STRUCTURAL ENGINEERING
FOR ARCHITECTS

STRUCTURAL ENGINEERING FOR ARCHITECTS

Kenneth R. Lauer

Professor of Civil Engineering
University of Notre Dame

McGraw-Hill Book Company

New York St. Louis San Francisco Auckland Bogotá Hamburg
Johannesburg London Madrid Mexico Montreal New Delhi
Panama Paris São Paulo Singapore Sydney Tokyo Toronto

This book was set in Times Roman by Science Typographers, Inc.
The editors were Diane D. Heiberg and J. W. Maisel;
the production supervisor was Charles Hess.
The drawings were done by ECL Art Associates, Inc.
The cover was designed by Anne Canevari Green.
R. R. Donnelley & Sons Company was printer and binder.

STRUCTURAL ENGINEERING FOR ARCHITECTS

1 2 3 4 5 6 7 8 9 0   D O D O   8 9 8 7 6 5 4 3 2 1

Library of Congress Cataloging in Publication Data

Lauer, Kenneth R
    Structural engineering for architects.

    Includes index
    1. Structural engineering. I. Title.
TA633.L38   624.1   80-23313
ISBN 0-07-036622-5

To My Mother
and
in Memory of My Father

CONTENTS

Appendix: Tables 490

Index 531

PREFACE

This book is written primarily for architecture students with a mathematical background in algebra, geometry, trigonometry, and preferably some calculus.

The book provides the mechanics and strength of materials required for an integrated course in the analysis and design of steel and timber structures. All topics are developed carefully to promote the student's understanding of these processes.

In developing the knowledge required to understand and design structural systems used in buildings and associated structures, the book considers tension members, cables, trusses, beams, columns, connections, and structural frames. The last chapter integrates this material into the design of buildings.

The coverage includes sufficient detail for the student to design simple structural members and systems and to understand the fundamentals involved in more complex systems. Approximate approaches are used for statically indeterminate structures.

One of the main features of the book is the presentation of a sufficient number of topics in mechanics and strength of materials to make a combined approach to the analysis and design of structures possible. This coordinated approach enables a student to design a roof truss and roof-cable system by the end of the first semester of work and helps generate and maintain interest on the part of the student. It also enables students to use their newly acquired knowledge immediately in other architectural courses.

The book puts the consideration of loads and reactions before the equilibrium of force systems. This makes the handling of force systems less abstract and more realistic for the student.

There is a chapter dealing with the properties of the main structural materials (steel, concrete, and wood). An understanding of these properties is necessary to understand the behavior and performance of the material under load.

The text provides all of the pertinent data required for the design of steel and timber members. This includes all of the information necessary for design according to the American Institute of Steel Construction (AISC) and the National Design Specifications for Wood Construction (NDS).

The text covers material sufficient for three semesters of work. The first semester includes the first eight or nine chapters, covering the basic requirements in mechanics and strength of materials as well as analysis and design of trusses and cables. The second semester would normally involve Chapters 10 to 12, covering the analysis and design of beams and columns. The third semester would pick up additional topics from Chapters 8, 9, 11, and 12 as well as Chapters 13 and 14 dealing with connections and elements of building design.

The author is, of course, indebted to a large number of people who over the years have been either directly or indirectly involved in the developement of this book. The general works and ideas of other authors and teachers have played an important role in the content and approach used. They include John Wilbur, Charles Norris, Ferdinand Singer, Charles Gaylord, Edwin Gaylord, Jr., D. Allan Firmage, Jack McCormac, German Gurfinkel, Mario Salvadori, Robert Heller, James Ambrose, and Henry Cowan to name a few. In particular, I would like to remember I. F. Morrison who made my initial contact with structures so interesting and vital. Thanks are also due to the architecture students at Notre Dame who, over several years, helped to shape the class notes from which this book was developed.

I would also like to acknowledge the contribution of my wife Ann not only for her help and patience over the years but in particular for her help in handling the page and galley proofs.

Kenneth R. Lauer

LIST OF SYMBOLS

A = area

A_{net} = required net area of member

A_s = effective shear area

A_1, A_2 = area of components of composite area

a = leg size of weld

a, b = leg sizes of unequal leg weld

a_1, a_2 = area of components of area

b = number of members in truss; width of beam or member

C = lateral-force coefficient for base shear

C_b = moment-gradient multiplier

C_c = upper limit for slenderness ratio for elastic buckling

C_{pe} = external pressure coefficient

C_{pi} = internal pressure coefficient

c = distance to outermost fiber of beam

D = depth of building; diameter of nail; shank diameter of lag bolt

d_e = effective diameter of fastener hole

d_s = diameter of fastener

E = modulus of elasticity

F = force acting on joint; friction force parallel to contact surface

F_b = allowable bending stress

F_c = allowable compressive stress parallel to grain

F_c' = allowable column stress

$F_{c\perp}$ = allowable compressive stress perpendicular to grain

F_p = allowable bearing stress

F_t = allowable tensile stress

F_u = ultimate strength; minimum ultimate tensile strength

F_v = allowable shear stress

F_w = load-carrying capacity of 1 in of weld

F_y = yield stress

FSP = fiber saturation point

f = coefficient of friction; stress

f_b = bending stress

f'_c = compressive strength of concrete

f_{cr} = column buckling stress

f_h = hoop stress

f_l = longitudinal stress

f_s = shear stress

G = modulus of rigidity; specific gravity of wood

g = gauge distance

H = elevation difference; height of building

h = depth of beam; vertical distance between cable supports

h_x = height of particular floor above ground level

I = moment of inertia

I_x = moment of inertia relative to x axis

I_{x0} = moment of inertia relative to x centroidal axis

I_y = moment of inertia relative to y axis

I_{y0} = moment of inertia relative to y centroidal axis

J = polar moment of inertia

$\bar{J}$ = polar moment of inertia through centroid

K = constant depending upon specific gravity, used in lateral-resistance formula for nails; numerical coefficient depending on ductility of structure

K_A = coefficient of active earth pressure

K_0 = coefficient of earth pressure at rest

k = distance from top of flange to web toe of fillet

L = length; span

L_b = unsupported length of steel beam

L_c = allowable unsupported length of beam based on flange buckling

L_u = allowable unsupported length of beam based on twisting

l = length; span

l_e = effective length of timber beam

l_u = unsupported length of timber beam

M = bending moment

N = allowable load at angle θ with direction of grain; bearing length; normal force between contact surfaces

n = number of fasteners; number of rows of fasteners; number of joints

P = allowable load acting parallel to grain; allowable normal lateral load per nail; allowable normal withdrawal load; applied load;

		concentrated load on beam; tensile load carried by member; total load on column including eccentric load P''
P'	=	allowable load on column
P''	=	eccentric load on column
P_c	=	load carried by concrete
P_s	=	load carried by steel
p	=	internal pressure in cylinder; pressure at a particular point below surface; spacing of rivets
p'	=	eccentric load on column
p_A	=	active earth pressure
p_e	=	external wind pressure
p_h	=	horizontal pressure
p_i	=	internal wind pressure
p_r	=	total force acting against 1-ft width of wall due to soil or water
p_v	=	vertical pressure
Q	=	allowable load acting perpendicular to grain; statical moment of area about neutral axis
q	=	wind-velocity pressure
q_{he}	=	external velocity pressure at height h
q_{hi}	=	internal velocity pressure at height h
R	=	reaction
R_{AH}	=	horizontal component of reaction A
R_{AV}	=	vertical component of reaction A
r	=	radius; radius of gyration
S	=	sag of cable; spacing of connectors; stagger distance
S_h	=	horizontal shear stress in beam
S_s	=	allowable shear stress in weld; shear stress in beam
S_v	=	transverse shear stress in a beam
T	=	fundamental period of vibration of structure; torque
t	=	thickness of member
t_e	=	effective throat dimension
V	=	shear force at a cross section of beam
V	=	total lateral shear force at base due to earthquake
v_z	=	velocity of wind at height z
v_{30}	=	velocity of wind at height 30 ft
W	=	total dead load; weight of body; weight of truss
W_x	=	weight of floor under consideration
w	=	distributed load on beam; unit weight of concrete
w_1, w_2	=	weight of elements of body
$\bar{x}, \bar{y}$	=	coordinates of center of gravity or centroid
x_0, y_0	=	coordinates of center of gravity or centroid

x_1, y_1 = coordinates of center of gravity of element; coordinates of element or components of area

y = deflection of beam

Z = depth in soil mass; numerical coefficient depending upon zone of seismic activity

α = coefficient of thermal expansion

γ = shear strain; unit weight of liquid, soil

Δt = change in temperature

δ = deformation

δ_s = shear deformation

ε = strain = deformation per unit length

θ = angle between direction of load and direction of grain; angle of twist

μ = Poisson's ratio

ρ = distance; radius; radius of curvature

ϕ = angle of internal friction of soil

CHAPTER 1

OBJECTIVES OF STRUCTURAL DESIGN

1.1 INTRODUCTION

A structure is created to serve a definite purpose. In order to fulfill this purpose a number of design objectives relating to safety, serviceability, feasibility, and aesthetics must be specified and satisfied. Meeting these design objectives requires a basic understanding of how construction materials behave and deform under load. Common construction materials are rock, earth, concrete, wood, and steel. Chapter 6 contains a brief discussion of some of the most important properties of these materials.

1.2 SAFETY

Safety represents one of the principal responsibilities of the structural engineer and the architect. The designer must understand the environment, the function of the structure, and the behavior of the materials of construction. The engineering judgment required to take into account the uncertainty involved in predicting load effects and structural response is basic to structural design.

The loading of a structure and its design can be considered in stages. The initial stage involves normal loads, those encountered on a day to day basis, which must be withstood indefinitely without distress. A second stage of loading consists of larger loads that might occur infrequently during the life of the

1

structure. These loads are often specified in design codes, and the structure must perform adequately with no sign of distress. The phenomenon of creep, in which a structural member continues to deform under constant load, and the rate of loading become important in this stage of design. A third stage of loading involves severe and unusual loads that might possibly occur. It is not economically reasonable to design for perfect performance under loads such as tornadoes and large earthquakes. Some damage can be expected under these conditions, but avoiding possible loss of life plays an important part in this stage of design. The final stage of loading that must be considered is that associated with failure. An understanding of this phase is based on tests carried out in the laboratory and the field. The probable mode of failure has important economic implications because if a structure collapses gradually, it permits the evacuation of people without loss of life.

A number of approaches to the problem of structural safety are possible. Two viewpoints are commonly found in specifications and codes: the first is based on allowable stress and the second on strength.

1.2.1 Allowable-Stress Design

This approach is based on the assumption that if the stresses (load per unit area) under working loads are kept substantially smaller than stresses corresponding to failure, safety will be assured. This design approach starts with the selection of appropriate working loads. Their nature and magnitude will depend upon the type of structure. The American National Standards Institute (ANSI) sets design working loads for buildings. The Basic Building Code (BOCA) adopts the loadings set by ANSI as design working loads. This Basic Building Code has been enacted into law by various states and cities. When applied, these loads become legal minimums and must be used in design.

Once these working loads have been established, the structure can be analyzed and, assuming elastic behavior, the members can be designed and the stresses calculated. Allowable maximum stresses are also specified by appropriate code-writing bodies. For example, the American Institute of Steel Construction (AISC) has recommendations for steel construction, the American Concrete Institute (ACI) provides a guide for reinforced concrete, and the National Forest Products Association provides National Design Specifications (NDS) for wood construction. This procedure is designed to ensure that in normal service the structure will behave elastically and avoid such modes of failure as buckling and fatigue.

1.2.2 Strength Design

Strength design is based on the use of failure loads in the calculation of the ultimate load-carrying capacity of structural members. Failure loads are obtained by multiplying working loads by appropriate load factors. In structural-steel design this approach is called *plastic design*. For reinforced concrete it is termed *ultimate-strength design*. The inherent variability of the materials and structural dimensions are taken into account by capacity-reduction factors, which are applied to the theoretical strength of the member and vary with the type of member. The ACI Building Code for Reinforced Concrete uses this approach.

1.3 SERVICEABILITY

The previously mentioned design approaches are based on strength. Two other important considerations are economy and rigidity. The cumulative effect of strain is displacement of the structure from its original unstressed position. The trend to stronger materials results in lighter sections and brings increasing problems from vibrations and deflections. The structural designer must understand and control this aspect of structural performance.

The incompatability of available construction materials also poses additional problems for the designer. Differences in thermal properties, stiffness, and response to moisture must be taken into account.

1.4 FEASIBILITY

The construction of a proposed structure must be economical as well as feasible. Familiarity with construction problems and techniques facilitates economical construction. The anticipated method of construction should be kept in mind during the design.

1.5 UNITS

The International System of Units (SI) is gradually being adopted on a worldwide basis. In the United States conversion is gradually taking place primarily on an industry-by-industry basis, depending on the need. As a result it is important for students to be trained in both systems.

At present design data (properties of materials and structural section details) are not readily available in SI units, but a straightforward conversion can be readily made. However, when a given industry converts to SI, the size of structural members may be altered to facilitate the change. For example, the nominal size of lumber may be dropped and the size of 4- by 8-ft sheets of plywood may be changed slightly to accommodate the conversion. On the other hand, ASTM Committee A6 has issued a soft conversion of the dimensional properties of some of the common rolled steel shapes into SI units. The conversion of the properties of structural steel A36 has been fairly well accepted at this time.

Table 1.1 The International System of units (SI)

Base and supplementary units		
Quantity	Name	Symbol
Length	Meter	m
Mass	Kilogram	kg
Time	Second	s
Plane angle	Radian	rad

Derived units			
Quantity	Derived unit	Special name	Symbol
Area	Square meter		m^2
Volume	Cubic meter		m^3
Density	Kilogram per cubic meter		kg/m^3
Force	Kilogram-meter per second squared	Newton	N
Moment of force	Newton-meter		$N \cdot m$
Pressure	Newton per square meter	Pascal	Pa or N/m^2
Stress	Newton per square meter	Pascal	Pa or N/m^2

Conversion factors		
Quantity	U.S. customary	SI to U.S. customary
Length	1 in = 25.40 mm	1 m = 39.37 in
Force	1 lb = 4.448 N	1 N = 0.2248 lb
Distributed load	1 lb/ft = 14.59 N/m	1 kN/m = 68.53 lb/ft
Moment of force	1 ft·lb = 1.356 N·m	1 N·m = 0.7376 ft·lb

To avoid proliferating the number of soft conversion tables of design data, both systems of units will be used in this book only where data are available from a recognized representative of the industry.

The International System is made up of a number of *basic units*, *supplementary units*, and *derived units*. Table 1.1 provides a list of important SI units with conversion factors.

1.5.1 Multiples of SI Units

Prefixes are used to form names and symbols of multiples of SI units. The choice of multiple unit is normally governed by convenience and should give numerical values between 0.1 and 1000. Approved prefixes with their names and symbols are given in Table 1.2.

Table 1.2 SI prefixes

Multiplier	Prefix Name	Prefix Symbol	Multiplier	Prefix Name	Prefix Symbol
10^{12}	tera	T	10^{-2}	centi	c
10^9	giga	G	10^{-3}	milli	m
10^6	mega	M	10^{-6}	micro	μ
10^3	kilo	k	10^{-9}	nano	n
10^2	hecto	h	10^{-12}	pico	p
10	deca	da	10^{-15}	femto	f
10^{-1}	deci	d	10^{-18}	atto	a

1.6 ACCURACY OF COMPUTATIONS

The number of significant figures considered necessary for engineering calculations depends on a number of factors. The common engineering materials used in structures (wood, steel, concrete, etc.) have ultimate strengths that can only be estimated. The loads applied to structures may be known within a few hundred pounds, but in some cases they are known only to a few thousand pounds. As will become evident, sometimes only partly true assumptions are made about structural behavior, and variations in dimensions are inevitable due to fabrication and construction. These deviations also introduce inaccuracies.

As a result, so-called slide-rule accuracy, usually three significant figures, is considered satisfactory for most engineering calculations. It is important not to develop a false sense of precision, particularly with the advent of the electronic calculator and the computer.

SELECTION OF STRUCTURAL SYSTEMS

2.1 INTRODUCTION

People build structures to enclose and support their activities. The first function, that of housing their activities, requires the creation of well-defined enclosed spaces. These spaces range in size and complexity from modular housing units to multistoried office buildings and from housing developments to large sports complexes.

The other important function is the requirement of facilities to support and encourage human activities. Such structures include the containment of material (tanks and walls), providing passageways for the movement of people and materials (highways, railroads, canals, pipelines), vehicles for the movement of people and materials (airplanes, ships, automobiles, trucks, trains), receiving and discharging facilities (ports and terminals), supporting a single load at a fixed point (communication antenna), and creating power (generating stations). These structures offer little opportunity for architectural embellishment, as their architectural aspects are integrated directly into the structural design.

The enclosure of space, on the other hand, is a fundamental objective of architectural design. As pointed out by Engel [1],† of all component elements contributing to the existence of rigid material form, structure is the most essential. Although structure is not architecture, it makes architecture possible.

†References will be found at the end of the chapter.

An architect must be able to formulate structural ideas and to propose structural systems. It is important that structural knowledge not only cover basic concepts of structural behavior and design but establish a concise relationship between a given structure and architectural form and space.

The essence of structural design from an architect's point of view is the development of a structural system that will direct loads in certain directions and bring them to the ground not only aesthetically but with a maximum of material efficiency and with a minimum obstruction of interior space.

There are two main types of forces involved in the design of structural systems. Gravity forces, directed downward, are in conflict with the development of roof systems and floor areas. Lateral forces, including wind and earthquakes, present problems in the vertical direction, e.g., in high-rise buildings. As a result it is important to understand the mechanisms available to redirect forces and the systems available for spanning space and resisting deformation.

2.2 STRUCTURAL FORM

The method of redirecting forces is a function of the geometric configuration of the load-resisting structure, i.e., the form of the structural system. It is important to understand the basic types of structural form and how they can be used in different structural situations. The selection of the structural form involves the highest level of structural engineering practice.

2.2.1 Tension or Compression Structures

One of the simplest structural systems available to us is that of a flexible cable. This tensile structure, which has many examples in nature, transmits the loads to the foundation by pure tension in the cable. The summation of the vertical components of cable tension must balance the applied load. This means that a cable must have a finite value of sag in order to carry a load. The shape acquired by a cable under concentrated loads is called a *funicular polygon* (from the Latin *funis*, "rope," and the Greek *gonia*, "angle"). It is the natural shape required to carry loads in tension. As the number of loads on the cable increases, the funicular polygon acquires an increasing number of sides, becoming in the extreme a funicular curve. The funicular polygon for equally spaced horizontal loads approaches the geometric curve known as a *parabola*. The optimal sag for such a cable is three-tenths of the span. When equal loads are distributed along the length of a cable, the funicular curve becomes a *catenary*, the natural shape acquired by a cable under its own weight. These basic aspects of cable systems are outlined in Fig. 2.1. Engel [1] classifies the tension structure as form-active. With the use of a high-tensile-strength material, such as steel, the form-active tensile structure is an efficient structural system. The limitations in the use of cables come primarily from their instability, i.e., their adaptability to

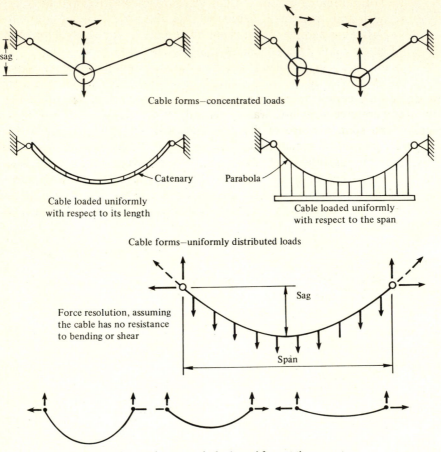

Cable forms—concentrated loads

Catenary

Parabola

Cable loaded uniformly
with respect to its length

Cable loaded uniformly
with respect to the span

Cable forms—uniformly distributed loads

Force resolution, assuming
the cable has no resistance
to bending or shear

Sag

Span

The less the sag, the greater the horizontal force at the supports

Figure 2.1 Basic aspects of cable systems [4].

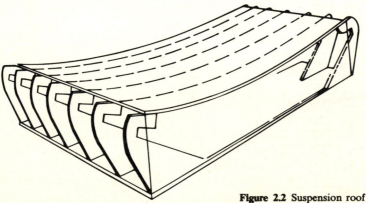

Figure 2.2 Suspension roof with parallel cables.

changing loads. Stiffening trusses and guy wires are used to increase the rigidity in the vertical and transverse directions. A cable is not a self-supporting structure unless ways and means are found to absorb the thrust developed at the supports. Figures 2.2 to 2.4 illustrate different anchorage systems.

When the parabolic shape assumed by a cable carrying horizontally distributed loads is turned up the other way, it gives the ideal shape of an arch, developing only compression under a specific loading system. The arch, essentially a compressive structure, was developed by the Romans. In a variety of shapes it can span a range of distances and is considered one of the basic structural elements of architecture. The shape of a masonry arch is usually chosen to be the funicular of the dead load. When moving loads are involved, a state of stress other than compression may develop since an arch can be funicular for only one set of loads. Figure 2.5 illustrates the basic aspects of arches. Figures 2.6 to 2.9 illustrate some interesting historical developments of the arch.

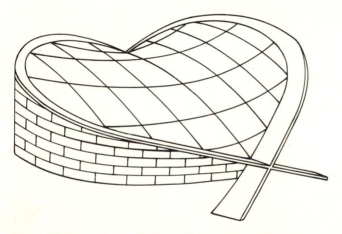

Figure 2.3 Suspension roof with parallel suspension cables using crossed arches.

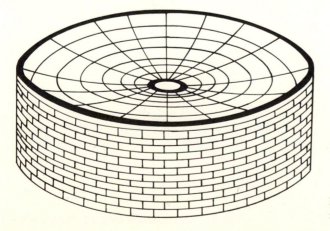

Figure 2.4 Suspension roof with radial cables.

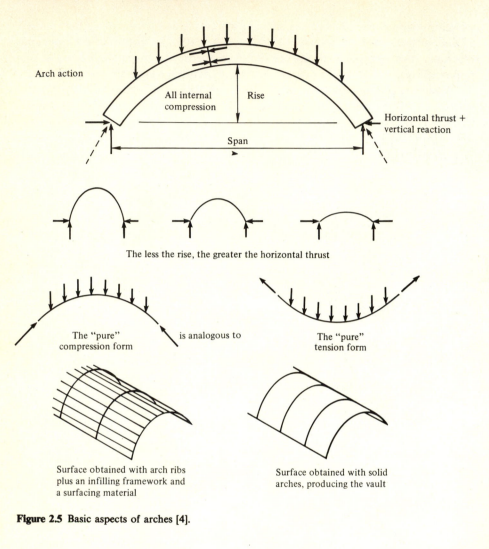

Arch action

All internal compression

Rise

Span

Horizontal thrust + vertical reaction

The less the rise, the greater the horizontal thrust

The "pure" compression form is analogous to The "pure" tension form

Surface obtained with arch ribs plus an infilling framework and a surfacing material

Surface obtained with solid arches, producing the vault

Figure 2.5 Basic aspects of arches [4].

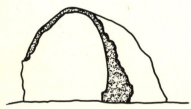

Figure 2.6 Ruins of the palace of Ctesiphon, built by the Persians in A.D. 550; catenary arch of mud bricks 112 ft high [5].

Figure 2.7 Primitive mud hut in central Africa [5].

Figure 2.8 Crossed arches of a gothic roof [5].

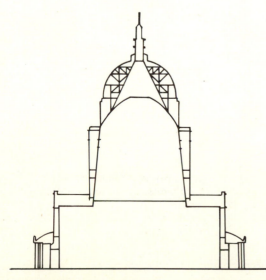

Figure 2.9 Outer dome of St. Paul's Cathedral, London [5].

2.2.2 Trusses

The horizontal forces exerted on the supports by a cable or an arch can be eliminated by adding a connecting member between the supports and making one of the supports movable (Fig. 2.10). The resulting structures, called *trusses*, are triangular in form and have both compressive and tensile members. The Romans are known to have built timber roof structures and bridges. Large timber trusses were revived during the Renaissance by Palladio. Although all the elements of mechanics required for the solution of this problem were available in the sixteenth century, it was not until the nineteenth century that the forces in truss members resulting from applied loads were solved analytically.

Trusses capable of spanning large distances are obtained by combining these elementary triangles. The truss does not have to change shape under different loadings. If loads are applied only at the joints, the members are not subject to bending and the only quantities that vary under variable loads are the magnitudes of the direct forces in the members. The truss form is capable of redirecting any conceivable system of joint loads to a given set of supports. The number of members and their configuration are design variables offering great flexibility to the engineer and architect in shaping a structure to fit a given problem.

For analytical purposes joints are normally considered to be pin-connected. Actually they are joined by being riveted, bolted, or welded to a gusset plate at their intersection. In smaller timber trusses nails and glue are used to connect the members to a gusset plate (Fig. 2.11). The restraint against relative rotation

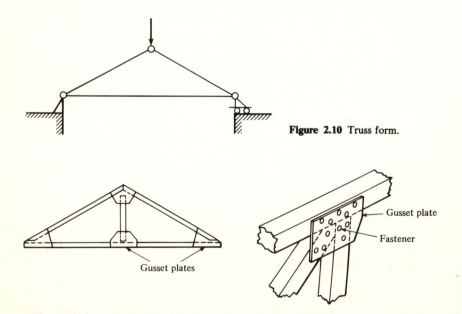

Figure 2.10 Truss form.

Gusset plates

Gusset plate

Fastener

Figure 2.11 Truss connection using a gusset plate.

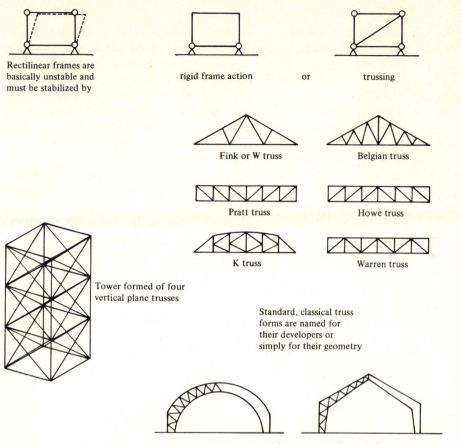

Rectilinear frames are basically unstable and must be stabilized by

rigid frame action   or   trussing

Fink or W truss

Belgian truss

Pratt truss

Howe truss

K truss

Warren truss

Tower formed of four vertical plane trusses

Standard, classical truss forms are named for their developers or simply for their geometry

Arches and rigid frames may be formed of trussed elements instead of solid members

Figure 2.12 Basic truss forms [4].

produced by the gusset plates introduces minor amounts of bending and so-called *secondary stresses*. Figure 2.12 illustrates some of the truss forms that have been developed over the years. Trusses can be used to span hundreds of feet between supports. They can be cantilevered from piers and in turn carry other simply supported trusses (*compound trusses*). Light trusses, called *open-web joists*, are used to span smaller distances, as roof or floor structures (Figs. 2.13 and 2.14).

When a roof or floor is supported by a number of parallel trusses, the load is first carried by a transverse system of beams and slabs to the trusses and then by the trusses to the supports (Figs. 2.13 and 2.14). The load-carrying capacity of each truss can be considered to be effective essentially in its own plane.

Better integrated and more efficient behavior can be obtained by connecting the parallel trusses by transverse trusses as rigid as the main trusses. In this case

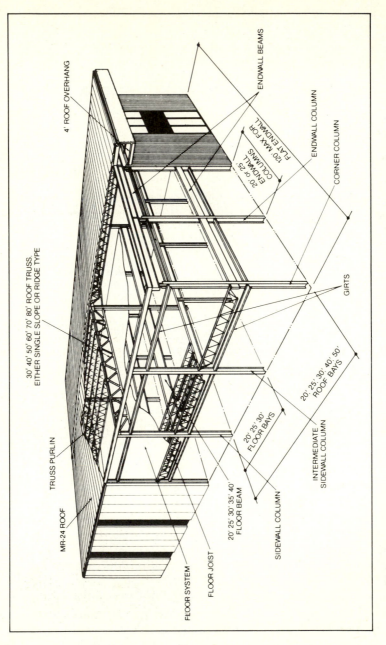

Figure 2.13 Modular-component system using trusses and columns [6]. (*Butler Manufacturing Company*.)

ENDWALL BEAMS

ENDWALL COLUMN

CORNER COLUMN

120' MAX FOR
FLAT ENDWALL

20' or 25'
ENDWALL
COLUMNS

GIRTS

20', 25', 30', 40', 50'
ROOF BAYS

INTERMEDIATE
SIDEWALL COLUMN

20', 25', 30'
FLOOR BAYS

SIDEWALL COLUMN

20', 25', 30', 35', 40'
FLOOR BEAM

FLOOR JOIST

FLOOR SYSTEM

MR-24 ROOF

TRUSS PURLIN

30', 40', 50', 60', 70', 80' ROOF TRUSS,
EITHER SINGLE SLOPE OR RIDGE TYPE

4' ROOF OVERHANG

14

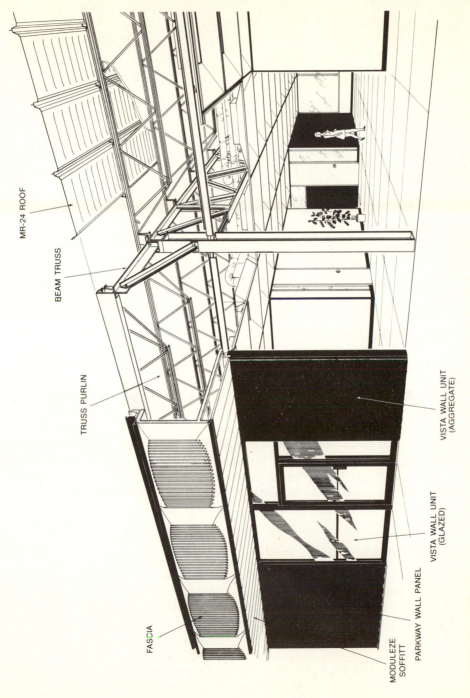

MR-24 ROOF

BEAM TRUSS

TRUSS PURLIN

FASCIA

MODULEZE SOFFITT

PARKWAY WALL PANEL

VISTA WALL UNIT (GLAZED)

VISTA WALL UNIT (AGGREGATE)

Figure 2.14 Details of beam trusses and columns [6]. (*Butler Manufacturing Company*.)

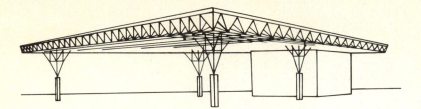

Figure 2.15 Space frame for flat roof [5].

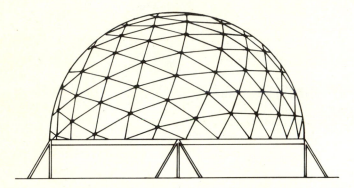

Figure 2.16 Geodesic dome [5].

the entire structure works more or less monolithically. Such spatial systems of members, called *space frames* (Fig. 2.15), are an economical solution to spanning large rectangular areas and are stiffer than systems of parallel trusses. The depth of parallel trusses is in the order of one-tenth the span. Space frames may have depths as small as one-twentieth or one-thirtieth of the span. When the area to be covered is a domed or vaulted surface, the surface can be subdivided into a number of triangles. A modern development of this type is the *geodesic dome* of Buckminster Fuller (Fig. 2.16). The *lamella roof* is a space-frame version of a diagonal grid requiring rigid joints for stability (Fig. 2.17).

2.2.3 Post and Lintel

The problem of sheltering human beings from the weather has been solved from very early times by an enclosure of walls and a roof. Separation of the protecting surface from the supporting system led to the simplest structural system known as the post and lintel. The *lintel* is a beam simply supported on posts (Fig. 2.18). Stonehenge is an excellent example of prehistoric post-and-beam construction (Fig. 2.19).

The transfer of load perpendicular to the axis of a long member is called *beam action*. The redirected perpendicular force is transmitted to the ends of the beam span by a complex system of stresses. This force transmission is not

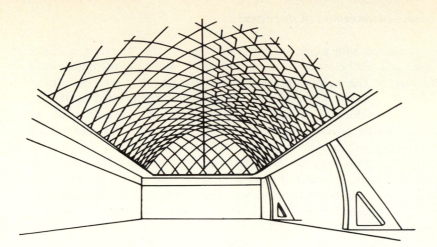

Figure 2.17 Lamella roof [5].

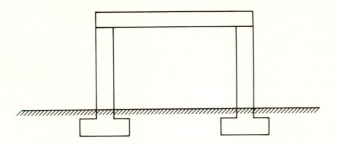

Figure 2.18 Post and lintel.

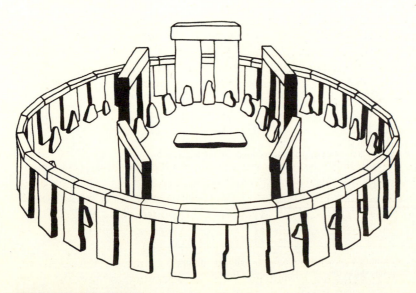

Figure 2.19 Stonehenge [5].

efficient compared with axial transmission because large amounts of material are stressed at levels below the maximum stress developed in the outer fibers of the beam section. The major advantage of beams is that they exert no horizontal loads on their supports. The posts are vertical struts compressed by the lintel. They must be capable of resisting some horizontal loads, such as wind. Some connection must be provided between the post and lintel. Post-and-lintel systems have been used to form multistory buildings but are not well suited for masonry construction because of a lack of bending resistance and strong connections between elements. Figure 2.20 illustrates some of the variations possible with post-and-beam construction. The problem of stabilizing this type of construction, illustrated in Fig. 2.21, leads to the development of framed and structural walls (Figs. 2.22 and 2.23).

When the action of the post-and-lintel system is changed by developing a rigid connection between the elements, the system becomes a rigid frame.

Overhangs, or cantilevers, introduce reverse bending in the beam

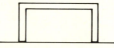

Rigid connection of posts and beam throws some bending into posts

Combination of cantilevers and rigid post-to-beam connection

Widened post tops; reduces length of span for beam but may induce eccentric load on post due to beam deflection

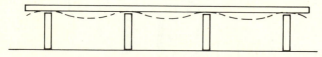

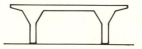

Combination of cantilevered ends, rigid connection, and widened post top

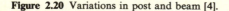

Reverse bending induced by making beam continuous through several spans; works the same as cantilevering the ends

Figure 2.20 Variations in post and beam [4].

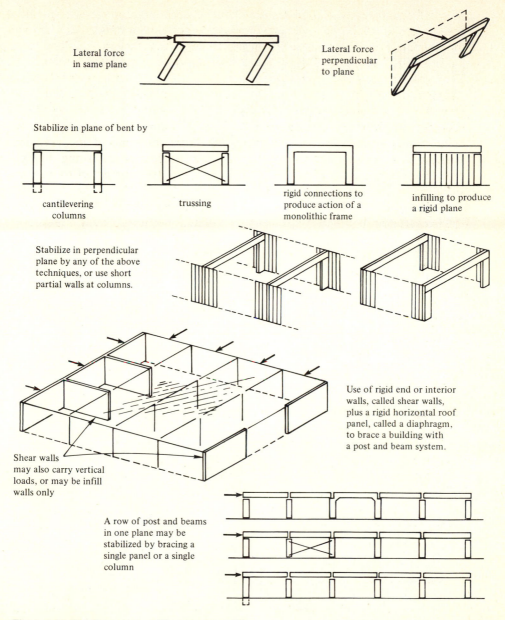

Lateral force in same plane

Lateral force perpendicular to plane

Stabilize in plane of bent by

cantilevering columns

trussing

rigid connections to produce action of a monolithic frame

infilling to produce a rigid plane

Stabilize in perpendicular plane by any of the above techniques, or use short partial walls at columns.

Use of rigid end or interior walls, called shear walls, plus a rigid horizontal roof panel, called a diaphragm, to brace a building with a post and beam system.

Shear walls may also carry vertical loads, or may be infill walls only

A row of post and beams in one plane may be stabilized by bracing a single panel or a single column

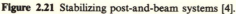

Figure 2.21 Stabilizing post-and-beam systems [4].

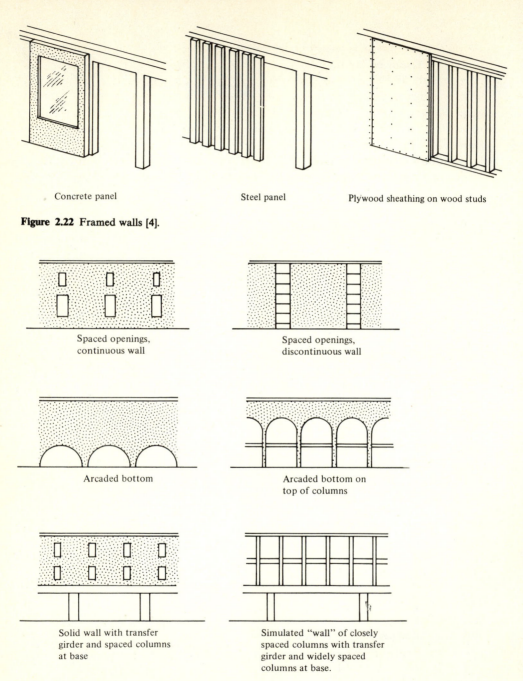

Concrete panel Steel panel Plywood sheathing on wood studs

Figure 2.22 Framed walls [4].

Spaced openings,
continuous wall

Spaced openings,
discontinuous wall

Arcaded bottom

Arcaded bottom on
top of columns

Solid wall with transfer
girder and spaced columns
at base

Simulated "wall" of closely
spaced columns with transfer
girder and widely spaced
columns at base.

Figure 2.23 Structural walls [4].

2.2.4 Rigid Frames

This new structure behaves monolithically and has greater strength to withstand both vertical and horizontal loads. Thanks to partially restrained ends, the beam is more rigid and capable of supporting heavier loads in bending. The columns, on the other hand, are subjected not only to compressive loads but also to bending stresses due to continuity with the beam. A new force, horizontal in nature and characteristic of frame action, becomes necessary to maintain the frame in equilibrium under vertical loads. This thrust is provided by the foundation's resistance to lateral displacements. The foot of the column may be *hinged* or *fixed* (see Chap. 3). Figure 2.24 illustrates some aspects of rigid-frame

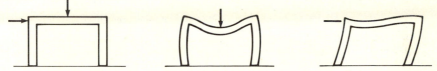

Rigid frame action: interaction of members through rigid jointing

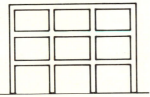

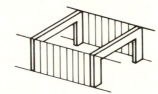

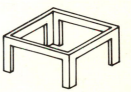

Multistory and/or
multispan frames

Rigid frame action in
one direction only

Rigid frame action
in two directions

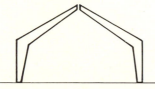

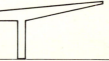

gabled frame

Cantilever frames

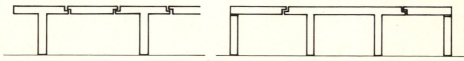

Partial rigid frames: combined with nonrigidly jointed elements

Figure 2.24 Basic aspects of rigid frames [4].

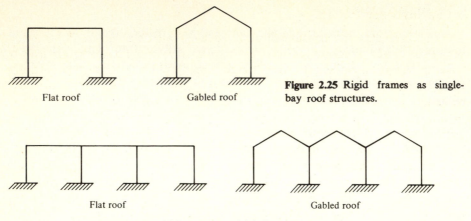

Flat roof   Gabled roof

Figure 2.25 Rigid frames as single-bay roof structures.

Flat roof   Gabled roof

Figure 2.26 Rigid frames as multiple-bay roof structures.

construction, and Fig. 2.25 shows the use of rigid frames as single-bay roof structures. The advantages of continuity can be extended by using multiple frames not only laterally (multiple bay) but also vertically in the form of multistory structures (Figs. 2.26 and 2.27). The first effective method for the design of rigid frames, developed by Alberto Castigliano in 1873, used strain-energy concepts and the principle of least work. The Vierendeel truss, named after its Belgian inventor, is used in bridge design and in buildings when the structure requires wide spans and unencumbered bays. The truss (Fig. 2.28) consists of the columns and floor beams of the building.

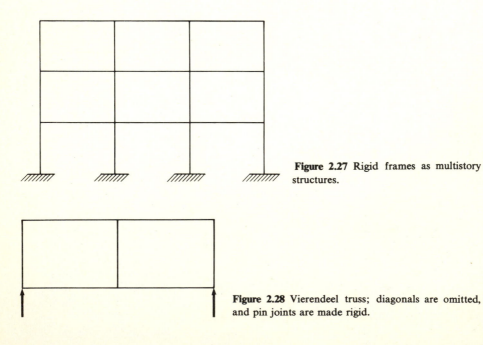

Figure 2.27 Rigid frames as multistory structures.

Figure 2.28 Vierendeel truss; diagonals are omitted, and pin joints are made rigid.

2.2.5 Plates and Shells

The structural elements considered so far have a common attribute, i.e., transferring loads in one direction. These one-dimensional resisting elements are used to cover or support a rectangular area, but they are impractical and inefficient. For example, a series of beams or trusses, all parallel to one side of the structure, can be used to support a floor. With a concentrated load, the member beneath it will be the primary supporting member, and the adjoining members become involved only to the extent that the floor develops two-dimensional resistance. The load transfer occurs only in the direction of the members to the supporting walls at their ends. This becomes uneconomical when four walls or girders are available to support the loads. Such dispersal is obtained by means of grids and plates, two-dimensional resisting structures acting in a plane. The basics of the grid system are illustrated in Fig. 2.29.

The two-way action of beam grids is due to the pointwise connection of the two-beam systems at their intersections. This action is even more pronounced in practice because the beam grid is covered by a slab, making it more of a monolithic structure. Extending this monolithic concept farther introduces plates and slabs. Any strip of plate parallel to one side of the supporting structure may be thought of as a beam acting in that direction. Any strip at right angles to the first may also be considered a beam (Fig. 2.30). Since any physical phenomenon in nature follows the easiest path, the beams may be considered to act in any

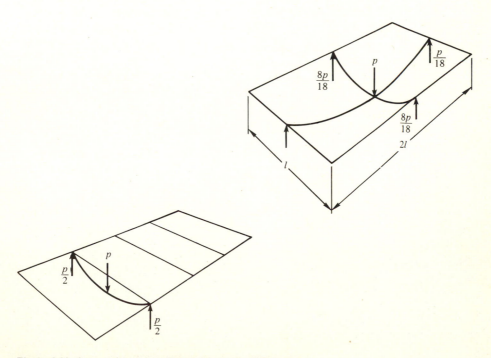

Figure 2.29 One- and two-way dispersal systems using beams.

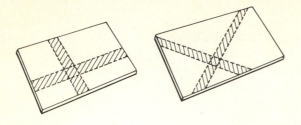

Figure 2.30 Equivalent grids in plate action.

direction. Plates may also be supported on columns. The connection between the column and plate must be designed to resist the so-called *punching shear* from the column and may require the use of capitals or thickened panels (Fig. 2.31).

The structural efficiency of plates is reduced by the fact that the allowable stress is developed only in the top and bottom plate fibers at the most stressed point in the most stressed direction. In beams this inefficiency is remedied to some extent by locating most of the material in the flange area. This approach has also been used with plate design by taking some of the material near the center of the section and locating it in ribs, which can run in one or more directions. Figure 2.32 shows examples of one-way ribbed plates and two-way, or waffle, plates.

The same result as ribbing plates can be obtained by folding the plate. Folded plates are made of wood, steel, aluminum, or reinforced concrete. Two plates at an angle are equivalent to a beam of rectangular cross section with a depth equal to that of the plates and a width equal to the combined horizontal width of the two plates. Figure 2.33 illustrates some folded-plate cross sections and an example of folded-plate construction.

Thin shells are considered form-resistant structures, thin enough not to develop appreciable bending stresses but thick enough to carry loads by compression, shear, and bending. Although thin shells have been built of wood, steel, and plastic materials, they are ideally suited to reinforced-concrete construction.

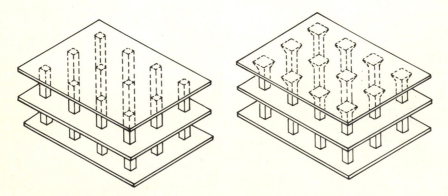

Figure 2.31 Flat-plate and flat-slab systems.

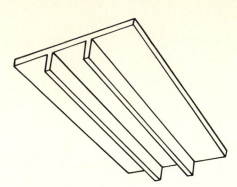
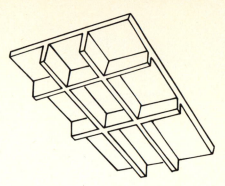

Figure 2.32 Ribbed-plate construction.

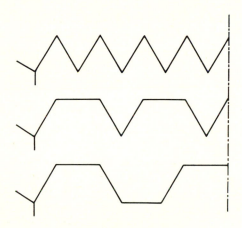

Folded-plate cross sections

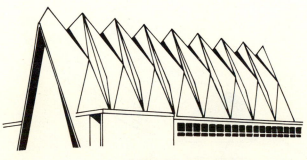

Folded-plate construction

Figure 2.33 Folded-plate construction [2].

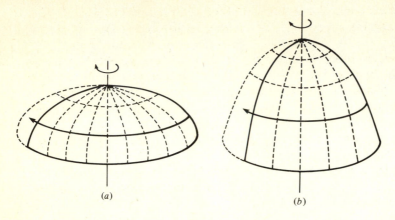

Figure 2.34 Rotational shell surfaces [2].

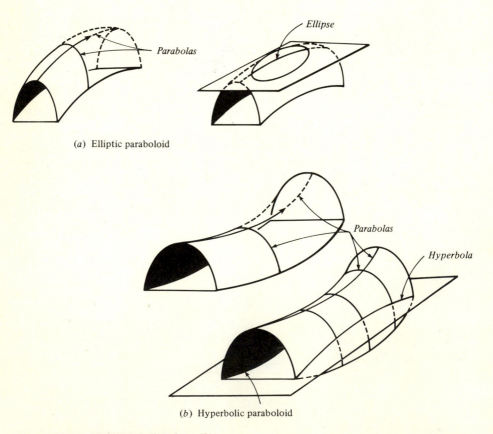

(a) Elliptic paraboloid

(b) Hyperbolic paraboloid

Figure 2.35 Translational shell surfaces [2].

Thin shells allow economical construction of beautiful curved surfaces of varied forms with remarkable strength. They are exceptional expressions of modern structural design.

Shell surfaces are often classified on the basis of their general shape, e.g., rotational, translational, and ruled surfaces. Rotational surfaces are described by the rotation of a plane curve around a vertical axis, giving rise to a variety of dome forms. The most commonly used are spherical, elliptical, and conical (Fig. 2.34).

The surface of a translational shell is obtained by translating or sliding a plane curve on another plane curve, usually at right angles to the first. For example, a cylinder is obtained by translating a horizonal straight line along a vertical curve or translating a vertical curve along a horizontal straight line. Both the elliptic paraboloid and the hyperbolic paraboloid (Fig. 2.35) are developed in this manner. A great variety of surfaces can be generated in this way, but not all of them will be capable of carrying loads.

Any surface developed by sliding the two ends of a straight-line segment on two separate curves is called a *ruled surface*. The cylinder is generated by sliding a horizontal line on two identical vertical surfaces. Conoidal surfaces are obtained by sliding a straight-line segment on two different parallel curves. Conoidal surfaces (Fig. 2.36) cover a rectangular area and allow more light to enter from one direction. All these geometrically defined surfaces can be combined in numerous ways to develop more complex surfaces (Fig. 2.37).

2.2.6 Membranes

A membrane is a sheet of thin material which can be designed to resist only tension (not compression, bending, or shear). For example, a tent requires compressive struts and guy wires for support. Pretensioning a membrane before loading is often desirable, as done, for example, in a trampoline or umbrella. Membranes can be prestressed by internal pressure to enclose a volume completely. The basic types of membrane are illustrated in Fig. 2.38.

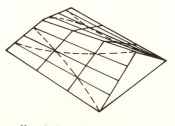

Hyperbolic paraboloidal conoid

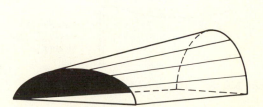

Figure 2.36 Ruled surface [2].

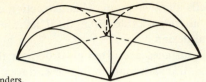

Intersecting cylinders.

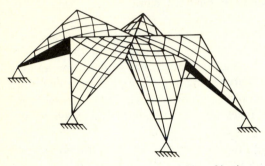

Skew hyperbolic paraboloid combination.

Figure 2.37 Combined surface [2].

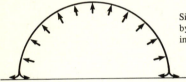

Single surface; tension maintained by pressure difference between interior of building and outside

Double surface; tension and stiffening of structure produced by inflation of the structure

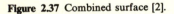

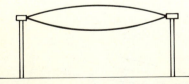

Double surface—pillow; bottom surface draped in tension from supports; top surface held up by inflation

Figure 2.38 Basic types of air-supported structures [4].

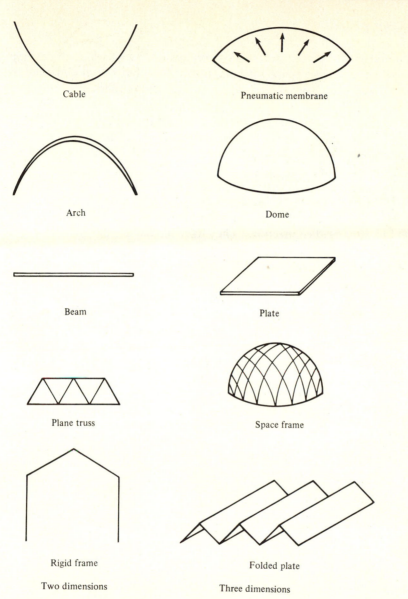

Cable

Pneumatic membrane

Arch

Dome

Beam

Plate

Plane truss

Space frame

Rigid frame

Folded plate

Two dimensions

Three dimensions

Figure 2.39 Comparison of structural types in two and three dimensions [5].

2.3 SUMMARY

There are a number of different structural forms available with which to develop structural systems to enclose space for human use. They have been classified in a number of ways but for convenience are grouped here as follows:

Tension and compression structures. Load is transmitted by a single state of stress. The tension structure (cables) takes on a specific geometry for a given loading. It is an efficient use of material. Compression structures (arches) cannot adjust shape with loading. They may be limited by buckling.

Trusses. They use compression and tension members. They are very efficient, stiff, and capable of spanning large distances.

Posts and lintels. These represent a combination of flexural and compression members without rigid connections. They have limited use in small one-story buildings.

Rigid frames. An extension or development of the post-and-lintel system, they use rigid connections to join the beams and columns. They are used to develop multibay and multistory structures.

Plates and shells. These are used extensively in floor and roof systems. Forces are transmitted primarily in a continuous surface. These systems are more difficult to analyze and design, but they provide economical use of materials. The shell is one of the most aesthetic structural forms available to the architect.

Membranes. This surface structure is capable of resisting only tension forces. Its primary use is in pneumatic structures.

Since the topic of structural form is of particular importance to the architect, the student is urged to do additional reading in the references. Critical analysis of existing structures is also very useful in appreciating form and understanding structural behavior. Figure 2.39 presents a comparison of structural forms in two and three dimensions.

REFERENCES

1. Heinrich Engel, "Structure Systems," Deutsche Verlags-Anstalt, Stuttgart, 1967.
2. Mario Salvadori and Robert Heller, "Structure in Architecture," 2d ed., Prentice-Hall, Englewood Cliffs, N.J., 1963.
3. Richard N. White, Peter Gergely, and Robert G. Sexsmith, "Introduction to Design Concepts and Analyses," vol. 1, Wiley, New York, 1972.
4. James E. Ambrose, "Building Structures Primer," Wiley, New York, 1967.
5. Henry J. Cowan, "Architectural Structures," Elsevier, New York, 1971.
6. Butler Manufacturing Company, "Landmark, a Beautifully Simple Building System," Kansas City, Missouri, 1979.

CHAPTER 3

LOADS AND SUPPORTS

3.1 ENGINEERING STRUCTURES

As we have seen, structures are composed of interconnected members and are supported in such a manner that they are able to resist the gravitational forces due to their own weight and other external forces expected to act on them. A building is acted upon by its own weight; wind, ice, and snow; and internal loads resulting from its use. A building must be so constructed that it will bring these forces into equilibrium by transferring their effects to the foundation of the building.

3.2 STRUCTURAL DESIGN

A structure is designed to perform a certain function. It must have sufficient strength and rigidity, be economical, and meet certain aesthetic requirements. The complete design of a structure is likely to require several phases.

One of these phases involves determining the loads acting on the structure. It is necessary to take into account a reasonable combination of loads that will produce maximum stresses or deflections in various parts of the structure. It is not feasible to design ordinary structures to resist all combinations of loads including such exceptionally large forces as tornadoes and major earthquakes. Thus a design, by necessity, is uncertain. While statistical and probabilistic evaluations of loads may be possible in the future, evaluations today rely on judgment based on experience, measurements, and logic. Load values given in codes, while legal minimums, often serve only as guides. They may not even be applicable to novel types of structure. As a result the determination of critical or controlling loads may be very time-consuming and may even require testing.

3.3 LOAD TYPES

Loads are normally classified into two broad categories, dead loads and live loads. *Dead loads* (DL) remain essentially constant during the life of the structure. They include the weight of the structural elements. It becomes necessary to assume the weights of members in order to design them. After the structure has been designed, the true weights can be computed. If the true weight is significantly different from the assumed weight, the design process must be repeated. *Live loads* (LL), which usually vary greatly, are considered to be the loads superimposed by use on the building or structure.

Another means of load classification is whether they result from human causes or from natural phenomena. The former can be regulated, but those resulting from natural phenomena require understanding and their prediction may be based on probabilistic studies.

In structural analysis it is convenient to idealize loads into three kinds: (1) concentrated loads, (2) line loads, and (3) distributed loads. *Concentrated loads* are single forces acting over a relatively small area, e.g., column loads. They are normally measured in pounds (lb), thousands of pounds (kips), or newtons (N). *Line loads* act along a line, e.g., the weight of a partition resting on a floor slab. They are usually designated as a force per unit length (pounds per foot or newtons per meter). A *distributed load* acts over a surface area, e.g., wind, soil pressure, and the weight of snow. Customary units are pounds per square foot and newtons per square meter (also called pascals).

Principal types of loads are discussed under the following headings:

1. Dead loads
2. Live loads
   a. Floor loads
   b. Roof loads
3. Snow loads
4. Wind loads
5. Earthquake loads
6. Water and earth pressure
7. Load combinations

3.4 DEAD LOADS

Dead loads include the weight of all permanent components of the structure—beams, columns, floor slabs, and roofing. They also include such architectural components as ceiling tile, window fixtures, and room partitions. Dead loads

Table 3.1 Unit weight of structural materials (adapted from [2])

Average values suitable for general use; specific products may differ considerably from these values; where available, actual weights from manufacturer's catalogs and reference books should be used

Material	lb/ft³	kN/m³
Metals:		
Aluminum	165	26
Iron, cast	450	70
Steel	490	77
Wood, soft:		
Douglas fir	32	5.0
Pine	30–44	4.7–6.9
Redwood	26	4.1
Spruce	27	4.2
Concrete, plain:		
Lightweight	32–110	5.0–17.0
Slag	130	20.7
Normal weight	144	22.6
Concrete, reinforced:		
Lightweight	65–120	5.0–17.0
Slag	140	21.7
Normal weight	150	23.6
Soils:		
Clay, dry	65	10
plastic	100	17.0
Earth, dry	75–95	
Sand and gravel	90–120	14.1–18.9
Liquids:		
Water	62.4	9.8
Masonry:		
Cement	90	14.1
Brick	100–140	15.7–22.0
Miscellaneous:		
Asphalt	80	12.7
Glass	156	24.5

can readily be computed from dimensions and the unit weight of materials. The unit weights of some common building and construction materials are given in Tables 3.1 and 3.2. Dead loads are usually computed on the high side to avoid redesign for minor changes in the final dimensions of the structural elements.

3.5 LIVE LOADS

Live loads can be defined as the weight superimposed by the use or occupancy of a building or structure but excluding such forces of nature as wind, snow, and earthquake, which usually act only during a fraction of the life of a structure. Live loads are often referred to as *normal loads*.

Table 3.2 Weights of structural materials per unit area (adapted from [2])

Material	lb/ft²	N/m²
Ceilings:		
Acoustical fiber tile	1.0	48
Plaster, 1-in	1.0	18.9†
Channel, suspended system	10.0	480†
Floors:		
Hardwood, 2-in	8.0	75†
Plywood, 1-in	3.0	56†
Asphalt mastic, 1-in	12.0	226†
Ceramic tile, $\frac{3}{4}$-in	10.0	250†
Flexicore, 6-in	46.0	2200
Linoleum, $\frac{1}{4}$-in	1.0	75†
Vinyl tile, $\frac{1}{8}$-in	1.4	210†
Roofs:		
Corrugated steel	1–5	48–250
3-ply roofing	1	48
3-ply felt and gravel	5.5	260
5-ply felt and gravel	6	290
Corrugated asbestos cement	3	145
Shingles:		
Wood, 1-in	3	56†
Asphalt, $\frac{1}{4}$-in	2	96
Clay tile	9–14	430–670
Slate, $\frac{1}{4}$-in	10	480
Sheathing:		
Wood, $\frac{3}{4}$-in	3	75†
Gypsum, 1-in	4	75†
Insulation:		
Loose	0.5	9
Poured	2.0	37
Rigid	2.5	47

† Per centimeter of thickness.

3.5.1 Floor Loads

To aid the designer, specifications usually prescribe uniformly distributed live loads. To simplify the analysis and for lack of a better approach, live loads are spread over the entire floor area though actual loads may be localized. Table 3.3 gives some of the requirements for minimum design loads in buildings and other structures. In warehouses the live load depends directly on the use of the structure, and it becomes important to avoid excessive loads as tenants change. Codes also specify concentrated loads to be used in conjunction with the uniform loads (Table 3.4). These concentrated loads are located in such a way that they develop maximum stresses in the structural elements.

The probability of having the full live load on an entire floor decreases as the area of the floor increases. Most codes take this into account with provisions

Table 3.3 Minimum uniformly distributed live loads (adapted from [2])

Category	Description	lb/ft^2	kN/m^2
Residential	Private rooms and apartments	40	1.9
Schools	Classrooms	40	1.9
	Corridors	100	4.8
Office buildings	Offices	80	3.8
	Lobbies	100	4.8
Hospitals	Rooms and wards	40	1.9
Libraries	Reading rooms	60	2.9
	Stacks	150	7.2
Stores, retail	First floor	100	4.8
	Upper floors	75	3.6
Wholesale		125	6.0
Theaters	Aisles, corridors and lobbies	100	4.8
	Orchestra	60	2.9
	Balconies	60	2.9
	Stage floors	150	7.2

Table 3.4 Concentrated live loads (adapted from [2])

	Load	
Location	lb	N
Elevator-machine-room grating (on area of 4 in^2)	300	1,335
Finish light floor-plate construction (on area of 1 in^2)	200	890
Office floors	2,000	8,896
Scuttles, skylights, ribs, and accessible ceilings	200	890
Sidewalks	8,000	35,584
Stair treads (on center of tread)	300	1,335

for reducing live loads. For example, for areas larger than 150 ft^2 the live load can be reduced at the rate of 0.08 percent per square foot up to a reduction of 60 percent. However, places of public assembly may receive full loading, and the reduction is not applicable to such structures. For multistory buildings it is unlikely that all floors will be fully loaded simultaneously. Reductions of 20 to 50 percent are allowed for columns that support several floors. The dynamic effects of live loads are not normally developed specifically in the codes. Special conditions require time-consuming analysis.

Example 3.1 What is the dead load of a 24-ft-long wood beam having a cross section of 4 by 8 in if the species of wood weighs 36 lb/ft^3?

SOLUTION Every foot of beam weighs the same amount. The weight of 1 ft of beam will be its volume per foot of length times its unit weight.

$$\text{Volume of beam per foot of length} = \frac{(12 \text{ in})(4 \text{ in})(8 \text{ in})}{(12 \text{ in/ft})(12 \text{ in/ft})(12 \text{ in/ft})} = \frac{2}{9} \text{ ft}^3$$

Weight of beam per foot of length = $(\frac{2}{9} \text{ ft}^3)(36 \text{ lb/ft}^3) = 8 \text{ lb}$

Total weight of beam = $(8 \text{ lb/ft})(24 \text{ ft}) = 192 \text{ lb}$

Example 3.2 What is the dead load of a masonry wall 2 m high, 10 m long, and 20 cm thick if the masonry weighs 14.2 kN/m³?

SOLUTION Consider a length of wall 1 m long. The volume of wall per meter length is $(1 \text{ m})(2 \text{ m})\frac{20 \text{ cm}}{100 \text{ cm/m}} = 0.40 \text{ m}^3$

The weight of the wall per meter length is $(0.40 \text{ m}^3/\text{m})(14.2 \text{ kN/m}^3) = 5.68 \text{ kN}$

Total weight of wall = $(5.68 \text{ kN/m})(10 \text{ m}) = 56.8 \text{ kN}$

3.5.2 Floor Framing

Probably the most common structural use of beams is to provide support for the floors and walls of buildings. The floor is supported by joists (called *floor beams* in steel construction). The joists in turn are supported by heavier beams, called *girders*; they are supported by columns that transmit the load to the foundation. The load on the joist will include the dead load of the joist itself and the weight of the flooring of width *b*. In addition to this it must carry the live load, which varies with the type of construction and is specified in pounds per square foot (see Table 3.3 and Fig. E3.3).

Example 3.3 Determine the loading on a typical joist shown in Fig. E3.3 with the joist spacing equal to 16 in. The joists are 2- by 10-in pine, and the flooring is hardwood $1\frac{1}{2}$ in thick. Assume residential construction.

SOLUTION Let us assume that pine weighs 38 lb/ft³ and oak weighs 54 lb/ft³. From Table 3.3 the live load for apartments is 40 lb/ft².

$$\text{Weight of joist} = \frac{(2)(10)(1)(38)}{(12)(12)} = 5.3 \text{ lb/ft}$$

$$\text{Weight of flooring} = \frac{(16)(1.5)(1)(54)}{(12)(12)} = 9.0 \text{ lb/ft}$$

Total dead weight = $5.3 + 9 = 14.3$ lb per foot of joist

$$\text{Live load} = \frac{16}{12}(1)(40) = 53.3 \text{ lb per foot of joist}$$

3.5.3 Roof Loads

If not covered by an existing building code, the live loads used in design should represent the designer's determination of the particular service requirements for the structure. ANSI recommends, as a minimum, a vertical live load of 20 lb/ft² of horizontal projection in addition to dead load and any special loads for all flat, pitched, or curved roofs regardless of their location. Where flat roofs are used as sundecks and roof gardens, floor live loads are in effect.

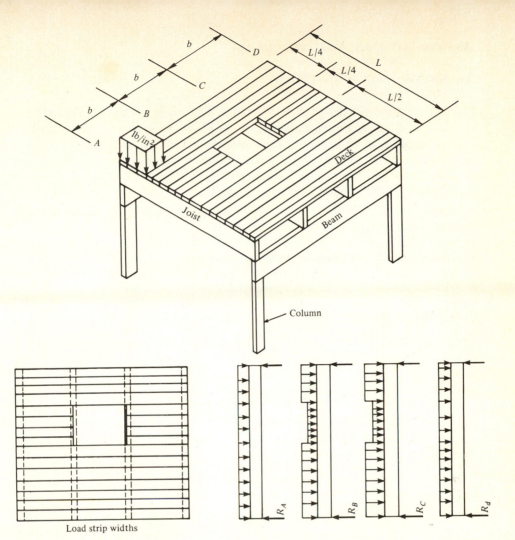

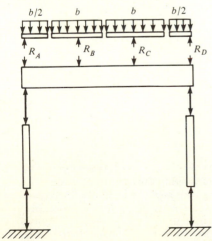

Figure E3.3 Structural framing.

3.6 SNOW AND RAIN LOADS

Snow loads affect the design of roof systems. The shape of the roof controls the magnitude of snow forces acting upon it. Many roof arrangements present special problems in designing for snow load. Some tend to accumulate the snow while others promote slide-off. Valleys formed by the intersection of multiple roofs and flat roofs promote the retention of snow. Figure 3.1 shows the greatest snow pack on the ground occurring from 1871 to 1944. A basic snow-load coefficient of 0.80 should be used to convert ground snow load to roof snow load. If this load is larger than the recommended live load of 20 lb/ft² for roofs, it should be used. If it is less than the minimum roof live load, it can be neglected. Snow loads are based on the horizontal projection of roof areas. Figure 3.2 indicates how they can be reduced as the slope of a roof increases.

Loads due to rain are not considered separately because they are usually less than the snow load. They can become important if the drains on flat roofs become clogged. AISC specifications contain provisions on ponding that specify stiffness of the roof beams to avoid progressive failure due to deflection.

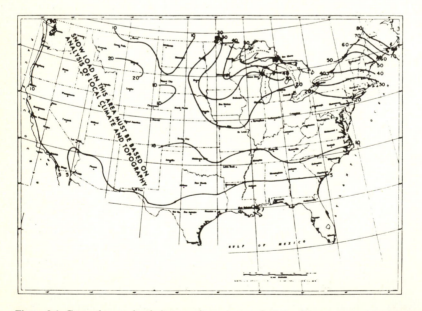

Figure 3.1 Ground snow loads in pounds per square foot for 50-year recurrence interval (from U.S. Weather Bureau data). Basic roof loads may be taken as 80 percent of the ground snow load. This map is intended to be used where there is no local building code and is not intended to supersede state or local code requirements. Special consideration should be given to areas (cross-hatched) where no design loads are shown and unusually high accumulations may occur. Variation of ground snow loads with elevation and exposure is not yet completely understood, and local differences in mountain regions are usually very significant [2].

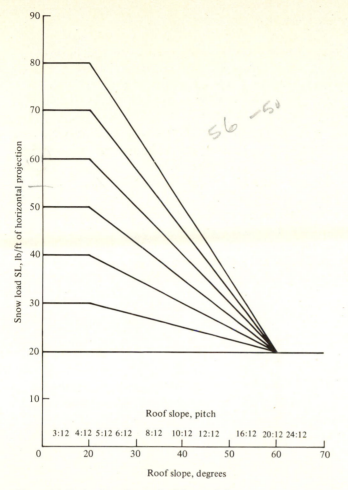

Figure 3.2 Reduction of snow load for roof slope.

3.7 WIND LOADS

Wind can develop significant forces on the exposed surfaces of structures, particularly tall buildings. The magnitude, frequency, and distribution of wind loads depend on several factors, including the velocity of the wind. The following material is based on Ref. 4. Wind velocities have been assembled from weather data and are based on probability of occurrence. Since maximum wind forces are dependent upon gusts (peak velocities), the quantity *fastest mile* has been developed for this purpose. It represents the maximum velocity of a 1-mi-long column of air passing a reference point. Such a wind-velocity map is illustrated in Figure 3.3. Approximate gust factors are required for structures that are small enough to be responsive to gusts involving less than a fastest mile. As a general guide, the gust factor depends on the size and height of the building

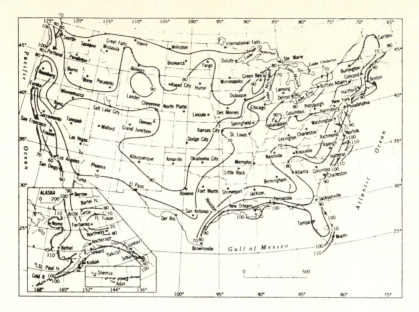

Figure 3.3 Fastest-mile wind velocities 30 ft above ground at airports, 50-year recurrence intervals [5].

as well as surface roughness and nearby obstructions. For small to medium-sized structures located in open, relatively level terrain and 100 ft or less in height a gust factor of 1.3 is commonly assumed. For taller buildings a gust factor of 1.1 is often used. Figure 3.3 provides basic wind velocities for observed airflows in open, level country at a height of 30 ft above ground. This value will require modification for other design heights. An accepted procedure is to apply an exponential profile law of the form

$$V_z = V_{30}(z/30)^{1/x} \tag{3.1}$$

where V_{30} = reference velocity at 30 ft elevation
V_z = velocity of wind at height z
x = exponent depending upon general site exposure conditions

For level or slightly rolling terrain with minimal obstructions, x may be taken as 7; for similar terrain with numerous obstructions, e.g., suburban areas, a value of 5 is recommended; and for urban areas, a value of 3 is suggested. Figure 3.4a illustrates this variation with height for rolling terrain.

For air at standard pressure (760 mmHg) and temperature (15°C) the wind-velocity pressure q in pounds per square foot is given by

$$q = 0.00256V^2 \tag{3.2}$$

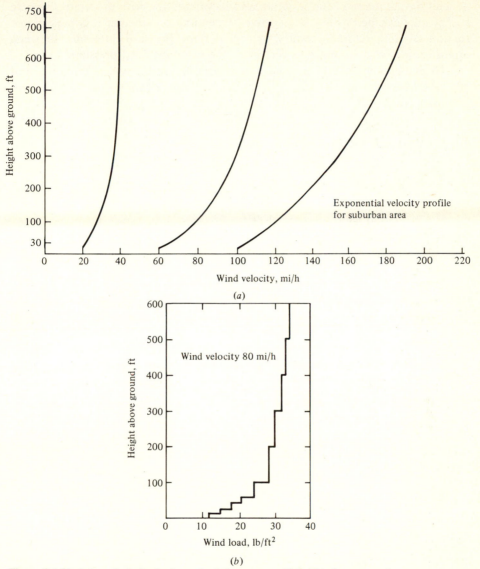

Figure 3.4 Variation of wind velocity and wind pressure with height above ground.

where V is wind velocity in miles per hour. In most municipal codes the continuous change in wind pressure with height is simplified into a stepped change (Fig. 3.4b).

3.7.1 Wind-Pressure Distribution on a Structure

To determine the design wind-pressure distribution on a building, the calculated wind-velocity pressure is multiplied by an appropriate coefficient. These coefficients define the pressure acting normally on the surface of the element and depend on the shape of the structure and its orientation with the wind. Pressure coefficients can be positive or negative, depending on whether it is pressure or suction. It is therefore necessary to consider the pressure difference between opposite sides of a structure to determine the resultant wind pressure.

Design-pressure coefficients have been established for typical buildings and enclosed structures with vertical walls that may have openings such as doors and windows.

3.7.2 Internal-Pressure Coefficients

The internal wind pressures acting on a structure can be determined by use of

$$p_i = q_{hi}C_{pi} \qquad (3.2a)$$

where p_i = internal wind pressure, lb/ft^2
q_{hi} = internal velocity pressure at height h, lb/ft^2
C_{pi} = internal-pressure coefficient

It is important to note that the internal velocity pressure q_{hi} does not need to include a gust factor because of the damping effect of the openings.

Figure 3.5 gives internal-pressure coefficients for walls and roofs regardless of the type of roof. The pressures determined by using these coefficients are assumed to be uniform on all internal surfaces at a particular building height. These values have application only for winds acting at right angles to the surface. The total force in any direction can be determined by summing up the pressures on appropriate surfaces (see Example 3.4).

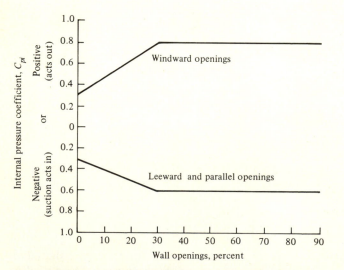

Figure 3.5 Internal-pressure coefficients for walls and roofs [6].

3.7.3 External-Pressure Coefficients

To calculate external wind pressures, use is made of

$$p_e = q_{he}C_{pe} \qquad\qquad (3.2b)$$

where p_e = external wind pressure, lb/ft^2
q_{he} = external velocity pressure at height h, lb/ft^2
C_{pe} = external-pressure coefficient

For wall surfaces of buildings, average external-pressure coefficients are given in Table 3.5, and the external-pressure coefficients for wind parallel to the surface of flat, pitched, or arched roofs can be assumed to be -0.8, as given in Table 3.6. For structures having a pitched or sloping roof with the direction of wind perpendicular to the ridge of the roof, the external-pressure coefficients are given in Fig. 3.6 for various slopes. For arched roofs with wind direction perpendicular to the axis of the arch, the coefficients can be obtained from Fig. 3.7.

Table 3.5 External-pressure coefficients for walls [4]

Wall location relative to wind direction	External-pressure coefficient C_{pe}
Windward wall	+ 0.8
Leeward wall:	
Height to width and height to length ≥ 2.5	− 0.6
Other dimensional ratios	− 0.5
Parallel to wall	− 0.8

Table 3.6 External-pressure coefficients for pitched roofs, wind parallel to roof surface [4]

Wall height to least width	External-pressure coefficient C_{pe}
< 2.5	− 0.7
≥ 2.5	− 0.8

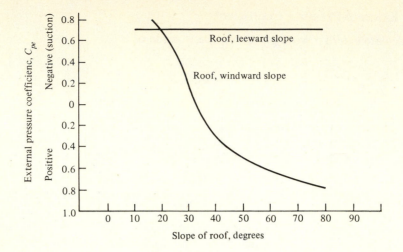

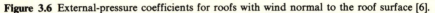

Figure 3.6 External-pressure coefficients for roofs with wind normal to the roof surface [6].

Example 3.4 Determine wind loads acting on the building in Fig. E3.4*a*.

SOLUTION The basic wind pressure can be obtained from Eq. (3.2).

$$q = (0.00256)(90)^2 = 20.7 \ \text{lb/ft}^2$$

External velocity pressure = $(20.7)(1.3) = 26.9 \ \text{lb/ft}^2$

Internal velocity pressure = $20.7 \ \text{lb/ft}^2$

Internal-pressure coefficients (Fig. 3.5):

      Leeward − 0.42   (acts in)

      Windward + 0.80   (acts out)

         Net internal load + 0.38   (acts out)

External-pressure coefficients for walls (Table 3.5):

    Leeward wall −0.50

    Windward wall − 0.80

External-pressure coefficients for roof (Fig. 3.6):

      Leeward slope   0.70 (suction)

      Windward slope   03.0 (push)

The results are shown in Fig. E3.4*b*.

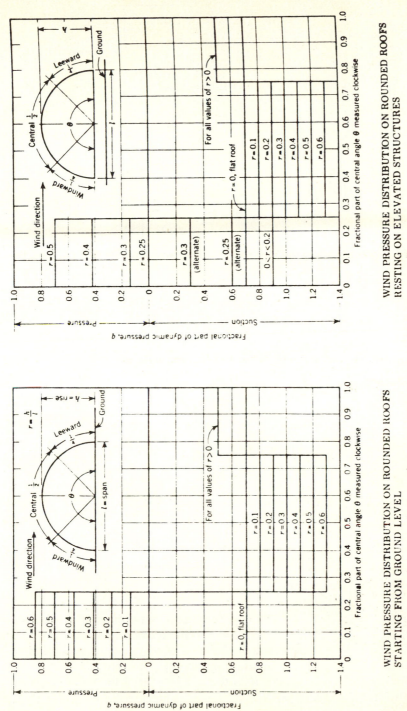

WIND PRESSURE DISTRIBUTION ON ROUNDED ROOFS
RESTING ON ELEVATED STRUCTURES

WIND PRESSURE DISTRIBUTION ON ROUNDED ROOFS
STARTING FROM GROUND LEVEL

Figure 3.7 External-pressure coefficients for arched roof when wind is perpendicular to the axis [7].

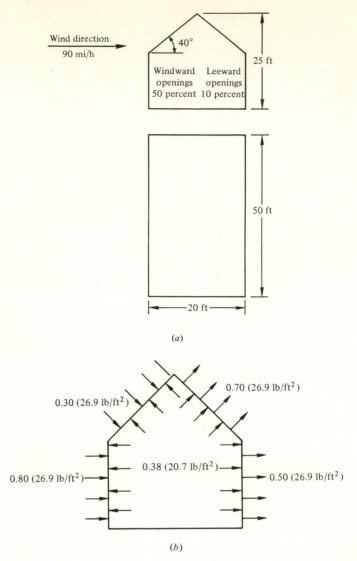

Figure E3.4

(a)

(b)

3.8 EARTHQUAKE LOADS

Earthquake loads must be taken into account in areas of great seismic activity, such as California, Alaska, and Japan. Seismic movement of the ground transmits accelerations to structures, and the mass of the structure resists the motion due to its inertia effects. This inertia force ranges from about $0.03W$ to $0.1W$ or more for most buildings, where W is the total weight of the building. This force is usually considered to act as a shear force (parallel to the surface of the earth) at the base of the structure.

The response of a structure depends on such factors as the stiffness and mass of the structure, foundation conditions, and amount of damping. As a result, the solution to this type of problem is very complex and can be solved only for artificially generated ground motion. This method may be desirable occasionally for special structures. Most common structures are designed by simpler methods usually detailed in codes. The seismic provisions of most codes are based on the recommendations of the Structural Engineers Association of California (SEACO) [8].

The dynamic forces are approximated by equivalent static forces. The total horizontal force is given by

$$V = ZKCW \tag{3.3}$$

In this equation V is the total lateral load or shear at base in pounds, and W is the total dead load in pounds. K is a numerical coefficient depending upon the ductility of the particular structure. It varies from about $\frac{2}{3}$ for a moment-resisting space frame to $1\frac{1}{2}$ for structures other than buildings. Z is a numerical coefficient depending on the zone of seismic activity. Figure 3.8 gives the zones of approximately equal seismic probability in the continental United States. C is a lateral-force coefficient for base shear which depends on the fundamental period of vibration T of the structure, where T is the time interval necessary for the structure to complete one fundamental mode of vibration. It depends on the mass and stiffness of the structure and on soil conditions. T varies from 0.1 s for a low, stiff building to about 5 s for a tall, flexible skyscraper. T varies from 0.2 to 0.4 s for most buildings. It can be determined from detailed analysis or

$$T = \frac{0.05H}{\sqrt{D}} \tag{3.4}$$

where H and D are the height and depth of the building in feet. The coefficient C is then given by

$$C = \frac{0.05}{\sqrt[3]{T}} \leq 0.10 \tag{3.5}$$

The lateral force V should be distributed over the height of the building (best done by considering the weight of the building to be concentrated at floor levels). This distribution can be computed from

$$F_x = V \frac{W_x h_x}{\sum W_i h_i} \tag{3.6}$$

where V = lateral force [Eq. (3.3)] due to earth movement
    W_x = weight of floor under consideration
    h_x = height of mass center of that floor from ground level
    $i = 1, \ldots, n$ = total number of floors
$$W = \sum_i^n W_i$$

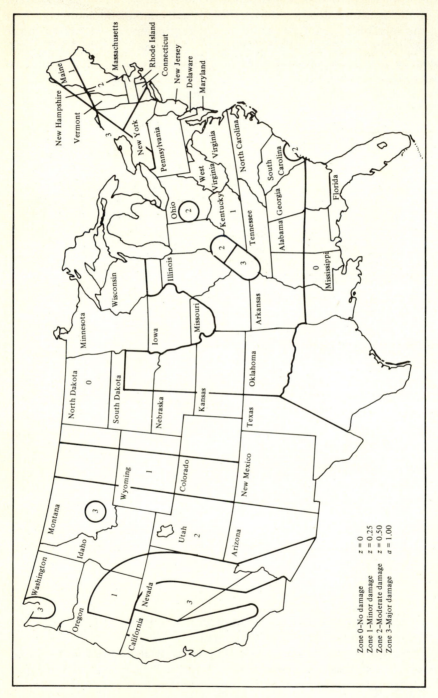

Figure 3.8 Zones of approximately equal seismic probability [9].

Zone 0–No damage $\quad z = 0$
Zone 1–Minor damage $\quad z = 0.25$
Zone 2–Moderate damage $\quad z = 0.50$
Zone 3–Major damage $\quad a = 1.00$

Additional guidelines [8] relate to such items as the lateral force on parts or portions of a structure, pile foundations, torsion, and overturning.

Example 3.5 A two-story building is designed to withstand a total lateral earthquake force $V = 0.06W$, where W = weight of building. The prescribed weights on each floor plus the heights of the mass centers above the base are given in Fig. E3.5. Compute the distribution of lateral force at each floor level.

SOLUTION

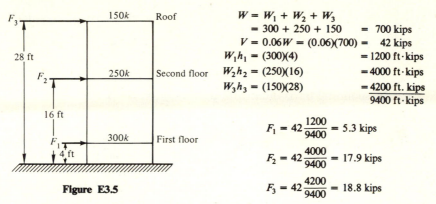

$$W = W_1 + W_2 + W_3$$

$$= 300 + 250 + 150 \qquad = 700 \text{ kips}$$

$$V = 0.06W = (0.06)(700) = \quad 42 \text{ kips}$$

$$W_1 h_1 = (300)(4) \qquad\qquad = 1200 \text{ ft} \cdot \text{kips}$$

$$W_2 h_2 = (250)(16) \qquad\qquad = 4000 \text{ ft} \cdot \text{kips}$$

$$W_3 h_3 = (150)(28) \qquad\qquad = \underline{4200 \text{ ft. kips}}$$

$$9400 \text{ ft} \cdot \text{kips}$$

Figure E3.5

$$F_1 = 42\frac{1200}{9400} = 5.3 \text{ kips}$$

$$F_2 = 42\frac{4000}{9400} = 17.9 \text{ kips}$$

$$F_3 = 42\frac{4200}{9400} = 18.8 \text{ kips}$$

3.9 WATER AND EARTH PRESSURES

The pressure exerted by a fluid is the same in all directions and is normal to a surface. The magnitude of the pressure at a given depth H is equal to the weight of fluid above the particular level,

$$p = \gamma H \qquad\qquad (3.7)$$

where p = pressure at particular point below surface

γ = unit weight of liquid

H = elevation difference between surface and point in question

This linear distribution occurs in tanks and vessels and on underwater structures (Fig. 3.9). The total force acting against a 1-ft width of wall is equal to the area in the pressure diagram,

$$P_r = \gamma H \frac{H}{2} \qquad\qquad (3.8)$$

and it acts in a horizontal direction at a depth of $2H/3$ below the surface. This assumption will be proved in Sec. 4.16.1.

Structures below ground, e.g., foundation walls, retaining walls, and tunnels, are subject to soil pressures, which depend on such factors as soil density, cohesion, angle of internal friction, rigidity of the structure, and the swelling of certain soils.

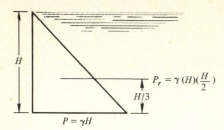

$$P_r = \gamma (H)\left(\frac{H}{2}\right)$$

$H/3$

$P = \gamma H$

Figure 3.9 Distribution of water pressure with depth.

According to a general theory of earth pressure, the vertical pressure at a depth Z in a level extensive mass of soil is equal to

$$p_v = \gamma Z \tag{3.9}$$

where γ is the unit weight of the soil.

The corresponding horizontal earth pressure p_h is given by

$$p_h = K_0 p_v \tag{3.10}$$

K_0 is designated the *coefficient of earth pressure at rest* and can be determined by field and laboratory tests. For rough estimation K_0 can be taken as 1 for clay and 0.5 for sand.

The total force per foot of width acting on a wall of height Z can be obtained from a pressure diagram similar to that in Fig. 3.9 and is equal to $K_0 \gamma Z^2/2$ acting through the centroid of the pressure diagram at a depth of $2Z/3$.

In practice it is unusual to have active soil pressures equivalent to those at rest. Any yield or movement of a structure away from the soil will be accompanied by a reduction in earth pressure to that of an active state. This minimum horizontal pressure, associated with a state of incipient shear failure at any depth Z for dry sands and gravels is

$$p_A = K_A \gamma Z \tag{3.11}$$

where $K_A = \tan^2(45 - \phi/2)$ is called the *coefficient of active earth pressure*, ϕ being the angle of internal friction of the soil. As already shown, the force per foot of wall width can be obtained from the triangular pressure diagram and is $K_A \gamma Z^2/2$. This analysis is based on dry sand. If the sand is saturated, ϕ remains unchanged but the static water pressure reduces the unit weight of the soil and consequently the effective stress. However, to this reduced effective stress must be added the pressure of the water as determined by Eq. (3.7). Cohesive soils do not remain in the active condition for long. Slow yield, called *creep*, tends to return the soil to an at-rest condition.

Example 3.6. A concrete swimming pool contains 10 ft of water. Determine the distribution of water pressure on the walls and the total force acting on a 1-ft strip of wall (Fig. E3.6).

SOLUTION

$$p_w = (62.4)(10) = 624 \text{ lb/ft}^2$$

$$P_W = \tfrac{10}{2} p_w = \frac{(624)(10)}{2} = 3120 \text{ lb}$$

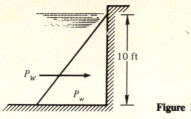

Figure E3.6

Example 3.7 (*a*) Compute the active earth pressure at a depth of 15 ft in a dry sand whose angle of internal friction is 38° and whose unit weight $\gamma_s = 108$ lb/ft^3. (*b*) Calculate the active pressure at the same depth in the same soil if the water table is at the ground surface. In this case the unit weight of the soil γ_s is 128 lb/ft^3, and the unit weight of water γ_w is 62.4 lb/ft^3.

SOLUTION (*a*) Dry sand

$$P_A = \gamma_s Z \tan^2\left(45 - \frac{\phi}{2}\right)$$

$$= (108)(15)(0.49)^2$$

$$= 389.0 \text{ lb/ft}^2$$

(*b*) Wet sand

$$P_A = Z(\gamma_s - \gamma_w) \tan^2\left(45 - \frac{\phi}{2}\right)$$

$$= (15)(128 - 62.4)(0.49)^2$$

$$= 236.2 \text{ lb/ft}^2$$

$$P_T = P_A + P_W = 236.2 + (15)(62.4)$$

$$= 1172 \text{ lb/ft}^2$$

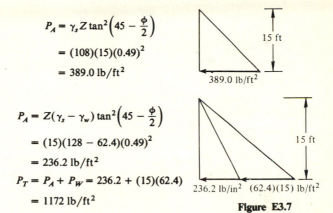

Figure E3.7

Example 3.8 Compute the active earth pressure acting against a wall at a depth of 6 m (19.7 ft) in a sand whose angle of internal friction is 34° and which weighs 15.4 kN/m^2 (98 lb/ft^3). What is the resultant force acting on a 1-m strip of wall?

SOLUTION See Fig. E3.8. The lateral pressure at depth of 6 m is

$$p_A = k_A \gamma Z$$

$$= \gamma Z \tan^2\left(\frac{45 - \phi}{2}\right)$$

$$= (15.4)(6)(0.532)^2 = 26.2 \text{ kN/m}^2$$

or

$$p_A = (98)(19.7)(0.532)^2 = 546 \text{ lb/ft}^2$$

The lateral force acting on a 1-m strip of wall is

$$p_A = K_A \gamma \frac{Z^2}{2}$$

$$= (0.532)^2(15.4)\frac{6^2}{2} = 78.5 \text{ kN} = \text{kPa}$$

$$= (0.532)^2(98)\frac{(19.7)^2}{2} = 5382 \text{ lb}$$

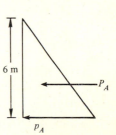

Figure E3.8

3.10 LOAD COMBINATION

Engineering judgment must be used when determining the critical combination of loads acting on a particular structure. It is unrealistic to expect a joint occurrence of all loadings. For example, the probability of maximum wind, snow, and earthquake loads occurring at the same time is negligible. It may be necessary to consider a combination of dead load, live load, and wind loads but not necessarily with a full snow load because it is unlikely that a full snow load would remain on a roof with a maximum wind velocity. The probability of critical load combinations, the safety of occupants, and the economic loss due to failure all vary from structure to structure and must be taken into account by the factor of safety.

3.11 CONVENTIONAL SUPPORTS

Most structures are either partly or completely restrained by supports connecting them to some stationary body like the earth. In restricting a structure's motion the supports resist the action of the forces acting upon the structure. These resistances are termed *reactions*. As a result, the effect of the supports can be replaced by the reactions. There are a few standard types of supports, described by conventional symbols.

A *hinge support* is represented by the symbol shown in Sketch 3.1a. It is normally assumed that the hinge is frictionless and that the support supplies a reactive force which has a line of action through the center of the hinge. The magnitude and direction of the reaction are unknown. These unknown elements can also be represented by the unknown magnitude of horizontal and vertical components that act through the pin (Sketch 3.1a). A *roller support* is normally represented by either of two symbols (Sketch 3.1b). The reactive force must be directed through the pin and be normal to the surface on which it rolls. It is applied at a known point in a known direction but is of unknown magnitude. Roller supports are usually designed so that they can provide a reactive force in either direction from the surface (Sketch 3.1b).

A *link support* reproduces the action of a roller support for situations requiring relatively small movement (Sketch 3.1c). Again, assuming that the pins are frictionless, the force must be transmitted along the link through the centers of the pins. As a result, a link supplies a reaction of known direction through a known point of application but of unknown magnitude (Sketch 3.1c).

Another common type of support, a *fixed support*, prevents not only the translation of the end of the member but also its rotation. This is normally accomplished by encasing the member in a solid support (Sketch 3.1d). A fixed support is therefore equivalent to a reaction, the magnitude, point of application, and direction being unknown. These three unknown elements are normally represented by a force of unknown magnitude and direction acting through a point and a couple or moment of unknown magnitude. The force may also be

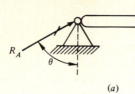
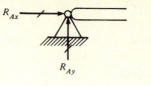

(a)

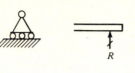

(b)

(c)

(d)

Sketch 3.1 Conventional supports and their symbols.

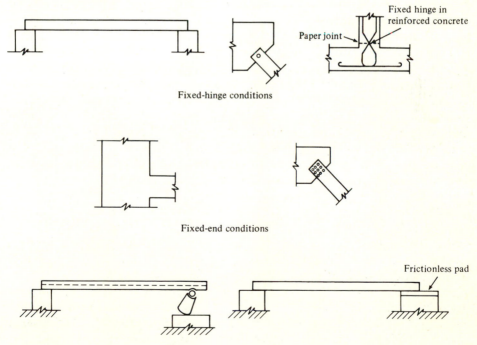

Fixed-hinge conditions

Fixed-end conditions

Frictionless pad

Moveable-hinge conditions

Sketch 3.2 Structural supports.

represented by vertical and horizontal reaction components of unknown magnitude (Sketch 3.1*d*).

In practice, a support is never completely free to move or rotate or totally prevented from moving or rotating. The definitions given above are only approximations of real behavior. Sketch 3.2 illustrates some of these approximations.

3.12 STRUCTURAL EQUILIBRIUM

The external forces acting on a structure, in terms of our previous discussion, consist of two distinct types, applied loads and reactions. Since most of the structures we consider are initially at rest and remain that way, the external forces must be in a state of static equilibrium. The implications of this condition and its importance in structural analysis and design will be developed later.

3.13 SUMMARY

The loads acting on a structure are categorized as dead, live, snow, wind, and earthquake loads. Live loads are considered in terms of floor and roof loads. The nature and magnitude of water and earth pressures are discussed because of their importance relative to basement walls and retaining structures.

Examples show how these loads act on a structure and how their magnitude and location can be determined.

Supports resist the action of loads acting on a structure. These supports are classified as fixed, hinged, and roller and are capable of supplying certain reaction forces to the structure.

PROBLEMS

3.1 A vertical wall 30 ft high moves outward enough to establish an active state in a dry-sand backfill. The sand density is 110 lb/ft^3, $\phi = 38°$, and $\gamma_{sat} = 135$ lb/ft^3.

(*a*) Compute the active earth-pressure distribution assuming a dry backfill. Draw the pressure diagram and compute P_A, the total earth pressure acting per foot of length of the wall.

(*b*) Compute the active earth pressure assuming that the water table rises to the surface of the soil behind the wall.

3.2 Water rises to level E in the pipe attached to tank $ABCD$. Neglecting the weight of the tank and pipe:

(*a*) Determine and locate the resultant force acting on area AB, which is 8 ft wide.

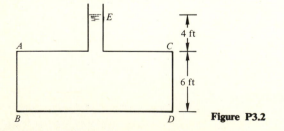

Figure P3.2

(*b*) Determine the total force on the bottom of the tank.

(*c*) Compare the total weight of the water with the result in (*b*).

3.3 A billboard 30 by 30 ft is supported at right angles to the surface of the earth. Determine the wind pressure on the billboard in a 90 mi/h wind blowing at right angles to the axis of the board.

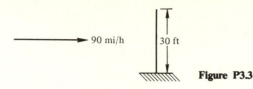

90 mi/h

30 ft

Figure P3.3

3.4 A tall building is subject to an earthquake giving a lateral force to the building of $0.10W$. Compute the lateral forces at the mass centers of various levels.

Floor level i	W_i, kips	h_i, ft
1	2000	2
2	1500	18
3	1500	32
4	1500	46
5	1500	60
6	1500	74
7	500	88

3.5 For a six-story building, a typical interior column supports a floor area of 20 by 20 ft on each floor. Assume that the dead weight on the roof is 20 lb/ft² and on the floors 40 lb/ft² per floor. The snow load on the roof is 30 lb/ft², and the live load on each floor is 60 lb/ft². Determine the load transmitted to the column at a level above the first floor considering (*a*) dead load and (*b*) live load with the following distribution:

Floor	Percent live load used in design
Roof	100
6	95
5	90
4	85
All others	80

3.6 An airtight mill building has a pitched roof, as shown in Fig. P3.6. Determine the external forces acting on the building in a 60 mi/h wind.

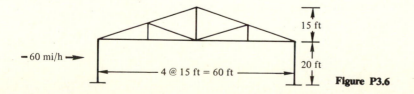

15 ft

60 mi/h

20 ft

4 @ 15 ft = 60 ft

Figure P3.6

3.7 The floor structure shown in Fig. P3.7 consists of beams which span 30 ft across the building and a floor slab which spans 14 ft between beams. Determine the load carried by the beams if the concrete slab is 6 in thick and the live load is 100 lb/ft^2.

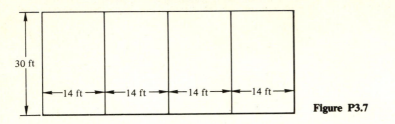

30 ft

—14 ft— —14 ft— —14 ft— —14 ft—

Figure P3.7

3.8 Figure P3.8 shows a reservoir holding a volume of water. What is the pressure on the reservoir wall at a depth of (*a*) 10 ft and (*b*) 50 ft? Determine the total force of water acting on a 1-ft length of reservoir wall.

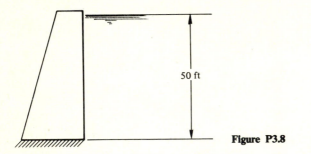

50 ft

Figure P3.8

REFERENCES

1. American Institute of Timber Construction, "Timber Construction Manual," 2d ed., Englewood, Colo., 1974.
2. Building Officials and Code Administration International, Inc., BOAC Basic Building Code/1978, Homewood, Ill., 1978.
3. International Conference of Building Officials, Uniform Building Code, Whittier, Calif., 1976.
4. Task Committee on Wind Forces, Committee on Loads and Stresses, Final Report, Wind Forces on Structures, *Trans. ASCE*, vol. 126, pt. II, p. 1124, 1961.
5. H. C. S. Thom, New Distributions of Extreme Winds in the United States, *Proc. Pap.* 6083, *J. Struct. Div. ASCE*, vol. 97-ST7, July 1968.
6. National Bureau of Standards, Strength of Houses, *BMS Rep.* 109, Washington, 1948.
7. ASCE Subcommittee 31, Committee on Steel, Final Report, Wind Bracing in Steel Buildings, *Trans. ASCE*, vol. 105, p. 1713, 1940.
8. Structural Engineers Association of California, Recommended Lateral Force Requirements Seismology Committee, 1967–1968 revisions, San Francisco.
9. S. T. Algermissen, Seismic Risk Studies in the United States, *Proc. 4th World Conf. Earthquake Eng., Santiago*, vol. 1, 1969.

CHAPTER 4

FORCE SYSTEMS

4.1 INTRODUCTION

Previous chapters discussed the nature of applied loads and the numbers and kinds of reactions required to support a structural system with these loads acting on it. This combination of load and support reactions constitute a *system of forces* with several unknowns. In order to design the structural system these unknowns must be evaluated. This chapter is important because it defines forces, classifies them into logical groups, and indicates ways of solving for the unknowns in such groups.

4.2 FORCE

According to Newton's first law of motion, a body will continue in its state of rest or of uniform motion unless acted upon by a force that changes or tends to change this state. We are primarily interested in bodies at rest. In this case a force can be defined as an action that changes the shape of the body upon which it acts. Newton's third law states that for every action there is an equal and opposite reaction. A force, being an action, is always opposed by an equal reaction. As a result forces always exist in pairs, equal and opposite. For example, the weight of a structural element is a force (gravitational) which acts vertically downward, and the reaction of the structure is equal and opposite, preventing the element from dropping. These forces are in equilibrium. Thus the loads acting on a structure can be considered forces. Force can be exerted only through the action of one physical body upon another, either through contact or at a distance.

4.3 TYPES OF FORCES

Forces can be classified under two general headings, *applied* (contact) *forces* and *nonapplied forces*. Examples of the latter are the gravitational pull of the earth on all physical bodies, magnetic force, and inertia forces.

An important distinction must be made between *external forces* acting on a structure and the resulting *internal forces* produced within the structure. For example, consider a truss acted upon by external forces (snow or wind loads and support reactions). These external forces produce internal forces in the members of the truss, which can be visualized in the sense that if a single member is cut, the truss will collapse.

External forces can be further divided into *acting* and *reacting* forces (Fig. 4.1). Gravity forces, wind forces, and the pressure of water and snow are examples of acting forces. Resisting forces developed at the supports of a structure are reacting forces.

Forces being resisted by material of limited strength must be distributed over an area. Nevertheless it is convenient to differentiate between a *concentrated force* and a *distributed force*. Examples of concentrated loads are the force exerted by the pull of a rope or the leg of a table. Distributed forces, on the other hand, are exemplified by the action of wind on the side of a building or the pressure of water on the side of a tank. The method of indicating such forces is shown in Fig. 4.2.

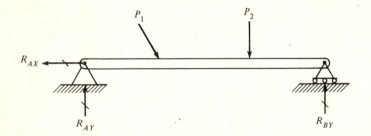

Figure 4.1 Acting and reacting forces. P_1 and P_2 are acting forces; R_{AX}, R_{AY}, and R_{BY} are reacting forces.

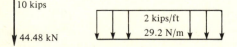

Concentrated load   Distributed load   **Figure 4.2** Representation of forces.

4.4 TYPES OF FORCE SYSTEMS

A structure is acted upon by a system of forces representing known applied loads and unknown reaction components. Systems of forces have been classified into a number of groups. When the forces of a given system are in a particular plane, they are said to be *coplanar*. When the system is three-dimensional, it is said to be *noncoplanar*. When the lines of action of all the forces intersect at a single point, the forces are said to be *concurrent*. When the lines of action do not intersect at a single point, they are said to be *nonconcurrent*. Forces with the same line of action are said to be *collinear*. With these definitions force systems can be classified as (1) coplanar parallel, (2) coplanar concurrent, (3) coplanar nonconcurrent, (4) noncoplanar parallel, (5) noncoplanar concurrent, and (6) noncoplanar nonconcurrent. These systems will be discussed more fully after a few basic concepts involving forces and moments have been considered.

4.5 CHARACTERISTICS OF A FORCE

A force can be completely described by its *magnitude*, *direction*, and *point of application*. The magnitude is normally expressed in pounds, kips, or newtons. The direction of a force is the direction of the line along which the force acts; it is expressed by an angle made with some coordinate system, normally a rectangular system with x and y axes. Unless otherwise specified, the angle is measured counterclockwise from the positive x axis. Since an applied force is transmitted from one body to another by contact, the point of application is the point of contact between the two bodies. The point of application of a force on a body is of importance when we study its effect on the body; it is a *vector* quantity, graphically represented by a vector (Sketch 4.1). The length of the line denotes the magnitude of the force P and its angle of inclination θ the direction. The sense is indicated by the vector head and the point of application by the location of the vector head.

4.6 TRANSMISSIBILITY OF A FORCE

This principle stipulates that the point of an external force acting on a body can be considered to act anywhere along its line of action without changing the other external forces acting on the body (Sketch 4.2). The reactions of the beam

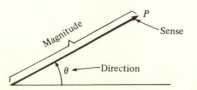

Sketch 4.1

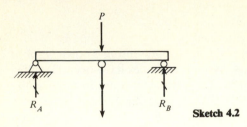

Sketch 4.2

are going to remain the same whether the force acts down on the beam, pulls at the ring, or is applied some distance down on a rope attached to the ring.

4.7 COMPONENTS OF A FORCE

A force can be replaced by two or more components that are the vector equivalent of the force they replace. This technique is useful in solving problems. Forces that are neither vertical nor horizontal can be resolved into vertical and horizontal components to facilitate the summation process. With rectangular coordinates the components can be evaluated analytically as follows. Angle θ in Sketch 4.3 defines the direction of force OC in the coordinate system. By trigonometry

$$OA = OC \sin \theta \quad \text{and} \quad OB = OC \cos \theta$$

If the direction of OC is defined by the slope triangle, abc and BC can be replaced by OA, making a force triangle which is similar to abc; then

$$\frac{a}{OA} : \frac{b}{OB} : \frac{c}{OC}$$

and by cross multiplication

$$OA = \frac{a}{c} OC \quad \text{and} \quad OB = \frac{b}{c} OC$$

where $a/c = \sin \theta$ and $b/c = \cos \theta$.

Since vectors can be added graphically (geometrically), the components of a force can be recombined to give the original force. This can be done by the triangle law; the y component is placed at the end of the x component and the resultant is then equal to the original force (Sketch 4.4). Algebraically the

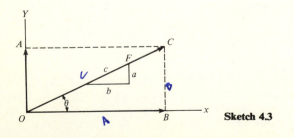

Sketch 4.3

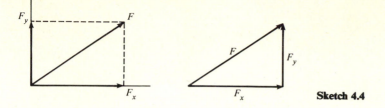

Sketch 4.4

components can be combined to form the original force by using the Pythagorean theorem

$$F^2 = F_x^2 + F_y^2 \tag{4.1}$$

Example 4.1 Determine the x and y components of the 100-lb (444.8-N) force in Fig. E4.1.

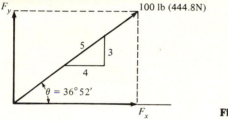

Figure E4.1

SOLUTION

$$F_x = F\cos\theta = (100)(0.800) = 80 \text{ lb}$$
$$= 444.8(0.800) = 355.8 \text{ N}$$
$$F_x = F \times 4/5 = (100)(4/5) = 80 \text{ lb}$$
$$= 444.8 \times 4/5 = 355.8 \text{ N}$$
$$F_y = F\sin\theta = (100)(0.600) = 60 \text{ lb}$$
$$= 444.8(0.600) = 266.9 \text{ N}$$
$$F_y = F \times 3/5 = (100)(3/5) = 60 \text{ lb}$$
$$= 444.8(3/5) = 266.9 \text{ N}$$
$$F^2 = F_x^2 + F_y^2$$
$$(100)^2 = (80)^2 + (60)^2$$
$$(444.8)^2 = (355.8)^2 + (266.9)^2$$

4.8 MOMENT OF A FORCE

The moment of a force is its tendency to rotate the body upon which it acts about some axis. A moment is quantified by the product of the force and the perpendicular distance between the axis of rotation and the force's line of

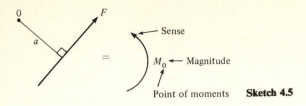

Sketch 4.5

action. This distance is called the *moment arm* (Sketch 4.5)

$$\text{Moment} = Fa \tag{4.2}$$

The units of a moment are the product of the units of force and distance. The basic units are inch-pounds (in·lb) and newton-meters (N·m). Other units are foot-pounds (ft·lb) and foot-kips (ft·kips).

It is important to realize that the moment of a force about an axis passing through its line of action is zero.

Varignon's theorem states that about any point, the algebraic sum of the moments of a force's components equals the moment of the force. This theorem can readily be proved and is based on the fact that a force can be replaced by its components without changing the total effect. This principal can be readily extended to a resultant force; i.e., about any point, the moment of the resultant of a system of forces equals the algebraic sum of the moments of the separate forces.

Example 4.2 Illustrate Varignon's theorem relative to the resultant 100-lb force and its two components about points A and B (Fig. E4.2).

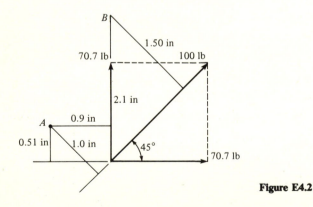

Figure E4.2

SOLUTION Take moments about A:

$$(1.00)(100) = (70.7)(0.90) + (70.0)(0.51)$$
$$100 = 66.63 + 36.07 \approx 100$$

because of accuracy of measurements. Take moments about B:

$$(1.50)(100) = 70.7 + (2.12)(70.7)$$
$$150 = 150$$

Note the advantage of having the center of moments on the line of action of one of the components.

Example 4.3 Determine the moment about A of the 100-N force shown in Fig. E4.3a.

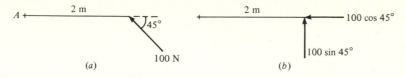

Figure E4.3

SOLUTION To find the moment about A resolve the force into vertical and horizontal components (Fig. E4.3b). The moment about A is

$$(-70.7 \text{ N})(2 \text{ M}) + (0)(70.7 \text{ N}) = -141.4 \text{ N} \cdot \text{m} \qquad \text{counterclockwise}$$

4.9 COUPLES

A couple can be defined as a system of two parallel forces equal in magnitude but opposite in direction (Sketch 4.6). The moment of this combination of forces is equal to the magnitude of one of the forces multiplied by the perpendicular distance between them. The moment of the couple in Sketch 4.6 is Fa.

4.10 CHARACTERISTICS OF A COUPLE

This combination of forces has a number of characteristics. The resultant force of such a system is zero. The moment of a couple is the same for all points in the plane of the couple, and a couple can be balanced only by an equal and opposite couple in the same plane or a parallel plane. Two couples are equivalent if their moments and directions are the same. The particular values of the forces and moment arm are not significant since only the product of the two determines the action of the couple. The forces of a couple can be rotated through any angle in their plane, translated to any position in the plane, or translated into any parallel plane without changing the motional effects of the couple on a body. Some of these properties are illustrated in Example 4.4.

F

a

$=$

$M = Fa$

F

Sketch 4.6

Example 4.4 Illustrate the principle that the moment of a couple is dependent upon the product of the magnitude of one of the forces and the distance between them and not on their location in space (Fig. E4.4).

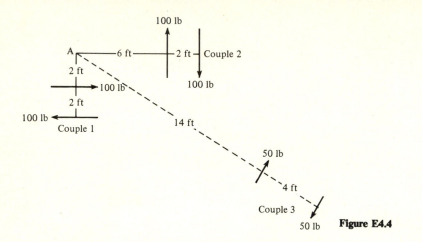

Figure E4.4

SOLUTION Counterclockwise moments are positive and clockwise moments are negative. The moments about A of forces in couple 1 are

$$(+100)(2) - (100)(4) = -200 \text{ ft·lb} = \text{moment of couple 1}$$

The moments about A of forces in couple 2 are

$$(+100)(6) - (100)(8) = (-100)(2) = -200 \text{ ft·lb} = \text{moment of couple 2}$$

$$\text{Moment of couple 1} = (-100)(2) = \text{moment of couple 2} = (-100)(2)$$

Couple 1 is equal to couple 2 because each has two forces, parallel, equal, and opposite, of the same magnitude and 2 ft apart.

The moments about A of forces in couple 3 are

$$(+50)(14) = (50)(18) = (-50)(4) = -200 \text{ ft·lb}$$

Couple 3 is also equal to couples 1 and 2 because the product of the forces and the distance between them are equal in each case.

4.11 RELOCATING A FORCE

Sometimes it is desirable to move a force from a given location to some new location. Consider the column bracket shown in Sketch 4.7*a*, which has a force P located at a distance e (called the *eccentricity*) from the centerline of the column. For design purposes it is preferable to have the force act along the centerline of the column. The transposition can be accomplished by adding and subtracting forces of magnitude P along the centerline of the column (see Sketch 4.7*b*). This does not change the loading on the column and bracket. This new combination of loads can be considered as an axial load P acting down the centerline of the column and two parallel forces of equal magnitude but opposite sign spaced a distance e apart. These two forces constitute a couple with a moment magnitude equal to Pe. Sketch 4.7*c* shows the final effect of transferring the load P to the centerline of the column. The column can be designed for this system of forces.

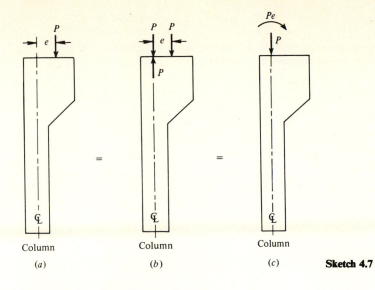

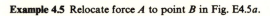

Column (a)   Column (b)   Column (c)   **Sketch 4.7**

Example 4.5 Relocate force A to point B in Fig. E4.5a.

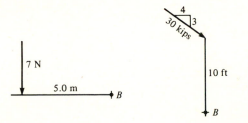

(a)

 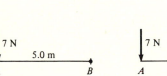 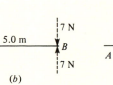

(b)

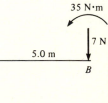

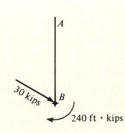

(c)

SOLUTION See Fig. E4.5b and c.

Example 4.6 Change each eccentrically loaded column shown in the structural diagrams in Fig. E4.6a into an axially loaded column with a moment.

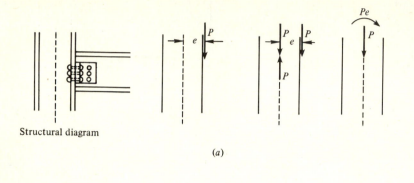

Structural diagram

(a)

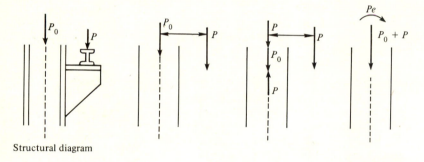

Structural diagram

(b)

Figure E4.6

SOLUTION See Fig. E4.6b.

4.12 FORCE SYSTEMS

It is not unusual to extract a single planar frame from a three-dimensional building system and consider the system of forces acting upon it. This simplifies the analysis and design of the structure. As a result, coplanar force systems are of considerable importance and will be discussed in detail in this chapter. Three-dimensional space structures which cannot be handled in this manner will be considered in some detail later in this chapter.

Coplanar force systems resolve themselves into concurrent and nonconcurrent systems. Parallel-force systems, which represent gravitational forces, are a special case of the nonconcurrent systems.

In considering force systems it may be convenient to replace a system of forces by a resultant, which produces the same effects as the forces it replaces.

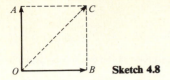

Sketch 4.8

4.13 RESULTANT OF TWO CONCURRENT FORCES

Consider the case where the two forces are at right angles (Sketch 4.8). We can use the Pythagorean theorem to find that the resultant force equals $OC = \sqrt{OA^2 + OB^2}$ since BC is equivalent to OA.

Example 4.7 Determine the magnitude and direction of the resultant of the two forces shown in Fig. E4.7.

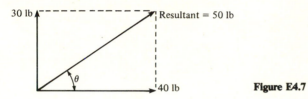

Figure E4.7

SOLUTION

$$R = \sqrt{30^2 + 40^2} = 50 \text{ lb}$$
$$\tan \theta = \tfrac{30}{40} = 0.75$$
$$\theta = 36°52'$$

When the forces are not at right angles to each other, the resultant can be found trigonometrically by means of the cosine law. In Fig. 4.3a the resultant becomes the diagonal OC of the parallelogram $OACB$. In Figure 4.3b OA is placed in the position of BC, and the resultant OC becomes part of the triangle OCB. In both representations the magnitude and direction of the resultant force

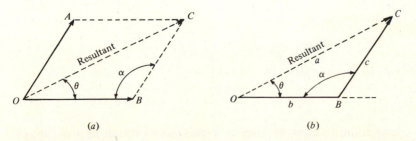

(a)

(b)

Figure 4.3 Resultants by use of cosine law: (a) parallelogram method; (b) triangle method.

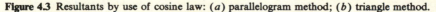

can be obtained by using the cosine law

$$a^2 = b^2 + c^2 - 2bc \cos \alpha \tag{4.3}$$

Example 4.8 Determine the magnitude and direction of the resultant of the two forces shown in Fig. E4.8.

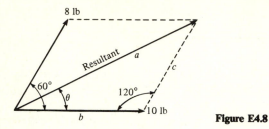

Figure E4.8

SOLUTION

$$\alpha = 180° - 60° = 120° \qquad \cos 120° = -0.50$$

To find R use the law of cosines:

$$R^2 = (10)^2 + (8)^2 - (2)(10)(8) \cos 120°$$
$$= 164 - (160)(-0.5) = 244$$
$$R = 15.6 \text{ lb}$$

To find θ use the law of sines:

$$\frac{\sin \theta}{c} = \frac{\sin 120}{R} \qquad \sin \theta = \frac{c \sin 120°}{R}$$

$$\sin \theta = \frac{(8)(0.866)}{15.6} = 0.444$$

$$\theta = 26°22'$$

4.14 RESULTANT OF A COPLANAR CONCURRENT FORCE SYSTEM

The resultant of a concurrent system of forces acts through the point of concurrence. Thus it will be fully defined when its magnitude, sense, and line of action are known. These characteristics can be determined either algebraically or graphically (geometrically).

The algebraic method involves a summation of rectangular components ΣF_x and ΣF_y, from which

$$R = \sqrt{(\Sigma F_x)^2 + (\Sigma F_y)^2} \tag{4.4}$$

and its direction in space is given by

$$\tan \theta = \frac{\Sigma F_y}{\Sigma F_x} \tag{4.5}$$

The graphical solution involves developing a polygon of forces by the triangle method. In Fig. 4.4b the forces P, Q, and S are laid out to scale end to end. The

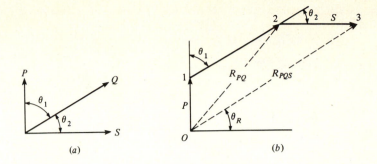

Figure 4.4 Resultant by graphical means: (*a*) space diagram; (*b*) triangle method.

diagonal $O2$ is the resultant of the forces P and Q and diagonal $O3$ is the resultant of forces P, Q, and S. The intermediate resultant R_{PQ} is not usually drawn.

Example 4.9 Determine the resultant R of the concurrent system of forces shown in Fig. E4.9*a* and its angle of inclination θ with the horizontal.

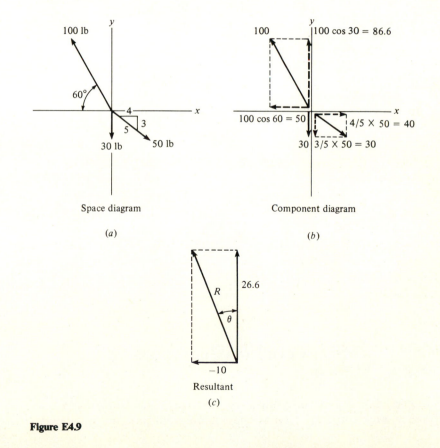

Figure E4.9

SOLUTION The vertical and horizontal components of each force are computed as described in Sec. 4.7 (see Fig. E4.9b):

$$\overrightarrow{\Sigma F_x} = -50 + 40 = -10$$

$$\uparrow \Sigma F_y = 86.6 - 30 - 30 = +26.6$$

For convenience, forces acting to the right and up are considered positive (Fig. E4.9c):

$$R = \sqrt{(+26.6)^2 + (-10)^2} = 28.4$$

$$\tan \theta = \frac{F_y}{F_x} = \frac{-10}{+26.6} = 0.375$$

$$\theta = 20°33'$$

It is important to realize which axis the angle θ uses as a reference. Here since we are dealing with $-\Sigma F_x$ and $+\Sigma F_y$, it is relative to the positive y axis in a counterclockwise direction.

Example 4.10 Determine the resultant of the concurrent force system of Example 4.9 graphically.

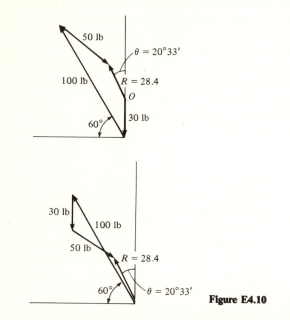

Figure E4.10

SOLUTION Starting with point O; applying the polygon law, add vectorially each force in turn. The resultant of the system is drawn from O to the arrowhead of the last force (Fig. E4.10). Note that the resultant R is independent of the sequence of vectorial addition.

4.15 RESULTANT OF A COPLANAR NONCONCURRENT SYSTEM OF FORCES

Since a coplanar nonconcurrent system of forces does not have a point of concurrence, each force will tend to rotate about a given point in the plane and

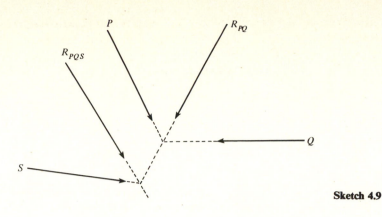

Sketch 4.9

the system of forces will have a net moment about this point. This moment is used to locate the resultant force in the plane by recognizing that the moment of the resultant about the given point must be equal to the moment of the system of forces about the same point. The magnitude and direction of the line of action of the resultant are determined in the same way as for the resultant of a concurrent system of forces. The resultant can be determined completely by either the algebraic or graphical method.

The resultant of a coplanar nonconcurrent system of forces can be determined graphically by determining the resultant of two of the forces. This resultant is then combined graphically with a third force, and this process is continued until a final resultant is obtained for the entire system, as illustrated in Sketch 4.9.

Another, more efficient method of determining the resultant of a coplanar nonconcurrent force system graphically involves the use of a force polygon (Sec. 4.14) to determine the magnitude and direction of the resultant and a funicular force polygon to locate the resultant in the space diagram of the force system.

Consider the system of four forces shown in Fig. 4.5. To identify the forces *Bow's notation* is used. The spaces between the forces in the space diagram are lettered with lowercase letters to designate the forces. Equivalent capital letters are used to designate the forces in the force polygon. The magnitude and direction of the resultant force *AE* can be measured directly from the force polygon.

To find the location of the resultant in the space diagram a funicular polygon is drawn. A pole *O* is selected adjacent to the force polygon. The exact location is unimportant. The pole *O* is connected by rays to the points *A*, *B*, etc. These rays can be considered components of the forces. For example, the rays *AO* and *OB* represent components of the force *AB*, and rays *BO* and *OC* those of force *BC*. The components *OB* and *BO* are equal, opposite, and collinear. This system of components is then transferred to the space diagram, where it replaces the force system. These components, called *strings*, are designated by lowercase letters associated with the spaces between the forces. This replacement

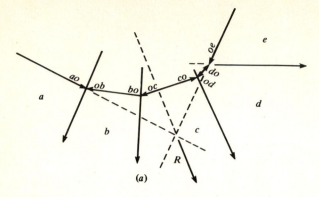

(a)

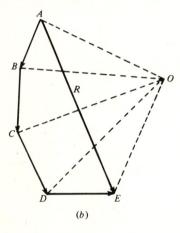

(b)

Figure 4.5 Resultant of a nonconcurrent system of forces: (*a*) space
diagram and funicular polygon;
(*b*) force polygon.

system is started by selecting any point *P* on the line of action of the force *ab* in
the space diagram and by drawing through this point lines (strings) parallel to
the rays *AO* and *OB*. These strings are designated *ao* and *ob*. String *ob* is
extended to intersect force *bc* and becomes component *bo* of force *bc*. A string
oc is then drawn from this point of intersection parallel to ray *OC* and becomes
a component *oc* of the force *bc*. This procedure is continued until the entire
force system has been replaced by its components. This diagram is called a
funicular or *string polygon* since it has the form which would be assumed by a
flexible string loaded by the original force system. In this particular string
polygon the components *ob* and *bo*, *oc* and *co*, and *od* and *do* cancel each other,
leaving components *ao* and *oe*, the components of the resultant. The point of
intersection of these two strings is on the line of action of the resultant force *ae*.

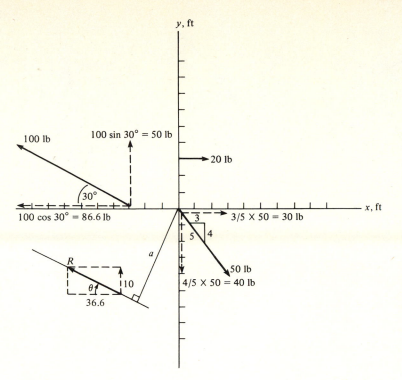

Figure E4.11

Example 4.11 Determine algebraically the magnitude, direction, and location of the resultant of the forces shown in Fig. E4.11.

SOLUTION This figure shows a nonconcurrent force system made up of three forces. The forces are resolved into x and y components

$$\overrightarrow{\Sigma F_x} = 20 + 30 - 86.6 = -36.6 \text{ lb}$$

$$\uparrow \Sigma F_y = 50 - 40 = 10 \text{ lb}$$

$$R = \sqrt{(36.6)^2 + (10)^2} = 37.9 \text{ lb}$$

$$\tan \theta = \frac{10}{36.6} = 0.273$$

$$\theta = 15.3° \quad \text{clockwise from negative } x \text{ axis}$$

The sum of the moments of the system about the origin is

$$\Sigma M_{(O,O)} = (20)(3) + (50)(3) = 210 \text{ ft·lb}$$

The moment of the resultant, 37.9 lb, about the origin must also be equal to 210 ft·lb

$$37.9a = 210 \text{ ft·lb}$$

where a is the perpendicular distance from the origin to the resultant

$$a = \frac{210}{37.9} = 5.54 \text{ ft}$$

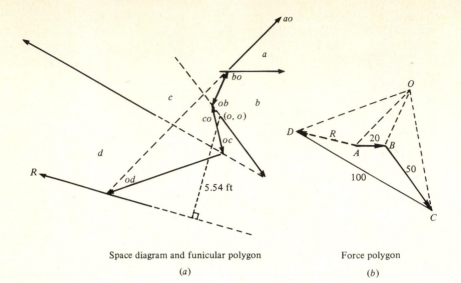

Space diagram and funicular polygon

(a)

Force polygon

(b)

Figure E4.12

Example 4.12 Solve Example 4.11 graphically.

SOLUTION See Fig. E4.12.

4.16 THE RESULTANT OF A COPLANAR PARALLEL-FORCE SYSTEM

Since vertical forces can be used to represent gravitational forces, they are often found in engineering practice as parallel-force systems. The resultant of such a system is often useful in the solution of problems. Both analytical and graphical solutions are available. This is a special case of a nonconcurrent system.

Example 4.13 Determine the magnitude and location of the line of action of the resultant of the four parallel forces shown in Fig. E4.13.

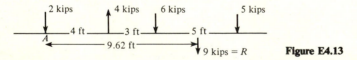

Figure E4.13

SOLUTION The resultant of the system is

$$\downarrow \Sigma F_y = R = 2 - 4 + 6 + 5 = 9 \text{ kips}$$

The moment of the system of forces about A is

$$\widehat{M_A} = -4(4) + (6)(7) + (12)(5) = 86 \text{ ft} \cdot \text{kips}$$

The moment of the resultant about A is

$$Ra = 9a = 86 \text{ ft} \cdot \text{kips}$$

$$a = \frac{86}{9} = 9.6 \text{ ft}$$

Example 4.14 Determine graphically the magnitude and location of the line of action of the resultant of the parallel system of forces in Fig. E4.14.

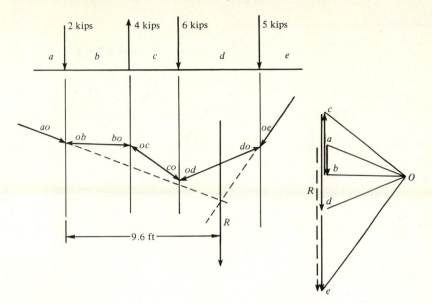

Figure E4.14

SOLUTION See the figure.

Example 4.15 Determine algebraically the resultant of the parallel system of forces shown in Fig. E4.15.

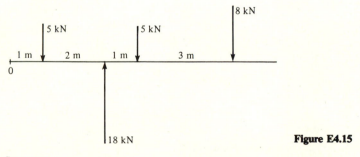

Figure E4.15

SOLUTION

$$\uparrow \Sigma F_y = -5 + 18 - 5 - 8 = 0$$

$$\Sigma M_o = (5)(1) - (18)(3) + (5)(4) + (8)(7) = 27 \text{ kN} \cdot \text{m}$$

The resultant of the system of forces is a clockwise moment of 27 kN·m. This moment can be represented by any couple whose moment is +27 kN·m.

4.16.1 Distributed Loads

Structural members like beams, foundations, and tanks containing liquids resist loads that are distributed instead of concentrated. The intensity of the load is

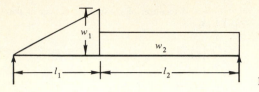

Figure 4.6 Distributed loads.

then expressed in terms of pounds per foot of length or some similar unit, and the loading is illustrated by a pressure or load diagram (Fig. 4.6 and Example 3.1).

When dealing with the equilibrium of external loads, it is often convenient to use the resultants of the distributed loads. The concept of a center of pressure through which the resultant pressure or load acts can be developed as follows. Consider a beam of uniform depth and homogeneous material. The beam can be divided into a number of small elements of equal size, as shown in Fig. 4.7. If the weight of each element is w concentrated at its center, a parallel-force system is formed whose resultant W is the weight of the beam.

Let the weights of these elements be designated by w_1, w_2, etc., and their coordinates be x_1, x_2, etc., and let the coordinates of the resultant weight W be $\overline{X}$. Now, about any point O, the moment of the resultant W equals the algebraic sum of the moments of the separate weights

$$W\overline{X} = w_1x_1 + w_2x_2 + \cdots = \Sigma wx \tag{4.6}$$

thus, in essence locating the center of gravity of the beam.

If we allow the beam to decrease in thickness until it becomes zero, the beam no longer has weight and only a surface or area remains. What was the center of gravity of a body has become the centroid of an area. When it represents a pressure or load diagram, this area has dimensions of length and load intensity (feet times pounds per foot = pounds).

Equation (4.6) now becomes

$$\overline{X} = \frac{\Sigma ax}{A} \tag{4.7}$$

where a represents the area of a component part and A is the total area, and it can be used to determine the centroid of any area and consequently the center of pressure of any pressure or load diagram.

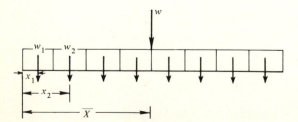

Figure 4.7 Distributed load divided into elements.

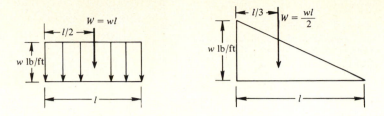

Figure 4.8 Common load distributions.

The centroids of common geometric areas have been found by geometry and calculus (Table 10.1). The areas most frequently used in pressure diagrams are given in Fig. 4.8. More complex pressure diagrams can be handled by resolving them into a number of simple areas or diagrams. The centroid of the original or composite area can then be determined by using Eq. (4.7).

Example 4.16 Determine the location and magnitude of the resultant of the distributed loads in Fig. E4.16a.

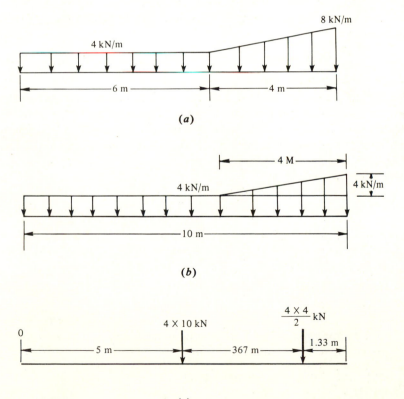

Figure E4.16

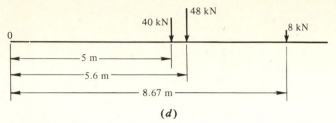

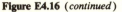

(d)

Figure E4.16 (*continued*)

SOLUTION The loads can be considered as a uniform loading of 4 kN/m over the entire 10-m length and a triangular loading varying from 0 to 4 kN/m over a 4-m length (Fig. E4.16b). The resultant of each loading is equal to the area of the load diagram and acts through the centroid of the diagram (Fig. E4.16c):

$$\downarrow \Sigma F_y = 40 + 8 = 48 \text{ kN} \qquad \Sigma M_o = (40)(5) + (8)(8.67) = 269.4 \text{ kN} \cdot \text{m}$$

The moment of the resultant of the system must be equal to the sum of the moments of the components (Fig. E4.16d):

$$48d = (40)(5) + (8)(8.67)$$

where d is the perpendicular distance from the center of moments o to the resultant force,

$$d = 5.6 \text{ m}$$

4.17 EQUATIONS OF STATIC EQUILIBRIUM

A body that is initially at rest and remains at rest when acted upon by a system of forces is considered to be in a state of static equilibrium. For such a condition to exist the net effect of the system of forces must be zero, neither a force nor a couple. As a result, the algebraic sum of all the horizontal components of the forces acting on the body must be equal to zero. Likewise, the algebraic sum of the vertical components must be equal to zero. The algebraic sum of the moments of all the forces about any axis normal to the plane of the structure must also be equal to zero. These three conditions, fulfilled simultaneously by the loads and reactions of a planar structure in a state of static equilibrium, can be designated by

$$\Sigma F_x = 0 \qquad \Sigma F_y = 0 \qquad \Sigma M_z = 0 \qquad \qquad (4.8)$$

called the *equations of static equilibrium* of a planar structure subjected to a general system of forces.

The equations for static equilibrium for a three-dimensional system of forces can readily be obtained by considering the equations of equilibrium for the three planes involved.

xy plane:

$$\Sigma F_x = 0 \qquad \Sigma F_y = 0 \qquad \Sigma M_z = 0$$

xz plane:

$$\Sigma F_x = 0 \qquad \Sigma F_z = 0 \qquad \Sigma M_y = 0$$

yz plane:

$$\Sigma F_y = 0 \qquad \Sigma F_z = 0 \qquad \Sigma M_x = 0$$

Eliminating the common equations leaves us with a set of six equations of equilibrium:

$$\Sigma F_x = 0 \qquad \Sigma F_y = 0 \qquad \Sigma F_z = 0$$

$$\Sigma M_x = 0 \qquad \Sigma M_y = 0 \qquad \Sigma M_z = 0 \qquad (4.9)$$

These equations enable us to solve for the unknown quantities or characteristics of the forces in a system that is in equilibrium. These characteristics are (1) the magnitude and sense of a force, (2) the direction of the line of action of a force, or (3) the location of the line of action of the force or of its point of application. Usually, it is the magnitude and/or the line of action of a given force that is used.

4.18 STATICALLY DETERMINATE AND INDETERMINATE STRUCTURES

Whenever there are enough equations of static equilibrium to solve for the unknowns in a system of forces acting on a structure, the structure is said to be *statically determinate*. There are many situations where the number of unknown forces in the system (usually reaction components) exceed the number of equations of static equilibrium. This type of structure is termed *statically indeterminate*. The degree of indeterminacy is indicated by the number of unknown forces in excess of the equations of static equilibrium. Examples are illustrated in Sketch 4.10. Statically indeterminate structures require additional relationships which depend on the compatability of elastic deformations for the solution of the unknowns.

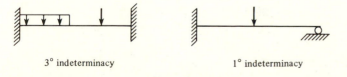

3° indeterminacy   1° indeterminacy

Sketch 4.10

4.19 SIGNS OF FORCES AND MOMENTS

No sign convention need be used in the summation of forces and moments. In any given direction, the sum of the forces acting in one way must be equal to the sum of the forces acting in the opposite way. Similarly the sum of the clockwise moments is equated to the sum of the counterclockwise moments. If necessary, signs can be conveniently assigned to the forces and moments. Forces acting upward or to the right are usually considered positive, and forces acting downward or to the left are negative. Clockwise moments are often considered positive and counterclockwise moments negative.

4.20 PRINCIPLES OF FORCE EQUILIBRIUM

A number of important principles can be developed for forces in equilibrium. These principles can be very helpful in solving problems.

4.20.1 The Two-Force Principle

When two forces are in equilibrium, they must be equal, opposite, and collinear. The proof of this principle is embodied in Newton's third law: to every action there is an equal and opposite reaction (Sketch 4.11).

4.20.2 The Three-Force Principle

When three nonparallel forces are in equilibrium, their lines of action must intersect at a common point. The proof of this principle can be developed from the fact that the resultant of two of the forces must be equal, opposite, and collinear with the third, according to the two-force principle. This, of course, means that the system of forces must be coplanar (Sketch 4.12).

4.20.3 The Four-Force Principle

When four forces are in equilibrium, the resultant of any two of the forces must be equal, opposite, and collinear with the resultant of the other two forces. The proof of this principle is an extension of the one used for the three-force

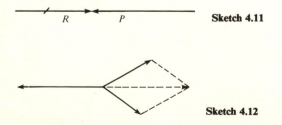

R P **Sketch 4.11**

Sketch 4.12

principle above. The two-force principle can be applied to any two resultants regardless of the number of forces involved in the system.

4.21 FREE-BODY DIAGRAMS

Methods of solving for unknowns in a system of forces in equilibrium are straightforward. The problem confronting the engineer and architect is primarily one of determining the system of forces representing a given physical situation. Two common problems exist: (1) to determine the external forces, called *reactions*, acting at the supports of a structure and caused by the loads it carries and (2) to determine the internal forces which the external forces produce in the various members or parts of the structure.

In order to isolate a particular system it is necessary to distinguish between external forces and internal forces. When isolating a portion of a structural system by cutting a member or members, the internal forces within the cut members become part of the external system. A diagram showing all the external forces acting on a structure or portion of a structure is called a *free-body diagram*. The preparation of a free-body diagram should be the first step in the solution of every problem.

A free-body diagram of the entire structure is drawn by isolating the structure from its supports and showing it acted upon by all the applied loads and possible reaction components that the supports may supply to the structure. Such a diagram is illustrated in Sketch 4.13*b*.

In a similar way, any portion of the structure can be isolated by passing a cutting plane through the desired members. The free-body diagram shows this portion of the structure acted upon by the applied loads and reactions together with any forces which may act on the faces of the members cut by the isolating section. Such a free-body diagram is illustrated in Sketch 4.14*b*.

Any force of unknown magnitude can be assumed to act in either sense along its line of action. For a cut two-force member (internal force is axial because loads are applied only at the joints of structure) it is convenient to indicate the internal force as *tensile*. Thus if subsequent calculations give a positive sign, the force not only acts in the assumed direction but also has the sign normally associated with tension. Conversely a negative sign indicates the force to be of opposite sign and therefore *compressive*.

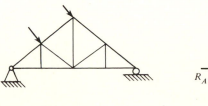

(*a*)

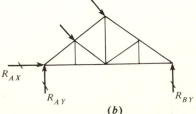

(*b*)

Sketch 4.13

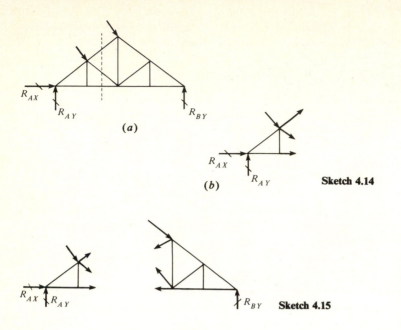

Sketch 4.14

Sketch 4.15

If free-body sketches or diagrams are drawn for two adjacent portions of the structure and the internal forces have been assumed to act in certain senses on an internal face of one portion, the corresponding forces must be assumed to act with the same numerical values but in opposite senses on the matching face of the adjacent portion (Sketch 4.15) since the action and reaction of one body on another must be numerically equal but opposite in sense. This consistency within free-body diagrams is necessary to obtain correct solutions from them. It must be remembered that a reaction assumed to act in one direction in a given free-body diagram must be shown to act in the same direction in the other diagrams.

Example 4.17 On the basis of the structural diagrams in Fig. E4.17a to c draw appropriate free-body diagrams.

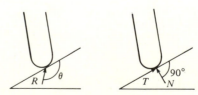

Structural diagram   Free body diagram

Figure E4.17   (a)

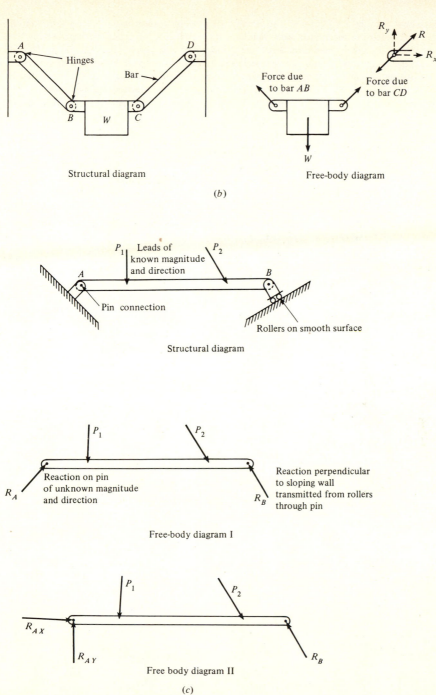

Structural diagram

Free-body diagram

(b)

Structural diagram

Free-body diagram I

Free body diagram II

(c)

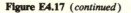

Figure E4.17 (*continued*)

SOLUTION See Fig. E4.17a to c. The reaction of a rough surface can be considered as one force of unknown magnitude and direction. This reactive force can be replaced by two components having known directions, one parallel and one perpendicular to the surface on which the body rests. A short pin-connected strut, link, or bar with no loads acting between pins can be assumed to develop an axial force along its longitudinal axis. This force can be tensile or compressive. Since a pin can resist motion in all direction but not rotation, it is customary to use rectangular components of the reaction in setting up free-body diagrams.

4.22 COPLANAR CONCURRENT FORCE SYSTEMS IN EQUILIBRIUM

Since all forces of a concurrent system acting upon a body at rest act through a point, they cannot cause rotation of the body. As a result, only two equations of equilibrium, $\Sigma F_x = 0$ and $\Sigma F_y = 0$, are available for solving unknowns in the system.

The force polygon is the graphical equivalent of the algebraic summation of forces in the x and y directions. For a system of concurrent forces in equilibrium the force polygon must close. This is known as the *polygon law* (Fig. 4.9). In such a polygon the arrowheads will always follow each other, regardless of the order in which the forces are laid out. As a result, it is possible to solve for two unknowns in closing the force polygon. Forces are normally considered in a clockwise order, starting with the known forces.

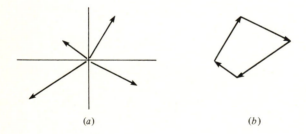

(a)   (b)

Figure 4.9 Concurrent system of forces in equilibrium: (a) space diagram; (b) closed-force polygon.

Example 4.18 Determine the tension in each cable supporting the 50-lb lighting fixture in Fig. E4.18a.

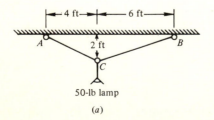

50-lb lamp

(a)

Figure E4.18

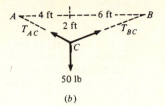

(b)

Figure E4.18 (*continued*)

SOLUTION A free-body diagram of the ring is drawn at C (Fig. E4.18b). Each cable is cut, exposing its internal force. These two forces, along with the weight of the lamp, are in equilibrium. It is a concurrent system of forces with two unknowns (magnitude of forces in cables). Two equations of equilibrium are available for any concurrent system of forces in equilibrium

$$\overrightarrow{\Sigma F_x} = 0: \quad -\frac{4}{\sqrt{20}} T_{AC} + \frac{6}{\sqrt{40}} T_{BC} \tag{1}$$

$$\uparrow \Sigma F_y = 0: \quad -50 + \frac{2}{\sqrt{20}} T_{AC} + \frac{2}{\sqrt{40}} T_{BC} \tag{2}$$

These two equations can be solved simultaneously for T_{AC} and T_{BC}. From Eq. (1)

$$T_{AC} = \frac{\sqrt{20}}{\sqrt{40}} \frac{6}{4} T_{BC} = 1.06 T_{BC}$$

Substituting into Eq. (2) gives

$$1.06 T_{BC} \frac{2}{\sqrt{20}} + \frac{2}{\sqrt{40}} T_{BC} = 50$$

from which $T_{BC} = 63.2$ lb and $T_{AC} = 67.0$ lb.

A solution could also be obtained by taking moments of the system of forces about some point. An advantage is obtained by taking moments about a point on the line of action of one of the forces (A or B). The moment equation replaces one of the summation equations.

Taking moments about A, we get

$$\widehat{\Sigma M_A} = 0: \quad (50)(4) - \frac{2}{\sqrt{40}} 4 T_{BC} - \frac{6}{\sqrt{40}} 2 T_{BC}$$

$$T_{BC} = 63.2 \text{ lb}$$

Example 4.19 Solve the problem of Example 4.18 graphically.

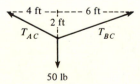

Free-body diagram

(a)

Figure E4.19 (*a*) Free-body diagram.

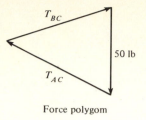

Force polygom

(b)

Figure E4.19 (b) Force polygon; by scaling T_{BC} = 63.2 lb and T_{AC} = 67.0 lb.

SOLUTION Draw to a particular scale the 50-lb vertical force. Starting at the tip of the 50-lb force, draw in a line parallel to the known direction of T_{AC}. Then draw a line parallel to T_{BC} through the origin of the 50-lb force. Where the two lines intersect is the head of T_{AC} and the tail of T_{BC}. Using the same scale, measure the lengths of T_{AC} and T_{BC}. These lengths represent the magnitude of the forces.

4.23 COPLANAR NONCONCURRENT SYSTEM OF FORCES IN EQUILIBRIUM

This kind of force system is common in practice and is involved in the determination of reactions at the supports of structures and the internal forces in structural members. As indicated in Sec. 4.17, the analytical solution of unknowns is based on the use of three equations of equilibrium, $\Sigma F_x = 0$, $\Sigma F_y = 0$, and $\Sigma M_z = 0$.

The graphical solution of unknowns in a nonconcurrent system of forces in equilibrium is based on an extension of the method developed in Sec. 4.15 for a resultant. Since the resultant of a nonconcurrent system of forces in equilibrium is zero, the force polygon must close and all the components of the forces in the system, represented by strings in the funicular polygon, cancel each other. Consequently the funicular polygon closes. The fact that both these polygons close can be used to solve for three unknowns in the system.

Example 4.20 Consider a steel beam supported by two columns carrying a column load of 20 kN, as shown in Fig. E4.20a. Draw a line diagram representing the structural diagram and determine the reactions.

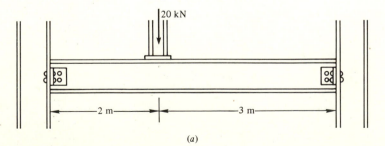

20 kN

←—2 m—→ ←————3 m————→

(a)

Figure E4.20

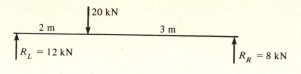

(b)

Figure E4.20 (*continued*)

SOLUTION See Fig. E4.20b.

$\overset{\frown}{\Sigma M_{R_L}} = 0:$

$$(20 \text{ kN})(2 \text{ m}) - (5 \text{ m})(R_R) = 0$$

$$R_R = \frac{(20 \text{ kN})(2 \text{ m})}{5 \text{ m}} = 8 \text{ kN}$$

$\uparrow \Sigma F_y = 0:$

$$R_L - 20 \text{ kN} + 8\text{kN} = 0$$

$$R_L = 12 \text{ kN}$$

$\overset{\frown}{\Sigma M_{R_R}} = 0:$

$$(-20 \text{ kN})(3 \text{ m}) + (12 \text{ kN})(5 \text{ m}) = 0 \quad \text{check}$$

Example 4.21 Determine algebraically the reactions for the beam shown in Fig. E4.21a.

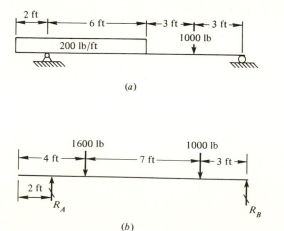

Figure E4.21

SOLUTION See Fig. E4.21b. The distributed load is replaced by a single concentrated load of 1600 lb acting at the centroid of the load diagram. This is permissible when determining the reactions (external load), but it is not permissible when determining internal stresses within the beam. The solution can best be obtained by taking moments about A and then about B. In this way each equation will yield one unknown. Each reaction is found independently, and then the summation of forces, which must equal zero, becomes an excellent check:

$$\overset{\frown}{\Sigma M_A} = (1600)(2) + (1000)(9) - 12 R_B = 0$$

$$R_B = \frac{3200 + 9000}{12} = 1017 \text{ lb}$$

$$\overset{\frown}{\Sigma M_B} = (-1000)(3) - (1600)(10) + 12 R_A = 0$$

$$R_A = \frac{16,000 + 3000}{12} = 1583 \text{ lb}$$

$$\uparrow \Sigma F_Y = 0: 1017 - 1600 - 1000 + 1583 = 0 \quad \text{check}$$

Example 4.22 Determine graphically the reactions of the loaded beam in Example 4.21.

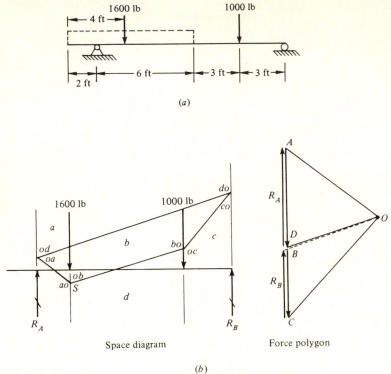

(a)

Space diagram   Force polygon

(b)

Figure E4.22

SOLUTION A free-body diagram is drawn of the beam with the uniformly distributed load replaced by a concentrated load at its midpoint (Fig. E4.22a). The fields in the space diagram are labeled. Letter d represents the field between the reactions. The force polygon is drawn as in Fig. E4.22b, starting with the known forces AB and BC. Since the force polygon, including the reactions, must close, the sum of the two reactions must be included from C to A. To determine the location of point D, which divides the line CA into two distances proportional to the magnitude of the reactions, it is necessary to complete a funicular polygon of the forces in the space diagram.

First choose a pole O and draw rays OA, OB, and OC. Select a convenient point S on the line of action of force ab in the space diagram, and through this point draw strings ao and ob parallel to rays AO and BO, respectively. Next draw oc through the point where bo intersects force bc and extend it to intersect force cd (R_B). In order for the substitute system of strings (components) completely to replace the system of space forces which are in equilibrium, it is necessary for a string od to connect the point where oc intersects the line of action of force cd (R_B) and the point where ao intersects the line of action of ed (R_A). This enables us to draw a ray DO in the force polygon parallel to this closing string do in the funicular polygon.

Then we can locate point D and thus establish the magnitude of the reactions R_B and R_A; their directions were already known.

Example 4.23 Determine algebraically the reactions of the beam loaded as shown in Fig. E4.23a.

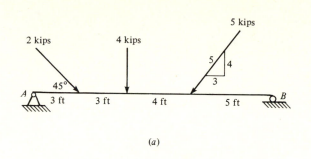

(a)

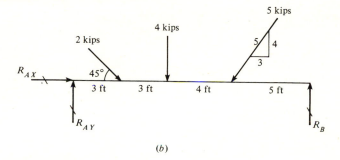

(b)

Figure E4.23

SOLUTION Draw a free-body diagram (Fig. E4.23b).

$$\overset{\curvearrowright}{\Sigma M_A} = 0: (2\sin 45°)(3) + (4)(6) + (5)(\tfrac{4}{5})(10) - 15R_B$$

$$R_B = \frac{4.2 + 25 + 40}{15} = 4.55 \text{ kips}$$

$$\overset{\curvearrowright}{\Sigma M_B} = 0: (5)(\tfrac{4}{5})(5) + (4)(9) + (2\sin 45°)(12) - 15R_{AY}$$

$$R_{AY} = \frac{20 + 36 + 16.9}{15} = 4.85 \text{ kips}$$

$$\overset{\rightarrow}{\Sigma F_X} = 0: R_{AX} + 2\sin 45° - (\tfrac{3}{5})(5) = 0$$

$$R_{AX} = 1.60 \text{ kips}$$

$$R_A = \sqrt{R_{AX}^2 + R_{AY}^2} = \sqrt{(4.85)^2 + (1.60)^2} = 5.1 \text{ kips}$$

$$\tan\theta = \frac{R_{AX}}{R_{AY}} = \frac{1.60}{4.85} = 0.330 \qquad \theta = 18.3°$$

Example 4.24 Determine graphically the reactions of the beam given in Example 4.23.

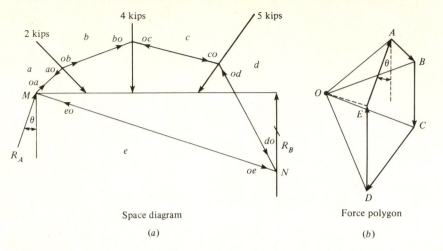

Space diagram

(a)

Force polygon

(b)

Figure E4.24 (a) Space diagram; (b) force diagram.

SOLUTION This constitutes a nonconcurrent nonparallel system of forces and as such requires a refinement of the graphical technique shown in Example 4.22. The space diagram (Fig. E4.24a) is labeled with small letters of Bow's notation. The right reaction R_A is vertical, and it is known that the left reaction R_A goes through point M.

The force polygon (Fig. E4.24b) is drawn and, since the right reaction is vertical, a vertical line is drawn through D. The point E which delineates the reactions must be on this line. Select the pole O and draw rays OA, OB, OC, and OD. It is important to realize that since the line of action of the left reaction (force ea) goes through point M, the funicular polygon is started at M by drawing string ao through M and parallel to AO. The string eo, as yet unknown, must also pass through M in order to close the funicular polygon. The rest of the funicular polygon is drawn until string do intersects force de (reaction R_B) at N. The closing string oe and eo must then join N and M. This means that ray OE can be drawn parallel to line MN; where it intersects the vertical line through D point E is located. The force polygon can then be closed by joining E and A. DE is now equivalent to reaction R_B, and EA gives the magnitude and direction of reaction R_A

$$R_A = 4.5 \text{ kips} \qquad R_B = 4.8 \text{ kips} \qquad \theta = 18.3°$$

4.24 THE MIDDLE-THIRD CONCEPT

Consider a retaining wall held in equilibrium by its weight W, the horizontal pressure of the soil H, and the reaction of the foundation under the wall R (Fig. 4.10). The horizontal force H acting at the center of gravity of the horizontal pressure diagram of the soil and the weight of the wall acting through the center of gravity of the wall combine to form a resultant R_A of the acting forces. This force must be brought into equilibrium by the resultant of the soil reactions. The horizontal component R_H is developed as a frictional force between the soil and the base of the retaining wall (see Sec. 4.11). The vertical component R_v represents the upward pressure of the soil and must act through the center of gravity of the upward soil reaction on the base. This reaction varies from a

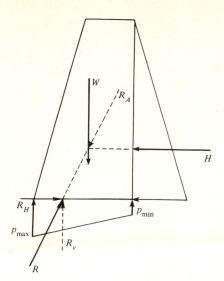

Figure 4.10 Equilibrium of forces acting on a retaining wall.

maximum pressure p_{max} to a minimum pressure p_{min}. The resultant of the soil reactions must be equal and opposite in magnitude and collinear with the resultant of the acting forces.

If p_{min} changes sign and becomes tensile, the base of the wall lifts off the soil since the interaction of soil with the base cannot develop tension. Although such uplift would not result in failure, it is undesirable. Retaining walls are normally designed to ensure compression in the soil below the base of the wall. A useful limit is reached when p_{min} becomes zero and the pressure diagram of the soil becomes triangular (Sketch 4.16).

For this limiting condition, the resultant of the upward soil pressure R_V is at a distance of $L/3$ from the edge of the retaining wall. This, then, is the basis for the rule that the resultant of a foundation reaction should remain within the middle third of the base. This concept has application in any situation where a contact surface cannot be relied upon to develop tension.

A famous application of this rule lies in the modern design of the Gothic buttress (Fig. 4.11). The thrust of the flying buttress is turned downward by the

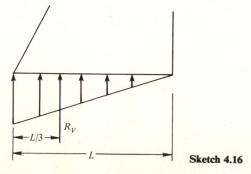

Sketch 4.16

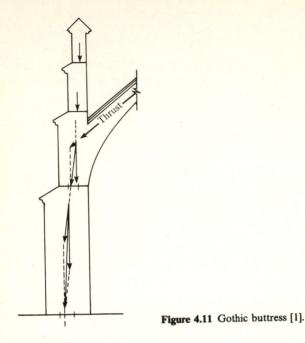

Figure 4.11 Gothic buttress [1].

weight of the pinnacle and turned farther downward by the weight in the buttress. If tension in the joints is to be avoided, the buttress must be made wide enough to keep the resultant within the middle third. The original Gothic builders were not familiar with the triangle of forces, but it was used in the neo-Gothic revival of the nineteenth century. The middle-third rule is conservative and has little application in modern architecture since tensile reinforcement can generally be used.

Example 4.25 Determine the vertical and horizontal reactions of the retaining wall in Fig. E4.25. The wall weighs 150 lb/ft^3, K_A for the soil retained by the wall is 0.25, and the soil weighs 120 lb/ft^3.

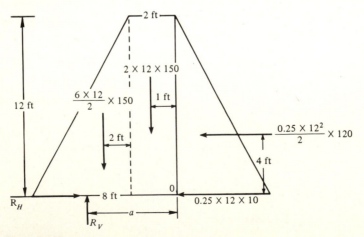

Figure E4.25

SOLUTION Consider a 1-ft length of wall. The horizontal force of the retained soil equals

$$K_A \frac{h^2}{2} \gamma = 0.25 \frac{(12)^2}{2} 120 = 2160 \text{ lb}$$

acting at a height of 4 ft from the bottom of the wall. The weight of the wall can be calculated most readily by considering the cross section to be made up of a rectangle and a triangle. Then

$$\text{Weight of rectangular portion} = (2)(12)(150) = 3600 \text{ lb}$$

acting vertically down 1 ft from the inner face.

$$\text{Weight of triangular portion} = \frac{(6)(12)(150)}{2} = 5400 \text{ lb}$$

acting vertically down 4 ft from the toe.

The reactions can be considered to be made up of a horizontal component R_H, which must be provided by friction or some mechanical anchorage, and the vertical reaction R_V, which must be provided by the upward pressure of the soil under the wall.

$$\overrightarrow{\Sigma H} = 0: \qquad\qquad R_H - 2160 = 0$$

$$R_H = 2160 \text{ lb}$$

$$\downarrow \Sigma V = 0: \qquad\quad R_V - 3600 - 5400 = 0$$

$$R_V = 9000 \text{ lb}$$

The location of the horizontal reaction component R_H is along the interface between the base of the wall and the soil. The location of R_V can be determined by taking moments of all the forces about some point O and equating them to zero.

$$\overset{\frown}{\Sigma M_O} = 0: \qquad (-2160)(4) - (3600)(1) - (5400)(4) + 9000a = 0$$

$$-8640 - 3600 - 21{,}600 - 9000a = 0$$

$$a = 3.76$$

The resultant of the reaction is within the middle third.

4.25 THREE-DIMENSIONAL FORCE SYSTEMS

By an extension of the principles developed for planar force systems, it is possible to resolve a force in space into three components parallel to three coordinate axes. The force F is represented by a vector whose origin is at the origin of the coordinate system (Sketch 4.17). As a result, the components of the force parallel with each axis are

$$F_x = F\cos\theta_x \qquad F_y = F\cos\theta_y \qquad F_z = F\cos\theta_z \qquad (4.10)$$

Since the force can be represented by a vector, these relationships can be developed by proportions rather than trigonometric functions. This is done by constructing a box relative to the force F, which is given a length d (Sketch 4.18). Realizing that the trigonometric relationships previously used can be developed from the dimensions of the box, we have

$$\cos\theta_x = \frac{x}{d} \qquad \cos\theta_y = \frac{y}{d} \qquad \cos\theta_z = \frac{z}{d}$$

and

$$F_x = F\frac{x}{d} \qquad F_y = F\frac{y}{d} \qquad F_z = F\frac{z}{d} \qquad (4.11)$$

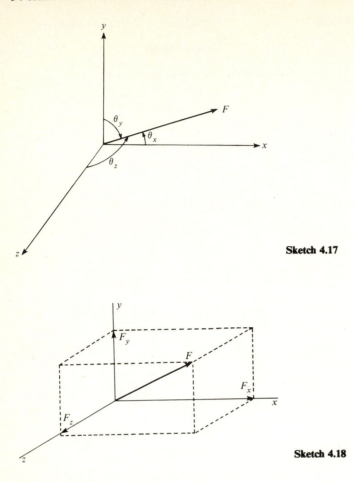

Sketch 4.17

Sketch 4.18

also
$$\frac{F_x}{x} = \frac{F_y}{y} = \frac{F_z}{z} = \frac{F}{d} \tag{4.12}$$

As developed earlier, if there are several forces acting on a rigid body, the following equations must be satisfied in order for the body to be in equilibrium

$$\Sigma F_x = 0 \qquad \Sigma F_y = 0 \qquad \Sigma F_z = 0$$

$$\Sigma M_x = 0 \qquad \Sigma M_y = 0 \qquad \Sigma M_z = 0 \tag{4.9}$$

In order to determine the moment of a force about an axis, it is normally easier to resolve the force into components parallel to the coordinate axis and to determine the perpendicular distances from the components to the various axes. According to Varignon's theorem, the moment is equal to the sum of the moments of the components.

Example 4.26 Determine the components of the force given in Fig. E4.26 and calculate the moments of the forces about the coordinate axes.

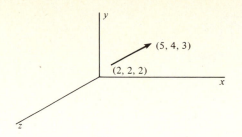

Figure E4.26

SOLUTION On the basis of Eqs. (4.11)

$$F_x = \frac{5-2}{\sqrt{(5-2)^2 + (4-2)^2 + (3-2)^2}}(100)$$

$$= \frac{3}{\sqrt{14}}(100) = 80 \text{ lb}$$

where $\sqrt{14}$ is the length of the 100-lb force,

$$F_y = \frac{2}{\sqrt{14}}(100) = 53.4 \text{ lb}$$

and

$$F_z = \frac{1}{\sqrt{14}}(100) = 26.7 \text{ lb}$$

The components of the force develop moments about some of the coordinate axes, as follows:

$$\Sigma M_x = +2F_y - 2F_z = (2)(53.4) - (2)(26.7) = 53.4 \text{ ft·lb}$$

$$\Sigma M_y = -2F_x + 2F_z = (-2)(80) + (2)(26.7) = -106.6 \text{ ft·lb}$$

and

$$\Sigma M_z = +2F_x - 2F_y = (2)(80) - (2)(53.4) = 53.2 \text{ ft·lb}$$

The moments were considered positive when, looking toward the origin, they were in a clockwise direction.

Example 4.27 Determine the forces in members AO, BO, and CO for a load of 2000 lb acting at O (Fig. E4.27a). Point O is at the same elevation as A and B.

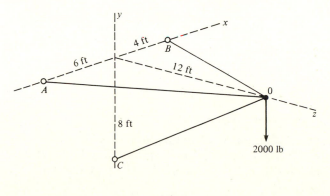

(*a*)

Figure E4.27

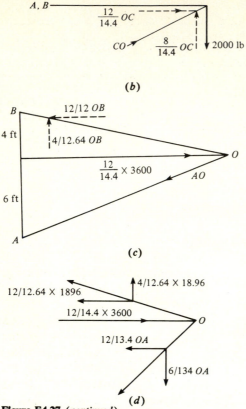

Figure E4.27 (*continued*)

SOLUTION Resolve the force in member CO into vertical and horizontal components at O and take moments about the X axis through AB (Fig. E4.27b).

$$\overset{\frown}{\Sigma M_{A,B}} = 0: (2000)(12) - \frac{8}{14.4}12OC$$

$$OC = \frac{(2000)(14.4)}{8} = 3600 \text{ lb}$$

Take moments about the Y axis through A. Resolve force OB into components at B (Fig. E4.27c).

$$\overset{\frown}{\Sigma M_{A_y}} = 0: \frac{12}{14.4}(3600)(6) - \frac{12}{12.64}10OB = 0$$

$$OB = 1896 \text{ lb}$$

Sum forces at O in the x direction (Fig. E4.27d)

$$\uparrow\Sigma F_x = 0: \frac{4}{12.64}(1896) - \frac{6}{13.4}OA = 0$$

$$OA = \frac{4}{12.64}(1896)\frac{13.4}{6} = 1340 \text{ lb}$$

$$\overrightarrow{\Sigma F_z} = 0: \frac{12}{14.4}(3600) - \frac{12}{12.64}(1896) - \frac{12}{13.4}(1340) = 0$$

$$= 3000 - 1800 - 1200 = 0 \quad \text{check}$$

Table 4.1

System	Number of unknowns that can be solved for in the system	Equations available for solution
Two-dimensional, concurrent	2	$\Sigma X = 0, \Sigma Y = 0$
Nonconcurrent	3	$\Sigma X = 0, \Sigma Y = 0, \Sigma M = 0$
Three-dimensional, concurrent	3	$\Sigma X = 0, \Sigma Y = 0, \Sigma Z = 0$
Nonconcurrent	6	$\Sigma X = 0, \Sigma Y = 0, \Sigma Z = 0$
		$\Sigma M_x = 0, \Sigma M_y = 0,$
		$\Sigma M_z = 0$

4.26 SUMMARY

Loads and reactions acting on a structure constitute a system of forces in equilibrium. Since forces can be described by their magnitude, direction, and point of application, they are vectors and can be characterized algebraically and graphically. The concept of the moment of a force and the moment of a couple are also developed. A system of forces can be classified as two- (coplanar) or three-dimensional, concurrent or nonconcurrent. Depending on the system of forces, there are a different number of equations of equilibrium available for solving unknowns in the system (see Table 4.1). The concept of a free-body diagram is developed as a basis for solving problems.

PROBLEMS

4.1 In a plane, add a 100-lb force at 30° and a 100-lb force at 90° (*a*) by the parallelogram method and (*b*) by the triangle law.

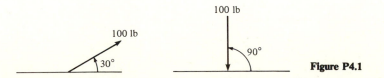

Figure P4.1

4.2 Determine the resultant of the coplanar forces of 40 lb at 20° and 60 lb at 190° (*a*) graphically and (*b*) algebraically.

4.3 In a plane, subtract the force 30 lb at 180° from the force 50 lb at 270°.

4.4 The resultant of two coplanar forces is 40 lb at 45°. If one of the forces is 24 lb at 0°, determine the other force by (*a*) algebra and (*b*) graphics.

4.5 Determine the moment of the 100-lb force about the point *O* in Fig. P4.5. It is in the *xy* plane.

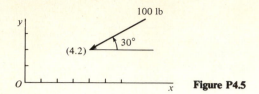

Figure P4.5

4.6 Solve Prob. 4.5 using Varignon's theorem.

4.7 A couple of moment $+60$ ft·lb acts in the plane of the paper. Indicate this couple with (*a*) 20-lb forces and (*b*) 60-lb forces.

4.8 Combine a couple of $+20$ ft·lb with a couple of -60 ft·lb, both in the same plane.

4.9 For a couple of $+30$ ft·lb in a horizontal plane and $+60$ ft·lb in a vertical plane determine the resultant graphically.

4.10 Combine a force of 30 lb at 90° with a $+60$-ft·lb couple in the same plane.

4.11 Determine the resultant of the concurrent system of forces in Fig. P4.11 (*a*) graphically and (*b*) algebraically.

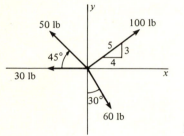

Figure P4.11

4.12 Determine the resultant of three forces originating at $(2, -2)$ and passing through the points indicated if they are 120 lb through $(8, 6)$, 180 lb through $(2, -4)$, and 270 lb through $(-5, 3)$.

4.13 Find the resultant of the three loads shown acting on the beam (*a*) algebraically and (*b*) graphically.

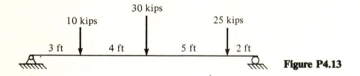

Figure P4.13

4.14 Determine the resultant of the four forces shown.

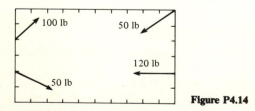

Figure P4.14

4.15 Find the resultant of forces acting on beam in Fig. P4.15 (*a*) algebraically and (*b*) graphically.

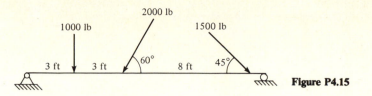

Figure P4.15

4.16 A bar *AB* weighs 10 lb/ft and is supported by a cable *AC* and a pin at *B*. Determine both algebraically and graphically the reaction at *B* and the tension in the cable.

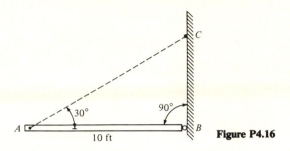

Figure P4.16

4.17 A beam, loaded as shown, weighs 50 lb/ft. Determine graphically and algebraically the reactions at *A* and *B*.

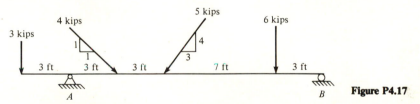

Figure P4.17

4.18 A cantilever beam 10 ft long weighing 50 lb/ft carries a concentrated load of 1000 lb at the free end. Determine the reactions.

4.19 Determine the compressive forces in the legs of the tripod.

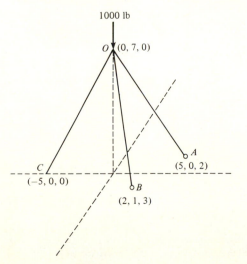

Figure P4.19

4.20 Determine the forces in the members of the tripod supporting the 100-kN force.

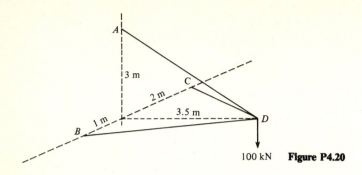

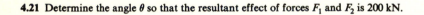

100 kN   **Figure P4.20**

4.21 Determine the angle θ so that the resultant effect of forces F_1 and F_2 is 200 kN.

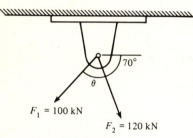

$F_1 = 100$ kN

$F_2 = 120$ kN   **Figure P4.21**

4.22 Determine the cable tensions T_1 and T_2.

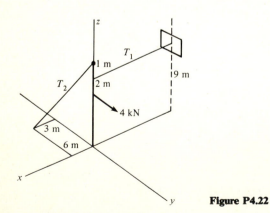

Figure P4.22

4.23 A welded joint of a truss is shown in Fig. P4.23. On the basis of equilibrium determine the forces in the remaining members. Indicate whether they are tensile or compressive forces.

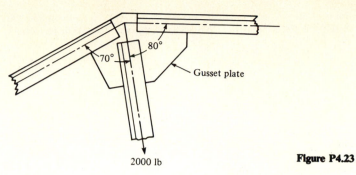

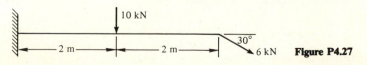

2000 lb

Figure P4.23

4.24 For the loading system in Fig. P4.24 determine W as a function of h.

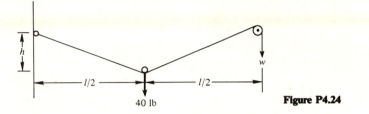

Figure P4.24

40 lb

4.25 Referring to Fig. P4.25, determine the reactions of the smooth floor and wall and the force exerted by the rope on a ladder weighing 50 lb.

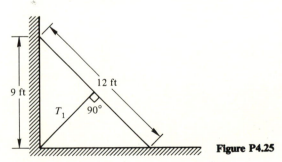

Figure P4.25

4.26 Consider the beam in Fig. P4.26 and determine its support reactions.

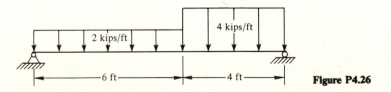

Figure P4.26

4.27 Determine the support reactions for the cantilever beam in Fig. P4.27.

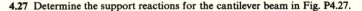

Figure P4.27

4.28 Determine the magnitude of the resultant force and the angles between the resultant and the coordinate axes for the three-dimensional concurrent force system shown. The 300-lb force is in plane XZ.

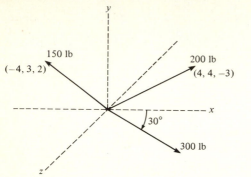

Figure P4.28

4.29 A concrete wall, of density 150 lb/ft^3, measures 5 by 4 ft by 8 in (Fig. P4.29). There is a horizontal pressure of 40 lb/ft^2 on one side (wind). Will the block overturn?

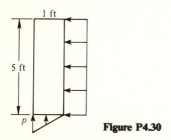

Figure P4.29

4.30 A masonry wall is 5 ft high and 1 ft thick. A uniform horizontal pressure of 10 lb/ft^2 acts on one side of the wall. Determine the minimum average density of the wall to limit the soil pressure to 0 lb/ft^2 on the heel of the wall (Fig. P4.30).

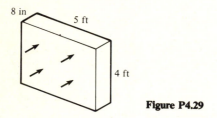

Figure P4.30

REFERENCES

1. Henry J. Cowan, "Architectural Structures," Elsevier, New York, 1971.

THE BEHAVIOR OF MATERIALS UNDER LOAD

5.1 INTRODUCTION

All materials deform under load. The deformation is usually elastic over a portion of the range of load required to cause failure. The term *elastic* refers to the ability of a material that has been deformed to return to its original shape after the deforming force is removed. *Brittle* materials are those which fail abruptly in tension and with little warning in compression. *Ductile* materials, on the other hand, exhibit large deformation before failure. The nonelastic portion of load-deformation curves provides useful margins of safety. Normally the allowable loads used in the design of structural members are within the elastic range of the material. Figure 5.1 illustrates the load-deformation curves of common brittle and ductile materials.

The elastic deformation of every structural material is directly proportional to the load it carries. This relationship was first enunciated by Hooke in 1678, and the constant of proportionality is termed the *modulus of elasticity*.

5.2 CONCEPT OF STRESS

Stress is defined as internal resistance to external forces. It is measured in terms of force per unit area

$$f = \frac{P}{A} \tag{5.1}$$

where f = force per unit area
P = applied load
A = cross-sectional area

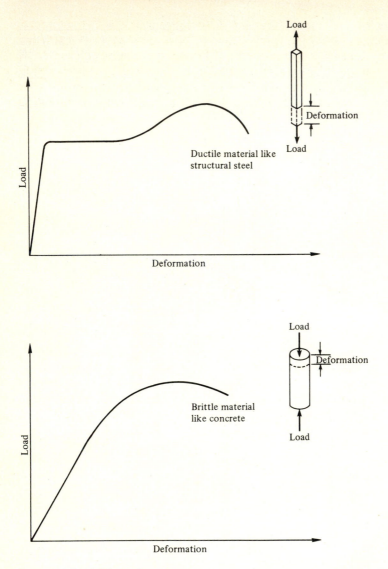

Figure 5.1 Load-deformation curves.

The applied load is normally measured in pounds or newtons. The area is usually given in square inches or square millimeters. As a result, the units of stress are pounds per square inch or newtons per square millimeter, normally converted into newtons per square meter, or pascals.

Any stress which acts perpendicular to the stressed surface is termed a *normal stress*. When the external force acting on the member is parallel to its major axis and the member is of constant cross section, the internal stresses are parallel to the axis and are referred to as *axial stresses*. When the axial forces pull on the member and tend to stretch it, they are called *tensile forces* and develop *tensile stresses* internally. When a pair of axial forces push on a member, they produce axial *compressive stresses* internally on a plane perpendicular to the axis. These two conditions are illustrated in Fig. 5.2. Compressive stress exerted on an external surface of a body is referred to as a *bearing stress*. In the other type of basic stress, a *shearing stress*, the stress is parallel to the surface (Fig. 5.3).

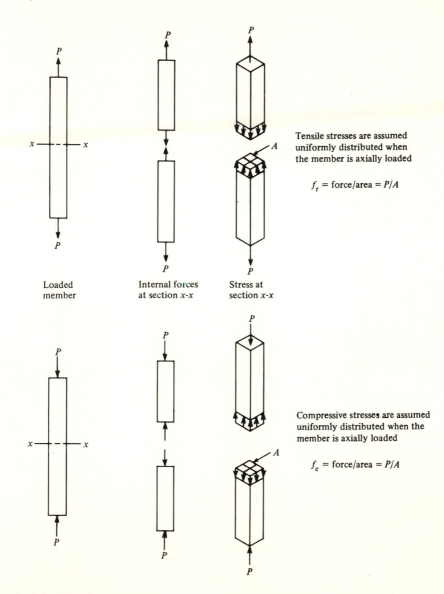

Tensile stresses are assumed uniformly distributed when the member is axially loaded

$$f_t = \text{force/area} = P/A$$

Loaded member

Internal forces at section *x-x*

Stress at section *x-x*

Compressive stresses are assumed uniformly distributed when the member is axially loaded

$$f_c = \text{force/area} = P/A$$

Figure 5.2 Normal stresses.

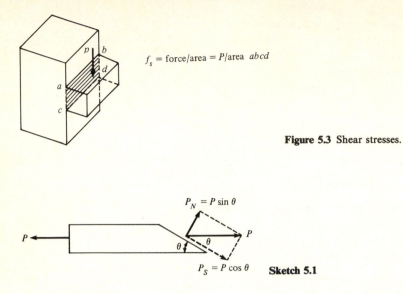

$$f_s = \text{force/area} = P/\text{area } abcd$$

Figure 5.3 Shear stresses.

$$P_N = P \sin \theta$$

$$P_S = P \cos \theta$$ **Sketch 5.1**

It is usually assumed that the stress is uniformly distributed over the cross section. In normal stresses this is considered to be the case when the resultant of the applied load acts through the centroid of the cross section. There are exceptions to this, termed *stress concentrations*, which develop at abrupt changes in cross sections and at points close to the applied loads. Shearing stresses are seldom uniformly distributed. Equation (5.1) can be considered as giving the average stress.

If the section is inclined at an angle to the axis of the member, both normal and shear stresses act on the section. A general relationship for this condition is illustrated in Sketch 5.1. The axial force P can be resolved into a component normal to the section, $P \sin \theta$, and a component parallel to the section, $P \cos \theta$. The area of the section becomes $A/(\sin \theta)$, where the area of the cross section is A. As a result, the normal stress on the inclined section becomes

$$f_n = \frac{P \sin \theta}{A/(\sin \theta)} = \frac{P \sin^2 \theta}{A} \tag{5.2}$$

and the corresponding shear stress is

$$f_s = \frac{P \cos \theta}{A/(\sin \theta)} = \frac{P \sin \theta \cos \theta}{A} \tag{5.3}$$

Note that for $\theta = 90°$, $f_n = P/A$ is a normal stress, as obtained by Eq. (5.1); $f_s = 0$ for this case. The shear stress can be maximized with $\theta = 45°$.

Example 5.1 Consider a bar with a cross-sectional area of $\frac{1}{2}$ in^2 supporting a load of 2000 lb (Fig. E5.1). Determine the average tensile stress in the bar.

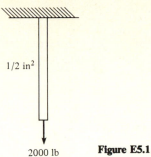

1/2 in²

2000 lb   **Figure E5.1**

SOLUTION The tensile stress is

$$f_n = \frac{P}{A} = \frac{2000}{\frac{1}{2}} = 4000 \text{ lb/in}^2$$

Example 5.2 Consider the riveted joint in Fig. E5.2. If the rivet diameter is 1/2 in, determine the average shearing stress in the rivet.

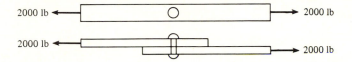

Figure E5.2

SOLUTION The area of the rivet is

$$A = \pi r^2 = \pi\left(\frac{1}{4}\right)^2$$

$$f_s = \frac{P}{A} = \frac{2000}{\pi/16} = 10{,}186 \text{ lb/in}^2$$

Example 5.3 Two blocks of wood 5 cm wide and 2.5 cm thick are glued together as shown in Fig. E5.3. Determine the normal and shearing stresses developed in the glued joint under the applied load of 5 kN.

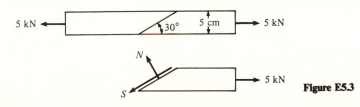

Figure E5.3

SOLUTION

$$N = P \sin 30° = (5 \text{ kN})(0.500) = 2.5 \text{ kN}$$

$$S = P \cos 30° = (5 \text{ kN})(0.866) = 4.33 \text{ kN}$$

$$f_n = \frac{P \sin^2 \theta}{A} = \frac{(5 \text{ kN})(0.5)^2}{(5)(2.5) \times 10^{-4} \text{ m}^2} = 1.0 \times 10^3 \text{ kN/m}^2 = 1.0 \text{ MPa}$$

$$f_s = \frac{P \sin \theta \cos \theta}{A} = \frac{(5 \text{ kN})(0.5)(8.66)}{(5)(25) \times 10^{-4} \text{ m}^2} = 1.73 \times 10^3 \text{kN/m}^2 = 1.73 \text{ MPa}$$

5.3 CONCEPT OF STRAIN

The deformation produced in a member is designated by the Greek letter δ (delta). If the length of the member is L, the deformation per unit length, termed *strain ε* (epsilon), is

$$\text{Strain} = \frac{\text{deformation}}{\text{length}} = \varepsilon = \frac{\delta}{L} \qquad (5.4)$$

Since deformation and member length are normally in the same units, strain has units of inches per inch or millimeters per millimeter.

5.4 MODULUS OF ELASTICITY

The proportionality of a load to deformation is normally expressed as the ratio of stress to strain, called the *modulus of elasticity E* of the material.

$$\text{Modulus of elasticity } E = \frac{\text{stress}}{\text{strain}} = \frac{f}{\varepsilon} \qquad (5.5)$$

Since the units of stress are pounds per square inch and the units for strain are inches per inch, the basic units for modulus of elasticity are pounds per square inch. Corresponding SI units are pascals (Table 5.1).

By combining Eqs. (5.1) to (5.3) it is possible to develop an equation to determine the deformation δ in a prismatic member of cross section A subject to an axial load P

$$\delta = \frac{PL}{AE} \qquad (5.6)$$

where $\delta =$ deformation
$P =$ axial load
$L =$ length of member
$A =$ cross-sectional area
$E =$ modulus of elasticity

Table 5.1 Modulus of elasticity of structural materials

Material	Tension E		Shear G	
	kips/in^2	MPa	kips/in^2	MPa
Steel	29,500	200,000	11,600	80,000
Aluminum	10,000	69,000	3,750	25,800
Cast iron	14,000	103,400	6,000	41,400
Copper (rods and bolts)	16,000	110,000		
Brass (hard)	15,000	103,500	5,500	37,900
Concrete	4,000	27,500		
Wood (southern pine)	1,500	10,000		

Example 5.4 Compute the elongation occurring in a steel bar 10 ft long with a cross-sectional area of 0.5 in² supporting a load of 10,000 lb (Fig. E5.4). Calculate the stress and strain in the member.

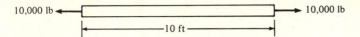

10,000 lb ← | 10,000 lb →

|← 10 ft →|

Figure E5.4

SOLUTION

$$\text{Stress in bar } f = \frac{P}{A} = \frac{10{,}000 \text{ lb}}{0.5 \text{ in}^2} = 20 \text{ kips/in}^2$$

$$\text{Strain } e = \frac{f}{E} = \frac{20 \text{ kips/in}^2}{29{,}500 \text{ kips/in}^2} = 0.678 \times 10^{-3} \text{ in/in}$$

Instead of leaving the strain as a unitless number it is customary to give it in inches per inch.

$$\text{Deformation } \delta = eL = (0.678 \times 10^{-3} \text{ in/in})(10)(12 \text{ in}) = 0.0814 \text{ in}$$

and

$$\delta = \frac{PL}{AE} = \frac{(10 \text{ kips})(10)(12 \text{ in})}{(0.5 \text{ in}^2)(29{,}500 \text{ kips/in}^2)} = 0.814 \text{ in}$$

Example 5.5 Compute the elongation occurring in a 5-mm-square steel bar 4 m long carrying a load of 5 kN (Fig. E5.5).

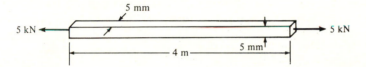

5 mm

5 kN ← 5 kN →

|← 4 m →| 5 mm

Figure E5.5

SOLUTION

$$\text{Stress in bar } f = \frac{P}{A} = \frac{5 \times 10^3 \text{N}}{(5)(5 \times 10^{-6} \text{ m}^2)} = 2.0 \times 10^8 \text{ N/m}^2 = 200 \text{ MPa}$$

$$\text{Strain } \varepsilon = \frac{f}{E} = \frac{200 \text{ MPa}}{200{,}000 \text{ MPa}} = 0.001 \text{ mm/mm}$$

$$\text{Deformation } \delta = \varepsilon L = (0.001 \text{ mm/mm})(4 \times 10^3 \text{ mm}) = 4 \text{ mm}$$

also

$$\delta = \frac{PL}{AE} = \frac{(5 \times 10^3 \text{ N})(4 \times 10^3 \text{ mm})}{(5)(5 \times 10^{-6} \text{ mm}^2)(200 \times 10^9 \text{ N/m}^2)} = 4 \text{ mm}$$

5.5 STATICALLY INDETERMINATE AXIALLY LOADED MEMBERS

In many structural systems the equations of equilibrium are not sufficient to determine the internal forces in the structural members uniquely. In these statically indeterminate cases (Sec. 4.17) the equation of equilibrium must be

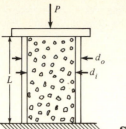

Sketch 5.2

supplemented. These equations depend upon the geometry of the structure and are often called *equations of compatibility*. At this point we shall be concerned only with indeterminate members subjected to axial loads.

For example, consider a pipe column filled with concrete (Sketch 5.2). It is loaded through a rigid plate P, which distributes the load to the concrete and steel. From statics we realize that

$$\uparrow \Sigma Y = 0: \qquad\qquad P_c + P_s - P = 0 \qquad\qquad (1)$$

where P_c is the force taken by the concrete and P_s is the force carried by the steel pipe. Under these circumstances the equations of statics supply only one equation with two unknowns, P_c and P_s. The additional equation is obtained by considering the geometry of the problem, namely, the compatibility of the deformations. Because of the rigid cap and concentric load the deformation in the concrete must be equal to the deformation in the steel

$$\delta_c = \delta_s$$

By using Eq. (5.6) we get

$$\frac{P_c L_c}{A_c E_c} = \frac{P_s L_s}{A_s E_s} \qquad\qquad (2)$$

Solving equations (1) and (2) simultaneously makes it possible to determine values of P_c and P_s in terms of the total load P. It is important to note that Eq. (2) does not apply if the stress exceeds the proportional limit.

Example 5.6 For the pipe column discussed above
$$L = 10 \text{ ft} \qquad d_o = 5.56 \text{ in} \qquad d_i = 5.05 \text{ in}$$

Determine the stresses in the steel and the concrete and the shortening under a load of 60 kips.

SOLUTION

$$\uparrow \Sigma Y = 0: \qquad P_c + P_s - 60 = 0 \qquad\qquad (1)$$

The deformation relationship, assuming it is elastic, yields

$$\frac{P_c L_c}{A_c E_c} = \frac{P_s L_s}{A_s E_s} \qquad\qquad (2)$$

Then we substitute

$$L_c = L_s = (10)(12) \text{ in}$$

$$A_c = \frac{\pi}{4}(5.05)^2 = 20.03 \text{ in}^2$$

$$A_s = \frac{\pi}{4}(5.56)^2 - A_c = 24.28 - 20.03 = 4.25 \text{ in}^2$$

$$E_s = 29{,}500 \text{ kips/in}^2 \qquad \text{Table 5.1}$$

$$E_c = 4000 \text{ kips/in}^2 \qquad \text{Table 5.1}$$

into Eq. (2), to get

$$\frac{120 P_c}{(20.03)(4 \times 10^3)} = \frac{120 P_s}{(4.25)(2.95 \times 10^2)} \qquad P_s = 1.56 P_c$$

Substituting this into Eq. (1), we get

$$P_c + 1.56 P_c - 60 = 0$$

$$2.56 P_c = 60 \text{ kips}$$

$$P_c = \frac{60}{2.56} = 23.44 \text{ kips/in}^2$$

$$P_s = 60 - 23.44 = 36.56 \text{ kips/in}^2$$

$$f_c = \frac{P}{A} = \frac{23.44}{20.03} = 1.17 \text{ kips/in}^2$$

$$f_s = \frac{P}{A} = \frac{36.56}{4.25} = 8.6 \text{ kips/in}^2$$

The stresses in both the concrete and steel are within the elastic range of the materials (see Chap. 6), and therefore our assumption is correct.

The total shortening of the column is

$$\delta = \delta_c = \delta_s = \frac{P_s L_s}{A_s E_s} = \frac{(36.56)(120)}{(4.25)(2.95 \times 10^2)} = 0.035 \text{ in}$$

Example 5.7 The load P is placed on a square rigid plate such that the plate will remain horizontal after loading (Fig. E5.7a). Determine the distance x.

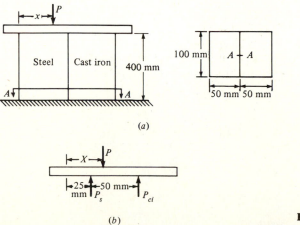

(a)

(b)

Figure E5.7

SOLUTION See Fig. E5.7b.

$$\Sigma M_0 = 0 \qquad PX - 25P_s - 75P_{ci} = 0 \qquad (1)$$

$$\downarrow \Sigma Y = 0 \qquad P - P_s - P_{ci} = 0 \qquad (2)$$

The required third equation is obtained from the compatability of deformations $\delta_s = \delta_{ci}$ (because cap remains horizontal)

$$\frac{P_s L_s}{A_s E_s} = \frac{P_{ci} L_{ci}}{A_{c_i} E_{c_i}} \qquad (3)$$

$$\frac{400 P_s}{(5000)(200,000)} = \frac{400 P_{ci}}{(5000)(103,400)}$$

$$P_s = 1.93 P_{ci}$$

Substituting the result from Eq. (3) into Eq. (2), we get

$$P - 1.93 P_{ci} - P_{ci} = 0 \qquad P = 2.93 P_{ci}$$

and substituting this into Eq. (1) along with $P_s = 1.93 P_{ci}$ from Eq. (3), we get

$$2.93 P_{ci} x - 25(1.93 P_{ci}) - 75 P_{ci} = 0$$

$$x = \frac{123.3 P_{ci}}{2.93 P_{ci}} = 42.1 \text{ mm}$$

5.6 MODULUS OF RIGIDITY

Shearing forces cause shearing deformation just as axial forces cause change in length. The difference is, however, that an element subject to shear does not change the length of its sides but undergoes a change in shape. As illustrated in Sketch 5.3, the change is from a rectangle to a parallelogram. In this simplified situation the deformation δ_s occurs in the length L, and the shear strain γ is given by

$$\gamma = \frac{\delta_s}{L} \qquad (5.7)$$

The relationship between shear stress and shear strain, according to Hooke's law, is

$$f_s = G\gamma \qquad (5.8)$$

where G represents the modulus of elasticity in shear, or modulus of rigidity (Table 5.1).

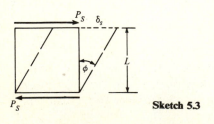

Sketch 5.3

The relationship between the shearing deformation and the applied shearing forces is similar to that given in Eq. (5.6) and is

$$\delta_s = \frac{P_s L}{A_s G} \tag{5.9}$$

where P_s is the shearing force acting upon the area A_s.

Sometimes the shear strain is expressed as an angular change such that

$$\tan \phi = \frac{\delta_s}{L} = \gamma \tag{5.10}$$

Since the angular change is usually very small, γ can be expressed in radians.

5.7 POISSON'S RATIO

Another type of deformation is the change in transverse dimensions accompanying axial tension or compression. Experiments show that if a bar is lengthened by axial tension, there is a reduction in the transverse dimension. The ratio of the strains in these directions is a constant within the proportional limit. It is designated μ and is named after Poisson, who first recognized this fact in 1811

$$\mu = -\frac{\varepsilon_y}{\varepsilon_x} = -\frac{\varepsilon_z}{\varepsilon_x} \tag{5.11}$$

where ε_x is the strain due to stress only in the x direction and ε_y and ε_z are the strains induced in the other directions. The minus sign indicates the decrease in transverse dimension associated with an elongation in the axial direction, or vice versa.

This particular phenomenon becomes very important when Hooke's law is extended to biaxial and triaxial stress conditions.

An important relationship exists between the constants E, G, and μ for a particular material.

$$G = \frac{E}{(2)(1 + \mu)} \tag{5.12}$$

This is a useful expression for calculating μ when E and G are known.

Common values of Poisson's ratio are 0.25 to 0.30 for steel, approximately 0.33 for most other metals, 0.25 to 0.50 for wood, and 0.085 to 0.125 for concrete.

5.8 TORSION

In the design of machinery and some structures a common problem is that of transmitting a torque (couple) from one plane to a parallel plane at right angles to the axis of the member. The simplest structural member for transmitting

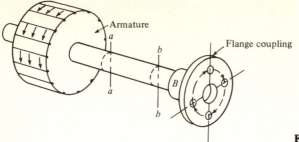

Figure 5.4 Drive shaft under torsion.

torque is a circular shaft, e.g., that connecting a motor to some other piece of equipment. The twisting of noncircular members becomes much more complicated and will not be considered, but principles that will be developed have also been applied to welded connections, among other problems.

A simplified diagram of a typical torsion problem is shown in Fig. 5.4. The basic problem is one of determining stresses in the shaft and its deformation. In deriving the torsion formulas we make the following assumptions:

1. Circular sections remain circular.
2. Plane sections remain plane.
3. Straight radial lines in a section remain straight.
4. Twisting couples are applied in planes perpendicular to the axis of the shaft.
5. Hooke's law applies to the behavior of the material.

Some of these assumptions can be demonstrated experimentally, and the others are reasonable in terms of loading and material properties. Visual examination of twisted rubber models indicates that some of these assumptions are apparently correct for circular sections (Fig. 5.5).

Consider the segment of the shaft between transverse planes a-a and b-b of Fig. 5.4. On the basis of the above assumptions, the distortion of the shaft will be as indicated in Fig. 5.6a. Points B and C on a common radius at section b-b move to B' and C' in the same plane and on the same radius. The angle θ is called the *angle of twist*. The surface ABB' is shown developed in Fig. 5.6b. The differential area at B assumes the distorted shape at B' due to shear stress (Sec. 5.5). If we consider an internal fiber at C' a radius distance ρ subtended by an angle of twist θ,

$$\delta = CC' = \rho\theta$$

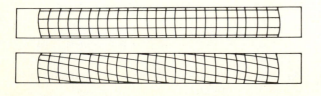

Figure 5.5 Plane sections remain plane during twisting.

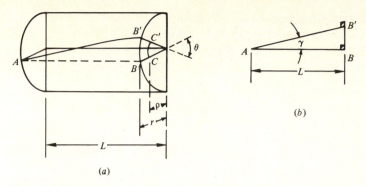

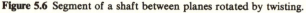

(b)

(a)

Figure 5.6 Segment of a shaft between planes rotated by twisting.

The shear strain in the fiber is

$$\gamma = \frac{\delta}{L} = \frac{\rho\theta}{L}$$

On the basis of Hooke's law the shearing stress in this typical fiber is

$$f_{s\rho} = G\gamma = \frac{G\theta}{L}\rho \tag{5.13}$$

It is important to realize that the term $G\theta/L$ is constant for any fiber on a given radius. Therefore, we can conclude that *the shearing-stress distribution along any radius varies linearly with the distance from the axis of the shaft* (Fig. 5.6).

Let us again consider the segment of shaft between sections *a-a* and *b-b* but this time in terms of the applied torque (Fig. 5.7). A differential area *dA* of section *b-b* located at a distance ρ from the centroidal axis develops the resisting force $dP = f_{s\rho}\,dA$. If we consider the area to be infinitesimally small, the stress

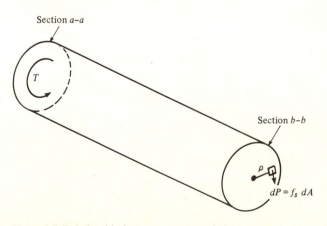

Figure 5.7 Relationship between torque and shear stress.

distributed on the area dA can be considered uniform. The resisting force dP, to be most efficient, acts perpendicular to the radius ρ. This is an axiom of structural mechanics. For equilibrium, the sum of the moments of all the differential resisting forces about the centroidal axis must be equal to the applied torque T:

$$\Sigma M_0 = T = \int \rho \, dP = \int \rho f_{sp} \, dA$$

Replacing f_{sp} by $G\theta\rho/L$ gives

$$T = \frac{G\theta}{L} \int \rho^2 \, dA$$

and since $\int \rho^2 \, dA = J$, the polar moment of inertia of the cross section (Sec. 10.5),

$$T = \frac{G\theta J}{L} \tag{5.14}$$

usually written

$$\theta = \frac{TL}{JG} \tag{5.15}$$

In order for θ to be in the proper units of radians, T must be in inch-pounds, L in inches, J in inches^4, and G in pounds per square inch. The SI units are T in newton-meters, L in meters, J in meters^4, and G in pascals (newtons per square meter).

By replacing $G\theta/L$ by its equivalent value T/J in Eq. (5.13) we obtain

$$f_{sp} = \frac{T}{J} \rho \tag{5.16}$$

This is called the *torsion formula*. A more common form gives the maximum shearing stress by replacing ρ with the radius r of the shaft

$$f_s = \frac{Tr}{J} \tag{5.17}$$

It is important to note that these relationships are applicable not only to solid shafts but also hollow ones (Sketch 5.4). The values of polar moments of inertia for circular shafts are

$$J = \frac{\pi r^4}{2} = \frac{\pi d^4}{32} \qquad J = \frac{\pi}{2}(R^4 - r^4) = \frac{\pi(D^4 - d^4)}{32}$$

Sketch 5.4

Example 5.8 A solid shaft 12 ft long is stressed to 10 kips/in^2 when twisted through 8°. If $G = 1.2 \times 10^4$ kips/in^2, determine the shaft diameter.

SOLUTION

$$T = \begin{cases} \dfrac{G\theta J}{L} & \text{from Eq. (5.14)} \\[2ex] \dfrac{f_s J}{r} & \text{from Eq. (5.17)} \end{cases}$$

Equating these two expressions on the basis of T, we get

$$\frac{G\theta J}{L} = \frac{f_s J}{r}$$

Solving for r gives

$$r = \frac{f_s L}{G\theta} = \frac{(10 \text{ kips/in}^2)(12)(12 \text{ in})}{(1.2 \times 10^4 \text{ kips/in}^2)(\frac{8}{180}\pi)} = 0.86 \text{ in} \qquad d = 1.72 \text{ in}$$

Example 5.9 A solid circular steel shaft 80 mm in diameter is subjected to a torque of 10 kN·m. The modulus of rigidity is 80 GPa. Determine the maximum shearing stress in the shaft and the magnitude of the angle of twist in a 2-m length.

SOLUTION Using Eq. (5.17), $f_s = Tr/J$, we have

$$f_s = \frac{(1 \times 10^4 \text{ N·m})(40 \times 10^{-3})}{\left[(80)^4 \pi \times 10^{-12} \text{ m}^4\right]/32} = 99.5 \times 10^6 \text{ N/m}^2 = 99.5 \text{ MPa}$$

Using Eq. (5.15), $\theta = TL/JG$, we have

$$\theta = \frac{(1 \times 10^4 \text{ M·m})(2 \text{ m})}{\left\{\left[(80)^4 \pi \times 10^{-12}\right]/32\right\}(8 \times 10^{10})} = 0.622 \text{ rad}$$

5.9 TEMPERATURE STRESSES

When a material body is subjected to changes in temperature, it expands with a temperature increase and contracts with a temperature decrease. This change in length per unit length of a bar of a given material for each degree of change in temperature is called the *linear coefficient of thermal expansion* α. Values of α for some of the more important structural materials are given in Table 5.2.

Table 5.2 Average coefficients of thermal expansion

Material	μin/in·°F	μm/cm·°C
Steel	6.5	0.138
Concrete	6.0	0.127
Wood	3.0	0.064
Cast iron	6.1	0.129
Aluminum	12.8	0.271

If a straight bar of material is free to change length with temperature change, there will be no change in internal stress, but if there is restraint to the change in length caused by a temperature change, internal stresses will be developed in proportion to the change in length prevented by the restraint.

The change in length of a bar as a result of a change in temperature can be represented by

$$\delta = \alpha \, \Delta t \, L \qquad (5.18)$$

where α = unit deformation per degree change in temperature, μ in/in·°F
   Δt = change in temperature, °F
   L = length of bar, in
   δ = change in length of member, in

The change in unit length $\alpha \, \Delta t$ for complete restraint is equal to a strain ε. Hence the stress f produced under these circumstances becomes

$$f = E\alpha \, \Delta t \qquad (5.19)$$

Example 5.10 A steel rod 10 ft long is attached between two walls. If the stress in the rod is zero at 70°F, calculate the stress in the rod when the temperature drops to 20°F. Solve assuming (*a*) that the walls are immovable and (*b*) that the walls move together 0.01 in as the temperature drops.

$$\alpha_{\text{steel}} = 6.5 \; \mu \text{in/in·°F} \qquad E = 2.95 \times 10^7 \; \text{lb/in}^2$$

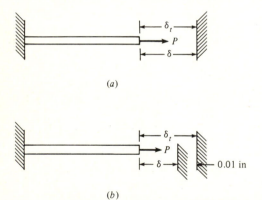

(*a*)

(*b*)   **Figure E5.10**

SOLUTION (*a*) Imagine the rod disconnected from the right wall (Fig. E5.10*a*). A temperature drop of 50°F causes the contraction δ, where $\delta = \alpha L \, \Delta t$. Reestablishing contact with the wall requires a force P sufficiently large to stretch the bar a distance δ, where $\delta = fL/E$. Consequently

$$\alpha L \, \Delta t = \frac{fL}{E}$$

and from Eq. (5.19), $f = E\alpha \, \Delta t = (6.5 \times 10^{-6})(50)(2.95 \times 10^7)$

$$= 9588 \approx 9590 \; \text{lb/in}^2$$

Note that the stress is independent of the length of the member.

(*b*) In this case, when the wall moves inward 0.01 in, the force P has to be large enough to elongate the bar a distance of only $\delta_t - 0.01$ (Fig. E5.10*b*). Consequently

$$\delta = \delta_t - 0.01$$

$$\frac{fL}{E} = \alpha L \Delta t - 0.01 \text{ in}$$

$$f = \alpha \Delta t E - 0.01 \frac{E}{L}$$

$$= 9590 - 0.01 \frac{2.95 \times 10^7}{(10)(12)}$$

$$= 9590 - 2460$$

$$= 7130 \text{ lb/in}^2$$

Note that in this case the length of the rod is important.

5.10 THIN-WALLED CYLINDERS

Consider a thin-walled cylinder containing a fluid or gas subjected to a pressure of p lb/in^2 (Fig. 5.8*a*). The internal pressure distribution in Fig. 5.8*a* is duplicated in Fig. 5.8*b*, where the cylinder is half full of a liquid. Since the fluid transmits pressure equally in all directions, the pressure distribution on the lower half of the cylinder is the same as in Fig. 5.8*a*.

A free-body diagram of the lower half of the cylinder formed by cutting plane $A - A$ is shown in Fig. 5.9. F equals the pressure p multiplied by the area dl on which it acts

$$\uparrow \Sigma V = 0: \qquad\qquad p\,dl - 2P = 0$$

where P is the force developed in each cylinder wall

$$P = \frac{p\,dl}{2} \tag{5.20}$$

The stress in the longitudinal section of a cylinder wall of thickness t is therefore

$$f_h = \frac{P}{A} = \frac{p\,dl}{2lt} = \frac{pd}{2t} \tag{5.21}$$

This stress is usually called *tangential*, *circumferential*, or *hoop stress*. It is the

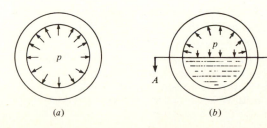

(*a*)   (*b*)

Figure 5.8 Pressure distribution in a cylinder.

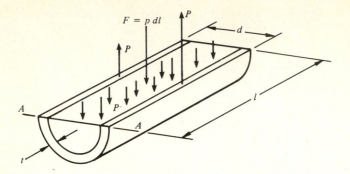

Figure 5.9 Free-body diagram of half of cylinder.

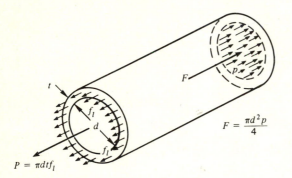

Figure 5.10 Transverse section of cylinder.

average stress. For cylinders having a wall thickness equal to one-tenth or less of the inner radius, it is practically equal to the maximum stress at the inside surface.

Consider the diagram of a transverse section (Fig. 5.10). The force acting over the end of the cylinder is resisted by the resultant of the forces acting on the transverse section of the cylinder wall

$$\overrightarrow{\Sigma H} = 0: \qquad\qquad -\pi \, dt f_l + \frac{\pi d^2}{4} p = 0$$

$$f_l = \frac{pd}{4t} \qquad\qquad (5.22)$$

The area of the transverse section is actually the wall thickness multiplied by the mean circumference, or $\pi(d + t)t$. If t is small compared with d, it can be approximated by $\pi \, dt$. f_l denotes what is called the *longitudinal stress*.

Example 5.11 A steel plate is used to cap the end of a pipe carrying steam at 500 lb/in². The pipe has an outside diameter of 12 in and a wall thickness of $\frac{1}{4}$ in. How many $1\frac{1}{2}$-in steel bolts are required to fasten the cap if the allowable stress in the bolts is 12,000 lb/in², of which 8000 lb/in² is used initially in tightening the bolts. What circumferential stress is developed in the pipe?

Figure E5.11

SOLUTION From Fig. E5.11,

Force acting against end plate $= 500\pi \dfrac{(11.5)^2}{4} = 51{,}934$ lb

Strength of one bolt $= \pi \dfrac{(1.5)^2}{4}(12{,}000 - 8000) = 7069$ lb

No. of bolts required $= \dfrac{\text{force acting on cap}}{\text{strength of bolt}} = \dfrac{51{,}934 \text{ lb}}{7069 \text{ lb/bolt}} = 7.35$ bolts

Use 8 bolts.
The longitudinal stress is

$$f_l = \frac{\text{force acting on cap}}{\text{solid cross section of pipe}}$$

$$= \frac{51{,}934}{\left[(12)^2\pi/4\right] - \left[(11.5)^2\pi/4\right]} = \frac{51{,}934 \text{ lb}}{113.1 - 103.9 \text{ in}^2}$$

$$= \frac{51{,}934}{9.2} = 5645 \text{ lb/in}^2$$

and the circumferential stress is

$$f_h = \frac{pd}{2t} = \frac{(500)(12)}{(2)\left(\frac{1}{4}\right)}$$

$$= 12{,}000 \text{ lb/in}^2$$

Example 5.12 A penstock 2 m in diameter, composed of wooden staves bound together by steel hoops each 10 cm² in area, is used to conduct water from a reservoir to a powerhouse (Fig. E5.12). If the maximum tensile stress permitted in the hoops is 250 MN/m², what is the maximum spacing of hoops under a 20-m head of water?

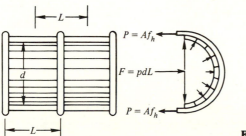

Figure E5.12

SOLUTION The water pressure corresponding to a 20-m head of water is

$$(20 \text{ m})(9.81 \text{ kN/m}^3) = 196 \text{ kN/m}^2$$

Each hoop must resist the bursting force exerted by water on a length L of pipe

$$(196 \text{ kN/m}^2)(2 \text{ m})(L \text{ m}) = 392L \text{ kN}$$

The tensile force developed by the hoop is

$$\frac{(2)(10 \text{ m}^2)}{(100)^2}(250 \text{ MN/m}^2) = 500 \text{ kN}$$

Σ forces = 0: $\qquad$ $500 \text{ kN} = 392\,L \text{ kN}$ $\qquad$ $L = \dfrac{500}{392} = 1.28 \text{ m}$

5.11 FRICTION

Friction is one of the most important phenomena in everyday life. Without friction it would be impossible to walk, roll on wheels, or use nails to hold pieces of wood together. Friction also has detrimental effects. Since it is a constant cause of wasted energy and wear of material, lubrication is used to minimize friction. Developing a qualitative explanation of friction is difficult in a text of this type. Our discussion will therefore be limited to a phenomenological treatment of so-called *dry friction*. It is important to note that friction involves material properties which require experimentation.

The main features of dry friction are best illustrated by a simple experiment. We place a block of weight W on a dry horizontal plane and apply to it a horizontal force P (Fig. 5.11). For a small force P our experience tells us that the block will not move because of a reaction force F developed at the contact surface which balances the applied force. This force is possible because the surfaces in contact are not smooth and because the small surface irregularities are capable of developing a reaction force with a component parallel to the surface. The load P can be increased to the point where the block will move because an increased frictional force is impossible. This force is called the *limiting friction* (static) *force*, and the system is said to be in a state of impending motion since any further increase in the applied force would cause motion. Once motion has been established, the resisting force is somewhat less than the limiting value but remains constant (sliding). The magnitudes of these frictional values have been determined experimentally for different kinds of contact surfaces, contact areas, and normal loads. These experimental facts are embodied in the laws of dry, or Coulomb, friction, first stated completely by Charles Coulomb (1736–1806):

1. The total amount of friction is independent of the area of contact.
2. The total amount of friction is proportional to the normal force.
3. For low velocities of sliding the total amount of friction is independent of velocity and is less than that for impending motion.

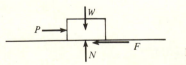

Figure 5.11 Block on plane with impending motion.

Table 5.3 Dry-friction coefficients

Material	Friction	
	Static f	Sliding f_s
Steel	0.15	0.10
Wood on wood	0.4–0.6	0.2–0.4
Wood on metal	0.6–0.7	0.4–0.5
Rubber on asphalt	0.7–1.0	0.5–0.6

These laws can be expressed analytically by

$$F = fN \qquad (5.23)$$

where F = frictional force parallel to surfaces in contact, at impending motion,
or during motion
N = normal force between contact surfaces
f = coefficient of friction = constant for given materials and conditions

Table 5.3 indicates the order of magnitude of the coefficients of friction. With changes in experimental conditions, particularly the quality of the contact surfaces, the given values can vary by over 100 percent.

Example 5.13 A 100-lb block rests on a horizontal plane. The coefficient of static friction between the block and plane is 0.30 (Fig. E5.13). What horizontal force P is required to put the block in a state of impending motion?

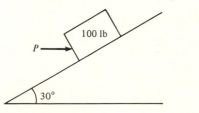

100 lb

100 × 0.30

100 lb

Figure E5.13

SOLUTION The static friction force that can be developed is equal to the normal force (100 lb) times the coefficient of friction (0.30), or (100)(0.30) = 30 lb. A horizontal force P = 30 lb would cause impending motion.

Example 5.14 A 100-lb block rests on a plane inclined at 30° to the horizontal. The coefficient of friction between the surfaces is 0.35. Determine the magnitude of the horizontal force P for impending motion up the incline (Fig. E5.14a).

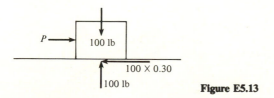

100 lb

P

30°

(a)   **Figure E5.14**

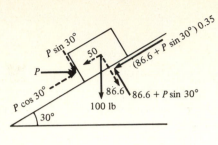

Figure E5.14 (*continued*)

(b)

SOLUTION Draw a free-body diagram of the block (Fig. E5.14*b*). Resolve all forces into components perpendicular and parallel to the plane.

$\overline{\Sigma F_\parallel} = 0$: $P\cos 30° - 50 - (86.6)(0.35) - (P\sin 30°)(0.35) = 0$

$$0.866P - 50 - 30.3 - 0.175P = 0$$
$$0.69P = 80.3$$
$$P = \frac{80.3}{0.69} = 116.4\ \text{lb}$$

Example 5.15 A block weighing 500 N rests on a horizontal plane. The coefficient of static friction between the two surfaces is 0.30. Determine the magnitude of the force required for impending motion if the force acts an angle of 30° from the horizontal.

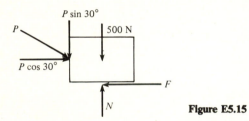

Figure E5.15

SOLUTION See Fig. E5.15.

$\uparrow \Sigma Y = 0$:	$N - 500 - P\sin 30° = 0$	(1)
$\overrightarrow{\Sigma X} = 0$:	$P\cos 30° - F = 0$	(2)
	$F = 0.30\,N$	(3)

Substitute (3) into (2) and solve simultaneously with (1):

$$P\cos 30° - (0.30)(500 + P\sin 30°) = 0$$
$$P = 210\ \text{N}$$

5.12 SUMMARY

This chapter introduces a series of concepts relative to the behavior of materials under load. Internal forces are considered in terms of the area upon which they act (stress) and the deformation they cause per unit length of material (strain),

$$f = \frac{P}{A} \qquad \varepsilon = \frac{\delta}{L}$$

Stresses are classified as normal (perpendicular to a surface) and shear (parallel to a surface). Normal stresses can be either tensile or compressive.

The development of strain in one direction results in strain of the opposite sign in a direction perpendicular to it. The ratio of these strains (Poisson's ratio) is different for different materials,

$$\mu = \frac{\varepsilon_y}{\varepsilon_x}$$

Most materials exhibit a linear relationship between stress and strain. This range is termed elastic and the ratio of stress to strain is called the modulus of elasticity,

$$E = \frac{f}{\varepsilon}$$

A similar relationship exists for shear and is termed modulus of rigidity,

$$G = \frac{f_s}{\gamma}$$

Torsion theory develops the relationship between shear stress f_s and angle of twist θ in a shaft subjected to a torque T,

$$f_s = \frac{Tr}{J} \qquad \theta = \frac{TL}{JG}$$

All materials change length with a change in temperature. The amount of change depends on the material and is evaluated by a coefficient of thermal expansion α μin/in·°F,

$$\delta = \alpha \, \Delta \, tL$$

If these changes in length are prevented or resisted, thermal stresses are developed.

Thin-walled cylinders develop longitudinal and hoop stresses as the result of an internal pressure p,

$$f_l = \frac{pd}{4t} \qquad f_h = \frac{pd}{2t}$$

The resistance to sliding of one surface relative to another is friction. It depends upon the materials involved and the smoothness of the surfaces (coefficient of friction, etc.). It also depends on the normal force acting on the surfaces,

$$F = fN$$

PROBLEMS

5.1 A short 8- by 8-in timber post supports an axial compressive load. Calculate (*a*) the compressive stress in the post if the load *P* is 50,000 lb and (*b*) the maximum allowable load *P* on the post if the allowable stress in compression is 1200 lb/in².

5.2 A timber post 10 in square rests on a steel bearing plate 12 in square, which rests on a concrete footing (Fig. P5.2). Determine the maximum load *P* if the allowable compressive stress in the timber is 1200 lb/in² and that of concrete is 800 lb/in². What dimension *d* of a square footing is necessary if the soil pressure must not exceed 4000 lb/ft²?

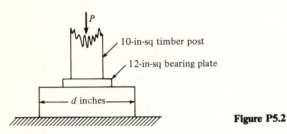

Figure P5.2

5.3 The end chord of a timber truss is framed into the bottom chord (Fig. P5.3). Neglecting friction, determine the required dimensions *a* and *b* if for wood the allowable shearing stress is 100 lb/in² and the allowable bearing stress 1000 lb/in².

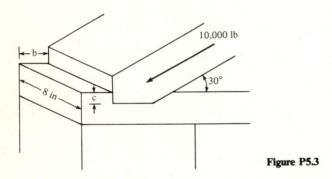

Figure P5.3

5.4 A 1-in diameter bolt with square head passes through a supporting steel plate (Fig. P5.4). Failure can occur in three ways: tension in the bolt, bearing in the plate under the bolt head, and shear in the bolt head. For a tensile load of 16,000 lb in the bolt calculate the average tensile, shear, and bearing stresses.

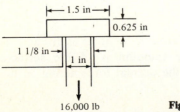

Figure P5.4

5.5 A 100-ft long surveyor's steel tape with a cross-sectional area of 0.004 in² must be stretched with a pull of 16 lb when in use. It is calibrated for a temperature of 70°F. Calculate (*a*) the elongation of the tape under the 16-lb pull, (*b*) the stress in the tape, and (*c*) the correction necessary if the temperature of the tape is 85°F.

5.6 A cast-iron column of 8 in OD, 6 in ID, and 15 ft long carries a compressive load of 225,000 lb. Calculate the average compressive stress in the column and the shortening of the column resulting from the load.

5.7 The ends of a laminated-wood roof arch are tied together with a horizontal steel rod 90 ft long. If the load carried by the rod is 52,000 lb, determine the required cross-sectional area for an allowable stress 20,000 lb/in². Determine the change in length resulting from the load.

5.8 A continuous concrete pavement was laid in 40-ft-long sections with $\frac{1}{2}$-in joints at 70°F. What width does the joint measure at (*a*) 35°F and (*b*) 100°F?

5.9 A steel wire 50 ft long is stretched between two fixed supports. If the stress in the wire is 4250 lb/in² at 70°F, what is the stress at 50°F? At what temperature would the stress in the wire be zero?

5.10 A steel bar has a cross-sectional area of 1.2 in² and carries the loads shown in Fig. P5.10. Compute the total deformation in the bar.

9 kips ← ... 21 kips ... 24 kips ... → 12 kips

|← 4 ft →|← 5 ft →|← 4 ft →|

Figure P5.10

5.11 A spherical shell of 6 ft OD and 2 in thick contains hydrogen at 1200 lb/in². Calculate the stress in the shell.

5.12 A water tank made of $\frac{1}{2}$-in-thick plate is 25 ft in diameter and 50 ft high. Determine the height of water in the tank that will cause a circumferential stress of 5000 lb/in². Assume that water weighs 62.5 lb/ft³.

5.13 A lap joint contains one high-strength bolt. Determine the shear strength and the frictional resistance of the joint if the bolt is 1 in in diameter and pretensioned to a stress of 50,000 lb/in². The coefficient of friction between the steel plates is equal to 0.15, and the allowable shear stress for the rivet is 15,000 lb/in².

5.14 An axial force *P* is hung on the end of the bar as shown in Fig. P5.14. This load causes the bar to elongate 0.04 in. Determine the load *P*.

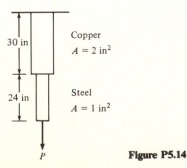

30 in — Copper $A = 2$ in²

24 in — Steel $A = 1$ in²

P

Figure P5.14

5.15 In Fig. P5.15 determine the strains in each of the rods. What is the total movement of the 500-lb force?

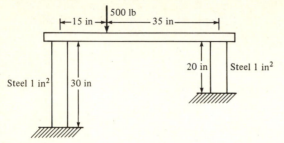

Figure P5.15

5.16 Determine the stresses in the bars in Fig. P5.16 when the temperature increases by 100°F.

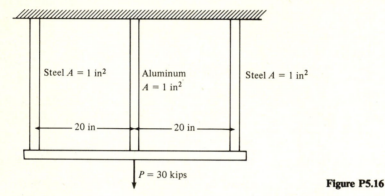

Figure P5.16

5.17 A steel band is to be shrunk-fit over a solid steel shaft which has a diameter 0.07 mm larger than the inside diameter of the band.

(*a*) At what temperature above normal must the band be raised to make this possible?

(*b*) What is the average stress in the steel band when normal temperature is regained? Neglect deformation in the shaft.

5.18 A reinforced concrete column (Fig. P5.18) is to support a load *P*. Determine the load *P* if the stress in the steel is not to exceed 20,000 lb/in^2 and the stress in the concrete is not to exceed 1350 lb/in^2.

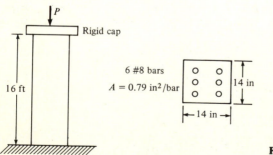

Figure P5.18

5.19 Determine the floor reactions R_1 and R_2 for the column shown in Fig. P5.19. Assume that the floors are unyielding.

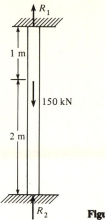

Figure P5.19

5.20 Determine the change in stresses in the steel and aluminum bars in Fig. P5.20 if the temperature increases by 50°C.

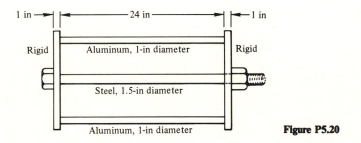

Figure P5.20

5.21 Figure P5.21 shows a bar composed of three rods fixed between two unyielding supports. Determine the stresses in each rod if the temperature increases by 75°F from that at which the rods have no stress in them.

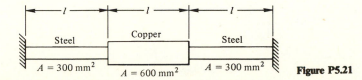

Figure P5.21

5.22 A tubular round aluminum shaft 1 m long fits over a concentric solid steel shaft 0.50 m long, as shown in Fig. P5.22. Determine (*a*) the maximum shearing stress in each material and (*b*) the total rotation of the free end relative to the fixed end.

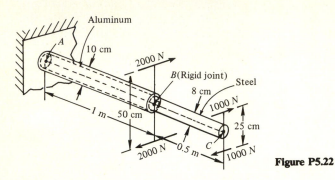

Figure P5.22

5.23 For Fig. P5.23 assume the following data. Material for *AB* is steel, for *BC* aluminum. When T_B = 10,000 in·lb, determine (*a*) the shearing stresses in each shaft and (*b*) the maximum rotation of each shaft in degrees. The sections are rigidly joined at *B*.

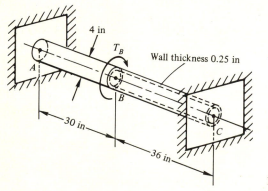

Figure P5.23

5.24 A tubular round shaft of aluminum is loaded as shown in Fig. P5.24. Determine (*a*) the maximum shearing stress and (*b*) the rotation of the free end relative to the fixed end.

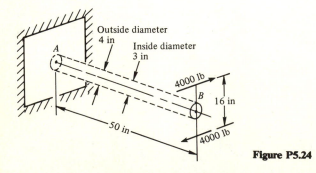

Figure P5.24

5.25 A cylindrical tank 2 m in diameter made of 20-mm-thick plate is to store oxygen under pressure. Determine the maximum allowable pressure for the tank if the allowable stress in the steel plate of the tank is 14 MPa.

5.26 Determine the maximum stress in the skin of a basketball, $t = 0.10$ in, when it is inflated to a pressure of 16 lb/in^2. Assume the diameter to be 12 in.

5.27 Determine the force P which will cause impending motion (*a*) up the plane and (*b*) down the plane in Fig. P5.27. The coefficient of friction f between the block and plane is 0.4.

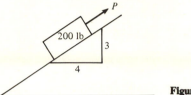

Figure P5.27

5.28 A block weighing 100 N rests on a horizontal surface. A force of 30 N pulling at an angle of 30° causes impending motion. Determine the coefficient of friction for the two surfaces (Fig. P5.28).

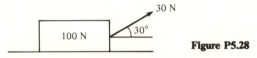

Figure P5.28

5.29 Block A weighs 200 lb, and block B weighs 300 lb. Find the force P required to move block B and the force T in the cord connecting block A to the wall. The coefficient of friction f for all surfaces is 0.30.

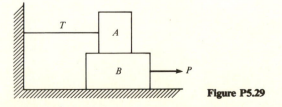

Figure P5.29

5.30 A force of 200 lb is applied to a crate weighing 1000 lb, as shown in Fig. P5.30. Determine whether the crate will slide, tip over, or remain stationary.

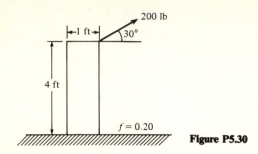

Figure P5.30

5.31 Two blocks resting against each other are placed on a surface inclined at an angle θ to the horizontal (Fig. P5.31). Block A, weighing 30 N, is placed down the incline from block B, which weighs 500 N. The coefficient of friction between block A and the surface is 0.3 and between block B and the surface is 0.2. Determine the maximum angle of inclination θ for which no slip will occur.

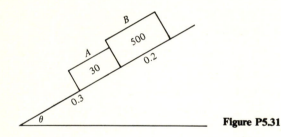

Figure P5.31

MATERIALS AND THEIR STRUCTURAL PROPERTIES

6.1 INTRODUCTION

The physical properties of construction materials are of importance to the designer since the behavior of a structure under load depends on these properties. Initially architecture was limited to compressive masonry forms for the larger spans, timber beams for intermediate spans, and heavy stone beams for the shorter spans.

The introduction of steel and reinforced concrete in the latter part of the nineteenth century made possible many innovations in structural form, e.g., multistoried buildings, long-span bridges, and large roof structures. Structural timber has become much more important with advances in lamination and connections. New high-strength steels, composites, and plastics are continuing to make even more sophisticated structural elements and forms possible.

The purpose of this chapter is to describe the basic properties and behavior of the most important structural materials. The coverage is brief and is not a substitute for more comprehensive works.

6.2 THE STRESS-STRAIN CURVE

Most of the important properties of a material are given in terms of its behavior during a simple uniaxial test. The nature of the loading depends on its function in a structure and on its physical properties. For example, structural steel is usually tested in tension and concrete in compression. Concrete is much stronger in compression and cannot readily be gripped for a tension test. In order to make comparisons between materials possible, the more basic stress-strain curve is used instead of the load-deformation curve (Sec. 5.1).

6.2.1 Tension Test

A typical stress-strain curve for structural steel is shown in Fig. 6.1. The stress is based on the original cross-sectional area of the specimen, and the curve is referred to as a *nominal stress-strain curve*. The slope of the elastic portion of the curve is E, the modulus of elasticity. The yield stress F_y and the ultimate strength (tensile strength) F_u are the most significant properties that differentiate various structural steels.

The yield point will vary somewhat with temperature, rate of loading, and specimen characteristics, which are standardized by the American Society of Testing and Materials (ASTM). After the initial yield, the specimen elongates in a plastic range without appreciable change in cross section. Yield actually occurs in localized regions which strain-harden (strengthen) and thus force yielding in new locations. After the elastic regions have been utilized, at strains of from 4 to 10 times the elastic strain, the stress starts to increase due to a more general strain hardening. This, along with inelastic deformation, continues until the maximum load is reached. The specimen then experiences a local constriction of cross section called *necking down*. After this point has been reached, the load-carrying capacity of the specimen decreases until failure occurs. The true stress, however, continues to increase until fracture.

Structural steels are unique in that they are tough. *Toughness* can be defined as a combination of strength and ductility which can be represented by the area

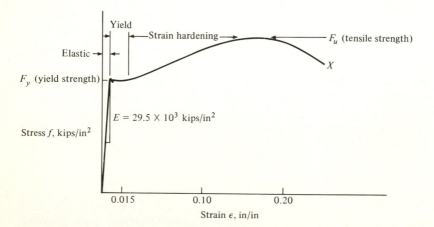

Figure 6.1 Stress-strain curve for a structural steel in tension.

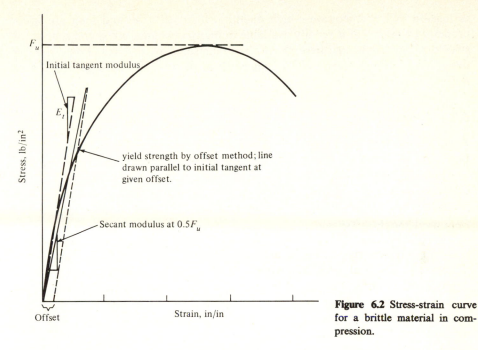

F_u

Initial tangent modulus

E_t

yield strength by offset method; line drawn parallel to initial tangent at given offset.

Secant modulus at $0.5F_u$

Stress, lb/in²

Offset

Strain, in/in

Figure 6.2 Stress-strain curve for a brittle material in compression.

under the stress-strain curve. *Ductility* can be defined as the permanent strain remaining in a specimen after failure.

6.2.2 Compression Test

The compression test is carried out on specimens of different size, depending upon the material. ASTM has standardized size and shape for such materials as concrete and wood. The stress is based on the original cross-sectional area. Figure 6.2 shows a typical stress-strain curve of a brittle material. The initial portion of the curve is linear and indicates the elastic behavior of the material. The upper limit of the elastic portion of the curve is termed the *proportional limit*. Since the proportional limit is difficult to locate and many materials do not have a yield stress, an offset method is used to locate a yield strength. The slope of the initial portion of the curve is termed E, the modulus of elasticity of the material. The ultimate strength F_u refers to the maximum stress developed by the material during the test. The shape of the curve beyond the ultimate strength is of little significance because it depends on rate of loading and machine characteristics.

6.3 STEEL

Steel is probably the most important structural material and along with concrete accounts for most engineered structures. Steel has essentially the same strength in tension and compression—high ductility, high stiffness, and long life when properly protected. It is readily available in many strengths, sizes, and shapes for structural use. Some of these shapes are illustrated in Fig. 6.3. Properties for designing and dimensions for detailing some of these sections are given in Table A.5.

Some of the disadvantages associated with steel as a construction material are its susceptibility to corrosion by water and chemicals, reduced strength and stiffness at elevated temperatures, and the possibility of fatigue in certain cyclic loading situations. Steel does not have the versatility of a cast-in-place material, but modern fabrication techniques facilitate the use of curved and variable section members.

The manufacture of steel is a highly developed technology, and many new grades of steel have been developed and adopted in recent years. The grade of steel used should be specified.

In general, *carbon steel* is the term applied to steels containing up to 1.7 percent carbon. These steels are divided into four categories: low carbon (less than 0.15 percent), mild carbon (0.15 to 0.29 percent), medium carbon (0.30 to 0.59 percent), and high carbon (0.60 to 1.70 percent). Structural carbon steels are in the mild-carbon range. Increased carbon content raises the yield stress but reduces ductility. A reduction in ductility increases problems associated with welding. Satisfactory economical welding without preheat, postheat, or special welding electrodes can normally be accomplished when the carbon content is less than 0.30 percent.

Two additional classes of structural steel have been developed, namely a high-strength steel using alloying elements and heat-treated low-alloy steels. The former include the addition of small amounts of alloying elements such as chromium, columbium, copper, manganese, molybdenum, nickel, phosphorus, vanadium, or zirconium. These steels have yield stresses from 40 to 70 kips/in^2. The yield stresses are well defined (Fig. 6.1). When quenched and tempered, low-alloy steels reach yield strengths of 80 to 110 kips/in^2. Since they are not well defined, they are determined at an offset strain. Typical stress-strain curves are illustrated in Fig. 6.4. Low-alloy steels come in various structural shapes but heat-treated low-alloy steels are available only as flat plates (Table 6.1).

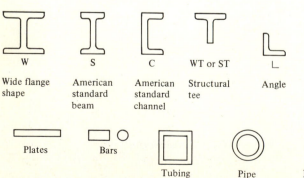

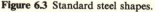

Figure 6.3 Standard steel shapes.

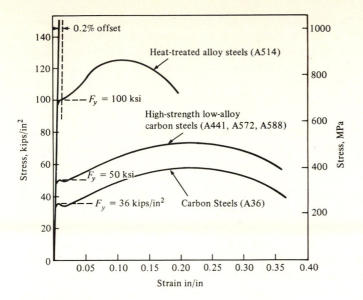

Figure 6.4 Stress-strain curves for steels used in structures.

Table 6.1 Structural steels and their properties

Code		Purpose		Minimum tensile strength F_u		Minimum yield point F_y	
				kips/in^2	MPa	kips/in^2	MPa
A36		General structural		58–80	400–550	36	250
A529		General structural, not over $\frac{1}{2}$-in thick		60–85	414–586	42	290
		Plates, in					
A242, A441		A588					
3/4 and under		4 and under		70	480	50	345
3/4–1$\frac{1}{2}$		4–5		67	460	46	315
1$\frac{1}{2}$–4		5–8		63	435	42	290
		Structural shapes					
A242, A441		A588					
Groups 1, 2		Groups 1–4		70	480	50	345
Group 3		Group 5		67	460	46	315
Groups 4, 5				63	435	42	290
Code	Grade	Group	Plate size, in				
A572	42	All	Up to 6	60	415	42	290
	50	All	Up to 2	65	450	50	345
	60	1, 2	Up to 1$\frac{1}{4}$	75	520	60	415
	65	1	Up to 1$\frac{1}{4}$	80	550	65	450
A514	Plates only		Up to 2$\frac{1}{2}$	110–130	760–895	100	690
			2$\frac{1}{2}$–4	100–130	690–895	90	620

Structural steels are referred to by ASTM designations as well as proprietary names. For design purposes the yield stress is the parameter which specifications, e.g., AISC, use as the material-property variable to establish allowable unit stresses for various types of members. The term *yield stress* is used to include either *yield point*, the well-defined deviation from elasticity, or *yield strength*, the stress at a given offset strain when no well-defined yield point exists. Figure 6.5 shows the various ASTM-designated structural steels and their range of yield stresses and tensile strengths. Two, A36 and A529, are carbon steels; four, A242, A441, A572, and A588, are high-strength low-alloy steels; and one, A514, is a high-strength quenched and tempered alloy steel. The high-strength steels A242, A441, and A588 are primarily for use in structural members where savings in weight and added durability are desirable. They have approximately twice the corrosion resistance of the carbon steels. The yield points and tensile strengths of plates and bars made of high-strength steels vary with thickness, and structural shapes made of these steels are classified in size groups according to their properties. The quenched and tempered alloy steel A514, is available only in plates. Table 6.1 outlines these characteristics, and Table 6.2 gives the grouping of structural shapes.

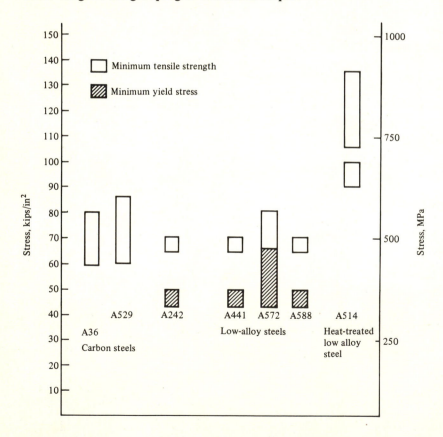

Figure 6.5 Range of yield and tensile strengths of structural steels.

Table 6.2 Grouping of structural shapes [1]†

Structural shape	Group 1	Group 2	Group 3	Group 4	Group 5
W shapes	W24 × 55 and 62	W36 × 135 to 210	W36 × 230 to 300	W14 × 233 to 550	W14 × 605 to 730
	W21 × 44 to 57‡	W33 × 118 to 152	W33 × 201 to 241	W12 × 210 to 336	
	W18 × 35 to 71	W30 × 99 to 211	W14 × 145 to 211		
	W16 × 26 to 57	W27 × 84 to 178	W12 × 120 to 190		
	W14 × 22 to 53	W24 × 68 to 162			
	W12 × 14 to 58	W21 × 62 to 147			
	W10 × 12 to 45	W18 × 76 to 119			
	W8 × 10 to 48	W16 × 67 to 100			
	W6 to 9 to 25	W14 × 61 to 132			
	W5 × 16 to 19	W12 × 65 to 106			
	W4 × 13	W10 × 49 to 112			
		W8 × 58 and 67			
M shapes	To 35 lb/ft	Over 35 lb/ft			
S shapes	To 35 lb/ft	Over 35 lb/ft			
HP shapes	To 20 lb/ft	To 102 lb/ft	Over 102 lb/ft		
American standard channels (C)		Over 20 lb/ft			
Miscellaneous channels (MC)	To 28.5 lb/ft	Over 28.5 lb/ft			
Angles (L) structural and bar size	To 1/2 in	Over $\frac{1}{2}$ to $\frac{3}{4}$ in	Over $\frac{3}{4}$ in		

†Structural tees from W, M, and S shapes fall in the same group as the structural shape from which they are cut.
‡All ranges are inclusive.

In addition to the hot-rolled plates and shapes, thin sheets of carbon and low-alloy steels may be *cold-rolled* into a wide variety of corrugated and folded shapes for floors, walls, and roofs of structures (Fig. 6.6). The thickness of these members ranges from about 0.01 in for panels to nearly $\frac{1}{4}$ in for structural shapes.

Structural cables for suspension bridges and suspended roof structures are normally made of high-strength steel wire, which is usually coated with zinc to provide corrosion resistance. Figure 6.7 illustrates the arrangement of the wires in cables and rope.

Prestressing tendons are made from high-strength wires or rods.

Reinforcing bars used in reinforced concrete have a variety of strengths and are made from high-carbon and low-alloy steels. Typical deformed bars are shown in Fig. 6.8.

The modulus of elasticity E of steel is nearly independent of steel type and is usually taken as 29,500 kips/in².

As mentioned in Chap. 1, structures and structural members are always designed to carry some reserve load beyond that expected under normal use. In

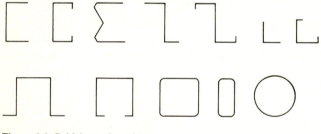

Figure 6.6 Cold-formed sections.

Strand   6 x 7 rope   6 x 19 rope

Figure 6.7 Structural cables and wire rope.

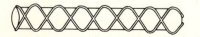
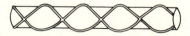

Figure 6.8 Deformed reinforcing bars.

Table 6.3 Allowable stresses [15]

Type	Notation	Comment
Tension	$F_t = 0.60F_y \leq 0.50F_u$	
Shear	$F_v = 0.40F_y$	
Bending	$F_b = 0.66F_y$	Tension and compression for compact adequately braced members symmetrical about, and loaded in, plane of minor axis
Bearing	$F_p = 0.90F_y$	Milled surfaces, including bearing, stiffness, and pins in reamed, drilled, or bored holes
	$F_p = 1.5F_u$	Projected areas of bolts and rivets in shear connections; F_u is miminum tensile strength of the connected parts
Compression		Because of buckling, allowable stress depends upon dimensional factors as well as material properties

the working-stress method, a factor of safety of 1.67 relative to the yield stress is often specified. For an A36 steel this gives a working stress of approximately 22 kips/in^2. In plastic or ultimate design, a corresponding load factor of 1.7 is often used relative to a combination of live plus dead loads. Table 6.3 presents some of the basic AISC recommendations relative to allowable stresses in terms of the yield and tensile strength of the steel.

6.4 CONCRETE

Structural concrete is a very popular construction material with an almost unlimited variety of structural shapes possible because the material can be easily formed. Its advantages include relatively low cost because of the use of local materials, good compressive strength, low maintenance, and durability. The disadvantages include low tensile strength, low compressive strength-to-weight ratio, the tendency to develop shrinkage cracks on drying out, and the possibility that variable quality control will result in a variable and low-strength material.

Concrete is a mixture of aggregates, cement, and water. The cement and water react chemically to bond the aggregate together. The aggregates, which are often naturally occurring sand and gravels or crushed rock but may also be synthetic, are normally graded in size from about $1\frac{1}{2}$ in down to 100 μm. Particles larger than about $\frac{1}{4}$ in are referred to as *coarse aggregates* and smaller ones as *fine aggregate* or *sand*. Portland cements, the most widely used, contain anhydrous calcium silicate and aluminate compounds and are classified as *hydraulic cements* because they set and harden under water to give a water-resistant product, unlike the lime and gypsum mortars.

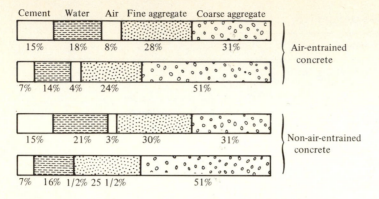

Figure 6.9 Range of concrete proportions by volume [2].

The typical range of mix proportions for normal concretes made with naturally occurring mineral aggregates and portland cements are given in Fig. 6.9.

Proportioning a mix for a given set of requirements is facilitated by previous experience and published design procedures [3]. A series of trial batches is normally required to check these proportions.

The compressive strength of a concrete mix depends primarily upon the water-cement ratio of the mix. At higher strengths (lower water-cement ratios) the strength also depends on the quality of the aggregate. The water-cement-ratio–strength relationship recommended for design by the ACI is given in Table 6.4.

6.4.1 Strength

The compressive strength of concrete is probably its most important property and is determined by testing 6- by 12-in cylinders in uniaxial compression. The ultimate strength is referred to as the *compressive strength* and is designated as f_c' in design theory. The test is normally carried out 28 days after curing under standard conditions of temperature (72°F) and humidity (100 percent).

Table 6.4 Relationship between water-cement ratio and compressive strength of concrete [3]

Compressive strength at 28 days, lb/in²	Water-cement ratio, by weight	
	Air-entrained concrete	Non-air-entrained concrete
6000	0.41	
5000	0.48	0.40
4000	0.57	0.48
3000	0.68	0.59
2000	0.82	0.74

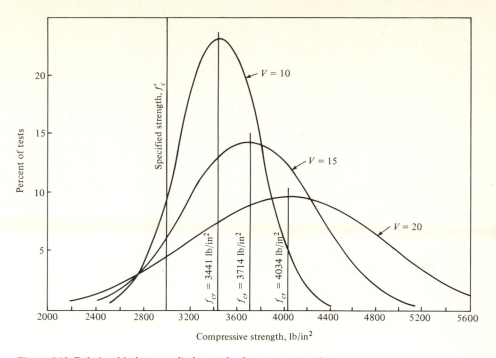

Figure 6.10 Relationship between f'_c, f_{ave}, and v for a concrete mix.

The average concrete strength for which a concrete mix must be designed must exceed f'_c by an amount which depends on the uniformity of production (V = coefficient of variation) and the percentage of test results that a particular code will allow to be below f'_c. These relationships are illustrated in Fig. 6.10.

A stress-strain curve gives more complete information on concrete. Figure 6.11 illustrates data of this type. As the load increases from zero, the curve remains linear up to about 35 percent of the ultimate load f'_c. The nonlinearity at higher loads is an indication of internal microcracking at aggregate-paste interfaces. At loads of 85 percent of ultimate these cracks become interconnected, the specimen has little additional load-carrying capacity, and the concrete has developed a strain of about 0.002 in/in. At higher strains the curve descends and is not well defined, as it depends on the rate of loading and stiffness of the testing machine. Final failure can involve splitting, shearing, and crushing of aggregate. The strain at ultimate load in a cylinder test is about 0.003 in/in for most concretes but it is larger for those containing lightweight aggregate.

The slope of the initial portion of the stress-strain curve defines the initial or tangent modulus of elasticity E_t of the concrete. The slope of the chord (up to about $0.5f'_c$) determines the secant modulus of elasticity E_c, which is generally used in straight-line stress calculations. The modulus of elasticity is sensitive to the same factors that affect strength. As a result E_c has been related to f'_c by

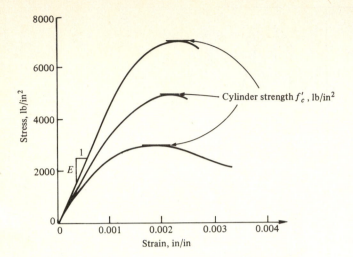

Figure 6.11 Stress-strain curves for concrete in compression.

various empirical relationships. The relationship recommended by ACI is

$$E = 33w^{3/2}\sqrt{f_c'}$$

where E and f_c' are expressed in pounds per square inch and w, the unit weight, is in pounds per cubic foot. This value of E should be used only for predicting short-term strains and deflections. Long-term observations must take creep into account.

The tensile strength of concrete ranges from about 8 to 15 percent of compressive strength. It is undependable and is not normally relied upon in design. Direct measurement of the tensile strength is difficult because of gripping problems. The split-cylinder test and beam test are widely used indirect testing methods. The test specimens and stress distributions are illustrated in Fig. 6.12. These tests will not give similar results because their distributions are not the same. They tend to be higher than the true tensile strength. The split-cylinder test is preferred.

In terms of the modulus of elasticity, creep is an important consideration when loads are being carried for long periods of time. Creep refers to a continued increase in deformation under constant loading (Fig. 6.13). Creep recovery is never complete. The structural implications of this phenomenon are very important. In a reinforced-concrete column the subsequent transfer of load from the concrete to the steel can be very significant. In reinforced-concrete beams the compression side of the beam continues to shorten with time, resulting in deflections that may be 2 or 3 times the elastic value. In prestressed beams a significant portion of the prestress can be lost with resulting losses in load-carrying capacity.

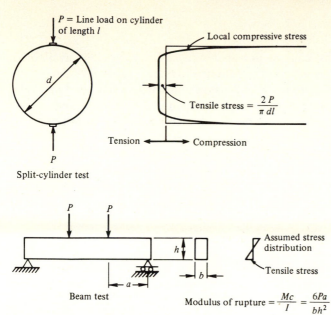

Split-cylinder test

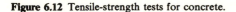

Beam test

Modulus of rupture $= \dfrac{Mc}{I} = \dfrac{6Pa}{bh^2}$

Figure 6.12 Tensile-strength tests for concrete.

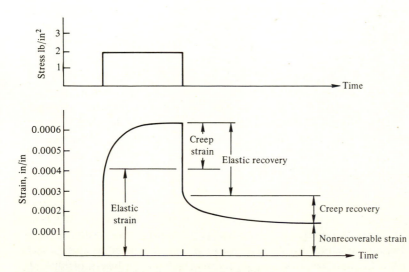

Figure 6.13 Creep behavior of concrete under load.

6.5 TIMBER

Wood was one of the first construction materials and is still readily available in most parts of the world. Some of its advantages include a relatively high ratio of strength to weight, similar strengths in tension and compression, and an ability to absorb considerable energy before failure. If kept moist or dry, timber is quite resistant to deterioration and in large solid members it is considered quite fire-resistant. Modern laminating techniques make the fabrication of members of nearly any shape and size possible. It also has natural aesthetic appeal.

Some of the disadvantages include a low modulus of elasticity, directional properties, and dimensional instability (shrinkage and swelling). Connections for both shear and tensile loads are expensive and complicated.

6.5.1 Physical Character

Wood is a cellular material composed of cellulose, lignin, and small quantities of other materials. The cell walls are made of cellulose, stiffened and cemented together by lignin. Cells vary in size and shape according to their function. Most are elongated and positioned vertically in the tree. They vary from $\frac{1}{8}$ to $\frac{1}{3}$ in long and about one-hundredth of these dimensions in width. New wood cells are formed in cambium, a thin layer located between the bark and the wood. In the spring, new thin-walled spring wood forms with large cavities. During the summer cell walls increase in thickness, and the cells become smaller toward the end of the season. In winter trees are normally dormant. These growth characteristics result in rings of annual growth (Fig. 6.14). This, along with the structure of the cells, is responsible for anisotropy. The directional properties of wood are associated with the principal axes (Fig. 6.15), i.e., the *tangential axis* (tangent to growth rings), *radial axis* (radius of tree), and *longitudinal axis* (vertical axis of tree).

The two main classes of trees are softwood (coniferous) and hardwood (deciduous). Softwood trees are the prime source of structural timber. Douglas fir and southern pine are among the most widely used species of softwood.

6.5.2 Moisture in Wood

In addition to the actual structure of the cell system, wood properties are affected by the environment, moisture being probably the most important factor. Wood is sensitive to moisture whether as vapor in air or as liquid water. Wood may contain moisture either *free water* in the cell cavities (lumina) or *absorbed water* in the cell walls. When green wood begins to lose moisture, the cell walls remain saturated and free water is lost from the lumina. When evaporation of free water is complete, the cell walls begin to lose moisture. This condition, termed *fiber saturation point* (FSP), occurs between a moisture content of 25 and 30 percent for most softwoods. The moisture content of wood in a living tree is approximately 30 to 90 percent in the heartwood portion, and it

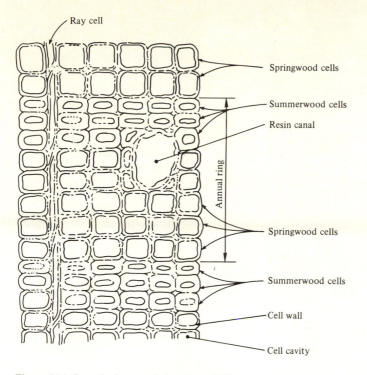

Figure 6.14 Growth characteristics of wood [5].

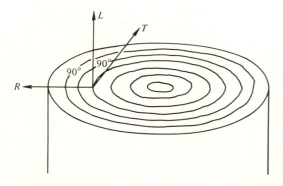

Figure 6.15 The three principal axes of wood: *L*, longitudinal (parallel to grain); *R*, radial (perpendicular to grain, radial to annual rings); *T*, tangential (perpendicular to grain, tangential to annual rings).

may be as high as 200 percent in the sapwood portion. The FSP is important because shrinkage, swelling, and strength change as moisture content decreases.

The moisture content of wood is the result of an equilibrium condition between the forces that hold water in wood and the vapor pressure of the surrounding environment. To remain dry, a completely dry piece of wood must be kept in a space free from water vapor. At 20 percent relative humidity and 70°F, a dry environment, wood comes to equilibrium at about 4 percent moisture content. At 75 percent relative humidity and 70°F, the equilibrium moisture content is about 14 percent, and for a relative humidity of 100 percent it approaches fiber saturation. Because there is always a time lag involved in attaining equilibrium conditions, the moisture content is generally close to that corresponding to the average daily temperature and relative humidity. It will fluctuate slowly throughout the year.

6.5.3 Dimensional Stability

Variations in moisture content from the FSP upward to complete saturation have no effect on the dimensional size of wood because these moisture changes are associated with the free water contained in the cell cavity. The moisture content of the cell-wall material remains unchanged throughout this range. Below the FSP a cell-wall moisture-content change results in a dimensional change. This is a reversible process; the addition of moisture causes swelling, and a reduction in moisture content causes shrinkage. The dimensional change is a straight-line function of moisture content. Shrinkage is also a directional property. Table 6.5 gives shrinkage factors for several common softwoods.

Longitudinal shrinkage is so small that it is usually neglected. Values of 0.1 to 0.2 percent from FSP to oven-dry are considered realistic. As lumber is often cut randomly with respect to the tangential and radial directions, the shrinkage can be estimated by averaging the tangential and radial values. The following example illustrates shrinkage calculations.

Shrinkage of Douglas fir from green to 12 percent moisture content is

$$7.6 \frac{30 - 12}{30} = 4.6\% \text{ of green size} \qquad \text{tangential direction}$$

$$4.8 \frac{30 - 12}{30} = 2.9\% \text{ of green size} \qquad \text{radial direction}$$

$$0.2 \frac{30 - 12}{30} = 0.12\% \text{ of green size} \quad \text{longitudinal direction}$$

For randomly cut lumber, using an average of the tangential and radial values,

Table 6.5 Shrinkage factors for several softwoods [7]

Shrinkage	Shrinkage FSP to oven-dry, % size at FSP	
	Tangential	Radial
Douglas fir	7.6	4.8
Western hemlock	7.8	4.2
Western larch	9.1	4.5
Southern pine	7.8	5.5
Engelman spruce	7.1	3.8
Redwood	4.9	2.2

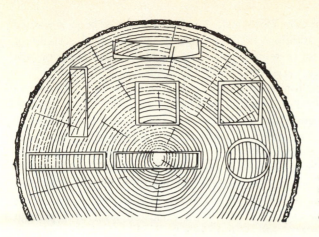

Figure 6.16 Characteristic shrinkage and distortion of structural shapes due to differences in tangential and radial shrinkage of wood [7].

we have

$$\frac{7.6 + 4.8}{2} \frac{30 - 12}{30} = 3.7\% \text{ of green size}$$

Swelling can be considered the reverse of shrinkage.

If wood had the same shrinkage coefficients in both the radial and tangential directions, the problem of warping and distortion of lumber on drying would be largely eliminated. Figure 6.16 illustrates characteristic distortions, which can be overcome by planing after seasoning to a moisture content close to that in service.

6.5.4 Strength

Like all other materials wood is characterized by variability, which can be illustrated by a frequency-distribution curve of strength-test data (Fig. 6.17). Such a distribution is described by the parameters σ and $\bar{x}$, the standard

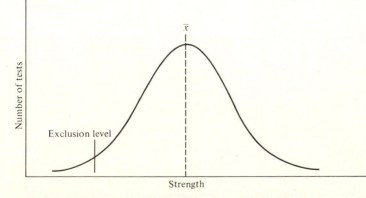

Figure 6.17 Frequency-distribution curve for wood-strength data,

deviation and average strength. This kind of information has been obtained under ASTM's clear wood standard [8]. The primary features of this standard are tables presenting the most reliable basic information developed on the strength of clear wood and its variability.

The properties evaluated for design purposes include extreme fiber stress in bending (modulus of rupture) F_b, tension parallel to the grain F_t, horizontal shear F_v, compression perpendicular to the grain $F_{c\perp}$, compression parallel to the grain F_c, and modulus of elasticity E. These test conditions are illustrated in Fig. 6.18. It is important to realize that $F_{c\perp}$ is based on a proportional limit strength while the other strength properties are based on failure conditions.

These properties are evaluated on small, clear green specimens. The first adjustment made to these average values, to arrive at a design stress, is the selection of an exclusion level. For example, an exclusion level of 1 percent indicates that only 1 piece in 100 is likely to have a lower strength value than the level selected (Fig. 6.17).

An exclusion level of 5 percent is used for framed wood structures because of the closely spaced structural elements. If one element has low strength, other

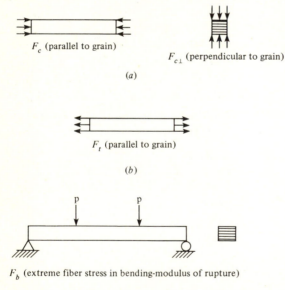

F_c (parallel to grain)

$F_{c\perp}$ (perpendicular to grain)

(a)

F_t (parallel to grain)

(b)

p   p

F_b (extreme fiber stress in bending-modulus of rupture)

(c)

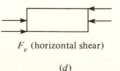

F_v (horizontal shear)

(d)

Figure 6.18 Test conditions for determining strength properties of wood: (a) loaded in compression, (b) loaded in tension, (c) loaded in bending, (d) loaded in shear.

closely located elements can take up the load. In laminated construction, an exclusion level of 1 percent is used because the structural members are more widely spaced.

After the exclusion-level adjustment three additional considerations are involved in determining allowable design stresses:

1. An increase in a property value due to the effect of seasoning
2. The effect of the strength-reducing defects permitted in grading
3. A general adjustment factor which is the composite result of considering other influences known to affect the property

As in dimensional stability, a change in moisture content below the FSP results in a change in the strength property of wood. A decrease in moisture content results in an increase in strength. The relationship is linear in a semilog plot (Fig. 6.19).

The slope of the relationship below the FSP depends on the property and applies to all the principal structural wood species. Moisture-content standards for grading rules provide for a MC-19 (19 percent maximum moisture content, 15 percent average) and MC-15 (15 percent maximum moisture content, 12 percent average). The effect of this seasoning (a reduction in moisture content from above the FSP) is indicated in Table 6.6.

By determining the effect of such defects as knot size, grain deviation and slope, end splits, checks, and shakes and by systematically codifying these characteristics it has been possible to establish structural grades and related allowable properties for visually graded lumber [9]. This involved establishing a

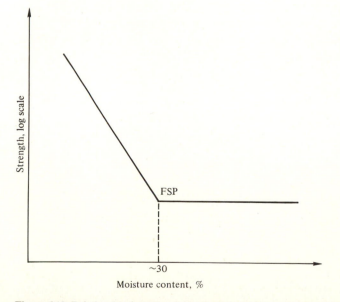

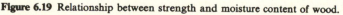

Figure 6.19 Relationship between strength and moisture content of wood.

Table 6.6 Allowable properties for a sample stress grade

Property	Clear-wood† strength value, lb/in²	Strength ratio	Seasoning increase for 19% moisture content	General adjustment factor	Allowable property, lb/in²‡
F_b	5500	0.54	1.25	1/2.1	1750
F_c	2575	0.62	1.50	1/1.9	1250
F_v	688	0.50	1.08	1/4.1	90‡
F_t	5500	0.30	1.25	1/2.1	1000
$F_{c\perp}$	382	1.00	1.50	1/1.5	1900
$E/1000$	1560	1.00	1.14	1/0.94	1900‡

†Unseasoned, 5% exclusion level, except E and $F_{c\perp}$ which are average values.

‡Values in the last column have been rounded off to the nearest 50 lb/in² except for shear (to the nearest 5 lb/in²) and the modulus of elasticity (to the nearest 100,000 lb/in²).

strength ratio for each property and grade, which represents the anticipated proportionate remaining strength after making allowance for the various defects permitted by the grade, compared with clear straight-grained lumber. The strength ratios for all properties of a grade are not the same.They represent the maximum effect of the defects in a particular grade. Within a given grade the strength ratio for any property will vary from the minimum permitted up to the minimum permitted by the next highest grade (Table 6.6).

Associations active in grading lumber include the Western Wood Products Association (WWPA) and the Southern Forest Products Association (SFPA). A new American Softwood Lumber Standard, PS20-70, promulgated by the U.S. Department of Commerce, has provided for a national grading rule [14], which has simplified grade names and new minimum sizes designed to assure greater uniformity, efficiency, and economy in the use of dimension lumber (Table 6.7).

Table 6.7 Grade names and sizes

Structural light framing, 2 to 4 in thick, 2 to 4 in wide Select structural No. 1 No. 2 No. 3	Studs, 2 to 4 in thick, 2 to 4 in wide, 10 ft and shorter Stud
Light framing, 2 to 4 in thick, 2 to 4 in wide Construction Standard Utility	Structural joists and planks, 2 to 4 in thick, 6 in and wider Select structural No. 1 No. 2 No. 3
Appearance framing, 2 to 4 in thick, 2 in and wider Appearance	

In addition to these grades and sizes the category of *timbers* includes sizes 5 by 5 in and larger and is of particular interest to engineers and architects in designing structural systems.

The third consideration in the development of allowable design stresses is the use of a general adjustment factor which takes into account several phenomena known to affect the properties of wood. These are summarized in Table 6.8. As is apparent from the table, some elements do not apply to every property.

The duration-load effect is a very important feature of timber design. The mechanical properties of wood are load- and time-dependent. The clear-wood strength properties are established from laboratory tests of close to 5 min duration from zero to failure load conditions. The "normal" duration of load condition for structural design properties is based on the concept that the maximum load condition will have a cumulative duration not to exceed one-tenth of an expected life of 100 years or more. The factors given in Table 6.6 are the ratio of laboratory test strengths to normal-duration strengths. The normal-load-duration adjustment of $F_{c\perp}$ is small because it is a proportional-limit value which is only about 60 percent of ultimate strength. The relationship between load duration and permissible load (Fig. 6.20) is fairly well defined. There is no load-duration factor for the modulus of elasticity. It is evident only under extremely rapid load application, rates which are not applicable to building structures and which would only be transient in terms of deflections.

The manufacture-and-use factors in Table 6.8 result from the consideration of such things as the effect of fastenings driven into members in use, machine skip in dressing, end splits occurring after construction, drilling holes for wiring and plumbing, error in grading, and variability in shrinkage. A separate stress-concentration factor is used for shear because of the influence of specimen shape and size. The l/d ratio for elastic modulus arises from the influence of internal shear deformation on the deflection of relatively short bending members. Table 6.6 illustrates the combination of these adjustments to produce design properties for a sample stress grade of lumber.

The purpose of this extended discussion is not only to realize the reliability of the legally grade-marked lumber but also to understand the factors involved

Table 6.8 Elements of the adjustment factor [5]

Property	Normal-duration-of-load factor	Manufacture-and-use factor	Stress-concentration factor	End position	l/d	Adjustment factor
F_b	10/16	10/13				1/2.1
F_c	2/3	4/5				1/1.9
F_v	10/16	8/9	4/9			1/4.1
F_t	10/16	10/13				1/2.1
$F_{c\perp}$	11/10	10/11		2/3		1/1.5
E	1.0				1.0/0.94	1.0/0.94

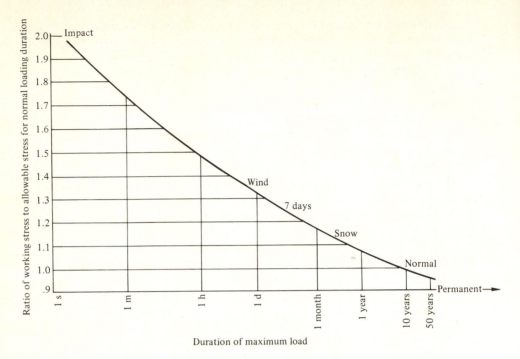

Figure 6.20 Relationship between working stress and load duration [10].

in arriving at such stresses. As a result, when service conditions change, appropriate changes can be made in the allowable design stress.

Allowable design stresses depend upon grade, species, and moisture content, Appendix Table A.1 provides allowable design stresses for some of the most common species and grades and Table A.2 provides information on section properties for various timber sizes. These properties will be useful in later chapters associated with the design of timber members.

6.6 GLUED-LAMINATED TIMBER

6.6.1 Introduction

Structural glued-laminated timber (glulam) refers to three or more layers of sawn lumber glued together with the grain direction of all layers approximately parallel. It has been used in engineering structures in the United States since 1935. The development of durable adhesives during World War II permitted the use of glued-laminated members under severe service and exposure conditions. With the use of proved adhesives, it is possible to produce joints that are durable and essentially as strong as the wood itself. The principal species used in the construction of laminated timber are Douglas fir and southern pine.

Glued-laminated members are classified as vertically or horizontally laminated beams according to the orientation of the glue line within the cross section. Vertically laminated members are limited in depth by the width of lumber available but can be fabricated in varying widths. Horizontally laminated members are, on the other hand, limited in width to the width of laminated stock available but generally are unlimited in depth. This latter type of member is the more common.

Significant advantages in the fabrication of glued-laminated structural members include a more efficient use of wood since lower-grade material can be used in the less highly stressed central laminations. Since the fabrication of large members is possible from smaller pieces, lumber from smaller trees can be used effectively. This kind of fabrication also means that the size and length of the member do not depend on the size and length of the tree. Spans of 200 ft with depths of 7 ft have been used. Since the members are laminated from thin pieces of wood, the strength-reducing characteristics can be more readily seen and controlled. Seasoned lumber can be used, which reduces seasoning checks and makes it possible to use the increased strength resulting from seasoning. Members of variable depth can be designed and fabricated, with additional economy and aesthetic bonuses. There is, of course, the cost of fabrication of laminated members, so that this type of construction does not compete economically with solid-sawn designs for short spans and small members.

6.6.2 Section Properties

Table A.3 gives section properties for structural glued-laminated timbers based on industry-recommended standard thicknesses and widths. Laminations are either $\frac{3}{4}$ or $1\frac{1}{2}$ in thick. Dimension lumber, surfaced to $1\frac{1}{2}$ in before gluing, is used in straight members and curved members having a radius of curvature of 27 ft 6 in or greater. Boards, surfaced to $\frac{3}{4}$ in before gluing, are used for curved members with a bending radius too short for $1\frac{1}{2}$-in-thick laminations but not less than 9 ft 4 in. Finished depths of laminated members are therefore generally multiples of the $1\frac{1}{2}$ or $\frac{3}{4}$-in net thicknesses.

6.6.3 Allowable Stresses

Most glued-laminated members currently being produced are manufactured from visually graded lumber. The strength is principally a function of two properties, the clear-wood strength, a species property, and the size and frequency of knots, a property of the lumber used in manufacture. Besides knots, cross grain and jointing efficiency can also influence strength. As these factors are not cumulative, and as the lowest determines strength, current practice is to control cross grain and end joints so that the effect of each is less than that of the knots. Table A.4 presents allowable design stresses for structural glued-laminated members using visually graded lumber.

REFERENCES

1. American Society for Testing and Materials, "Annual Book of ASTM Standards," Philadelphia, 1979.
2. Portland Cement Association, "Design and Control of Concrete Mixtures," 12th ed., Skokie, Ill., 1979.
3. American Concrete Institute, Recommended Practice for Selecting Proportions for Normal and Heavy-Weight Concrete, Standard 211.1-77, Detroit, 1977.
4. American Concrete Institute, Recommended Practice for Evaluation of Strength Test Results of Concrete, Standard 214-77, Detroit, 1977.
5. R. J. Hoyle, Jr., "Wood Technology in the Design of Structures," Mountain Press, Missoula, Mont., 1973.
6. National Forest Products Association, Design Specifications for Wood Construction, Washington, 1977.
7. U.S. Department of Agriculture, "Wood Handbook: Wood as an Engineering Material," Agriculture Handbook 72, Washington, 1974.
8. American Society for Testing Materials, Standard Methods for Establishing Clear Wood Strength Values, ASTM D2555-78, Philadelphia, 1978.
9. American Society for Testing and Materials, Standard for Establishing Structural Grades and Related Allowable Properties for Visually Graded Lumber, ASTM D245-74, Philadelphia, 1974.
10. National Forest Products Association, Supplement of National Design Specifications for Wood Construction: Design Values for Wood Construction, Washington, 1977.
11. Southern Forest Products Association, *Tech. Bull* 2, New Orleans, n.d.
12. G. Gurfinkel, "Wood Engineering," Southern Forest Products Association, New Orleans, 1973.
13. Bethlehem Steel Company, Structural Steel Data for Architectural and Engineering Students, Bethlehem, Pa., 1978.
14. U.S. Department of Commerce, American Softwood Lumber Standard PS20-70, Washington, 1970.
15. American Institute of Steel Construction, "Manual of Steel Construction," 8th ed., Chicago, Ill., 1978.

TENSION MEMBERS AND THEIR CONNECTIONS

7.1 INTRODUCTION

Many types of structures contain members that are loaded primarily in tension. Tension members may be of various sizes and shapes, depending on the nature of the material. Because a tensile member in a structure must be connected at its ends to other members, connection behavior is an integral part of tensile-member performance. For example, in a riveted steel connection or a bolted timber connection, the undesirable effects are a reduction in the effective cross-sectional area of a member and the development of stress concentration in the material adjacent to the holes. Bending effects may also be induced when the load resultant at the connection is not coincident with the centroid of the member.

7.2 STRENGTH AS A DESIGN CRITERION

The design of a tension member is one of the simplest and most straightforward problems in structural engineering. The problem is basically one of providing a member· with sufficient cross-sectional area to resist the applied load without exceeding allowable tensile stress. Thus

$$A_{net} = \frac{P}{F_t} \qquad (7.1)$$

where A_{net} = required net area of member
P = tensile load to be carried by member
F_t = allowable tensile stress

Computing the required net area is not difficult, but the proportioning and arrangement of the member so that it is compact and efficient can become involved. Difficulties arise from the connections, which may develop eccentricities, cause reduced areas, and develop stress concentrations. The allowable stresses to be used in the design of tension members are presented in Table 6.3 for steel and Table A.1 for timber.

7.3 STEEL TENSION MEMBERS

7.3.1 Introduction

In addition to appearance, the two most important considerations for tension members are slenderness and connection details. While buckling is not involved, slenderness must be considered if the members are subject to lateral loads, vibration, or fluttering in the wind. Most design specifications recommend that slenderness be controlled by limiting the L/r ratio, the ratio of unbraced length to least radius of gyration (Chap. 10). For example, AISC suggests a limit of 240 for main members and 300 for secondary members.

Figure 7.1 illustrates some typical steel cross sections. Of the shapes shown the simplest one is a solid bar, or rod. For larger areas and more stiffness a tube or pipe becomes more effective, but the connection details become more difficult. The use of single angles introduces the problem of eccentric loading, which results in bending stresses, which are often neglected in design but which should be minimized where possible. A common section is a double angle with a gusset plate between two adjacent legs. For larger members, standard I and W sections can be used, as well as built-up sections and tee sections.

Steel structural members are usually connected by *welding*, *riveting*, or *bolting*. The principles discussed concerning riveted connections have direct application to bolted connections.

7.3.2 Tension Rods

Threaded rods, a common and simple tension member, are generally used as secondary members, e.g., sag rods to help support purlins in industrial buildings,

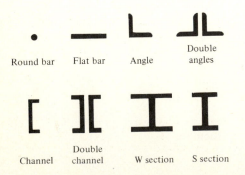

Figure 7.1 Cross sections of typical steel tension members.

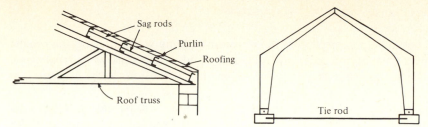

Figure 7.2 Use of tension rods.

Table 7.1 Screw thread sizes [8]

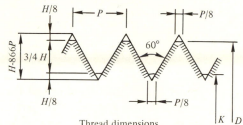

Thread dimensions

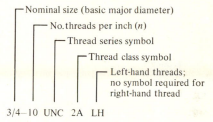

Standard designations

Diameter		Area				Diameter		Area			
Basic major D	Root K	Gross A_D	Root A_K	[a] Tensile stress	Threads per inch[b]	Basic major D	Root K	Gross A_D	Root A_K	[a] Tensile stress	Threads per inch[b]
in	in	in^2	in^2	in^2	n	in	in	in^2	in^2	in^2	n
$\frac{1}{4}$	0.185	0.049	0.027	0.032	20	$2\frac{3}{4}$	2.425	5.940	4.62	4.93	4
$\frac{3}{8}$	0.294	0.110	0.068	0.078	16	3	2.675	7.069	5.62	5.97	4
$\frac{1}{2}$	0.400	0.196	0.126	0.142	13	$3\frac{1}{4}$	2.925	8.296	6.72	7.10	4
$\frac{5}{8}$	0.507	0.307	0.202	0.226	11	$3\frac{1}{2}$	3.175	9.621	7.92	8.33	4
$\frac{3}{4}$	0.620	0.442	0.302	0.334	10	$3\frac{3}{4}$	3.425	11.045	9.21	9.66	4
$\frac{7}{8}$	0.731	0.601	0.419	0.462	9	4	3.675	12.566	10.6	11.1	4
1	0.838	0.785	0.551	0.606	8	$4\frac{1}{4}$	3.925	14.186	12.1	12.6	4
$1\frac{1}{8}$	0.939	0.994	0.693	0.763	7	$4\frac{1}{2}$	4.175	15.904	13.7	14.2	4
$1\frac{1}{4}$	1.064	1.227	0.890	0.969	7	$4\frac{3}{4}$	4.425	17.721	15.4	16.0	4
$1\frac{3}{8}$	1.158	1.485	1.05	1.16	6	5	4.675	19.635	17.2	17.8	4
$1\frac{1}{2}$	1.283	1.767	1.29	1.41	6	$5\frac{1}{4}$	4.925	21.648	19.1	19.7	4
$1\frac{3}{4}$	1.490	2.405	1.74	1.90	5	$5\frac{1}{2}$	5.175	23.758	21.0	21.7	4
2	1.711	3.142	2.30	2.50	$4\frac{1}{2}$	$5\frac{3}{4}$	5.425	25.967	23.1	23.8	4
$2\frac{1}{4}$	1.961	3.976	3.02	3.25	$4\frac{1}{2}$	6	5.675	28.274	25.3	26.0	4
$2\frac{1}{2}$	2.175	4.909	3.72	4.00	4						

[a] Tensile stress area = $0.7854\left(D - \dfrac{.9743}{n}\right)^2$.

[b] For basic major diameters of $\frac{1}{4}$ to 4 in inclusive, thread series is UNC (coarse); for $4\frac{1}{4}$ in diameter and larger, thread series is 4UN.

[c] 2A denotes Class 2 fit external thread. 2B denotes Class 2 fit internal thread.

159

vertical ties to help support girts in industrial building walls, hangers, tie rods to resist the thrust of an arch, and diagonal wind bracing in walls, roofs, and towers. In the last case they may have some initial tension. Some of these uses are illustrated in Fig. 7.2.

Tension rods must be threaded so that they can be fastened to other members. The effective diameter of the rod is at the root of the threads, and this *root area* must be used in design. To overcome this inefficiency in long rods upset ends are often used. In this case the threaded part is enlarged enough to ensure that the root-diameter threading will not be smaller than the diameter of the unthreaded portion of the rod. Table 7.1 provides information on standard threaded fasteners.

Example 7.1 Determine the size of a threaded round steel tension rod to carry 10 kips using A36 steel.

SOLUTION Refer to Tables 6.1 and 6.3.

$$F_y = 36 \text{ kips/in}^2 \qquad\qquad F_t = 0.60 F_y = 21.6 \text{ kips/in}^2$$

AISC uses 22 kips/in^2;

$$A_{\text{net}} = \frac{10}{22} = 0.455 \text{ in}^2$$

The tensile-stress area in Table 7.1 should be used as a basis for design. This is an empirical quantity which takes into account the added strength resulting when loads bear against the threads at an angle to the axis of the member.

Select a 7/8-in-diameter rod (A_{net} tensile stress $= 0.462$ in^2).

7.3.3 Plates and Angles

Whenever a tension member is to be fastened by bolting or riveting, holes must be provided for the connection. As a result the cross-sectional area is reduced and hence the allowable tensile load. The amount of this reduction depends upon the size and spacing of the holes.

There are a number of methods used to make holes, depending upon the requirements of the connection. The most common and least expensive is to punch holes full size, that is $\frac{1}{16}$ in larger than the rivet or bolt. A second method consists of subpunching an undersize hole and then reaming it to the finished size after the pieces being joined have been assembled. This method is more expensive but has the advantage of accurate alignment of holes. A third method consists of drilling the holes to a diameter of the bolt or rivet plus $\frac{1}{32}$ in. This method is used for thick members and is most expensive. For computing net areas, the diameter of a rivet or bolt hole should be taken as $\frac{1}{16}$ in greater than the normal dimension of the hole.

Whenever there is more than one row of holes in a tension member, a number of potential transverse failure lines are frequently possible. The designer must determine the failure line yielding the minimum net section.

Figure 7.3a indicates the only failure plane associated with one row of holes. For two or more rows of holes without stagger (Fig. 7.3b) failure will again

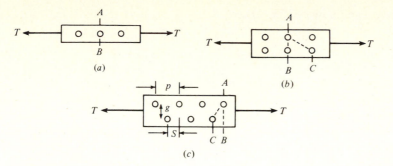

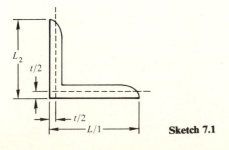

Figure 7.3 Different failure lines.

occur along section A-B because the length of section A-C is longer than that of A-B and both require the deduction of two holes. Figure 7.3c shows two rows of staggered holes with a spacing p and separated by a gauge distance g. In this case it is not immediately evident whether section A-B or A-C is critical. It is important to note that section A-C requires the deduction of two holes compared with one for section A-B.

AISC specifications [1] recommend a correction factor of $S^2/4g$ to account for the difference in length between paths A-B and A-C.

$$\text{Net length } A\text{-}B = \text{length } A\text{-}B - \text{diam of connector plus } \tfrac{1}{8} \text{ in}$$

$$\text{Net length } A\text{-}C = \text{length } A\text{-}B - (2)\left(\text{diam of connector plus } \tfrac{1}{8} \text{ in}\right)$$

$$+ \frac{S^2}{4g}$$

where S is the stagger distance and g is the gauge distance. The development of the $S^2/4g$ correction attempts to take into account the maximum principal stress on the inclined portion of the section.

For an angle the net area is determined on the basis of its thickness and net length. The net length is assumed to be along the centerline of the angle shown in Sketch 7.1. As a result, the net length of an angle is $L_1 + L_2 - t$. Every rolled angle has prescribed gauge lines for the location of holes depending upon the size of the angle. Table 7.2 gives the usual gauges for angles. Gauge distances other than standard should normally be avoided because of higher fabrication costs.

Sketch 7.1

Table 7.2 Usual gauges in inches for angles [1]

Leg	g	g_1	g_2	Leg	g
8	$4\frac{1}{2}$	3	3	$2\frac{1}{2}$	$1\frac{3}{8}$
7	4	$2\frac{1}{2}$	3	2	$1\frac{1}{8}$
6	$3\frac{1}{2}$	$2\frac{1}{4}$	$2\frac{1}{2}$	$1\frac{3}{4}$	1
5	3	2	$1\frac{3}{4}$	$1\frac{1}{2}$	$\frac{7}{8}$
4	$2\frac{1}{2}$			$1\frac{3}{8}$	$\frac{7}{8}$
$3\frac{1}{2}$	2			$1\frac{1}{4}$	$\frac{3}{4}$
3	$1\frac{3}{4}$			1	$\frac{5}{8}$

The AISC also has recommendations for spacing and end distances. The distance between centers of fastener holes should not be less than $2\frac{2}{3}d$, where d is the diameter of the fastener, nor less than

$$\frac{2P}{F_u t} + \frac{d}{2}$$

along a line of transmitted force, where P equals the force transmitted by one fastener, F_u equals the specified minimum tensile strength of the critical connected part, and t equals the thickness of the critical part.

The distance from the center of a standard hole to an edge of a connected part depends on whether the edge is sheared or rolled. Recommended values are given in Table 7.3. Along a line of transmitted force in the direction of the force

Table 7.3 Recommended edge distances for holes [1]

	Minimum edge distance for punched, reamed, or drilled holes, in	
Rivet or bolt diameter, in	At sheared edges	At rolled edges of plates, shapes, or bars or gas-cut edges†
$\frac{1}{2}$	$\frac{7}{8}$	$\frac{3}{4}$
$\frac{5}{8}$	$1\frac{1}{8}$	$\frac{7}{8}$
$\frac{3}{4}$	$1\frac{1}{4}$	1
$\frac{7}{8}$	$1\frac{1}{2}$‡	$1\frac{1}{8}$
1	$1\frac{3}{4}$‡	$1\frac{1}{4}$
$1\frac{1}{8}$	2	$1\frac{1}{2}$
$1\frac{1}{4}$	$2\frac{1}{4}$	$1\frac{5}{8}$
Over $1\frac{1}{4}$	$1\frac{3}{4}$ × diameter	$1\frac{1}{4}$ × diameter

† All edge distances in this column may be reduced $\frac{1}{8}$ in when the hole is at a point where stress does not exceed 25 percent of the maximum allowed stress in the element.

‡ These may be $1\frac{1}{4}$ in at the ends of beam-connection angles.

the edge distance should not be less than

$$\frac{2P}{F_u t}$$

where P, F_u, and t are as defined above. Additional details for slotted and oversized holes can be found in the AISC manual [1].

Experiments indicate that a tension member will have about a 15 percent reduction in ultimate strength even if the area removed by the hole is less. As a result, an additional criterion imposed by AISC specifications is that the effective net area shall not exceed 85 percent of the gross area whenever there is a hole (or holes) in a tension member.

Example 7.2 Determine the net section of the angle shown in Fig. E7.2a if $\frac{7}{8}$-in-diameter bolts are used.

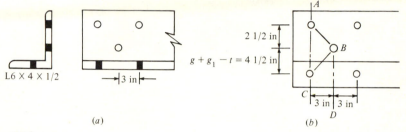

$g + g_1 - t = 4\ 1/2$ in

L6 × 4 × 1/2

(a)   (b)

Figure E7.2

SOLUTION Assuming that the net length is along the centerline of the angle, the section is as shown in Fig. E7.2b

$$A_{net} = A_{gross} - Dt + \frac{S^2}{4g}t$$

Path $AC = 4.75 - (2)(0.875 + 0.125)(0.5) = 3.75$ in^2

Path $ABC = 4.75 - (3)(0.875 + 0.125)(0.5) + \frac{(3)^2}{(4)(2.50)}0.5 + \frac{(3)^2}{(4)(4.25)}0.5$

$\qquad = 3.97$ in^2

Path $ABD = 4.75 - (2)(0.875 + 0.125)(0.5) + \frac{(3)^2}{(4)(2.50)}0.5 = 4.20$ in^2

85% of gross area $= (0.85)(4.75) = 4.04$ in^2

Path AC governs with area of 3.75 in^2

7.4 RIVET AND BOLT CONNECTIONS

7.4.1 Introduction

A riveted connection is a mechanical connection made by upsetting or distorting the ends of special pins known as *rivets* (Fig. 7.4). This forming process can be

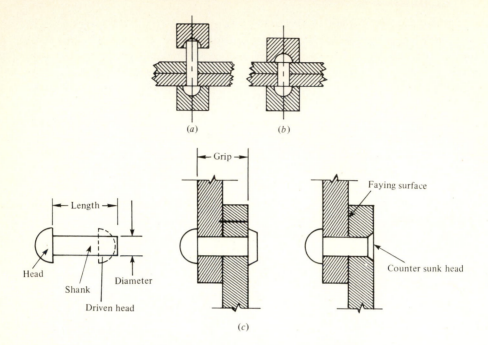

Figure 7.4 Riveted joint; setting up (a) (b); (c) dimensions and details.

carried out with either a hot or cold rivet, but usually hot ones (1600 to 1900°F) are used. Rivets are made from steel conforming to the specifications for rivet steel, ASTM A502. This specification covers two grades of steel, grade 1 for general purposes and grade 2, a carbon-manganese steel for high-strength structural steels. Commercially available rivets come in sizes from $\frac{3}{8}$-in shank diameter to $1\frac{1}{4}$-in diameter in increments of $\frac{1}{8}$ in. Rivets installed hot shrink upon cooling and develop a clamping force between the joined surfaces, but this clamping force cannot be relied upon in design.

High-strength *bolts* have to a large extent replaced rivets as the principal means of making nonwelded connections. In 1934 a report by Batho and Bateman to the Steel Structures Committee of Scientific and Industrial Research of Great Britain indicated that bolts could be tightened enough to prevent slip in structural joints. In 1947 the Research Council on Riveted and Bolted Structural Joints was formed. The Council became the major force in the development of high-strength bolting. In 1951 the first specification permitted the replacement of rivets by bolts on a one-for-one basis. Because slip into bearing is not always objectionable, a bearing-type connection was recognized in 1954. On the basis of continued research and development designers were permitted to take advantage of the superior characteristics of high-strength bolts. Bearing and friction connections were clearly defined. Minimum bolt tensions were standardized at 70 percent of the specified minimum tensile strength, and installation procedures permitted using the turn of the nut and calibrated wrench tightening.

At present specifications recognize three types of A325 high-strength carbon-steel bolts, which include a bolt with improved atmospheric corrosion resistance and weathering characteristics, and an **A490** higher-strength alloy-steel bolt (Fig. 7.5). Bolt diameters range from $\frac{1}{2}$ to $1\frac{1}{2}$ in, the most common sizes for buildings being $\frac{3}{4}$ and $\frac{7}{8}$ in. The same installation procedures are required for both types of *shear connections*, i.e., *friction* type and *bearing* type. Friction connections are recommended for joints subjected to stress reversals, impact, vibration, and other situations where slip is undesirable. Recent research has shown that where greater latitude is needed in meeting dimensional tolerances during erection, oversized holes, as well as short- and long-slotted holes, can be permitted for bolts $\frac{5}{8}$ in diameter or larger (Table 7.4). Since an increase in hole size generally reduces the net area of a connected part, the use of oversize holes is subject to approval by the designer.

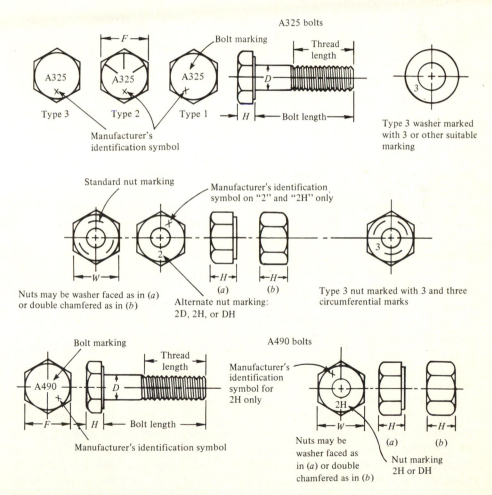

Figure 7.5 Bolt dimensions and details [1].

Table 7.4 Oversized and slotted holes [1]

Nominal fastener diameter d	Maximum size† of fastener holes, in			
	Standard hole diameter	Oversized‡ hole diameter	Short-slotted‡ hole dimensions	Long-slotted‡ hole dimensions
$\leq \frac{7}{8}$	$d + \frac{1}{16}$	$d + \frac{3}{16}$	$(d + \frac{1}{16}) \times (d + \frac{1}{4})$	$(d + \frac{1}{16}) \times 2\frac{1}{2}d$
1	$1\frac{1}{16}$	$1\frac{1}{4}$	$1\frac{1}{16} \times 1\frac{5}{16}$	$1\frac{1}{16} \times 2\frac{1}{2}$
$\geq 1\frac{1}{8}$	$d + \frac{1}{16}$	$d + \frac{5}{16}$	$(d + \frac{1}{16}) \times (d + \frac{3}{8})$	$(d + \frac{1}{16}) \times 2\frac{1}{2}d$

† Sizes are nominal.
‡ Not permitted for riveted connections.

Oversized and slotted holes should not be used in riveted connections. Oversized holes may be used in any or all plies of friction-type connections but should not be used in bearing-type connections. Short-slotted holes may be used in any or all plies of friction-type or bearing-type connections. Long-slotted holes may be used in only one of the connected parts of either a friction-type or bearing-type connection. Extensive research has also shown that for the same clamping force, frictional resistance varies widely but predictably with surface treatment. Surface treatments on contact surfaces are in no way detrimental to the performance of bearing-type connections.

7.4.2 Failure Modes

A riveted or bolted connection can fail in several ways. The most common are illustrated in Fig. 7.6. Failure in which the fastener is sheared along planes of slip is shown in Fig. 7.6a. The area of shear is the cross-sectional area of the fastener. For a lap joint, failure is on one plane (*single shear*), and for a butt joint, failure occurs on two planes (*double shear*). Bearing failure, caused by compression between the cylindrical surface of the plate hole and the fastener, is illustrated in Fig. 7.6c. The variation of the compressive stresses around the perimeter of the hole is unknown. For design purposes it is assumed to be uniform over a rectangular area equal to the thickness of the plate times the diameter of the fastener. Failure may also occur in the plate between the hole and the end of the plate (Fig. 7.6b and d). To avoid this type of failure in structural steel the required distance from the center of the hole to the end of the plate is of the order of 1.5 to 2 times the diameter of the fastener (Table 7.3). Tension failure of the connector is unusual, but it can occur in rivets and overtightened bolts. These fasteners are normally replaced (Fig. 7.6e). Bending of a fastener (Fig. 7.6f) occurs when the fastener is long compared with its diameter. This action subjects the fastener not only to bending but also to tension. The use of multiple rows and larger-diameter fasteners helps alleviate this condition. The eccentricity of the applied loads in a single lap joint also tends to bend the fastener. Nevertheless tests indicate that the bearing strength

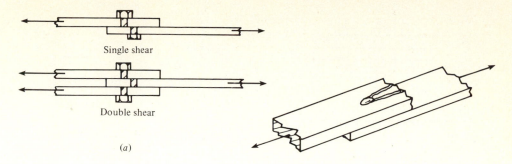

Single shear

Double shear

(a)

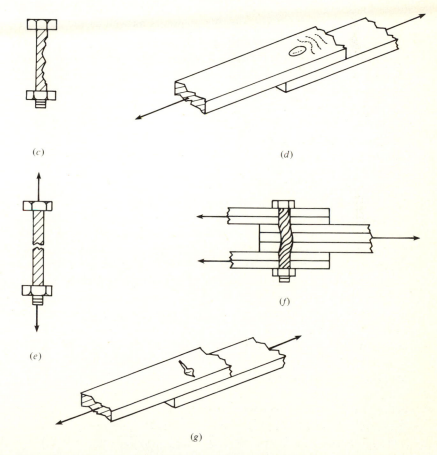

(b)

(c)

(d)

(e)

(f)

(g)

Figure 7.6 Possible modes of failure connections: (*a*) shear failure of bolts, (*b*) shear failure of plate, (*c*) bearing failure of bolt, (*d*) bearing failure of plate, (*e*) tensile failure of bolt, (*f*) bending failure of bolt, and (*g*) tensile failure of plate.

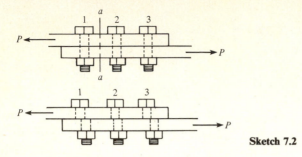

Sketch 7.2

for single shear is about the same as for double shear. Some specifications allow larger unit stresses for bearing in double shear.

Bolts and rivets decrease the area available in a given member to resist direct load. Figure 7.6g shows this type of failure. For more complex butt joints this type of failure can occur in the members as well as the strap plates.

In the behavior of rivets and bolted joints for tension members, the general assumption is that the fasteners share the load equally if it is not eccentric. This assumption is in error when there are more than two fasteners in a given gauge line. Consider the connection shown in Sketch 7.2. Assuming that each bolt carries $P/3$ and that stress is proportional to strain, we see that the force in the upper plate at section a-a is $2P/3$, while in the lower plate it is $P/3$. As a result, the upper plate elongates twice as much as the lower, and the shear deformations of bolts 1 and 2 cannot be equal. Analysis of the joint between bolts 2 and 3 shows that here the lower plate elongates twice as much as the upper one, so that the shear deformation of bolt 3 equals that of 1. Thus the end bolts carry larger shears than the middle bolt. This means that bolts 1 and 3 reach the proportional limit first, after which the center bolt picks up a larger portion of the load. Thus at failure, the bolt forces may be nearly equal, depending on the ability of the fasteners to tolerate large shear deformations. As a result, it is usual to assume that each fastener carries an equal share of the load, or the load is distributed in proportion to the shear areas of the fasteners.

7.4.3 Design of Riveted and Bolted Connections

As a consequence of the discussions in Sec. 7.4.2, it is usual to consider that a joint can fail in one of three possible modes and these failure modes can be expressed mathematically as follows.

Failure by shear See Sketch 7.3.

$$f_s = \frac{F}{A_s} = \frac{F}{2n\pi(d^2/4)} \tag{7.2}$$

where f_s = shear stress
F = force acting on joint
A_s = effective shear area
n = number of fasteners
d = diameter of fastener

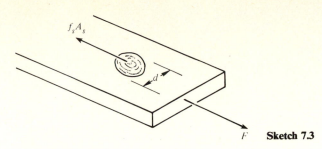

F **Sketch 7.3**

The coefficient 2 refers to double-shear conditions.

Bearing failure See Sketch 7.4.

$$f_b = \frac{F}{A_b} = \frac{F}{ntd} \tag{7.3}$$

where f_b = bearing stress
$\quad\quad F$ = force acting in member
$\quad\quad t$ = thickness of member
$\quad\quad d$ = diameter of fastener
$\quad\quad n$ = number of fasteners

Tension failure See Sketch 7.5.

$$f_t = \frac{F}{A_t} = \frac{F}{(b - nd_e)t} \tag{7.4}$$

where f_t = tensile stress
$\quad\quad F$ = force acting in member
$\quad\quad t$ = thickness of member
$\quad\quad b$ = width of member
$\quad\quad d_e$ = diameter of fastener $+ \frac{1}{8}$ in
$\quad\quad n$ = number of rows of fasteners

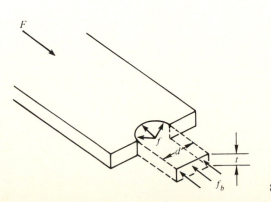

Sketch 7.4

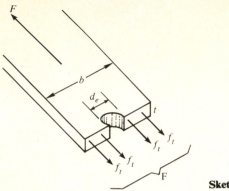

Sketch 7.5

Table 7.5 Allowable stress on fasteners, kips/in² [1]

| | | Allowable shear F_v | | | |
| | | Friction-type connections[a] | | | |
Description of fasteners	Allow-able tension F_t	Standard size holes	Oversized and short-slotted holes	Long-slotted holes	Bearing-type connec-tions
A502, grade 1, hot-driven rivets	23.0[b]				17.5[c]
Grades 2 and 3, hot-driven rivets	29.0[b]				22.0[c]
A307 bolts	20.0[b]				10.0[cd]
Threaded parts, threads not excluded from shear planes	$0.33F_u$[be]				$0.17F_u$
Threads excluded from shear planes	$0.33F_u$[b]				$0.22F_u$
A325 bolts, threads not excluded from shear planes	44.0	17.5	15.0	12.5	21.0[c]
Threads excluded from shear planes	44.0	17.5	15.0	12.5	30.0[c]
A490 bolts, threads not excluded from shear planes	54.0	22.0	19.0	16.0	28.0[c]
Threads are excluded from shear planes	54.0	22.0	19.0	16.0	40.0[c]

[a] When specified by the designer, the allowable shear stress F_v for friction-type connections having special faying surface conditions may be increased to the applicable value given in Table 7.6.

[b] Static loading only.

[c] When bearing-type connections used to splice tension members have a fastener pattern whose length, measured parallel to the line of force, exceeds 50 in, tabulated values shall be reduced by 20 percent.

[d] Threads permitted in shear planes.

[e] The tensile capacity of the threaded portion of an upset rod, based upon the cross-sectional area at its major thread diameter A_b, shall be larger than the nominal body area of the rod before upsetting times $0.60F_y$.

Table 7.6 Allowable stresses based upon surface conditions of bolted parts for friction-type shear connections, kips/in² † [1]

Class	Surface condition of bolted parts	Standard holes		Oversized holes and short-slotted holes		Long-slotted holes	
		A325	A490	A325	A490	A325	A490
A	Clean mill scale	17.5	22.0	15.0	19.0	12.5	16.0
B	Blast-cleaned, carbon and low alloy steel	27.5	34.5	23.5	29.5	19.5	24.0
C	Blast-cleaned, quenched and tempered steel	19.0	23.5	16.0	20.0	13.5	16.5
D	Hot-dip galvanized and roughened	21.5	27.0	18.5	23.0	15.0	19.0
E	Blast-cleaned, organic zinc-rich paint	21.0	26.0	18.0	22.0	14.5	18.0
F	Blast-cleaned, inorganic zinc-rich paint	29.5	37.0	25.0	31.5	20.5	26.0
G	Blast-cleaned, metallized with zinc	29.5	37.0	25.0	31.5	20.5	26.0
H	Blast-cleaned, metallized with aluminum	30.0	37.5	25.5	32.0	21.0	26.5
I	Vinyl wash	16.5	20.5	14.0	17.5	11.5	14.5

† Values from this table are applicable *only* when they do not exceed the lowest appropriate allowable working stresses for *bearing-type* connections, taking into account the position of threads relative to shear planes and, if required, the 20 percent reduction due to joint length (see Table 7.5).

The maximum load a given member and connection will carry depends upon the mode of failure. Ideally each failure mechanism should be capable of developing the same allowable load. In order to accomplish this, allowable stresses have to be used for values of f_s, f_b, and f_t. They are given in Tables 7.5, 7.6, and 6.3.

Example 7.3 Determine the tensile capacity of the bearing-type connection shown in Fig. E7.3 if threads are excluded from the shear planes. Assume the plate to be A572 grade 50 steel and A325 bolts.

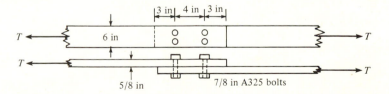

Figure E7.3

SOLUTION The allowable tensile stress is

$$F_t = 0.6F_y > 0.50F_u = (0.60)(50) = 30 \text{ kips/in}^2$$

From Eq. (7.4) the plate capacity in tension is

$$T_t = \left(\tfrac{5}{8}\right)\left[6 - (2)\left(\tfrac{7}{8} + \tfrac{1}{8}\right)\right](0.60)(50) = 75.0 \text{ kips}$$

and from Table 7.5 the allowable shear stress is

$$F_s = 30 \text{ kips/in}^2$$

Using Eq. (7.2), we find the shear capacity to be

$$T_s = (4)\frac{\pi}{4}\left(\frac{7}{8}\right)^2(30) = 72.2 \text{ kips}$$

From Table 6.3 the allowable bearing stress is

$$F_b = 1.5F_u = (1.5)(65) = 97.5 \text{ kips/in}^2$$

and from Eq. (7.3) the bearing capacity is

$$T_b = (4)(97.5)\left(\frac{7}{8}\right)\left(\frac{5}{8}\right) = 213.3 \text{ kips}$$

Since shear governs, the allowable load for the joint is 72.2 kips.

Example 7.4 A tension diagonal of a roof truss carries a load of 68 kips. The gusset plates at the ends of the diagonal are $\frac{3}{8}$ in thick. Select a pair of angles (A36 steel) to carry the load and determine the number of A325 $\frac{3}{4}$-in bolts required for the connection (Fig. E7.4).

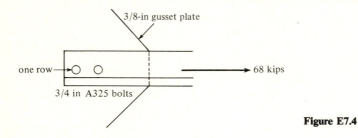

Figure E7.4

SOLUTION The allowable tensile stress is

$$0.60F_y < 0.50F_u = (0.60)(36) = 22 \text{ kips/in}^2 \qquad \text{(See Ex. 7.1)}$$

$$\text{Net area of angles required} = \frac{68}{22} = 3.10 \text{ in}^2$$

Assuming one row of bolts, the reduction in area of two angles $\frac{3}{8}$ in thick would be $(2)\left(\frac{7}{8}\right)\left(\frac{3}{8}\right) = 0.66 \text{ in}^2$.

Gross area of angles required is

$$3.10 + 0.66 = 3.76 \text{ in}^2$$

Checking the structural-steel data for angles (Table A.5), we find that two angles 3 by 2 by $\frac{3}{8}$ in $= (2)(1.92) = 3.84 \text{ in}^2$.

The number of bolts required in the single row is based on the double-shear capacity of $\frac{3}{4}$-in bolts (Table 7.5)

$$68 = \frac{2n\pi}{4}\left(\frac{3}{4}\right)^2(30) \qquad \text{Eq. (7.2)}$$

$$n = 5.13$$

Consider six bolts. Check to see whether the connection is adequate in bearing. The allowable bearing stress equals $1.5F_u$.

$$T_b = (6)(1.5)(58)\left(\frac{3}{4}\right)\left(\frac{3}{8}\right) = 146.8 \text{ kips} > 68 \qquad \text{OK}$$

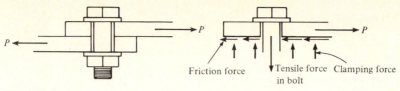

Friction force   Tensile force   Clamping force
                 in bolt

Sketch 7.6

7.4.4 Design of Friction-Type Connections

In a friction-type connection, the bolts are not actually stressed in shear or bearing since no slip occurs at allowable loads, but an allowable shear stress is developed by friction. This stress is used in determining the number of bolts required in the same way as for a bearing-type connection [Eq. (7.2)]; see Sketch 7.6.

Example 7.5 Two $\frac{1}{4}$- by 8-in A36 plates are to be connected using $\frac{3}{4}$-in A325 bolts in a friction-type connection. How many bolts are required to transfer a load of 30 kips?

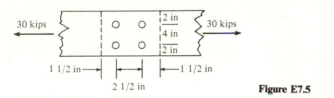

Figure E7.5

SOLUTION From Table 7.5 the allowable shear stress for a friction-type connection with standard holes is 17.5 kips/in². The shear resistance of one bolt is

$$F_v A = 17.5 \frac{\pi}{4} \left(\frac{3}{4} \right)^2 = 7.73 \text{ kips}$$

No. of bolts required $= \dfrac{30}{7.73} = 3.38$ use 4 (Fig. E7.5).

Check section for tensile strength

$$T_t = \left(\tfrac{1}{4} \right) \left[8 - (2)\left(\tfrac{3}{4} + \tfrac{1}{8} \right) \right] (0.60)(36) = 33.8 \text{ kips} > 30 \text{ kips} \qquad \text{OK}$$

7.5 WELDED CONNECTIONS

7.5.1 Introduction

Welding is a metal-joining process involving fusion or melting of the parts being joined. Welding offers the possibility of more economical joints, which do not require punching or drilling, are lighter, and yield a more uniform distribution of stress than riveted or bolted connections.

The most common welding processes, particularly for structural steel, are arc welding, resistance welding, and gas welding. Arc welding generates heat

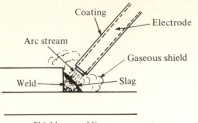

Shield-arc welding

Sketch 7.7

from the resistance of an electric current across a gas gap. Filler metal may be added to the joint, and an inert gas is generally used to envelop the area being welded. Resistance welding obtains the required heat from the electrical resistance at the interface of the parts being joined. Gas welding generates the required heat by the combustion of a gas such as acetylene with air. The electric arc is used for most structural welding. The basis of the arc-welding process is illustrated in Sketch 7.7. Using a gaseous shield and different electrodes (coated, uncoated, consumable, nonconsumable, and flux-cored) results in a number of distinct processes—shielded metal-arc, submerged-arc, gas metal-arc, and flux-cored arc welding.

Most ASTM specification structural steels can be welded without special precautions or procedures. The weldability of a steel is a measure of the ease of producing a crackfree and sound structural joint.

The economy of welded connections must be considered from a broad point of view. Welded connections are not only neater but offer the designer more freedom to be innovative in the entire design concept. The designer is not bound by standard sections but may build up configurations not only for aesthetic reasons but for efficient transmittal of loads. Welded connections generally eliminate the need for holes in members except for construction. As a result, the efficiency of welded joints can approach 100 percent and the members can usually be smaller.

Another advantage of welded connections is the possibility of reducing construction costs by being able to make minor length adjustments in the field.

7.5.2 Classification of Welds

The three separate classifications of welds are based upon type, position during welding, and type of joint. The two main types of welds are *fillet* and *butt welds* (also known as *groove welds*). *Plug* and *slot welds* are not common in structural work. These types are illustrated in Fig. 7.7. Butt, or groove, welds can be used only if the members are lined up in the same plane. Figure 7.8 shows the common types of groove welds and the end preparations required. When welded from both sides or from one side with a backup strip on the far side, they may be said to achieve *complete penetration*. *Incomplete-penetration* groove welds are used only when the plates need not be fully stressed and full continuity is not

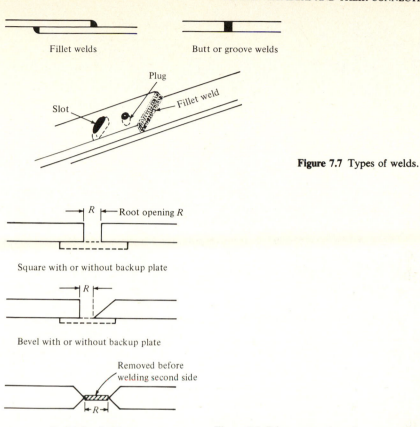

Fillet welds   Butt or groove welds

Figure 7.7 Types of welds.

Square with or without backup plate

Bevel with or without backup plate

Double bevel with spacer   **Figure 7.8** Edge preparation of groove welds.

required. Overlapping the members allows for larger tolerances during erection and the use of fillet welds. Figure 7.9 shows some typical uses of fillet welds.

A plug is a circular weld passing through one member to another and joining the two together. The slot weld differs from the plug only in the shape of the weld formed in the slot or elongated hole. These two types of weld have use when the overlap is insufficient for the desired length of fillet welds.

Welds are also referred to in terms of their position during welding as *flat, horizontal, vertical,* and *overhead* (Fig. 7.10). The position of welding is important relative to cost. The list above is in order of their economy, the flat position being the most economical and overhead the most expensive.

Welds are also classified according to the type of joint formed. The most common ones (*butt, lap, tee, edge,* and *corner*) are shown in Fig. 7.11.

Figure 7.12 presents the method of making welding symbols developed by the American Welding Society. It is a form of shorthand which enables a large amount of information to be presented in a small space on engineering plans and drawings with a relatively few lines and numbers. Figure 7.13 illustrates the use of these symbols.

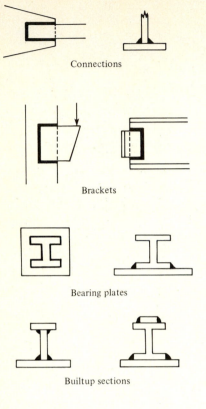

Connections

Brackets

Bearing plates

Builtup sections

Figure 7.9 Uses of fillet welds.

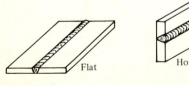

Flat

Horizontal

Vertical

Overhead

Figure 7.10 Welding positions.

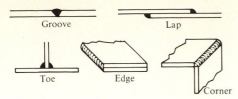

Groove   Lap   Toe   Edge   Corner

Figure 7.11 Weld joints.

Type of weld							
Bead	Fillet	Plug or slot	Groove				
			Square	V	Bevel	U	J
⌒	△	▽	||	∨	∨	∪	∪

Basic arc and gas weld symbols

Weld all around	Field weld	Contour	
		Flush	Convex
○	●	—	⌒

Supplementary symbols

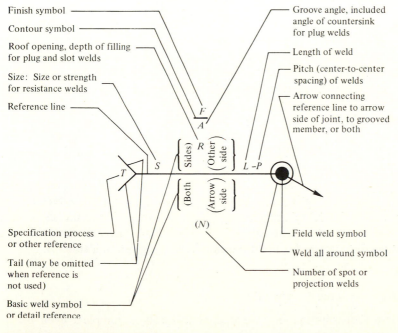

Finish symbol

Contour symbol

Roof opening, depth of filling for plug and slot welds

Size: Size or strength for resistance welds

Reference line

Groove angle, included angle of countersink for plug welds

Length of weld

Pitch (center-to-center spacing) of welds

Arrow connecting reference line to arrow side of joint, to grooved member, or both

Specification process or other reference

Tail (may be omitted when reference is not used)

Basic weld symbol or detail reference

Field weld symbol

Weld all around symbol

Number of spot or projection welds

F

A

R (Other side)

(Sides)

(Both)

(Arrow side)

(N)

T S L –P

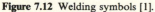

Figure 7.12 Welding symbols [1].

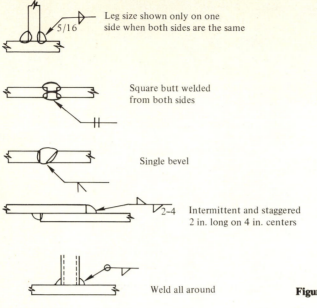

Leg size shown only on one side when both sides are the same

Square butt welded from both sides

Single bevel

Intermittent and staggered 2 in. long on 4 in. centers

Weld all around

Figure 7.13 Use of welding symbols.

7.5.3 Allowable Stresses in Welds

Welding electrodes are classified on the basis of the mechanical properties of the weld metal, welding position, type of coating, and type of current required. Electrodes for shielded metal-arc welding are designated by a code number Exxxxx, where the first two or three numbers indicate the tensile strength of the weld metal. Combinations of flux and electrodes for submerged-arc welding are designated by F followed by the tensile strength of the weld material. Electrodes for gas metal-arc welding are identified by an E plus digits for tensile strength and S or U for solid or emissive electrodes, respectively. The mechanical properties of electrodes are given in Table 7.7.

Since welds must transmit the entire load from one member to another, they must be sized correctly and made from the right electrode material. *Fillet welds are assumed, for design purposes, to transmit loads through shear stress on the effective area no matter how the fillets are oriented on the structural connections.* Groove welds transmit loads exactly as in the pieces they join. As a result, the required electrode material for groove welds depends on the base metals in the members. These combinations are given in Table 7.8 for the different welding processes. Allowable stresses for the different welds are given in Table 7.9.

7.5.4 Design of Welded Connections

The actual stress distribution in a welded connection is complex, even for a simple joint. Welds and the members they join must deform together; otherwise

Table 7.7 Mechanical properties of electrodes [2]

Welding process				Minimum mechanical properties		
Shielded metal-arc	Submerged arc	Gas metal-arc	Flux-cored arc	F_u, kips/in^2	F_y, kips/in^2	Elongation in 2 in, %
E60				62–67	50–55	17–25
	F60			60–80	45	25
		E60S		62	50	22
E70				72	60	17–22
	F70			70–95	50	22
		E70S		70	60	20–22
E80				80	67	16–19
	F80	E80S	E80T	80	65	18
E90				90	77	14–17
	F90	E90S	E90T	90	78	17
E100				100	87	13–16
	F100	E100S	E100T	100	90	16
E110				110	97	15
	F110	F110S	F110T	110	98	15

Table 7.8 Electrodes for use with various steels [1]

Base metal†	Welding process			
	Shielded metal-arc	Submerged arc	Gas metal-arc	Flux-cored arc
A36, A53 grade B, A375, A500, A501, A529, A570 grades D, E	E60, E70	F60, F70	E70S, E70U	E60T, E70T
A242, A441, A572 grades 42–60, A588	E70	F70	E70S, E70U	E70T
A572 grade 65	E80	F80	E80S	E80T
A514, $2\frac{1}{2}$ in	E100	F100	E100S	E100T
Under $2\frac{1}{2}$ in	E110	F110	E110S	E110T

† Use of same type of filler metal having next higher mechanical properties is permitted.

a separation would occur. True stresses are also altered by the existence of residual stresses due to cooling of the welds, warping stresses arising from poor welding procedures, and stress relieving in the pieces joined. As a result, exact analyses are not used in the design of welded connections. Instead, criteria of safe nominal stresses are used. If a welded joint is subjected to pure shear, compression, or tension, the stresses are assumed to be uniform over the length of the welds. If it is subjected to pure moment or torsion, the stresses are assumed to vary linearly with distance from the neutral axis. If two or more of

Table 7.9 Allowable stresses in welds [1]

Weld type	Kind of stress	Permissible stress
Groove weld	Tension and compression parallel to axis of complete penetration weld	Same as base metal
	Tension normal to complete penetration weld	Same as allowable tensile stress for base metal
	Compression normal to complete or partial-penetration weld	Same as allowable compressive stress for base metal
	Tension normal to throat of partial-penetration weld	$0.30F_u$ electrode
	Shear on effective throat of complete and partial-penetration welds	Same as allowable shear stress for base metal
Fillet weld	Shear	$0.30F_u$ electrode
Plug or slot welds	Shear	$0.30F_u$ electrode

these conditions exist simultaneously, the stresses are assumed to add vectorially. These allowable nominal stresses resulting from service loads are related to ultimate strength by a factor of safety. The allowable stresses depend upon their effective areas, which vary with the type of weld and expected failure plane.

Groove weld The effective area can be considered to be the product of the effective throat dimension t_e and the length of the weld. The effective throat dimension of a full penetration weld is the thickness of the thinner part joined as shown in Fig. 7.14a and b.

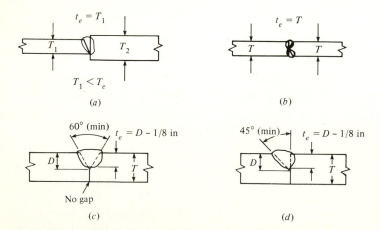

Figure 7.14 Effective throat dimensions for groove welds.

The effective throat dimension of single and double partial-penetration welds is the depth of the groove, except for a bevel joint made by the manual shielded metal-arc process. In the latter case the effective throat dimension is taken as the depth of the groove less $\frac{1}{8}$ in, but it cannot be less than $\sqrt{t_e/6}$, where t_e is the thickness of the thinner piece being joined, as shown in Fig. 7.14a and d

$$F = S_s t_e L \qquad (7.5)$$

where S_s = allowable stress (tension, compression or shear depending on condition), lb/in^2
t_e = effective throat dimension, in
L = length of weld, in
F = allowable load on joint, lb

Groove welds can be reinforced. Reinforcement is added weld metal which causes the throat dimension to be greater than the thickness of the welded material. These welds are referred to as 100, 125, 150 percent, etc., depending on the added thickness and strength. Reinforcement affords greater strength under static loads but is undesirable for vibratory and repeated loads because of stress concentrations. For these conditions the reinforcement is normally ground off flush with the connected members.

Fillet weld Tests have shown that fillet welds are stronger in tension and compression than in shear. As a result controlling fillet-weld stresses are considered to be shear stresses. Where practical, it is desirable to arrange fillet connections that will be subjected to shearing stresses only.

The effective throat dimension of fillet welds is the shortest distance from the root to the face of the weld, as shown in Fig. 7.15.

For the equal-leg fillet, the throat dimension $t_e = 0.707$ times the leg of the weld. For the unusual weld with unequal legs, the throat dimension

$$t_e = \frac{ab}{\sqrt{a^2 + b^2}}$$

where a and b are the legs of the weld. The effective throat dimensions for fillet welds made by the submerged-arc process are modified because of the larger throat dimension (Fig. 7.16). For leg sizes $\frac{3}{8}$ in or less the throat dimension is taken to be equal to leg size a. For leg sizes larger than $\frac{3}{8}$ in the effective throat

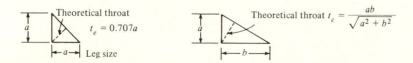

Figure 7.15 Effective throat dimensions.

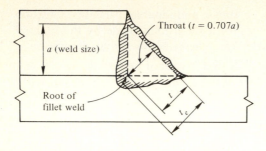

Figure 7.16 Throat for submerged-arc process.

dimension $t_e = 0.707a + 0.11$ in. As a result the strength of a fillet weld becomes

$$F = S_s t_e L \qquad (7.5)$$

where S_s = allowable shear stress, lb/in²
$\quad$ t_e = effective throat dimension, in
$\quad$ L = length of weld, in
$\quad$ F = allowable load carried by weld, lb

Since welding involves heating the metal pieces, it is important to prevent the welds from cooling too rapidly and to consider the effect of plate thickness. The thicker the plate the faster the dissipation of heat and the lower the temperature of the weld. Unless a proper temperature is maintained, a lack of fusion will result. Table 7.10 shows *minimum sizes* of fillet welds depending on the plate thickness.

The *maximum size* of fillet welds used along the edges of pieces joined is limited by the thickness of the thinner piece. The maximum permitted by AISC is shown in Fig. 7.17.

Large fillet welds made manually require two or more passes, as indicated in Fig. 7.18. The largest size that can be made in one pass depends on the welding

Table 7.10 Minimum fillet weld sizes [1]

Thickness of thicker part joined, in	Minimum size of fillet weld, in
To $\frac{1}{4}$	$\frac{1}{8}$
Over $\frac{1}{4}$ to $\frac{1}{2}$	$\frac{3}{16}$
Over $\frac{1}{2}$ to $\frac{3}{4}$	$\frac{1}{4}$
Over $\frac{3}{4}$ to $1\frac{1}{2}$	$\frac{5}{16}$
Over $1\frac{1}{2}$ to $2\frac{1}{4}$	$\frac{3}{8}$
Over $2\frac{1}{4}$ to 6	$\frac{1}{2}$
Over 6	$\frac{5}{8}$

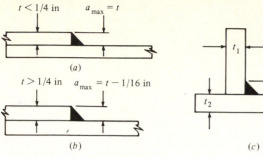

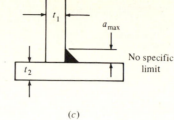

Figure 7.17 Maximum weld size.

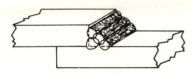

Figure 7.18 Multiple-pass fillet weld.

position and should not exceed

$\frac{5}{16}$ in in the horizontal or overhead position

$\frac{3}{8}$ in in the flat position

$\frac{1}{2}$ in in the vertical position

The most commonly used fillet welds increase in size by sixteenths of an inch from $\frac{1}{8}$ to $\frac{1}{2}$ in and by eights of an inch for sizes greater than $\frac{1}{2}$ in. The efficiency of welding is affected by the amount of filler material used. For example, a $\frac{5}{16}$-in fillet weld has only a 25 percent larger throat than a $\frac{1}{4}$-in weld but uses 56 percent more material. The need for multiple passes will also decrease efficiency.

The *minimum effective length* of a fillet weld should not be less than 4 times the nominal size, or the size of the weld should be considered only one-fourth of the effective length. The effective length of any segment of intermittent fillet welding should not be less than 4 times the weld size, with a minimum of $1\frac{1}{2}$ in. The recommended *minimum* amount of lap on lap joints should be 5 times the thickness of the thinner part joined and not less than 1 in. The ends of lapped joints should be welded if deflection under loading will cause opening of the joint.

AISC also recommends the use of end returns since there is a slight tapering off in the region where they are started and where they end. AISC limits the minimum effective length of fillet welds to 4 times their nominal size and end returns equal to $2a$, where a is the size of the weld (see Fig. 7.19).

Plug and slot welds The effective shearing area of plug or slot welds is their nominal area in the shear plane. As a result the resistance of a plug or slot weld

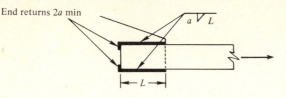

End returns 2a min

Figure 7.19 Use of end returns.

is the product of the nominal cross section times the allowable stress

$$F = A_e S_s \qquad (7.6)$$

where A_e = nominal cross-sectional area, in²
S_s = allowable shear stress, lb/in²
F = resistance of plug or slot, lb

Example 7.6 Determine the allowable shear resistance of a $\frac{3}{8}$-in fillet weld produced by a shielded metal-arc process; use E70 electrode.

SOLUTION

$$t_e = 0.707a = (0.707)(0.375) = 0.265 \text{ in}$$

$$S_s = 0.30 F_u = (0.30)(70) = 21 \text{ kips/in}^2$$

$$F_w = S_s t_e L = (0.265 \text{ in})(21 \text{ kips/in}^2)(1 \text{ in}) = 5.57 \text{ kips per inch of weld}$$

Example 7.7 Determine the allowable capacity of a $\frac{3}{4}$-in-diameter plug weld using E70 electrode material.

SOLUTION

$$A_e = \frac{\pi d^2}{4} = \frac{\pi}{4}\left(\frac{3}{4}\right)^2 = 0.44 \text{ in}^2$$

$$S_s = (0.30)(70) = 21 \text{ kips/in}^2$$

Allowable shear capacity $= A_e S_s = (0.44)(21) = 9.24 \text{ kips}$

Example 7.8 Select the required thickness of the plates (A572 grade 55) and the proper electrode material, assuming a bevel weld and the submerged-arc process (Fig. E7.8a).

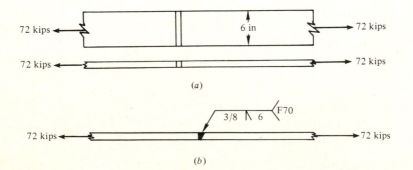

(a)

(b)

Figure E7.8

SOLUTION See Fig. E7.8b.
The allowable tensile stress is

$$F_t = 0.6F_y = (0.6)(55) = 33 \text{ kips/in}^2$$

The required plate thickness is

$$t_{\text{req}} = \frac{72}{(6)(33)} = 0.364 \text{ in}$$

Use 6- by $\frac{3}{8}$-in plates. From Table 7.8 use an F70 electrode and full-penetration bevel weld.

Example 7.9 Determine the capacity of the tee connection shown in Fig. E7.9 for a shielded metal-arc process. Assume that the flange does not control design.

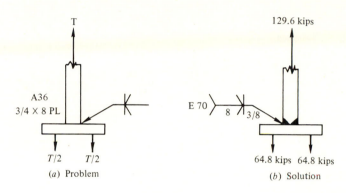

(a) Problem   (b) Solution

Figure E7.9

SOLUTION Use double bevel weld to avoid warping, and use $\frac{3}{8}$-in welds for complete penetration. Use E60 electrodes. Then

$$T = S_s t_e L = (0.6)(36)(2)\left(\tfrac{3}{8}\right)(8) = 129.6 \text{ kips}$$

Example 7.10 Determine the size and length of the fillet weld for the lap joint in Fig. E7.10. Use submerged-arc process. Plates are A36 steel.

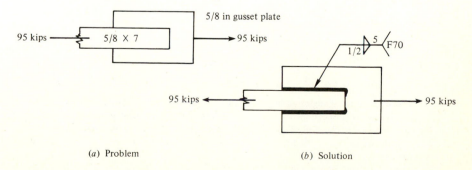

(a) Problem   (b) Solution

Figure E7.10

SOLUTION

$$\text{Maximum size} = \tfrac{5}{8} - \tfrac{1}{16} = \tfrac{9}{16} \text{ in} \qquad \text{Fig. 7.17}$$

$$\text{Minimum size} = \tfrac{1}{4} \text{ in} \qquad \text{Table 7.10}$$

Use $\tfrac{1}{2}$-in fillet weld and F70 electrode (Table 7.8)

$$t_e = (0.707)\left(\tfrac{1}{2}\right) + 0.11 = 0.464 \text{ in}$$

$$F_w = (0.464)(0.30)(70) = 9.74 \text{ kips/in}$$

$$\text{Length of weld } L = \frac{95}{9.74} = 9.8 \text{ in}$$

Use 5-in fillet on each side.

Example 7.11 A tension diagonal of a roof truss carries a load of 80 kips. The gusset plates at the ends of the diagonal are $\tfrac{3}{8}$ in thick. Select a pair of angles to carry the load and determine the size and length of fillet welding required for the connection.

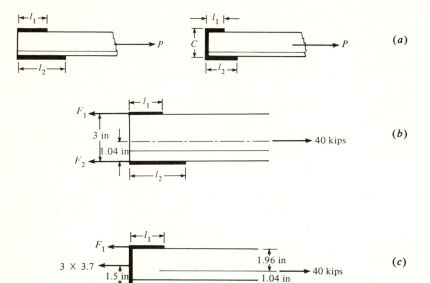

Figure E7.11

SOLUTION The allowable tensile stress is

$$0.60 F_y < 0.50 F_u = (0.60)(36) = 22 \text{ kips/in}^2 \qquad \text{see Example 7.1}$$

The net area is

$$80/22 = 3.64 \text{ in}^2. \quad \text{Use two angles 3 by 2 by } \tfrac{3}{8} = 3.84 \text{ in}^2 \quad \text{(see Table A.5)}.$$

For single and double angles under static tension, AISC does not require the fillet welds in end connections to be proportioned to balance the forces about the neutral axis of the member, but for members in compression or subject to repeated stress variations, it is recommended that the fillet welds be placed so that they balance the forces about the neutral axis and eliminate eccentricity.

Two common cases are illustrated in Fig. E7.11a.

Consider the first case for the problem at hand (Fig. E7.11b).

Angles are made of A36 steel; use E60 electrode. The angles are $\frac{3}{8}$ in thick; use $\frac{1}{4}$-in fillet welds.

$$\text{Shear strength } F_w \text{ of 1 in of weld} = \left(\tfrac{1}{4}\right)(0.707)(0.30)(70) = 3.71 \text{ kips}$$

The total length of weld required per angle is

$$l = P/F_w = \frac{40}{3.71} = 10.8 \text{ in} = l_1 + l_2$$

Consider the conditions for equilibrium of the forces of shear resistance provided by the welds and the axial force of 40 kips in the member acting along its neutral axis

$$\overrightarrow{\Sigma X} = 0: 40 - F_1 - F_2 = 0$$

where
$$F_1 = 3.71 l_1 \text{ kips and } F_2 = 3.71 l_2 \text{ kips.}$$

In addition to the equilibrium of forces there must be equilibrium in terms of moments

$$\widehat{\Sigma M_{F_2}} = 0: (40)(1.04) - 3F_1 = 0$$

$$F_1 = \frac{(40)(1.04)}{3} = 13.9 \text{ kips}$$

$$F_1 = 3.71 l_1 = 13.9 \text{ kips}$$

$$l_1 = 3.8 \text{ in}$$

Since $l_1 + l_2 = 10.8$ in,

$$l_2 = 10.8 - 3.8 = 7.0 \text{ in}$$

In the second case, where there is a 3-in end fillet, we have the free-body diagram shown in Fig. E7.11c, and

$$\overrightarrow{\Sigma F} = 0: -F_1 - F_2 - (3)(3.71) + 40 = 0$$

$$\widehat{\Sigma M_{F_2}} = 0: (40)(1.04) - (3)(3.71)(1.5) - 3F_1 = 0$$

$$F_1 = \frac{24.9}{3} = 8.3 \text{ kips}$$

$$F_1 = 3.71 l_1 = 8.3 \text{ kips}$$

$$l_1 = \frac{8.3}{3.71} = 2.2 \text{ in}$$

$$l_2 = 10.8 - 3 - 2.2 = 5.6 \text{ in}$$

7.6 MECHANICAL FASTENERS FOR TIMBER JOINTS

7.6.1 Introduction

In the design of wood structures, fastening one member to another is a critical feature. Many devices have been used, from wooden pegs and wrought-iron nails to modern shear connectors. Figure 7.20 illustrates the common mechanical devices used to connect wood members. The formulas and safe working loads are the results of tests conducted by the Forest Products Laboratory and other research institutes. They have been presented in Ref. 6 and are recommended as specifications in Ref. 4.

Factors to be taken into account in determining the allowable loads for fasteners are (1) lumber species (specific gravity), (2) critical section, (3) angle

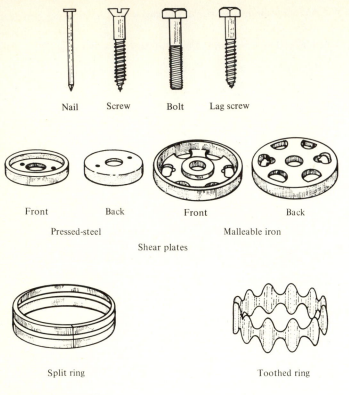

Figure 7.20 Wood fasteners.

of load to grain, (4) spacing of fasteners, (5) edge and end distances, and (6) eccentricity.

Lumber species The allowable load for fasteners varies with the species of wood. The species are divided into four load groups based on their specific gravity. Table 7.11 lists various species of wood for fastener design.

Critical section The critical section of a timber joint is that section taken at right angles to the direction of the load which gives the maximum stress based on the net area. The net area is equal to the cross-sectional area of the member less the projected area of the fastener within the member and the projected area of associated holes not within the fastener's projected area. This provision may vary for different connectors.

Angle of load to grain Since the wood has a greater bearing value parallel to the grain than perpendicular to the grain, the angle of load to grain becomes important. It is the angle between the resultant of the load exerted by a mechanical fastener on a member and the longitudinal axis of the member. Figure 7.21 illustrates this relationship. It is important to remember that this angle can be different for the members involved in the joint. For example, in

Table 7.11 Species load groups for fastenings [4]

Species	Specific gravity†	Timber-connector load groups	Lag bolt and driven fastening‡ load groups
Ash, commercial white	0.62	A	I
Beech	0.68	A	I
Birch, sweet and yellow	0.66	A	I
Cedar, northern white	0.31	D	IV
Western	0.36	D	IV
Cottonwood, eastern	0.41	D	IV
Cypress, southern	0.48	C	III
Douglas fir–larch§	0.51	B	II
South	0.48	C	III
Fir, balsam	0.38	D	IV
Subalpine	0.34	D	IV
Hem-fir	0.44	C	III
Hemlock, eastern–tamarack	0.45	C	III
Mountain	0.47	C	III
Hickory and pecan	0.75	A	I
Maple, black and sugar	0.66	A	I
Oak, red and white	0.67	A	I
Pine, eastern white	0.38	D	IV
Idaho white	0.42	C	III
Lodgepole	0.44	C	III
Northern	0.46	C	III
Ponderosa–sugar	0.42	C	III
Southern	0.55	B	II
Poplar, yellow	0.46	C	III
Redwood, California, close grain	0.42	C	III
Open grain	0.37	D	IV
Spruce, eastern	0.43	C	III
Engelmann	0.37	D	IV
Sitka	0.43	C	III
Sweetgum and tupelo	0.54	B	II

† Based on weight and volume when oven-dry.
‡ Nails, spikes, wood screws, and spiral dowels.
§ When graded for density, these species qualify for group A connector loads.

Fig. 7.21 the angle of load to grain is θ for member B and zero for member A. The allowable load acting at some angle θ to the grain can best be determined by use of the *Hankinson formula*

$$N = \frac{PQ}{P \sin^2 \theta + Q \cos^2 \theta} \tag{7.7}$$

where N = allowable load at angle θ with direction of grain, lb
   P = allowable load acting parallel to grain, lb
   Q = allowable load acting perpendicular to grain, lb
   θ = angle between direction of load and direction of grain, deg

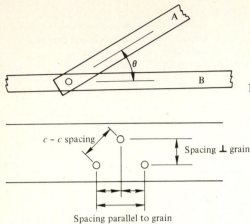

Figure 7.21 Angle of load to grain.

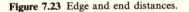

Figure 7.22 Spacing of fasteners.

This relationship has application not only with fasteners but also with working stresses where wood bears against wood.

Spacing of fasteners The spacing of fasteners is considered to be the center-to-center distance measured on a line joining them. This distance can also be expressed in terms of distances measured perpendicular and parallel to the grain (Fig. 7.22). The spacing in a group should be sufficient to develop the full strength of each fastener.

Edge and end distances The edge distance is the perpendicular distance from the edge of a member to the center of the fastener nearest that edge. The loaded edge is the edge toward which the load induced by the fastener acts. The unloaded edge is the edge away from which the load acts. The end distance is the distance measured parallel to the grain from the center of a fastener to the square-cut end of the member. If the end of the member is cut at an angle, the end distance is measured parallel to the length of the piece on a parallel line that is one-fourth the fastener diameter from the center of the connector on the side of the larger angle of the end cut. These distances are illustrated in Fig. 7.23.

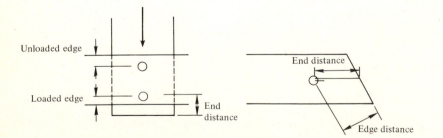

Figure 7.23 Edge and end distances.

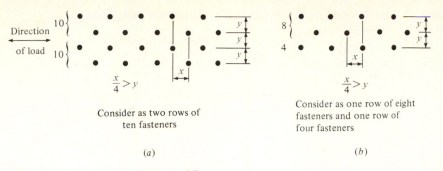

Consider as two rows of
ten fasteners

(a)

Consider as one row of eight
fasteners and one row of
four fasteners

(b)

Figure 7.24 Group action of fasteners [4].

Group action of fasteners Research has indicated that a load carried by a row of fasteners (*connectors* and *bolts*) is not equally divided among the fasteners. The end fasteners tend to carry a larger portion of the load. The distribution of load depends on the relative stiffness of the main and side members.

A group of fasteners is considered to be one or more rows of fasteners aligned in the direction of the load. When fasteners in adjacent rows are staggered and the distance between the rows y is less than one-fourth the spacing between the closest fasteners in adjacent rows x, as shown in Fig. 7.24a, the fasteners in adjacent rows should be considered as one row for purposes of the

Table 7.12 Modification factors for fastener groups with wood side plates [5]

A_1/A_2†	A_1, in.²‡	\multicolumn{11}{c}{Number of fasteners in row}
		2	3	4	5	6	7	8	9	10	11	12
0.5§,¶	< 12	1.00	0.92	0.84	0.76	0.68	0.61	0.55	0.49	0.43	0.38	0.34
	12–19	1.00	0.95	0.88	0.82	0.75	0.68	0.62	0.57	0.52	0.48	0.43
	> 19–28	1.00	0.97	0.93	0.88	0.82	0.77	0.71	0.67	0.63	0.59	0.55
	> 28–40	1.00	0.98	0.96	0.92	0.87	0.83	0.79	0.75	0.71	0.69	0.66
	> 40–64	1.00	1.00	0.97	0.94	0.90	0.86	0.83	0.79	0.76	0.74	0.72
	> 64	1.00	1.00	0.98	0.95	0.91	0.88	0.85	0.82	0.80	0.78	0.76
1.0§,¶	< 12	1.00	0.97	0.92	0.85	0.78	0.71	0.65	0.59	0.54	0.49	0.44
	12–19	1.00	0.98	0.94	0.89	0.84	0.78	0.72	0.66	0.61	0.56	0.51
	> 19–28	1.00	1.00	0.97	0.93	0.89	0.85	0.80	0.76	0.72	0.68	0.64
	> 28–40	1.00	1.00	0.99	0.96	0.92	0.89	0.86	0.83	0.80	0.78	0.75
	> 40–64	1.00	1.00	1.00	0.97	0.94	0.91	0.88	0.85	0.84	0.82	0.80
	> 64	1.00	1.00	1.00	0.99	0.96	0.93	0.91	0.88	0.87	0.86	0.85

† A_1 = cross-sectional area of main member(s) before boring or grooving. A_2 = sum of the cross-sectional areas of side members before boring or grooving.
‡ When A_1/A_2 exceeds 1.0, use A_2 instead of A_1.
§ When A_1/A_2 exceeds 1.0, use A_2/A_1.
¶ For A_1/A_2 between 0 and 1.0, interpolate or extrapolate from the tabulated values.

Table 7.13 Modification factors for fastener groups with metal side plates [5]

A_1/A_2†	A_1, in^2	Number of fasteners in row										
		2	3	4	5	6	7	8	9	10	11	12
2–12	25–39	1.00	0.94	0.87	0.80	0.73	0.67	0.61	0.56	0.51	0.46	0.42
	40–64	1.00	0.96	0.92	0.87	0.81	0.75	0.70	0.66	0.62	0.58	0.55
	65–119	1.00	0.98	0.95	0.91	0.87	0.82	0.78	0.75	0.72	0.69	0.66
	120–199	1.00	0.99	0.97	0.95	0.92	0.89	0.86	0.84	0.81	0.79	0.78
12–18	40–64	1.00	0.98	0.94	0.90	0.85	0.80	0.75	0.70	0.67	0.62	0.58
	65–119	1.00	0.99	0.96	0.93	0.90	0.86	0.82	0.79	0.75	0.72	0.69
	120–199	1.00	1.00	0.98	0.96	0.94	0.92	0.89	0.86	0.83	0.80	0.78
	> 200	1.00	1.00	1.00	0.98	0.97	0.95	0.93	0.91	0.90	0.88	0.87
18–24	40–64	1.00	1.00	0.96	0.93	0.89	0.84	0.79	0.74	0.69	0.64	0.59
	65–119	1.00	1.00	0.97	0.94	0.92	0.89	0.86	0.83	0.80	0.76	0.73
	120–199	1.00	1.00	0.99	0.98	0.96	0.94	0.92	0.90	0.88	0.86	0.85
	> 200	1.00	1.00	1.00	1.00	0.98	0.96	0.95	0.93	0.92	0.92	0.91
24–30	40–64	1.00	0.98	0.94	0.90	0.85	0.80	0.74	0.69	0.65	0.61	0.58
	65–119	1.00	0.99	0.97	0.93	0.90	0.86	0.82	0.79	0.76	0.73	0.71
	120–199	1.00	1.00	0.98	0.96	0.94	0.92	0.89	0.87	0.85	0.83	0.81
	> 200	1.00	1.00	0.99	0.98	0.97	0.95	0.93	0.92	0.90	0.89	0.89
30–35	40–64	1.00	0.96	0.92	0.86	0.80	0.74	0.68	0.64	0.60	0.57	0.55
	65–119	1.00	0.98	0.95	0.90	0.86	0.81	0.76	0.72	0.68	0.65	0.62
	120–199	1.00	0.99	0.97	0.95	0.92	0.88	0.85	0.82	0.80	0.78	0.77
	> 200	1.00	1.00	0.98	0.97	0.95	0.93	0.90	0.89	0.87	0.86	0.85
35–42	40–64	1.00	0.95	0.89	0.82	0.75	0.69	0.63	0.58	0.53	0.49	0.46
	65–119	1.00	0.97	0.93	0.88	0.82	0.77	0.71	0.67	0.63	0.59	0.56
	120–199	1.00	0.98	0.96	0.93	0.89	0.85	0.81	0.78	0.76	0.73	0.71
	> 200	1.00	0.99	0.98	0.96	0.93	0.90	0.87	0.84	0.82	0.80	0.78

† A_1 = cross-sectional area of main member before boring or grooving. A_2 = sum of cross-sectional areas of metal side plates before drilling.

modification factor K. This principle applies to groups with either even or odd numbers of rows, as shown in Fig. 7.24b.

The load for each row of fasteners is determined by summing the individual loads for each fastener in the row and then multiplying this value by the modification factor K (Tables 7.12 and 7.13). The allowable load for the group is the sum of the allowable loads on the rows in the group.

For a member loaded perpendicular to the grain, the equivalent cross-sectional area is the product of the thickness of the member and the overall width of the fastener group. In general, long rows of fasteners perpendicular to the grain should be avoided.

Eccentricity Timber joints should be designed to be concentric. If this is impossible, the effect of the induced moment must be taken into account in calculating fiber stresses.

7.7 NAILS AND SPIKES

7.7.1 Introduction

The basic factors controlling the holding strength of a nailed joint are the specific gravity of the wood, diameter of the nail, depth of penetration, and moisture-content changes in the wood. The ultimate load for nails and spikes is about 4 or 5 times the values determined by the given formulas and tables. If conditions warrant it, on the basis of good engineering judgment, these values can be modified. For a joint involving more than one nail or spike, the total resistance is the sum of the allowable loads for the individual nails or spikes, providing the edge and end distances are such that they avoid objectionable splitting. Loads may be applied to nails in a lateral or withdrawal direction. The withdrawal loading of nails should be avoided if possible. Specifications recommend that nails never be used in withdrawal from the end grain. In general if holes no larger than three-fourths the diameter of the nail or spike are prebored, splitting is reduced substantially and the allowable loads are not decreased. The sizes and dimensions of common wire nails and spikes are given in Table 7.14. Note that spikes are similar to nails but have larger diameters for the same sizes.

7.7.2 Lateral Resistance

The lateral resistance of nails and spikes in the side grain of seasoned wood for normal load duration is given by

$$P = KD^{3/2} \tag{7.8}$$

Table 7.14 Sizes of nails and spikes [4]

	Nails			Spikes		Threaded, hardened-steel nails and spikes		
Penny-weight	Length, in	Diam, in	Penny-weight	Length, in	Diam, in	Penny-weight	Length, in	Diam, in
6	2	0.113	10	3	0.192	6	2	0.120
8	$2\frac{1}{2}$	0.131	12	$3\frac{1}{4}$	0.192	8	$2\frac{1}{2}$	0.120
10	3	0.148	16	$3\frac{1}{2}$	0.207	10	3	0.135
12	$3\frac{1}{4}$	0.148	20	4	0.225	12	$3\frac{1}{4}$	0.135
16	$3\frac{1}{2}$	0.162	30	$4\frac{1}{2}$	0.244	16	$3\frac{1}{2}$	0.148
20	4	0.192	40	5	0.263	20	4	0.177
30	$4\frac{1}{2}$	0.207	50	$5\frac{1}{2}$	0.283	30	$4\frac{1}{2}$	0.177
40	5	0.225	60	6	0.283	40	5	0.177
50	$5\frac{1}{2}$	0.244	$\frac{5}{16}$ in	7	0.312	50	$5\frac{1}{2}$	0.177
60	6	0.263	$\frac{3}{8}$ in	$8\frac{1}{2}$	0.375	60	6	0.177
						70	7	0.207
						80	8	0.207
						90	9	0.207

where P = allowable normal lateral load, lb per nail
K = constant depending on specific gravity
D = diameter of nails, in

The values for K for the various species groups (Table 7.11) are

Group	IV	III	II	I
K value	1080	1350	1650	2040

The allowable design values for different sizes (Table 7.15) are valid whether the load is parallel to the grain or at some angle to it. When the side piece and the main member are of different specific gravities, the smaller value controls. The depths of penetration required to develop the lateral loads given by Eq. (7.8) depend on the species group of the wood (Table 7.11). The depth of penetration into the member holding the point should be

Not less than 10 diameters for group I
Not less than 11 diameters for group II
Not less than 13 diameters for group III
Not less than 14 diameters for group IV

For a penetration less than that specified, the allowable load can be determined by straight-line interpolation between zero and the tabulated load. The penetration should not be less than three diameters.

For double shear, the allowable load may be increased by one-third if the side members are not less than one-third the thickness of the main member and by two-thirds if the side members are equal in thickness to the main member. When nails are used to fasten metal plates to wood, the value of P can be increased by 25 percent. For unseasoned wood exposed to the weather, using 75 percent of the tabulated value is realistic. For nails or spikes driven into end grain, the allowable load is two-thirds that of the value for side grain. The allowable lateral load in a toe-nailed joint is five-sixths that given by Eq. (7.8). The end distance for nailed joints should be no less than 15 times the diameter of the nail if the end is stressed or 12 times the nail diameter if unstressed. Edge distances should be no less than 10 times the nail diameter.

7.7.3 Withdrawal Resistance

When necessary, the withdrawal resistance of nails and spikes in side grain of seasoned wood for normal load duration can be determined by

$$P = 350G^{5/2}D \tag{7.9}$$

where P = allowable normal load per inch of penetration, lb/in
G = specific gravity of wood
D = diameter of nail or spike, in

Table 7.15 Allowable lateral loads for nails and spikes [4]†

Allowable lateral loads (shear) in pounds for nails and spikes penetrating 10 diameters in group I species, 11 diameters in group II species, 13 diameters in group III species, and 14 diameters in group IV species into the member holding the point

Size, pennyweight	Group			
	I	II	III	IV
Common nail				
6	78	63	51	41
8	97	78	64	51
10	116	94	77	62
12	116	94	77	62
16	132	107	88	70
20	171	139	113	91
30	191	154	126	101
40	218	176	144	116
50	249	202	165	132
60	276	223	182	146
Threaded nail‡				
30–60	171	139	113	91
70–80	218	176	144	116
Common spike				
10	171	139	113	91
12	171	139	113	91
16	191	155	126	101
20	218	176	144	116
30	249	202	165	132
40	276	223	182	146
50	306	248	202	162
60	306	248	202	162
$\frac{5}{16}$ in	357	289	236	189
$\frac{3}{8}$ in	469	380	310	248

  † For species load groups see Table 7.11.
  ‡ Loads for threaded hardened-steel nails in 6d to 20d sizes are the same as for common nails.

These values of allowable load incorporate a factor of safety of 4 and are adjusted for normal loading conditions. The allowable withdrawal load in toe-nailed joints is equivalent to two-thirds of that indicated above.

Example 7.12 A 2- by 6-in plank is nailed to a seasoned Douglas fir–larch 6- by 6-in post with four 40d nails. Determine the load allowed by the connection.

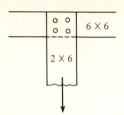

Figure E7.12

Solution (Fig. E7.12)   Douglas fir–larch, group II (Table 7.11)

$$\text{Penetration} = 5 - 1\,1/2\ \text{in} = 3\,1/2\ \text{in} > (11)(0.225)$$

$$\text{Allowable load} = 4P = (4)(176) \qquad \text{Table 7.15}$$

$$= 704\ \text{lb}$$

Example 7.13 A scaffold is made of rough unseasoned southern pine, as shown in Fig. E7.13. Determine the number of 20d nails required in joint *B*.

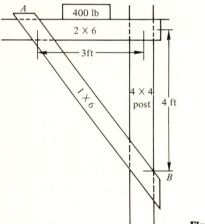

Figure E7.13

Solution Southern pine, group II (Table 7.11)

$$\text{Load on nails} = \frac{400}{2}\frac{5}{4} = 250\ \text{lb}$$

$$\text{Penetration}\ 4 - \tfrac{5}{8}\ \text{in} = 3\tfrac{3}{8} > (11)(0.192)$$

From Table 7.15

$$\text{Allowable lateral load for 20d nail} = 91\ \text{lb} \qquad \text{seasoned}$$

For unseasoned conditions

$$\left(\tfrac{3}{4}\right)(91) = 68\ \text{lb/nail}$$

$$\text{Nails required} = \frac{250}{68} = 3.7 \qquad \text{use four nails}$$

7.8 LAG BOLTS

7.8.1 Introduction

Wood screws are seldom used in structural applications of timber. Lag bolts, on the other hand, are used in places where the use of a bolt would be difficult or impossible or a nut on the surface would be objectionable. Lag bolts should be screwed into a prebored hole with a wrench. Washers are used unless metal side plates are being used. Commonly available lag bolts range from about 0.2 to 1 in in diameter and from 1 to 16 in long. The length of the threaded part varies from $\frac{3}{4}$ in for 1- and $1\frac{1}{4}$-in bolts to half the length for all lengths greater than 10 in (Table 7.16).

7.8.2 Lateral Resistance

The experimentally determined lateral loads for *lag screws* or *bolts* inserted in the side grain and loaded *parallel to the grain* of a piece of seasoned wood can be computed from

$$P = KD^2 \qquad (7.10)$$

where P = allowable normal lateral load parallel to grain, lb
           K = coefficient depending upon specific gravity
           D = shank diameter of lag bolt, in

Values of K for a number of specific-gravity ranges of soft wood are

Specific gravity	K
0.29–0.42	1650
0.43–0.47	1870
0.48–0.52	2090

The loads given by Eq. (7.10) apply when the thickness of the attached member is 3.5 times the shank diameter and the depth of penetration in the main member is 11 times the diameter in soft wood. For other thicknesses of attached members the computed loads should be multiplied by the following factors:

Ratio of thickness of attached member to shank diameter	2	$2\frac{1}{2}$	3	$3\frac{1}{2}$	4	$4\frac{1}{2}$	5	$5\frac{1}{2}$	6	$6\frac{1}{2}$
Factor	0.62	0.77	0.93	1.00	1.07	1.13	1.18	1.21	1.22	1.22

Table 7.16 Dimensions for standard lag bolts or lag screws [4]

(All dimensions in inches)
D = Nominal diameter.
Ds = D = Diameter of shank.
Dr = Diameter at root of thread.
W = Width of bolt head across flats.
N = Number of threads per inch.

H = Height of bolt head.
L = Nominal length of bolt.
S = Length of shank.
T = Length of thread.
F = Length of tapered tip.

Nominal length of bolt (L) in inches*	Item	Dimensions of lag bolt with nominal diameter (D) of–									
		$\frac{1}{4}$	$\frac{5}{16}$	$\frac{1}{8}$	$\frac{7}{16}$	$\frac{1}{2}$	$\frac{9}{16}$	$\frac{5}{8}$	$\frac{3}{4}$	$\frac{7}{8}$	1
All lengths	Ds = D	0.250	0.3125	0.375	0.4375	0.500	0.5625	0.625	0.750	0.875	1.000
	Dr	.173	.227	.265	.328	.371	.435	.471	.579	.683	.780
	E	$\frac{3}{16}$	$\frac{1}{4}$	$\frac{1}{4}$	$\frac{9}{32}$	$\frac{5}{16}$	$\frac{3}{8}$	$\frac{3}{8}$	$\frac{7}{16}$	$\frac{1}{2}$	$\frac{9}{16}$
	H	$\frac{11}{64}$	$\frac{13}{64}$	$\frac{1}{4}$	$\frac{19}{64}$	$\frac{21}{64}$	$\frac{3}{8}$	$\frac{27}{64}$	$\frac{1}{2}$	$\frac{19}{32}$	$\frac{21}{32}$
	W	$\frac{3}{8}$	$\frac{1}{2}$	$\frac{9}{16}$	$\frac{5}{8}$	$\frac{3}{4}$	$\frac{7}{8}$	$\frac{15}{16}$	$1\frac{1}{8}$	$1\frac{5}{16}$	$1\frac{1}{2}$
	N	10	9	7	7	6	6	5	$4\frac{1}{2}$	4	$3\frac{1}{2}$
1	S	$\frac{1}{4}$	$\frac{1}{4}$	$\frac{1}{4}$	$\frac{1}{4}$	$\frac{1}{4}$	...	...	...	...	...
	T	$\frac{3}{4}$	$\frac{3}{4}$	$\frac{3}{4}$	$\frac{3}{4}$	$\frac{3}{4}$	...	...	...	...	...
	T-E	$\frac{9}{16}$	$\frac{1}{2}$	$\frac{1}{2}$	$\frac{15}{32}$	$\frac{7}{16}$	...	...	...	...	...
$1\frac{1}{2}$	S	$\frac{3}{8}$	$\frac{3}{8}$	$\frac{3}{8}$	$\frac{3}{8}$	$\frac{3}{8}$	...	...	...	...	...
	T	$1\frac{1}{8}$	$1\frac{1}{8}$	$1\frac{1}{8}$	$1\frac{1}{8}$	$1\frac{1}{8}$	...	...	...	...	...
	T-E	$\frac{15}{16}$	$\frac{7}{8}$	$\frac{7}{8}$	$\frac{27}{32}$	$\frac{13}{16}$	...	...	...	...	...
2	S	$\frac{1}{2}$	$\frac{1}{2}$	$\frac{1}{2}$	$\frac{1}{2}$	$\frac{1}{2}$	$\frac{1}{2}$	$\frac{1}{2}$	...	...	...
	T	$1\frac{1}{2}$	$1\frac{1}{2}$	$1\frac{1}{2}$	$1\frac{1}{2}$	$1\frac{1}{2}$	$1\frac{1}{2}$	$1\frac{1}{2}$	...	...	...
	T-E	$1\frac{5}{16}$	$1\frac{1}{4}$	$1\frac{1}{4}$	$1\frac{17}{32}$	$1\frac{3}{16}$	$1\frac{1}{8}$	$1\frac{1}{8}$	...	...	...
$2\frac{1}{2}$	S	1	$\frac{7}{8}$	$\frac{7}{8}$	$\frac{3}{4}$	$\frac{3}{4}$	$\frac{3}{4}$	...	...	...	...
	T	$1\frac{1}{2}$	$1\frac{5}{8}$	$1\frac{5}{8}$	$1\frac{3}{4}$	$1\frac{3}{4}$	$1\frac{3}{4}$	$1\frac{3}{4}$	...	...	...
	T-E	$1\frac{5}{16}$	$1\frac{3}{8}$	$1\frac{3}{8}$	$1\frac{15}{32}$	$1\frac{7}{16}$	$1\frac{3}{8}$	$1\frac{3}{8}$	...	...	...
3	S	1	1	1	1	1	1	1	1	1	1
	T	2	2	2	2	2	2	2	2	2	2
	T-E	$1\frac{13}{16}$	$1\frac{3}{4}$	$1\frac{3}{4}$	$1\frac{23}{32}$	$1\frac{11}{16}$	$1\frac{5}{8}$	$1\frac{5}{8}$	$1\frac{5}{16}$	$1\frac{1}{2}$	$1\frac{7}{16}$
4	S	$1\frac{1}{2}$	$1\frac{1}{2}$	$1\frac{1}{2}$	$1\frac{1}{2}$	$1\frac{1}{2}$	$1\frac{1}{2}$	$1\frac{1}{2}$	$1\frac{1}{2}$	$1\frac{1}{2}$	$1\frac{1}{2}$
	T	$2\frac{1}{2}$	$2\frac{1}{2}$	$2\frac{1}{2}$	$2\frac{1}{2}$	$2\frac{1}{2}$	$2\frac{1}{2}$	$2\frac{1}{2}$	$2\frac{1}{2}$	$2\frac{1}{2}$	$2\frac{1}{2}$
	T-E	$2\frac{5}{16}$	$2\frac{1}{4}$	$2\frac{1}{4}$	$2\frac{7}{32}$	$2\frac{3}{16}$	$2\frac{1}{8}$	$2\frac{1}{8}$	$2\frac{1}{16}$	2	$1\frac{15}{16}$
5	S	2	2	2	2	2	2	2	2	2	2
	T	3	3	3	3	3	3	3	3	3	3
	T-E	$2\frac{13}{16}$	$2\frac{3}{4}$	$2\frac{3}{4}$	$2\frac{23}{32}$	$2\frac{11}{16}$	$2\frac{5}{8}$	$2\frac{5}{8}$	$2\frac{9}{16}$	$2\frac{1}{2}$	$2\frac{7}{16}$
6	S	$2\frac{1}{2}$	$2\frac{1}{2}$	$2\frac{1}{2}$	$2\frac{1}{2}$	$2\frac{1}{2}$	$2\frac{1}{2}$	$2\frac{1}{2}$	$2\frac{1}{2}$	$2\frac{1}{2}$	$2\frac{1}{2}$
	T	$3\frac{1}{2}$	$3\frac{1}{2}$	$3\frac{1}{2}$	$3\frac{1}{2}$	$3\frac{1}{2}$	$3\frac{1}{2}$	$3\frac{1}{2}$	$3\frac{1}{2}$	$3\frac{1}{2}$	$3\frac{1}{2}$
	T-E	$3\frac{5}{16}$	$3\frac{1}{4}$	$3\frac{1}{4}$	$3\frac{7}{32}$	$3\frac{3}{16}$	$3\frac{1}{8}$	$3\frac{1}{8}$	$3\frac{1}{16}$	3	$2\frac{15}{16}$
7	S	3	3	3	3	3	3	3	3	3	3
	T	4	4	4	4	4	4	4	4	4	4
	T-E	$3\frac{13}{16}$	$3\frac{3}{4}$	$3\frac{3}{4}$	$3\frac{23}{32}$	$3\frac{11}{16}$	$3\frac{5}{8}$	$3\frac{5}{8}$	$3\frac{9}{16}$	$3\frac{1}{2}$	$3\frac{7}{16}$
8	S	$3\frac{1}{2}$	$3\frac{1}{2}$	$3\frac{1}{2}$	$3\frac{1}{2}$	$3\frac{1}{2}$	$3\frac{1}{2}$	$3\frac{1}{2}$	$3\frac{1}{2}$	$3\frac{1}{2}$	$3\frac{1}{2}$
	T	$4\frac{1}{2}$	$4\frac{1}{2}$	$4\frac{1}{2}$	$4\frac{1}{2}$	$4\frac{1}{2}$	$4\frac{1}{2}$	$4\frac{1}{2}$	$4\frac{1}{2}$	$4\frac{1}{2}$	$4\frac{1}{2}$
	T-E	$4\frac{5}{16}$	$4\frac{1}{4}$	$4\frac{1}{4}$	$4\frac{7}{32}$	$4\frac{3}{16}$	$4\frac{5}{8}$	$4\frac{1}{8}$	$4\frac{1}{16}$	4	$3\frac{15}{16}$
9	S	4	4	4	4	4	4	4	4	4	4
	T	5	5	5	5	5	5	5	5	5	5
	T-E	$4\frac{13}{16}$	$4\frac{3}{4}$	$4\frac{3}{4}$	$4\frac{23}{32}$	$4\frac{11}{16}$	$4\frac{5}{8}$	$4\frac{5}{8}$	$4\frac{9}{16}$	$4\frac{1}{2}$	$4\frac{7}{16}$
10	S	$4\frac{3}{4}$	$4\frac{3}{4}$	$4\frac{3}{4}$	$4\frac{3}{4}$	$4\frac{3}{4}$	$4\frac{3}{4}$	$4\frac{3}{4}$	$4\frac{3}{4}$	$4\frac{3}{4}$	$4\frac{3}{4}$
	T	$5\frac{1}{4}$	$5\frac{1}{4}$	$5\frac{1}{4}$	$5\frac{1}{4}$	$5\frac{1}{4}$	$5\frac{1}{4}$	$5\frac{1}{4}$	$5\frac{1}{4}$	$5\frac{1}{4}$	$5\frac{1}{4}$
	T-E	$5\frac{1}{16}$	5	5	$4\frac{31}{32}$	$4\frac{15}{16}$	$4\frac{7}{8}$	$4\frac{7}{8}$	$4\frac{13}{16}$	$4\frac{3}{4}$	$4\frac{11}{16}$
11	S	$5\frac{1}{2}$	$5\frac{1}{2}$	$5\frac{1}{2}$	$5\frac{1}{2}$	$5\frac{1}{2}$	$5\frac{1}{2}$	$5\frac{1}{2}$	$5\frac{1}{2}$	$5\frac{1}{2}$	$5\frac{1}{2}$
	T	$5\frac{1}{2}$	$5\frac{1}{2}$	$5\frac{1}{2}$	$5\frac{1}{2}$	$5\frac{1}{2}$	$5\frac{1}{2}$	$5\frac{1}{2}$	$5\frac{1}{2}$	$5\frac{1}{2}$	$5\frac{1}{2}$
	T-E	$5\frac{9}{32}$	$5\frac{1}{4}$	$5\frac{1}{4}$	$5\frac{7}{32}$	$5\frac{3}{16}$	$5\frac{1}{8}$	$5\frac{1}{8}$	$5\frac{1}{16}$	5	$4\frac{15}{16}$
12	S	6	6	6	6	6	6	6	6	6	6
	T	6	6	6	6	6	6	6	6	6	6
	T-E	$5\frac{13}{16}$	$5\frac{3}{4}$	$5\frac{3}{4}$	$5\frac{23}{32}$	$5\frac{11}{16}$	$5\frac{5}{8}$	$5\frac{5}{8}$	$5\frac{9}{16}$	$5\frac{1}{2}$	$5\frac{7}{16}$

*Length of thread (T) on intervening bolt lengths is the same as that of the next shorter bolt length listed. The length of thread (T) on standard bolt lengths (L) in excess of 12 inches is equal to $\frac{1}{2}$ the bolt length (L/2).

The thickness of the attached member should be about one-half the depth of penetration in the main member. When the lag bolt is inserted into the side grain of wood and the load is *applied perpendicular to the grain*, the load given by Eq. (7.10) should be multiplied by the following factors:

Shank diameter of lag bolt, in	$\frac{3}{16}$	$\frac{1}{4}$	$\frac{5}{16}$	$\frac{3}{8}$	$\frac{7}{16}$	$\frac{1}{2}$	$\frac{5}{8}$	$\frac{3}{4}$	$\frac{7}{8}$	1
Factor	1.00	0.97	0.85	0.76	0.70	0.65	0.60	0.55	0.52	0.50

For other angles of loading the Hankinson formula (Sec. 7.6.1) is used. When the unthreaded portion of the shank penetrates the main member, the following increases in loads are permitted:

Ratio of penetration of shank into main member to shank diameter	1	2	3	4	5	6	7
Increase in load, %	8	17	26	33	36	38	39

When lag bolts are used with metal plates, the lateral loads parallel to the grain may be increased by 25 percent. Lag bolts should not be used in *end grain* because of possible splitting. If necessary, the load should be taken as two-thirds of the load perpendicular to the grain. Spacings, end and edge distances, and net sections for lag-bolt joints should be the same as for bolted connections of similar size (Table 7.20). The allowable normal loads, according to accepted practice, are the proportional-limit load with a factor of safety of 2.25 and adjustments for normal loading conditions.

7.8.3 Withdrawal Resistance

The results of withdrawal tests have shown that the maximum load in direct withdrawal from seasoned wood can be calculated from

$$P = 400G^{3/2}D^{3/4}L \tag{7.11}$$

where P = allowable normal withdrawal load, lb
G = specific gravity of wood
D = shank diameter, in
L = length of penetration, in

In determining the withdrawal resistance, the allowable tensile strength of the lag bolt should not be exceeded. Penetrations of the threaded part, about 7 times the shank diameter in the denser species and 10 to 12 times the diameter in the less dense species, will normally develop the tensile load of the lag bolt. Resistance to withdrawal from end grain is about three-fourths that from side grain. The allowable normal withdrawal load [Eq. (7.11)] was obtained by using a factor of safety of 5 and a 10 percent increase for normal loading conditions.

Example 7.14 A $\frac{1}{4}$-in metal gusset plate is fastened to a seasoned Douglas fir post with five $\frac{1}{4}$- by 4-in lag screws (Fig. E7.14). Determine the load the connection will hold.

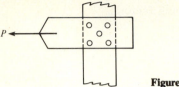

Figure E7.14

SOLUTION From Table 7.11 the specific gravity of Douglas fir is 0.48; then

$$K = 2090.$$

$$P = KD^2 = 2090\left(\tfrac{1}{4}\right)^2 = 131 \text{ lb/bolt}$$

In view of the fact that the load is perpendicular to the grain,

$$P = (131)(0.97) = 127 \text{ lb}$$

The shank penetration into post is $1\frac{1}{4}$ in, or 5 diameters,

$$P = (127)(1.36) = 173 \text{ lb}$$

$$P_{\text{allowable}} = 173 \text{ lb/bolt}$$

$$\text{Total allowable load} = (5)(173) = 865 \text{ lb}$$

7.9 BOLTS

7.9.1 Introduction

The results of many tests of joints having a seasoned-wood member and two steel splice plates show that bearing stress parallel to the grain at proportional limit loads approaches 60 percent of the crushing strength of softwoods and 80 percent of that of hardwoods. The crushing strengths are those obtained for clear, straight-grained specimens. Figure 7.25 indicates the reduction in bearing

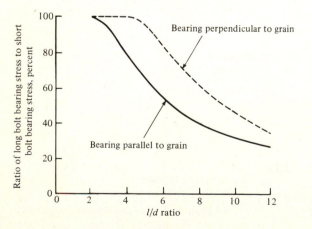

Figure 7.25 Variation of bearing stress with l/d ratio [6].

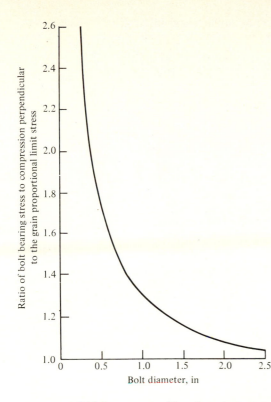

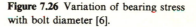

Figure 7.26 Variation of bearing stress with bolt diameter [6].

stress as l/d increases. For bearing stresses perpendicular to the grain in seasoned wood, the stresses for short bolts (small l/d ratios) depend upon bolt diameters (Fig. 7.26). As a result, the bolt-bearing stress perpendicular to the grain can be obtained from the proportional-limit stress perpendicular to the grain of small, clear specimens. This is increased by an appropriate factor from Fig. 7.26 and then reduced by a factor from Fig. 7.25.

7.9.2 Design Considerations

On the basis of the above considerations, tables have been developed giving the allowable loads various sized bolts can carry. Table 7.17 shows the allowable load for one bolt loaded in double shear in a timber joint under normal duration of loading and dry-use conditions. Loads for more than one bolt are the sum of the loads permitted for each bolt provided that spacings and end and edge distances are sufficient to develop the full strength of each bolt in a joint. Figure 7.27 illustrates examples of bolted joints for which full allowable loads can be used. For other types of joints the loads should be modified for group action (Sec. 7.6.1). For other than normal durations of load, the allowable loads should be multiplied by the appropriate adjustment factor from Fig. 6.20. For joints in use under other service or seasoning conditions, the allowable loads determined from Table 7.17 are multiplied by the appropriate factor from Table 7.18. For

Table 7.17 Allowable loads for one bolt in double shear [4]

Length of bolt in main member l	Diameter of bolt d	$\dfrac{l}{d}$	Projected area of bolt $A = ld$	Douglas fir–larch (dense) or southern pine (dense)		Eastern hemlock–tamarack, California redwood (open grain), hem-fir, western hemlock		Oak, red and white		Douglas fir, Douglas fir south, southern pine (open grain)	
				‖ to grain P	⊥ to grain Q	‖ to grain P	⊥ to grain Q	‖ to grain P	⊥ to grain Q	‖ to grain P	⊥ to grain Q
$1\frac{1}{2}$	$\frac{1}{2}$	3.00	.750	1120	500	810	280	830	650	820	370
	$\frac{5}{8}$	2.40	.938	1420	570	1010	310	1050	730	1030	420
	$\frac{3}{4}$	2.00	1.125	1700	630	1210	350	1260	820	1240	470
	$\frac{7}{8}$	1.71	1.313	1990	700	1410	380	1470	900	1440	520
	1	1.50	1.500	2270	760	1610	420	1690	980	1650	570
2	$\frac{1}{2}$	4.00	1.000	1400	670	1050	370	1040	870	1020	500
	$\frac{5}{8}$	3.20	1.250	1860	760	1340	410	1380	980	1350	560
	$\frac{3}{4}$	2.67	1.500	2260	840	1610	460	1680	1090	1640	630
	$\frac{7}{8}$	2.29	1.750	2640	930	1880	510	1960	1200	1920	690
	1	2.00	2.000	3030	1010	2150	550	2250	1310	2200	750
$2\frac{1}{2}$	$\frac{1}{2}$	5.00	1.250	1510	840	1190	460	1120	1080	1100	620
	$\frac{5}{8}$	4.00	1.563	2190	950	1640	520	1620	1220	1590	710
	$\frac{3}{4}$	3.33	1.875	2780	1060	2010	580	2060	1360	2020	790
	$\frac{7}{8}$	2.86	2.188	3290	1160	2350	630	2440	1500	2390	860
	1	2.50	2.500	3770	1270	2690	690	2800	1640	2740	940
3	$\frac{1}{2}$	6.00	1.500	1530	1010	1220	550	1130	1130	1110	750
	$\frac{5}{8}$	4.80	1.875	2350	1140	1830	620	1740	1470	1710	850
	$\frac{3}{4}$	4.00	2.250	3150	1270	2360	690	2340	1630	2290	940
	$\frac{7}{8}$	3.43	2.625	3860	1390	2810	760	2870	1800	2810	1040
	1	3.00	3.000	4500	1520	3230	830	3340	1960	3270	1130
$3\frac{1}{2}$	$\frac{1}{2}$	7.00	1.750	1530	1140	1220	640	1130	1130	1110	870
	$\frac{5}{8}$	5.60	2.188	2380	1330	1900	730	1760	1690	1730	990
	$\frac{3}{4}$	4.67	2.625	3360	1480	2600	810	2440	1910	2440	1100
	$\frac{7}{8}$	4.00	3.063	4290	1630	3210	890	3180	2100	3120	1210
	1	3.50	3.500	5120	1770	3740	970	3800	2290	3720	1320
4	$\frac{1}{2}$	8.00	2.000	1530	1180	1220	700	1130	1130	1110	960
	$\frac{5}{8}$	6.40	2.500	2380	1510	1910	830	1770	1770	1730	1130
	$\frac{3}{4}$	5.33	3.000	3420	1690	2710	920	2540	2170	2490	1260
	$\frac{7}{8}$	4.57	3.500	4560	1860	3510	1020	3380	2400	3310	1380
	1	4.00	4.000	5600	2030	4190	1110	4160	2620	4070	1510

Table 7.17 (*Continued*)

Length of bolt in main member l	Diameter of bolt d	$\frac{l}{d}$	Projected area of bolt $A = ld$	Douglas fir–larch (dense) or southern pine (dense) $\parallel$ to grain P	$\perp$ to grain Q	Eastern hemlock–tamarack, California redwood (open grain), hem-fir, western hemlock $\parallel$ to grain P	$\perp$ to grain Q	Oak, red and white $\parallel$ to grain P	$\perp$ to grain Q	Douglas fir, Douglas fir south, southern pine (open grain) $\parallel$ to grain P	$\perp$ to grain Q
$4\frac{1}{2}$	$\frac{5}{8}$	7.20	2.813	2380	1640	1910	930	1770	1770	1730	1270
	$\frac{3}{4}$	6.00	3.375	3430	1900	2750	1040	2550	2360	2500	1410
	$\frac{7}{8}$	5.14	3.938	4640	2090	3660	1140	3450	2690	3370	1560
	1	4.50	4.500	5910	2280	4540	1250	4390	2940	4300	1700
	$1\frac{1}{4}$	3.60	5.625	8170	2670	5990	1460	6070	3450	5940	1990
$5\frac{1}{2}$	$\frac{5}{8}$	8.80	3.438	2380	1650	1910	1010	1770	1770	1730	1380
	$\frac{3}{4}$	7.33	4.125	3430	2200	2750	1270	2540	2490	2490	1720
	$\frac{7}{8}$	6.29	4.813	4680	2550	3750	1400	3470	3110	3400	1900
	1	5.50	5.500	6080	2790	4860	1520	4510	3560	4420	2080
	$1\frac{1}{4}$	4.40	6.875	9160	3260	7000	1780	6800	4210	6660	2430
$7\frac{1}{2}$	$\frac{5}{8}$	12.00	4.688	2380	1480	1910	950	1770	1560	1730	1290
	$\frac{3}{4}$	10.00	5.625	3430	2130	2750	1320	2550	2260	2490	1800
	$\frac{7}{8}$	8.57	6.563	4670	2840	3750	1730	3460	3110	3390	2360
	1	7.50	7.500	6100	3550	4890	2060	4520	3980	4430	2800
	$1\frac{1}{4}$	6.00	9.375	9540	4450	7640	2430	7080	5530	6930	3310
$9\frac{1}{2}$	$\frac{3}{4}$	12.67	7.125	3430	1920	2740	1250	2550	2020	2490	1700
	$\frac{7}{8}$	10.86	8.313	4680	2660	3750	1660	3470	2810	3400	2260
	1	9.50	9.500	6100	3460	4890	2130	4520	3730	4430	2900
	$1\frac{1}{4}$	7.60	11.875	9530	5210	7640	3040	7070	5820	6920	4140
	$1\frac{1}{2}$	6.33	14.250	13740	6480	11000	3540	10200	7840	9990	4830
$11\frac{1}{2}$	$\frac{7}{8}$	13.14	10.062	4680	1980	3750	1590	3470	2550	3400	2170
	1	11.50	11.500	6090	3240	4900	2040	4520	3420	4430	2780
	$1\frac{1}{4}$	9.20	14.375	9530	5110	7640	3130	7070	5520	6930	4270
	$1\frac{1}{2}$	7.67	17.250	13730	7200	10960	4210	10190	8000	9980	5740
$13\frac{1}{2}$	1	13.50	13.500	6090	2410	4900	1970	4520	3120	4430	2680
	$1\frac{1}{4}$	10.80	16.875	9530	4860	7660	3030	7070	5140	6920	4130
	$1\frac{1}{2}$	9.00	20.250	13730	7070	11020	4340	10190	7680	9980	5920

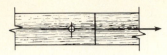

Single bolt loaded parallel to grain

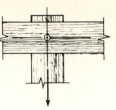

Single bolt loaded
perpendicular to grain

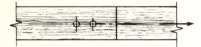

Single row of bolts loaded parallel to grain

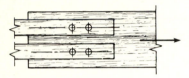

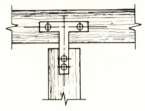

Multiple rows of bolts loaded
parallel to grain

Steel gusset plate connection

Figure 7.27 Joints with allowable loads fully developed [7].

other than the three-member joints of Table 7.17 appropriate length and load factors from Table 7.19 are applied. Recommended spacing and end and edge distances are given in Table 7.20 and illustrated in Fig. 7.23. When steel plates are used, the allowable load parallel to the grain can be increased 25 percent.

The net area at the critical section of a bolted joint is equal to the full cross-sectional area of the timber less the projected area of bolt holes at that section. Where bolts are staggered, parallel-to-grain spacing should be greater than eight bolt diameters.

Table 7.18 Adjustment of bolt-load values for service and seasoning conditions [7]

In service	Condition of lumber when installed	Percentage bolt-load values
Dry	Seasoned	100
Seasoned	Joints, unseasoned (above FSP)	100
	All other	40
Exposed to weather	Seasoned or unseasoned	75
Always wet	Seasoned or unseasoned	67

Table 7.19 Length and load factors of bolted joints [7]

Used in conjunction with Table 7.17

Type of joint	Length and load factors
Three-member: 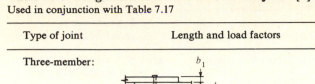	
$b_1 = b_2 = \dfrac{b}{2}$	Use $l = b$
$b_1, b_2 > \dfrac{b}{2}$	Use $l = b$
$b_1 \le b_2 < \dfrac{b}{2}$	Use $l = 2b_1$
Two-member: 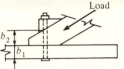	
Wood side member $b_1 \le b_2$	Use one-half of allowable load for $l = 2b_1$
Multiple-member: 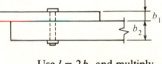	
$b_3 < b_1, b_2 \text{ or } b_4$	Use $l = 2b_3$ and multiply by $\frac{1}{2}$ and number of shear planes in joint
Two-member: 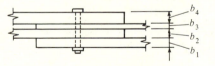	
Load acting at angle with axis of bolt	Allowable load component acting perpendicular to bolt equals one-half that for $l = 2b$ or $l = 2b_2$ whichever is smaller

Table 7.20 Recommended spacing and distance values for bolts [7]

See Figs 7.22 and 7.23	Parallel to grain loading	Perpendicular to grain loading
A, center-to-center spacing	Minimum of 4 times bolt diameter	4 times bolt diameter unless design load is less than bolt bearing capacity of side members; then spacing may be reduced proportionately
Staggered bolts	Adjacent bolts considered to be placed at critical section unless spaced at a minimum of 8 times bolt diameter	Staggering not permitted unless design load is less than bolt bearing capacity of side members
B, row spacing	Minimum of $1\frac{1}{2}$ times bolt diameter	$2\frac{1}{2}$ times bolt diameter for l/d ratio of 2; 5 times bolt diameter for l/d ratios of 6 or more; use straight-line interpolation for l/d between 2 and 6
	Spacing between rows paralleling a member not to exceed 5 in unless separate splice plates used for each row	
C, end distance	In tension, 7 times bolt diameter for softwoods, 5 times bolt diameter for hardwoods In compression, 4 times bolt diameter	Minimum of 4 times bolt diameter; when members abut at a joint (not illustrated), evaluate strength also as a beam supported by fastenings
D, edge distance	$1\frac{1}{2}$ times bolt diameter, except that for l/d ratios of more than 6, use one-half row spacing B	Minimum of 4 times bolt diameter at edge toward which load acts; minimum of $1\frac{1}{2}$ times bolt diameter at opposite edge

Example 7.15 Two seasoned 6 by 6 S4S southern pine (no. 1 SR) members are to be connected by gusset plates as shown in Fig. E7.15. Determine the capacity of the vertical member and connection.

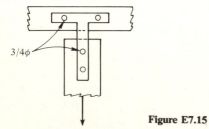

$3/4\phi$

Figure E7.15

SOLUTION Full allowable loads are permitted. Bolts acting perpendicular to grain will control;

$$\frac{l}{d} = \frac{5\frac{1}{2}}{\frac{3}{4}} = \frac{22}{3}$$

from Table 7.17, 2200 lb/bolt. No increase for metal gusset plates.

$$\text{Allowable load for connection} = (2)(2200) = 4400 \text{ lb}$$

Consider critical section of member (Tables A.1 and A.2):

$$\text{Net area} = 30.25 - 4.125 = 26.125 \text{ in}^2$$
$$\text{Allowable load in member} = (26.125)(875) = 22{,}860 \text{ lb}$$

Bolt connection controls.

Example 7.16 Design a splice for a seasoned S4S, 4 by 12 southern pine select structural grade member using metal side plates and carrying 40,000 lb under normal loading conditions (Fig. E7.16a).

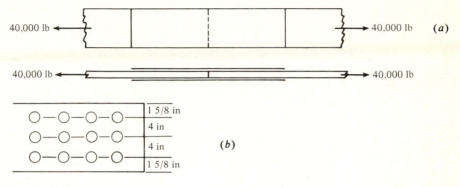

Figure E7.16

SOLUTION Try $\frac{3}{4}$-in bolts

$$l = 3\tfrac{1}{2}$$

$$P = 3360 \text{ lb/bolt} \qquad \text{Table 7.17}$$

For steel side plates

$$P = (3360)(1.25) = 4200 \text{ lb/bolt}$$

$$\text{No. of bolts} = \frac{40{,}000}{4200} = 9.5$$

Try 12. Consider three rows of four bolts for group action (Fig. E7.16b). From Table 7.13

$$A_1 = 39.38 \text{ in}^2 \qquad A_2 = (11.5)\left(\tfrac{3}{8}\right)(2) = 8.6 \text{ in}^2$$

$$\frac{A_1}{A_2} = 4.60$$

Consider the group as being made up of three rows. From Table 7.13 the modification factor is 0.92. Then the load-carrying capacity of the group is

$$(12)(4200)(0.92) = 46{,}370 \text{ lb}$$

Use 12 bolts as detailed. Consider the net section

$$39.38 - (3)(2.625) = 31.50 \text{ in}^2$$

The allowable load on the member is

$$(31.5)(1400) = 44{,}100 > 40{,}000 \text{ lb} \qquad \text{OK}$$

The connection and member are adequate.

Table 7.21 Allowable loads for normal loading conditions for one split ring and bolt in single shear [4]

Split-ring diam, in	Bolt diam, in	No. of faces of piece with connectors on same bolt	Net thickness of lumber, in	Loaded parallel to grain (0°)					Loaded perpendicular to grain (90°)						
				Min edge distance, in	Allowable load per connector unit and bolt, lb, for wood group				Edge distance, in		Allowable load per connector unit and bolt, lb, for wood group			
					A	B	C	D	Min unloaded	Loaded	A	B	C	D
$2\frac{1}{2}$	$\frac{1}{2}$	1	1 min	$1\frac{3}{4}$	2630	2270	1900	1640	$1\frac{3}{4}$	$1\frac{3}{4}$ min	1580	1350	1130	970
										$2\frac{3}{4}$ +	1900	1620	1350	1160
			$1\frac{1}{2}$ +		3160	2730	2290	1960		$1\frac{3}{4}$ min	1900	1620	1350	1160
										$2\frac{3}{4}$ +	2280	1940	1620	1390
		2	$1\frac{1}{2}$ min		2430	2100	1760	1510		$1\frac{3}{4}$ min	1460	1250	1040	890
										$2\frac{3}{4}$ +	1750	1500	1250	1070
			2 +		3160	2730	2290	1690		$1\frac{3}{4}$ min	1900	1620	1350	1160
										$2\frac{3}{4}$ +	2280	1940	1620	1390

			2370	2030	1700	1470
		2¾ min	2840	2440	2040	1760
		3¾ +	3490	2990	2490	2150
		2¾ min	4180	3590	2990	2580
		3¾ +	2480	2040	1700	1470
		2¾ min	2980	2450	2040	1760
		3¾ +	2870	2470	2050	1770
		2¾ min	3440	2960	2460	2120
		3¾ +	3380	2900	2410	2080
		2¾ min	4050	3480	2890	2500
		3¾ +	3560	3050	2540	2190
		2¾ min	4270	3660	3050	2630
		3¾ +				

	2¾			
	4090	3510	2920	2520
1 min	6020	5160	4280	3710
1½ +	4110	3520	2940	2540
1½ min	4950	4250	3540	3050
2	5830	5000	4160	3600
2½	6140	5260	4380	3790
3 +				

4 ¾ 1 2 2¾

209

Table 7.22 Allowable loads for one shear plate and bolt in single shear [4]

Shear-plate diam, in	Bolt diam, in	No. of faces of connectors on same piece with connectors on same bolt	Net thickness of lumber, in	Loaded parallel to grain (0°) edge dist, in	Loaded parallel to grain (0°) — Allowable load per connector unit and bolt, lb, for wood group A	B	C	D	Edge distance, in Min unloaded	Loaded perpendicular to grain (90°) Edge distance, in Loaded	Allowable load per connector unit and bolt, lb, for wood group A	B	C	D
$2\frac{5}{8}$	$\frac{3}{4}$	1	$1\frac{1}{2}$ min	$1\frac{3}{4}$	3110†	2670	2220	2010	$1\frac{3}{4}$	$1\frac{1}{4}$ min	1810	1550	1290	1110
										$2\frac{3}{4}$ +	2170	1860	1550	1330
		2	$1\frac{1}{2}$ min		2420	2080	1730	1500		$1\frac{1}{4}$ min	1410	1210	1010	870
										$2\frac{3}{4}$ +	1690	1450	1210	1040
			2		3190†	2730	2270	1960		$1\frac{1}{4}$ min	1850	1590	1320	1140
										$2\frac{3}{4}$ +	2220	1910	1580	1370
			$2\frac{1}{2}$ +		3330†	2860	2380	2060		$1\frac{1}{4}$ min	1940	1660	1380	1200
										$2\frac{3}{4}$ +	2320	1990	1650	1440
4	$\frac{3}{4}$ or $\frac{7}{8}$	1	$1\frac{1}{2}$ min	$2\frac{3}{4}$	4370	3750	3130	2700	$2\frac{3}{4}$	$2\frac{3}{4}$ min	2540	2180	1810	1550
										$3\frac{3}{4}$ +	3040	2620	2170	1860
			$1\frac{3}{4}$ +		5090†	4360	3640	3140		$2\frac{3}{4}$ min	2950	2530	2110	1810
										$3\frac{3}{4}$ +	3540	3040	2530	2200

	1¾ min	2¾				2¾ min			
						3¾ +			
2	3390	2910	2420	2090		1970	1680	1400	1250
						2360	2020	1680	1410
2	3790	3240	2700	2330		2200	1880	1570	1360
						2640	2260	1880	1630
2½	4310	3690	3080	2660		2500	2140	1780	1540
						3000	2550	2140	1850
3	4830	4140	3450	2980		2800	2400	2000	1720
						3360	2880	2400	2060
3½ +	5030†	4320	3600	3110		2920	2500	2090	1800
						3500	3000	2510	2160

† Loads exceed those permitted by note 2 but are needed for proper determination of loads for other angles of load to grain. Note 2 limitations apply in all cases.

NOTES: 1. For metal side plates, tabulated loads apply except that, for 4-in shear plates, the parallel-to-grain (not perpendicular) loads for wood side plates shall be increased 18, 11, 5 and 0% for groups A, B, C and D woods, respectively, but loads shall not exceed those permitted by note 2.

2. The allowable loads for all loadings, except wind, shall not exceed 2900 lbs for $2\frac{5}{8}$-in shear plates; 4970 and 6760 lb for 4-in shear plates with $\frac{3}{4}$- and $\frac{7}{8}$-in bolts, respectively; or, for wind loading, shall not exceed 3870, 6630, and 9020 lb, respectively.

3. Metal side plates, when used, shall be designed in accordance with accepted metal practices. For steel, the following unit stresses, in pounds per square inch, are suggested for all loadings except wind: net section in tension, 20,000; shear, 12,500; double-shear bearing, 28,125; single-shear bearing, 22,500; for wind, these values may be increased one-third; if bolt threads are in bearing, reduce the preceding shear and bearing values by one-ninth.

211

Table 7.23 Projected area of connectors and bolts [4]

Connector				Member thickness, in									
Type	Size, in	Bolt diam, in	Placement of connectors, faces	$1\frac{1}{2}$	$2\frac{1}{2}$	$3\frac{1}{8}$	$3\frac{1}{2}$	$5\frac{1}{8}$	$5\frac{1}{2}$	$6\frac{3}{4}$	$8\frac{3}{4}$	$10\frac{3}{4}$	$12\frac{1}{4}$
				Total projected area of connectors and bolts, in²									
Split rings:													
1	$2\frac{1}{2}$	$\frac{1}{2}$	1	1.73	2.29	2.65	2.86	3.77	3.98	4.69	5.81	6.94	7.78
			2	2.62	3.18	3.54	3.75	4.66	4.87	5.58	6.70	7.82	8.67
2	4	$\frac{3}{4}$	1	3.05	3.86	4.37	4.68	6.00	6.30	7.32	8.94	10.56	11.79
			2	4.88	5.69	6.20	6.51	7.83	8.13	9.15	10.77	12.40	13.62
Shear plates:													
1	$2\frac{5}{8}$	$\frac{3}{4}$	1	2.03	2.85	3.35	3.66	4.98	5.28	6.30	7.92	9.55	10.77
			2	2.84	3.66	4.16	4.47	5.79	6.09	7.11	8.73	10.36	11.58
1 LG	$2\frac{5}{8}$	$\frac{3}{4}$	1	1.91	2.72	3.23	3.53	4.85	5.16	6.18	7.80	9.43	10.64
			2	2.60	3.41	3.92	4.22	5.54	5.85	6.87	8.49	10.11	11.33
2	4	$\frac{3}{4}$	1	3.26	4.07	4.58	4.89	6.21	6.51	7.53	9.15	10.78	12.00
			2	—	6.11	6.62	6.93	8.25	8.55	9.57	11.19	12.82	14.04
2-A	4	$\frac{7}{8}$	1	3.37	4.30	4.89	5.24	6.77	7.12	8.29	10.16	12.04	13.45
			2	—	6.26	6.85	7.20	8.73	9.08	10.25	12.12	14.00	15.41

Table 7.24 Connector spacing and end distances for parallel-to-grain loading with corresponding percentages of tabulated loads [4]

Connector and diameter, in	Spacing parallel to grain†		Spacing perpendicular to grain		End distance		
	Spacing, in	Percentage of tabulated load	Minimum, in	Percentage of tabulated load	Tension member, in	Compression member, in	Percentage of tabulated load
Split rings:							
$2\frac{1}{2}$	$6\frac{3}{4}$	100	$3\frac{1}{2}$	100	$5\frac{1}{2}$	4	100
$2\frac{1}{2}$	$3\frac{1}{2}$	50	$3\frac{1}{2}$	100	$2\frac{3}{4}$	$2\frac{1}{2}$	62.5
4	9	100	5	100	7	$5\frac{1}{2}$	100
4	5	50	5	100	$3\frac{1}{2}$	$3\frac{1}{4}$	62.5
Shear plates:							
$2\frac{5}{8}$	$6\frac{3}{4}$	100	$3\frac{1}{2}$	100	$5\frac{1}{2}$	4	100
$2\frac{5}{8}$	$3\frac{1}{2}$	50	$3\frac{1}{2}$	100	$2\frac{3}{4}$	$2\frac{1}{2}$	62.5
4	9	100	5	100	7	$5\frac{1}{2}$	100
4	5	50	5	100	$3\frac{1}{2}$	$3\frac{1}{4}$	62.5

† No reduction in end distance is permitted for compression members loaded parallel to grain.

7.10 TIMBER CONNECTORS

7.10.1 Introduction

Shear developers, which act with a bolt or other fastener in the transfer of load from one wood piece to another in a timber joint, have been in use since the turn of the century. Many variations have been patented and used. Two common ones are *split rings* and *shear plates*. National Design Specifications [4] cover their use. They are illustrated in Fig. 7.20.

7.10.2 Design Considerations

The density of wood affects the allowable load for timber connectors. The species groupings are given in Table 7.11. The allowable loads in Tables 7.21 and 7.22 apply for one timber connector unit which consists of (1) one split ring with its bolt in single shear or (2) two shear plates used back-to-back in the contact faces of a timber-to-timber joint with their bolts in single shear, or (3) one shear plate with its bolt in single shear in conjunction with a metal plate.

The critical or net section of a connector joint will probably be at the centerline of a bolt and connector and is equal to the full cross-sectional area of the timber less the projected area of the connector and that portion of the projected area of the bolt not within the area of the connector. Table 7.23 gives the projected area of connectors and bolts. For staggered rows of connectors the parallel-to-grain spacing should exceed the diameter of the connectors to avoid the more complicated and less efficient critical section which passes through adjoining connectors. Appropriate percentages of tabulated loads (Tables 7.21 and 7.22) are given in Table 7.24 (see Figs. 7.22 and 7.23). Adjustments of allowable connector loads for different duration of loads are given in Fig. 6.20.

National Design Specifications do not provide for the use of connectors in end grain. Experience has shown, however, that connectors installed in the end grain of a member can be designed for an allowable load of 60 percent of that loaded perpendicular to the grain [6].

Example 7.17 Design a spliced tension connection for the following conditions:

Timber: group B species, $F_t = 1200$ lb/in^2 seasoned and dry, 4 by 6 main member, 2 by 6 side members
Connectors: $2\frac{1}{2}$-in split ring
Design load: 15,000 lb, normal duration

SOLUTION Check net section. Table 7.23 projected area connector and bolt is 1.73 in^2.

$$\text{Net area side member} = (8.25 - 1.73)(2) = 13.04 \text{ in}^2$$

$$\text{Net area main member} = 19.25 - 3.75 = 15.50 \text{ in}^2$$

$$\text{Allowable load} = (13.04)(1200) = 15,600 \text{ lb} > 15,000 \text{ lb} \qquad \text{OK}$$

Connector spacing details From Tables 7.21 and 7.24,

End distance $5\frac{1}{2}$ in   Edge distance $2\frac{3}{4}$ in   Spacing $6\frac{3}{4}$ in

Number of connectors From Table 7.21 the full allowable load is 2730 lb/bolt.

$$\text{No. of connectors required} = \frac{15,000}{2730} = 5.4$$

Try three on each side and check group action.

Group action Two rows of three.

$$\text{Area of side members } A_2 = (2)(8.25) = 16.50 \text{ in}^2$$

$$\text{Area of main member } A_1 = 19.25 \text{ in}^2$$

$$\frac{A_2}{A_1} = 0.86$$

since $A_1/A_2 > 1.0$. Table 7.12 interpolates between $A_1/A_2 = 0.5$ and $A_1/A_2 = 1.0$. Use factor of 0.97.

$$\text{Allowable load per connector} = (0.97)(2730) = 2648$$

$$\text{Allowable load} = (2)(3)(2648) = 15,890 \text{ lb} > 15,000 \text{ lb} \qquad \text{OK}$$

7.11 SUMMARY

This chapter covers the design of steel and timber tension members. This requires not only providing sufficient cross-sectional area

$$A_{\text{net}} = \frac{P}{F_t}$$

but also designing the connection, which involves possible failure of the fastener.

The most common steel tension members are rods, angles, and plates. Angles and plates are normally connected to gusset plates by rivets, bolts, or welds.

Rivets and bolts can be used in bearing-type connections, which mean considering the shear and bearing capacity of the fastener

$$P_b = ndtF_b \qquad P_s = 2n\frac{\pi d^2}{4}F_s$$

Bolted connections can also be used to develop a friction-type connection by putting the bolt in tension. The connection is designed on the basis of an allowable shear stress developed by friction.

Welded connections use fillet, butt, and plug welds. Fillet welds, the most common, are designed to fail in shear at the minimum or throat distance. A number of factors have to be taken into account in determining the size and length of weld, including the strength and thickness of the member and gusset plate.

The design of timber tension members has the added complication of the directional properties of wood, i.e., strength parallel and perpendicular to the grain. The most popular fasteners are nails, bolts, and connectors. The latter two reduce the cross-sectional area of the member. The load-carrying capacity of most timber connectors has been determined experimentally, and while the parameters influencing these values have been determined, most design values are available in tables.

PROBLEMS

7.1 Determine the allowable force P which can be applied to the riveted connection shown in Fig. P7.1 if A36 steel plates and A502 G1 rivets are used.

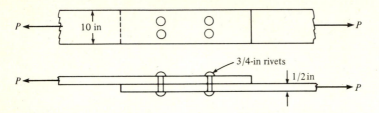

Figure P7.1

7.2 Determine the number of $\frac{7}{8}$-in A325 bolts required for the connection in Fig. P7.2. The plates are of A36 steel.

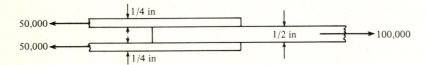

Figure P7.2

7.3 The butt splice shown in Fig. P7.3 is to be designed for a tensile load of 120 kips. Determine the required dimensions of the plate and connection using $\frac{7}{8}$-in A502 G1 rivets. The plates are A36 steel.

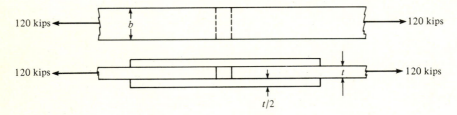

Figure P7.3

7.4 A truss member (Fig. P7.4) consists of two A36 10-in channels at 15.3 lb/ft connected to a $\frac{1}{2}$-in gusset plate. How many $\frac{7}{8}$-in A490 bolts are required to develop the full tensile capacity of the member? The channels are 0.240 in thick.

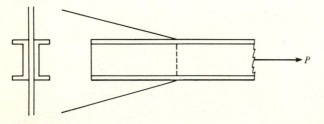

Figure P7.4

7.5 Fillet welds as indicated in Fig. P7.5 are used in the connection. Determine the maximum load which can be applied.

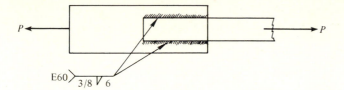

Figure P7.5

7.6 Determine the maximum load which can be applied to the member in Fig. P7.6.

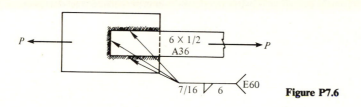

Figure P7.6

7.7 Design a connection using fillet welds to develop the full strength of the A441 bar in Fig. P7.7.
(*a*) Select the proper electrode and fillet-weld size and determine the length L.
(*b*) Redesign the connection to reduce the overlap length L by using plug welds in addition to the fillet welds.

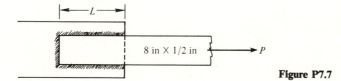

Figure P7.7

7.8 Determine the maximum force the $\frac{1}{2}$-in plate of A36 steel can carry in tension (Fig. P7.8). Rivet diameter is $\frac{7}{8}$ in.

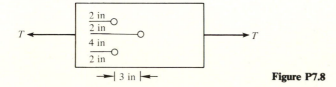

Figure P7.8

7.9 A pair of 6- by 4- by $\frac{1}{2}$-in angles constitute a tension member of a truss (Fig. P7.9). They are attached to a $\frac{5}{8}$-in gusset plate by two rows of $\frac{3}{4}$-in A325 bolts in the 6-in legs of the angle. Bolt pitch is 2 in. Angles are A36 steel. Determine the maximum load that can be carried according to AISC specifications.

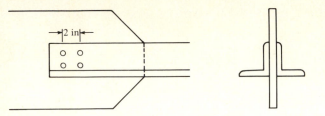

Figure P7.9

7.10 The tension member of a roof truss carries a load of 76 kips. Gusset plates are $\frac{3}{8}$ in thick. Use A36 steel.

  (*a*) Select a pair of angles and detail the welds required.

  (*b*) Select a pair of angles and determine the number of $\frac{3}{4}$-in rivets required. Detail connection.

7.11 The tension member shown in Fig. P7.11 is made up of 7- by $\frac{1}{2}$-in plates of A36 steel connected by $\frac{7}{8}$-in A325 bolts. Determine the maximum load that can be carried according to AISC specifications.

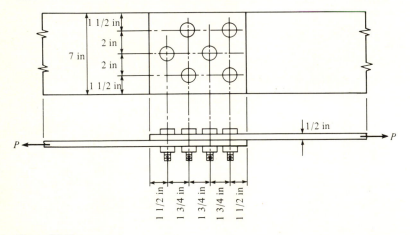

Figure P7.11

7.12 A tension member of a truss is composed of two angles of A36 steel and carries a load of 83 kips. What size angles are required and how many $\frac{7}{8}$-in-diameter rivets (A502 grade 1) are required?

7.13 An overhead bin is suspended by four steel rods. The maximum load carried by each rod is 20 kips. Use A36 steel. What diameter should be used (*a*) when the rods have upset ends and (*b*) when the rods do not have upset ends?

7.14 A 2- by 6-in wind brace is nailed to a 6- by 6-in southern pine post (Fig. P7.14). The wind load on the brace is 1200 lb. Determine the size and number of nails required.

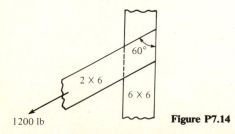

Figure P7.14

7.15 Determine the load-carrying capacity of the hook attached to a southern pine beam by four $\frac{1}{2}$-in lag screws 4 in long (Fig. P7.15).

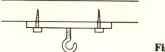

Figure P7.15

7.16 A 2 by 6 is attached to an 8 by 8 post with four $\frac{1}{2}$-in lag bolts 7 in long (Fig. P7.16). What is the allowable load P? Wood is Douglas fir no. 1 grade.

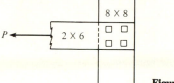

Figure P7.16

7.17 A 6 by 8 southern pine select structural grade timber is spliced with 3 by 8 side members and ten $\frac{3}{4}$-in bolts. What is the capacity of the structural member?

7.18 Determine the number of $\frac{3}{4}$-in bolts required to carry the 10-kip load shown in Fig. P7.18. Show spacing details. Material is Douglas fir no. 1 grade.

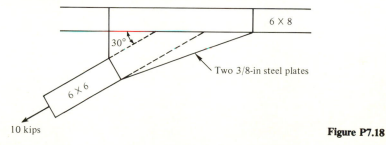

6 X 8

30°

Two 3/8-in steel plates

6 X 6

10 kips

Figure P7.18

7.19 Determine the safe load for the bolted connection in Fig. P7.19. Material is Douglas fir no. 1 grade used under continuously damp conditions. Loading is dead plus snow load.

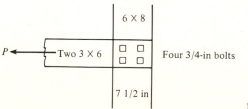

6 X 8

P ← Two 3 X 6   Four 3/4-in bolts

7 1/2 in

Figure P7.19

7.20 Determine the safe normal load P that can be developed by the illustrated splice. Material is southern pine no. 1 grade. Detail joint to provide for maximum allowable load on split rings.

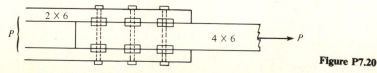

2 X 6

P

4 X 6   → P

Figure P7.20

7.21 Determine the size of the members and the number and placement of 4-in split rings required for the heal joint in Fig. P7.21. Timber is Douglas fir select structural grade; normal loading; thickness of all pieces is $2\frac{1}{2}$ in.

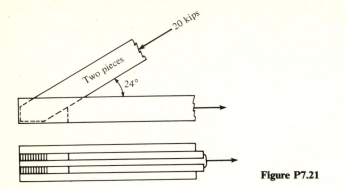

Figure P7.21

REFERENCES

1. American Institute of Steel Construction, "Steel Construction Manual," 8th ed., Chicago, 1978.
2. E. Gaylord and C. Gaylord, "Design of Steel Structures," McGraw-Hill, New York, 1972.
3. G. Gurfinkel, "Wood Engineering," Southern Forest Products Association, New Orleans, 1973.
4. National Forest Products Association, National Design Specifications for Wood Construction, Washington, 1977.
5. Canadian Standards Association, "Code for the Engineering Design of Wood," Rexdale, Ontario, 1976.
6. U.S. Department of Agriculture, "Wood Handbook: Wood as an Engineering Material," Agriculture Handbook 72, Washington, 1974.
7. American Institute of Timber Construction, "Timber Construction Manual," 2d ed., Englewood, Colo., 1974.
8. American Society of Mechanical Engineers, American Standards Association, B1.1-1960, New York, 1960.

CHAPTER 8

TRUSSES

8.1 INTRODUCTION

A *truss* can be defined as a structure composed of a number of bars all lying in one plane and connected at their ends by pins in such a manner as to form a rigid framework. To facilitate the calculation of forces in a truss due to applied loads the following ideal conditions are assumed to exist: (1) the members are connected together at their ends by frictionless pin joints; (2) loads and reactions are applied only at the joints of the truss; (3) the centroidal axis, i.e., an axis perpendicular to the cross section and passing through its center, of each member is straight, coincides with a line joining the centers of the joints, and lies in a plane containing the lines of action of the loads and reactions. The stresses determined under these circumstances are called *primary stresses*. It is a physical impossibility for an actual truss to fulfill all these assumed ideal conditions.

The pins of an actual pin-jointed truss are never frictionless, and most modern trusses have bolted or welded joints. This prevents angle change at the joints and consequently induces transverse loads and couples at the ends of the members. In addition, the weight of the member is distributed along the length, contributing to the bending of the member. These sources of stress are minimized by good detailing practice, which ensures that the centroidal axes coincide with joint centerlines and the members are relatively slender. These differences between the stresses in the members of an actual truss and the primary stresses computed for the corresponding ideal truss are called *secondary stresses*. The primary stresses computed on the basis of an ideal truss are usually satisfactory for practical design purposes, the secondary stresses being neglected.

8.2 ARRANGEMENT OF TRUSS MEMBERS

Members of a truss must be hinged together to form a rigid framework. A framework is said to be rigid if there is no relative movement other than that caused by the elastic deformation of the members of the framework.

In order to develop a rigid framework it is necessary to select three joints that do not lie along a straight line and to connect them by three bars pinned together to form a triangle. Each additional joint can then be formed by connecting two bars to each other and to two joints of the previous structure. This new joint and the two joints to which it is connected cannot lie in a straight line. Each of the trusses shown in Fig. 8.1 has been formed in this manner by starting with a rigid triangle *abc* and using two more bars to connect each of the other joints in alphabetical order.

Trusses arranged in this manner are called *simple trusses*. The entire framework can then be supported like a beam. When two or more simple trusses are connected together to form a rigid framework, the composite framework is called a *compound truss*. Examples of compound trusses are given in Fig. 8.2.

8.3 STABILITY AND DETERMINACY

In discussing the arrangement of members in a simple truss, it was shown that a rigid truss is formed by using three bars to connect three joints together in the form of a triangle and then using two bars to connect each additional joint to the framework already constructed.

As a result, to form a rigid simple truss of n joints, it is necessary to use the three bars of the original triangle plus two additional bars for each of the remaining $n - 3$ joints. If b denotes the total number of bars required, then

$$b = 3 + (2)(n - 3) = 2n - 3 \tag{8.1}$$

This is the minimum number of bars required to form a rigid simple truss. More bars are unnecessary, and fewer bars result in a nonrigid unstable truss.

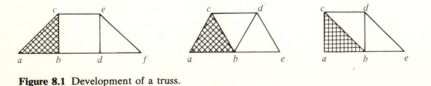

Figure 8.1 Development of a truss.

Figure 8.2 Compound trusses.

If a simple truss is supported in a manner equivalent to a hinge and roller, the structure is stable under a general condition of loading and the reactions are statically determinate.

If there are more than three reaction elements, the structure is statically indeterminate with respect to its reactions. If there are more than $2n - 3$ bars but only three reaction components, it is indeterminate with respect to the bar forces. If there is an excess of both bars and reaction elements, the structure is statically indeterminate with respect to both reactions and bar forces.

8.4 NOTATION AND SIGN CONVENTION

It is important to establish a definite notation and sign convention for designating the forces in the members of a truss. For an algebraic solution it is convenient to designate lower chord joints by L_0, L_1, L_2, etc., and correspond-

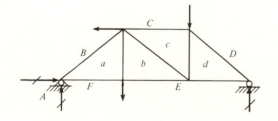

Sketch 8.1

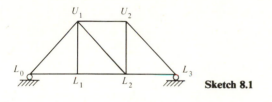

(a)

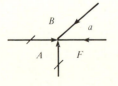

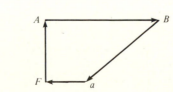

(b)

Sketch 8.2

ing upper chord joints by U_0, U_1, U_2, etc. (Sketch 8.1). Each member is then designated by the letters associated with the joints at the end of the member.

It is convenient to associate tension with a plus sign and compression with a minus sign. To facilitate this convention always assume an unknown force in a member to be tensile in character. Then if subsequent calculations give a negative sign, the member is in compression.

In solutions using a graphical method, Bow's notation (Sec. 4.15) is used to identify members and forces. In this system capital letters are placed in the spaces between external forces generally in a clockwise order around the truss. Lowercase letters are placed in each triangular panel of the truss (Sketch 8.2a). Each force and member will then be between two letters and is identified by them (Sketch 8.2b).

8.5 ANALYSIS OF TRUSSES

In order to design a truss to support a given condition of loading, it is necessary to determine the forces that must be carried by the members of the truss.

Either algebraic or graphic methods may be used in the analysis of trusses. The algebraic method consists essentially of isolating certain parts of a truss by cutting through some members and treating the forces in these cut members as external forces acting on a free body. If the force system acting on the free body is concurrent, as in a joint, two unknown forces can be determined. If the force system acting on the free body is nonconcurrent, three unknown forces can be found. Such a free body is usually obtained by cutting a section through three members and treating either side of the truss as a free body. As a result, the former procedure is known as the *method of joints* and the latter as the *method of sections*.

8.5.1 Method of Joints

The external reaction components are first determined by considering the entire truss as a free body. The two equations of equilibrium ($\Sigma F_x = 0$ and $\Sigma F_y = 0$) are then applied to the free-body diagram of each joint in succession in such a way that not more than two unknown bar forces are involved at any given joint. It is advisable to start at the left support and proceed joint by joint to the center of the truss and then to start at the right support and work back to the center. This makes it possible to check calculations at the center of the truss. It is also advisable to indicate on the truss diagram not only the force in the member but also the horizontal and vertical components, so that the equilibrium of any joint can be checked at a glance.

Example 8.1 Using the method of joints, determine the force in each member of the truss in Fig. E8.1a.

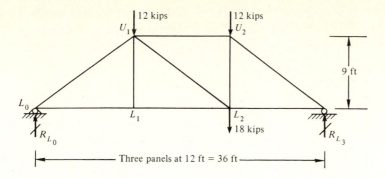

Figure E8.1*a*

SOLUTION By applying the equations of statics to the truss, R_{L_0} and R_{L_3} are determined to be 18 and 24 kips, respectively.

$$\overset{\frown}{\Sigma M_{L_0}} = 0: \qquad\qquad (12)(12) + (24)(12 + 18) - 36R_{L_3} = 0$$

$$R_{L_3} = 24 \text{ kips}$$

$$\downarrow \Sigma V = 0: \qquad\qquad 12 + 12 + 18 - 24 - R_{L_0} = 0$$

$$R_{L_0} = 18 \text{ kips}$$

$$\overset{\frown}{\Sigma M_{L_3}} = 0: \qquad (12)(18 + 12) + (12)(24) - (18)(36) = 0 \qquad \text{check}$$

Consider joint L_0 (Fig. E8.1*b*)

$$\uparrow \Sigma V = 0: \qquad\qquad \tfrac{3}{5}L_0U_1 + 18 = 0$$

$$L_0U_1 = \frac{(-18)(5)}{3} = -30 \text{ kips} \qquad \text{compression}$$

Note that the negative sign indicates compression because the force was assumed tensile.

$$\overrightarrow{\Sigma H} = 0: \qquad\qquad (4/5)(-30) + L_0L_1 = 0$$

$$L_0L_1 = 24 \text{ kips} \qquad \text{tension}$$

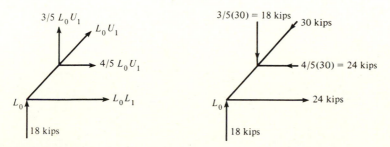

Figure E8.1*b*

Consider joint L_1 (Fig. E8.1c). Note that joint U_1 could not be considered next because it contains three unknowns (members U_1L_1, U_1L_2, and U_1U_2).

$\uparrow \Sigma V = 0$: $\qquad\qquad\qquad\qquad U_1L_1 = 0$

$\overrightarrow{\Sigma H} = 0$: $\qquad\qquad\qquad\qquad 24 - L_1L_2 = 0$

$$L_1L_2 = 24 \text{ kips} \qquad \text{tension}$$

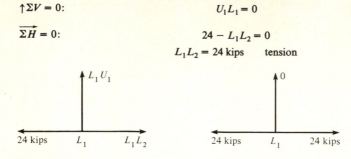

Figure E8.1c

Consider joint U_1 (Fig. E8.1d)

$\uparrow \Sigma V = 0$: $\qquad\qquad\qquad 18 - 0 - 12 - \frac{3}{5}U_1L_2 = 0$

$$U_1L_2 = (6)(5/3) = 10 \text{ kips} \qquad \text{tension}$$

$\overrightarrow{\Sigma H} = 0$: $\qquad\qquad\qquad 24 + (4/5)(10) + U_1U_2 = 0$

$$U_1U_2 = -32 \text{ kips} \qquad \text{compression}$$

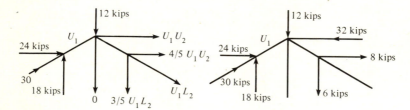

Figure E8.1d

Note that it is simpler to sum in the vertical direction first since there is only one unknown, which can be solved for directly. Consider joint L_3 (Fig. E8.1e)

$\uparrow \Sigma V = 0$: $\qquad\qquad\qquad\qquad 24 + \frac{3}{5}U_2L_3 = 0$

$$U_2L_3 = (-5/3)(24) = -40 \text{ kips} \qquad \text{compression}$$

$\overleftarrow{\Sigma H} = 0$: $\qquad\qquad\qquad\qquad (4/5)(-40) + L_2L_3 = 0$

$$L_2L_3 = 32 \text{ kips} \qquad \text{tension}$$

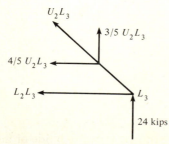

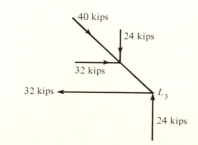

Figure E8.1e

Consider joint U_2 (Fig. E8.1f)

$\overrightarrow{\Sigma H} = 0$: $32 - 32 = 0$ check

$\downarrow\Sigma V = 0$: $12 + U_1 L_2 - 24 = 0$

$$U_2 L_2 = 12 \text{ kips} \quad \text{tension}$$

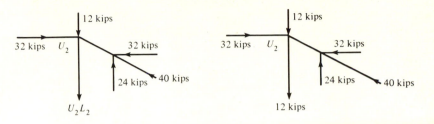

Figure E8.1 f

Consider joint L_2 (Fig. E8.1g)

$\overrightarrow{\Sigma H} = 0$: $24 + 8 - 32 = 0$ check

$\downarrow\Sigma V = 0$: $18 - 12 - 6 = 0$ check

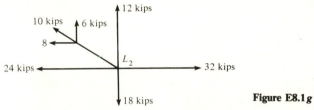

Figure E8.1 g

Figure E8.1h is an answer diagram on which the member forces are indicated. The arrows represent the direction of action of the member on the joint, not the joint on the member.

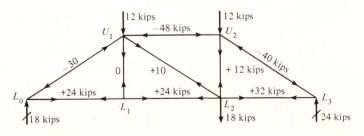

Figure E8.1 h

8.5.2 Method of Sections

The method of sections involves passing a section through three members and applying the three equations of equilibrium to the free body on either side of the section. In cutting a member the internal force becomes an external force in the free-body diagram. Usually the force system acting on such a free body is a nonconcurrent system in which three unknowns can be found. It is advisable always to assume that a cut member has a tensile force acting in it. As a result, if the answer has a negative sign associated with it, it is an indication of compression.

The method of sections is particularly useful when only the forces in some members are desired. Taking moments about the point of intersection of two of the unknowns will solve directly for the other unknown.

Example 8.2 Using the method of sections, determine the force in members U_1U_2, U_1L_2, and L_1L_2 in the truss shown in Fig. E8.2a.

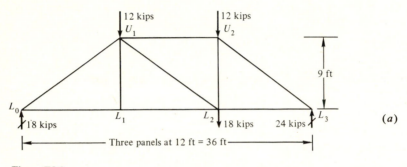

(a)

Figure E8.2

SOLUTION As in Example 8.1, reactions R_{L_0} and R_{L_3} are found to be 18.0 and 24.0 kips, respectively. A section is then passed through the truss cutting the members in question. It is not always possible to cut all the members in question with a single plane (Fig. E8.2b). Consider the left-hand portion of the truss as a free body. The internal forces in the cut members now become external forces which keep the free body in equilibrium. If these are indicated as tensile forces, the appearance of a negative sign with a force indicates that it is compressive (Fig. E8.2c). By taking moments about L_2 it is possible to determine the force in

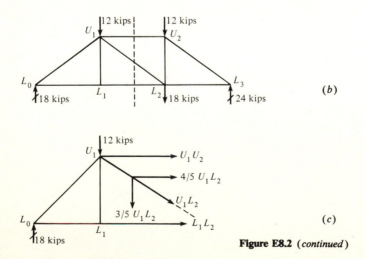

(b)

(c)

Figure E8.2 (*continued*)

U_1U_2 directly because forces U_1L_2 and L_1L_2 pass through the center of moments.

$$\overset{\curvearrowright}{\Sigma M_{L_2}} = 0: \qquad\qquad (18)(24) - (12)(12) + 9U_1U_2 = 0$$

$$U_1U_2 = \frac{-432 + 144}{9} = -32 \text{ kips} \qquad \text{compression}$$

By taking moments about U_1 it is possible to determine the force in L_1L_2 directly.

$$\overset{\curvearrowright}{\Sigma M_{U_1}} = 0: \qquad\qquad (18)(12) - 9L_1L_2 = 0$$

$$L_1L_2 = \frac{(18)(12)}{9} = 24 \text{ kips} \qquad \text{tension}$$

The force in member U_1L_2 can be determined by summing forces in the vertical direction.

$$\uparrow \Sigma V = 0: \qquad\qquad 18 - 12 - \tfrac{3}{5}U_1L_2 = 0$$

$$U_1L_2 = \frac{(6)(5)}{3} = 10 \text{ kips} \qquad \text{tension}$$

By summing forces in the horizontal direction a check can be made:

$$\overrightarrow{\Sigma H} = 0: \qquad\qquad -32 + 24 + \tfrac{4}{5}(10) = 0 \qquad \text{check}$$

In using the method of sections it is useful to remember that the sum of the moments of the components of a force about a point is equal to the moment of the force about that point. In taking moments of the components, they can be located anywhere along the line of action of the force providing they are considered to act at the same point.

8.5.3 The Graphical Method

Two steps are involved in the graphical method of analysis of trusses: (1) the external reaction components are determined by either the algebraic or graphical methods developed in Chap. 4, and (2) the forces in the members are obtained by a force diagram which involves the superposition of the individual force polygons for the concurrent force systems acting at each joint. In order to facilitate a systematic approach to the second phase, Bow's notation is used.

Example 8.3 Determine graphically the force in each member of the truss shown in Fig. E8.3a.

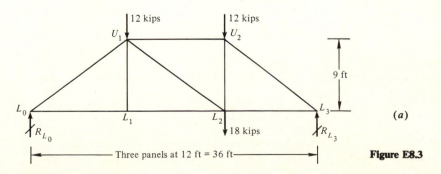

(a)

Figure E8.3

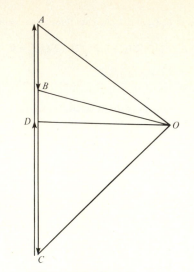

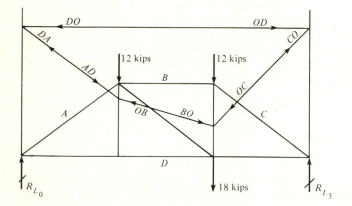

(b)

Figure E8.3 (*continued*)

SOLUTION The external reaction components are determined graphically in Fig. E8.3*b*. Note that the 12- and 18-kip loads have been combined in one force *BC*. The balanced external-force system acting on the truss includes 12, 12, 24, 18, and 18 kips named in a clockwise order around the truss. These forces and the members are identified by Bow's notation in Fig. E8.3*c*.

The force diagram shown in Fig. E8.3*j* is actually a composite of the closed force polygons representing the concurrent force systems of each joint. As in the method of joints, each force polygon can be closed only if there are no more than two unknowns in the concurrent system of forces.

Consider first joint L_0 (Fig. E8.3*d*). The force polygon is constructed by drawing first, to scale, a vector representing force $EA = 18$ kips. Forces of known magnitude and direction are drawn first, and then the force polygon is closed by working through the system in a clockwise manner. The direction of force Aa is known but not its sense (tension or compression). As a result, a line is drawn through point A of the force polygon parallel to the known direction of

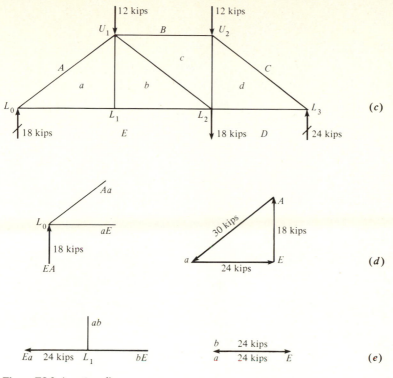

Figure E8.3 (*continued*)

force Aa. The force polygon is closed by a force aE which must pass through point E of the force polygon and intersect with vector Aa at a. The only solution for closing the force polygon in this manner is indicated in Fig. E8.3d. The values of forces Aa and aE can be obtained by scaling their lengths from the force polygon. The sense of the forces can be determined from the fact that the force EA acts up. As a result, again to close the polygon, force Aa acts down and to the left and force aE acts to the right, closing the force polygon. These are the directions of the forces as they act on joint L_0. Consequently, force Aa represents a compressive force in member L_0U_1, and force aE represents a tensile force in member L_0L_1.

This procedure is then repeated for joints L_1, U_1, U_2, L_2, and L_3. The force polygons are illustrated in Fig. E8.3e to j. It should be emphasized that each force polygon is started with the known forces and then completed by working clockwise through the force system.

In Fig. E8.3e, a and b are the same point, as this is the only way that the force polygon can be closed. This indicates that the force ab is zero.

Careful scrutiny of the individual force polygons will indicate that some forces are common to more than one force polygon. As a result, a more efficient diagram can be developed by using the initial force polygon of joint L_0 as a basis for the force polygon of joint L_1, and so on. Not all the forces are used in subsequent polygons. It is also important to realize that the direction of a force in a member changes as the joint at the other end of the member is considered. For example, force aE becomes force Ea when joint L_1 is considered instead of joint L_0.

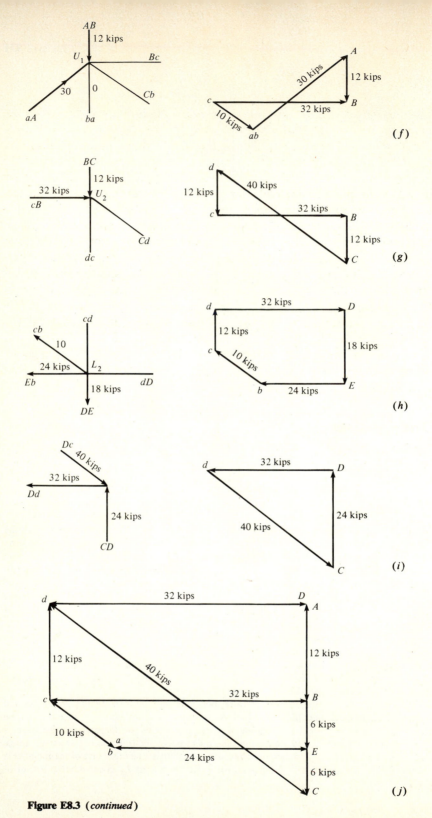

Figure E8.3 (continued)

8.6 ROOF TRUSSES

In building design where unobstructed space with a width of more than 40 to 50 ft is required, the roof is normally supported by roof trusses spaced 15 to 25 ft apart. These trusses normally rest on columns or on masonry walls along the sides of the building. If the span of the roof truss is small, less than 40 to 50 ft, the truss may be anchored to the wall at both ends or allowed to slide in slotted holes on a bearing plate at one end. For longer spans trusses should be hinged on one end and supported on rollers or rockers on the other end.

The distance between adjacent trusses is called a *bay*. *Purlins* are longitudinal beams which rest on the top chord, normally at the joints of the truss. The roof covering may rest directly on the purlins or on *rafters* which are supported by the purlins. The *span* of a truss is the horizontal distance between the supports. The *rise* is the vertical distance from the eaves to the ridge. The *pitch* is the ratio of the rise to span. Additional lateral bracing is normally required. This diagonal bracing, which runs from truss to truss, may be in the plane of the upper or lower chords. This terminology is illustrated in Fig. 8.3.

The selection of the type of roof truss generally depends upon the length of span, amount of loading, and choice of materials. Some common trusses are illustrated in Fig. 8.4.

8.6.1 Truss Loadings

The loads on roof trusses generally consist of dead, snow, and wind loads. The dead load includes (1) the weight of the roof covering, rafters, and purlins, (2)

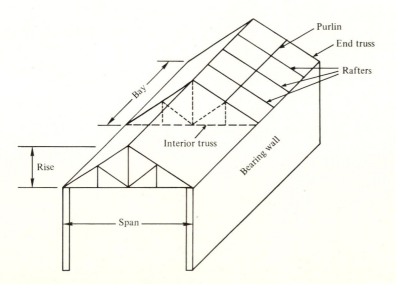

Figure 8.3 Truss terminology.

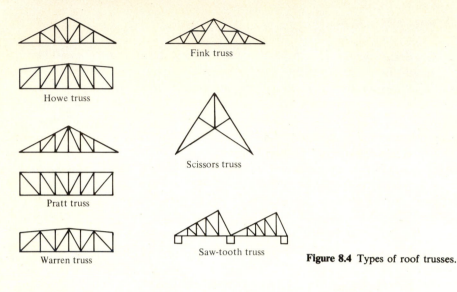

Figure 8.4 Types of roof trusses.

the weight of the bracing system, (3) the weight of the truss, and (4) ceiling and other suspended loads. Items 1 and 4 can be determined before the analysis, but items 2 and 3 have to be estimated and then reviewed after design calculations have been made. Fortunately, these items usually represent only a small portion of the total dead load.

The weight of a roof truss is often estimated by empirical formulas. Two are given here, and others can be found in handbooks.

$$W = \begin{cases} 0.5 + 0.075L & \text{for timber roof trusses} \\ 0.4 + 0.04L & \text{for steel roof trusses} \end{cases}$$

where W is the weight of the truss in pounds per square foot of horizontal surface and L is the span of the truss in feet.

Roof coverings include corrugated steel, asphalt and asbestos, shingles, tiles, slates, concrete slabs, and tar and gravel. The weight of these items can vary from 5 to 25 lb per square foot of surface. The weight of a plastered ceiling may be from 8 to 10 lb per square foot of horizontal surface. The weight of bracing systems vary from $\frac{1}{2}$ to $1\frac{1}{2}$ lb per square foot of roof surface.

8.6.2 Combinations of Loads

The primary purpose of analysis is to determine maximum and minimum forces occurring in truss members. As a result, consideration must be given to possible combinations of dead, snow, and wind loads. The usual combinations include

1. Dead plus full snow on both sides
2. Dead plus wind on either side
3. Dead plus half snow on both sides plus wind from either side
4. Dead plus full snow on the leeward side and wind on the windward side

5. Dead plus ice on both sides plus wind on either side
6. Dead plus ice on both sides plus full snow on the leeward side plus wind on the windward side

It is important that the designer use judgment in selecting the critical load combination. Because maximum winds occur only occasionally and for relatively short durations, most specifications allow a $33\frac{1}{3}$ percent increase in allowable stresses for cases where wind effect is included. Ordinarily, a consideration of load combinations 1 to 3 will provide adequate design data. If a stress reversal is to occur in a member, it will be caused by wind from the opposite side and is normally possible only with small dead loads and large wind loads. The following example will illustrate the general procedure of analysis of a roof truss.

Example 8.4 Consider the truss in Fig. E8.4a. The location is northern Indiana, bay length is 15 ft, and the trusses are supported on masonry walls. Wall openings are 30 percent. The dead load is

Weight of roofing, rafters, and purlins = 16 lb/ft^2 roof surface

Weight of bracing = 1 lb/ft^2 roof surface

Weight of truss = $0.5 + 0.075L = 4.1$ lb/ft^2 horizontal surface

Weight of ceiling = 10 lb/ft^2 horizontal surface

From Fig. 3.2 the greatest snow load on the ground is 20 lb/ft^2; therefore

Roof load = $(20)(0.80) = 16$ lb/ft^2

The live load of 20 lb/ft^2 controls. The wind load is determined as follows. Figure 3.4 indicates a fastest-mile velocity of 90 mi/h. The basic wind pressure, from Eq. (3.2), is

$$q = (0.00256)(90)^2 = 20.7 \text{ lb/ft}^2$$

External velocity pressure = $(20.7)(1.3) = 26.9$ lb/ft^2

Internal velocity pressure = 20.7 lb/ft^2

Internal-pressure coefficients are obtained by using Fig. 3.6.

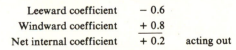

Leeward coefficient	-0.6	
Windward coefficient	$+0.8$	
Net internal coefficient	$+0.2$	acting out

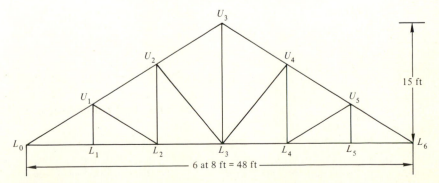

Figure E8.4a

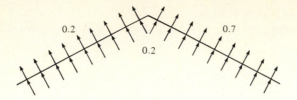

0.2   0.7

0.2

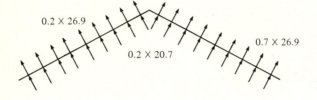

0.2 × 26.9

0.7 × 26.9

0.2 × 20.7

Figure E8.4*b* and *c*

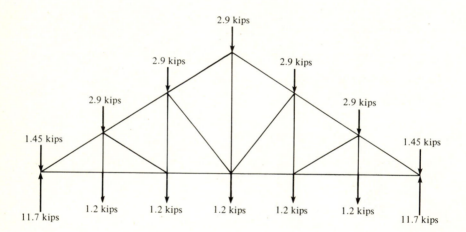

2.9 kips

2.9 kips   2.9 kips

2.9 kips   2.9 kips

1.45 kips   1.45 kips

11.7 kips   1.2 kips   1.2 kips   1.2 kips   1.2 kips   1.2 kips   11.7 kips

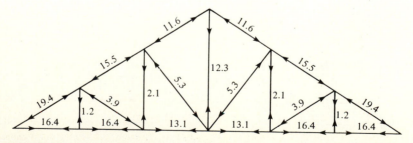

11.6   11.6

15.5   12.3   15.5

19.4   5.3   5.3   19.4

2.1   2.1

1.2   3.9   3.9   1.2

16.4   16.4   13.1   13.1   16.4   16.4

Figure E8.4*d* and *e* Dead loads and resulting forces.

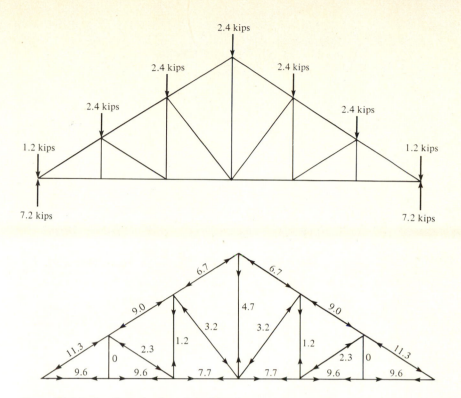

Figure E8.4 f and g Snow loads and resulting forces.

External-pressure coefficients for the roof are obtained using Fig. 3.7:

Leeward slope $= -0.70$ windward slope $= -0.20$

The net pressure coefficients are indicated in Fig. E8.4b. When they are used with the internal and external velocity pressures, the wind pressures shown in Fig. E8.4c are developed. Find (a) the dead-load forces, (b) the snow-load forces, and (c) the wind loads.

SOLUTION (a)

$$\text{Length } L_0 L_1 = \sqrt{(8)^2 + (5)^2} = 9.44 \text{ ft}$$

The area of roof surface associated with a joint is

$$(9.44)(15) = 141.6 \text{ ft}^2$$

and the area of horizontal surface associated with a joint is

$$(8)(15) = 120 \text{ ft}^2$$

The panel load on the upper chord joint is

$$(16 + 1)(141.6) + (4.1)(120) = 2900 \text{ lb}$$

and on the bottom chord joint is

$$(10)(120) = 1200 \text{ lb}$$

The weight of the truss has been assumed to act at the panel points of the top chord. It could have been divided between the upper and lower panel points. The panel loads are shown

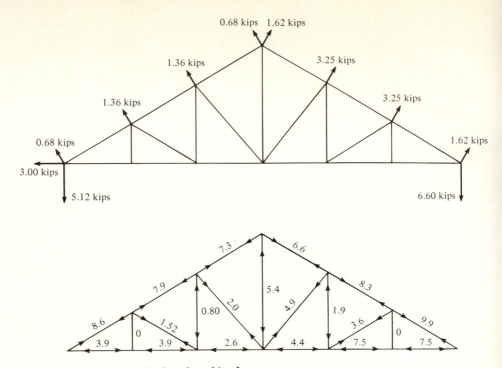

Figure E8.4h and i Wind loads and resulting forces.

in Fig. E8.4d, and the associated member forces are given in Fig. E8.4e. They can be computed by either the algebraic or graphic methods.

(b) Panel loads at top chord points = (20)(120) = 2400 lb
The live loading is shown in Fig. E8.4f, and the corresponding live-load forces are given in Fig. E8.4g.

(c) Panel loads can be calculated as follows:

$$[(0.2)(20.7) + (0.2)(26.9)](141.6) = 1360 \text{ lb} = 1.36 \text{ kips}$$

$$[(0.2)(20.7) + (0.7)(26.9)](141.6) = 3250 \text{ lb} = 3.25 \text{ kips}$$

Figures E8.4h and i shows the wind loads on the truss with the wind from the left. Reactions are considered to be a hinge on the left and a roller or expansion joint on the right.

Table 8.1 lists the member forces for the various loadings. These forces are combined in the indicated ways to determine the maximum load each member must be capable of carrying. These become design loads. Note that the AISC provision for a $33\frac{1}{3}$ percent increase in allowable stresses for wind loads is used by taking three-fourths of any combination load including wind. Consideration is given to the wind blowing from both directions. This is necessary when the truss is supported on what is equivalent to a hinge and roller. It is not unusual to neglect expansion for shorter trusses. In this case, the horizontal reactions are split between supports to avoid a statically indeterminate situation. In the above example there are stress reversals because of the wind loads. In most instances the critical force combination is that of dead load plus snow load.

Table 8.1 Maximum member forces

Member	Forces due to					Force combination			
	Dead load	Live load	Half live load	Wind from left	Wind from right	DL + LL	$\frac{3}{4}$(DL +WL)	$\frac{3}{4}$(DL $+\frac{1}{2}$(LL) +WL)	Design force
L_0L_1	+ 16.4	+ 9.6	+ 4.8	− 3.9	− 10.5	+ 26.0	+ 9.4	+ 13.0	+ 26.0
L_1L_2	+ 16.4	+ 9.6	+ 4.8	− 3.9	− 10.5	+ 26.0	+ 9.4	+ 13.0	+ 26.0
L_2L_3	+ 13.0	+ 7.7	+ 3.9	− 2.6	− 7.4	+ 20.8	+ 7.8	+ 10.7	+ 20.8
L_0U_1	− 19.4	+ 11.3	− 5.7	+ 8.6	+ 9.9	− 30.7	− 8.1	− 12.4	− 30.7
U_1U_2	− 15.5	− 9.0	− 4.5	+ 7.9	+ 8.3	− 24.5	− 5.7	− 9.1	− 24.5
U_2U_3	− 11.6	− 6.7	− 3.4	+ 7.3	+ 6.7	− 18.3	− 3.7	− 6.3	− 18.3
U_1L_1	+ 1.2	0	0	0	0	+ 1.2	+ 0.9	+ 0.9	+ 1.2
U_1L_2	− 3.9	+ 2.3	+ 1.2	+ 1.5	+ 3.6	− 1.6	− 0.2	+ 0.7	$\left\{\begin{matrix} -\ 1.6 \\ +\ 0.7 \end{matrix}\right.$
U_2L_2	+ 2.1	− 1.2	− 0.6	− 0.8	− 1.9	+ 0.9	+ 0.2	− 0.3	$\left\{\begin{matrix} +\ 0.9 \\ -\ 0.3 \end{matrix}\right.$
U_2L_3	+ 5.3	− 3.2	− 1.6	+ 2.0	+ 4.9	+ 2.1	+ 7.7	+ 6.5	+ 7.7
U_3L_3	+ 12.3	+ 4.7	+ 2.4	− 5.4	− 5.4	+ 17.0	+ 5.2	+ 7.0	+ 17.0
L_3U_4	+ 5.3	− 3.2	− 1.6	+ 4.9	+ 2.0	+ 2.1	+ 7.7	+ 6.5	+ 7.7
L_4U_4	+ 2.1	− 1.2	− 0.6	+ 1.9	− 0.8	+ 0.9	+ 3.0	+ 2.6	+ 3.0
L_4U_5	− 3.9	+ 2.3	+ 1.4	+ 3.6	+ 1.5	− 1.6	− 1.8	− 0.8	− 1.6
L_5U_5	+ 1.2	0	0	0	0	+ 1.2	+ 0.9	+ 0.9	+ 1.2
U_3U_4	− 11.6	− 6.7	− 3.4	+ 6.7	+ 7.2	+ 18.3	− 3.7	− 6.3	+ 18.3
U_4U_5	− 15.5	− 9.0	− 4.5	+ 8.3	+ 7.9	− 24.5	− 5.7	− 9.1	− 24.5
U_5L_6	− 19.4	− 11.3	− 5.7	+ 9.9	+ 8.6	− 30.7	− 8.1	− 12.4	− 30.7
L_3L_4	+ 13.1	+ 7.7	+ 3.9	− 4.4	− 5.6	+ 20.8	+ 6.5	+ 9.4	+ 20.8
L_4L_5	+ 16.4	+ 9.6	+ 4.8	− 7.5	− 6.9	+ 26.0	+ 7.1	+ 10.7	+ 26.0
L_5L_6	+ 16.4	+ 9.6	+ 4.8	− 7.5	− 6.9	+ 26.0	+ 7.1	+ 10.7	+ 26.0

8.7 DESIGN CONSIDERATIONS

8.7.1 Steel Trusses

Some general remarks concerning the selection of members for a steel roof truss follow. For riveted and bolted trusses a pair of angles back to back is the most common type of member. For larger trusses WF or I sections may be used for some members. A minimum-sized member may be required by specifications. The minimum-sized member often specified consists of two angles 2 by 2 by $\frac{1}{4}$ in. The web members are commonly made of angles, channels, or WF sections. It is desirable to limit the width of members as this reduces secondary stresses.

Chord members are often continuous through several panel points. Upper chords may be continuous from the eave to the ridge. Bottom chords are often made continuous for half the span. Since a member like this has to be designed for the maximum stress, some of the members will be overdesigned. The use of more frequent splices is not necessarily more economical. Using purlins between the joints of the top chord produces bending in the chord member as well as direct stress. The chord member then becomes a continuous beam. For welded roof trusses, members are usually made from structural tees, two angles, or WF sections. If a member is subject to stress reversal, it should be designed to resist both tension and compression.

8.7.2 Wood Trusses

With the choice of connectors available, it is possible to fabricate almost any kind of truss. The choice is usually made on the basis of architectural requirements. Joint details are a major factor in truss cost. Simplicity in details leads to economy.

Gusset plates are not needed when split rings and toothed rings are used, and these connectors are suitable for most spans of average length. The 4-in ring is used for the longer spans.

It is not unusual to have joints and purlins located between the panel points, particularly in flat trusses. Member size is normally the same throughout the entire length of the top chord. Column action must be considered about both parallel axes—length of panel for one axis and spacing of joists for the other. In computing bending moments for the upper chord $Wl/8$ can be used even though continuity makes $Wl/10$ realistic (see Chap. 11).

The thickness of the end diagonal and vertical determines the spacing of chord members. Other web members are normally the same thickness. The stresses in these members are so low that the other dimensions are determined by the requirements of connector spacing and end distances.

8.8 DEFLECTION OF TRUSSES

The basic requirements of structural design include not only the consideration of allowable stresses and economics but also deflection. Nearly all design specifications contain limits of deflection for particular types of structures. Ceiling beams have deflections limited by the cracking of plaster. Bridges have an inherent upward deflection (camber) fabricated into them so that any dead-load deflection will be zero or slightly upward.

Deflections are also important during erection. The different loading conditions which can occur during construction must be anticipated, particularly for their effect upon deflection. Excessive flexibility is a very real mechanism of failure.

The methods of calculating deflections also form the foundation of analysis of statically indeterminate structures. When the equations of static equilibrium are not sufficient to solve for all the unknown forces acting on a structure, deflection and slope conditions can be used in developing additional equations.

Many different methods can be used to compute deflections. No one method is best for all applications.

8.8.1 The Williot-Mohr Method

In 1877 the French engineer Williot developed a graphical method for determining the deflection of structures made up of two-force members. The original method developed by Williot is limited. In 1887 a German, Otto Mohr, suggested a supplement or correction which makes Williot's method one of general application. It has the particular advantage of giving the deflection of all the joints. This method, like all graphical methods, requires care in scaling and drawing lines.

The fundamentals of the Williot-Mohr method can be developed by considering a sequence of examples using simple trusses. Assume that the change of length Δl of all members has been computed for the given conditions, $\Delta l = \alpha l \Delta t$ for temperature changes, length changes due to applied loads $\Delta l = Fl/AE$, and mistakes in member lengths. From these changes in length the deflected shape of the structure can be obtained.

Consider the two-bar truss in Fig. 8.5a. Under the applied load P member AB elongates a distance ΔAB and member BC is shortened an amount ΔCB. Figure 8.5b shows by the heavy-line segments the exaggerated changes in length of the members. It must be remembered that these changes in length are very small compared with the length of the members (a fraction of an inch compared with many feet). They would not be visible on a scaled drawing. Consider the members to be disconnected at joint B with the changes in length as indicated in Fig. 8.5b. With these changes in length to reconnect the members at B' it is necessary to rotate the members about their connected ends, as indicated in Fig. 8.5c. It is permissible to draw perpendiculars from the stressed ends instead of striking arcs because in an actual structure the rotation of the members would at most be only a few minutes of arc. B' and B_1' would coincide for all practical purposes. Williot realized the possibility of using only the change in lengths of the members, as illustrated in Fig. 8.5d, called the *Williot diagram*. A somewhat more complicated truss will be used to illustrate the Williot diagram with its Mohr correction.

Consider the truss in Fig. 8.6a. It is symmetrical about the center member with respect to shape, size of member, and loading. The changes in length of the members due to the load P are indicated. It is necessary, as will be shown, to start the Williot diagram with a member that does not rotate. Consider member BD. Fix joint B and disconnect joint D. Figure 8.6b shows by heavy-line segments the exaggerated change in length of the members. D' is the new location of joint D. To locate the deflected position of C disconnect joint C. Again, the heavy-line segments indicate the change in length of the members. To reconnect members BC and DC at C' the intersection of perpendiculars to the ends of the members is used instead of rotating the members about B and D and using the intersection of the corresponding arcs. In a similar manner, the deflected position of joint A is found. The deflected shape of the truss is now given by $A'B'C'D'$. It would be impossible to draw the above diagram to scale, but the exaggerated sketch indicates some important aspects. It is apparent that

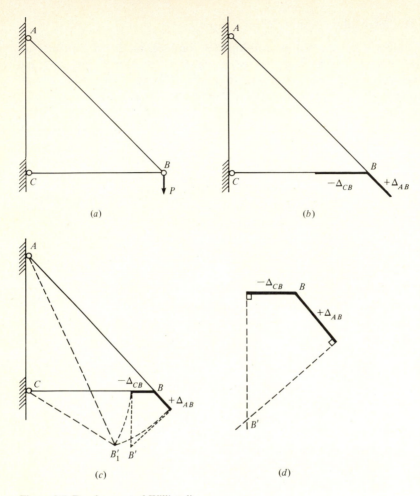

Figure 8.5 Development of Williot diagram.

the movements of the joints can be isolated in terms of the changes in length of the members. It is equivalent to allowing the member lengths to become zero (see the circled portions of Fig. 8.6b). These are combined in Fig. 8.6c and constitute the Williot diagram, which can be drawn to a sufficiently large scale to give engineering accuracy. In order to use the Williot diagram it must be remembered that joint A is a fixed joint and that joint C can move laterally only because it is a roller support. As a result, because the Williot diagram was started with member BD, which does not rotate, A' can be considered fixed and B', D', and C' give the deflected position of the corresponding joints relative to A'.

If the truss had not been symmetrical, both physically and in loading, member BD would not have remained vertical. The rotation would have been

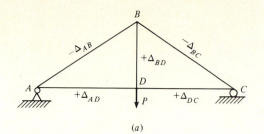

(a)

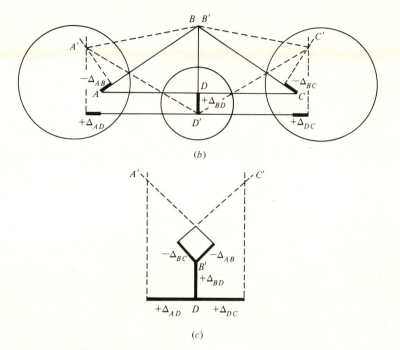

(b)

(c)

Figure 8.6 Williot diagram.

indeterminate, and in drawing the displacement diagram the location of A' and C' would be unknown.

Instead of using a different unsymmetrically loaded truss in which there is no member which does not rotate, we shall use the same truss but start with member AD, a member which does rotate under the applied loads. Consider the truss in Fig. 8.7a, which is the same as the truss in Fig. 8.6a. Member AD is considered fixed at A and in its direction AD. It is able to move horizontally at D. Figure 8.7b gives the position of the deflected structure (dashed lines) relative to the original structure (light solid lines). As described previously, the movements of the joints can be obtained independently by means of the Williot diagram (Fig. 8.7c). This diagram does not give true deflections because of the

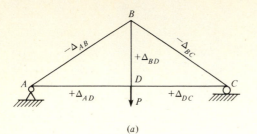

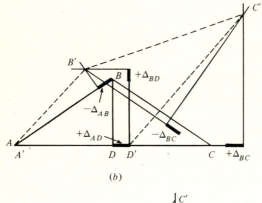

(a)

(b)

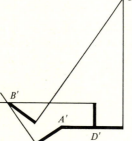

(c)

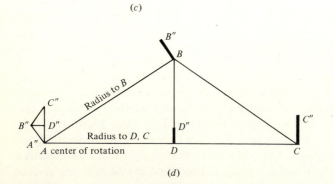

(d)

Figure 8.7 Development of Williot diagram and Mohr correction. (*d*) Evolution of Mohr diagram.

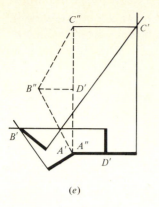

(e)

Figure 8.7 (*continued*) (*e*) Williot-Mohr diagram.

rotation of the truss. The true deflected position of the truss could be obtained by rotating the deflected truss back about reaction *A* until the right-hand end of the truss rests again at the original level of support *C*. If Fig. 8.7c is to represent true deflections instead of relative ones, *C'* must lie on a horizontal line through *A'*. This correction was first developed by Mohr and is often referred to as the *Mohr rotation diagram*. The Mohr rotation diagram is developed in Fig. 8.7d. A small counterclockwise rotation of the truss about *A* produces joint movements *B-B''*, *C-C''*, and *D-D''*, which can be drawn as short straight lines perpendicular to radii through *A*. These joint movements can be grouped together in a small diagram at *A* representing the joint movements with the truss members eliminated from Fig. 8.7d, where *A''-B''* represents *B-B''* on the truss, *A''-C''* represents *C-C''* and *A''-D''* represents *D-D''*. This small diagram is a scale drawing of the truss (turned counterclockwise through 90°) drawn upon *A''-C''* (which is equal to *C-C''*) as a base. Since there is no translation of the joint *A*, the points *A*, *A'*, and *A''* are coincident.

This Mohr rotation diagram is then combined with the Williot diagram to give true deflections of the truss joints (Fig. 8.7e) by leaving the truss in its deflected position (Williot diagram) and rotating the undeflected truss (Mohr diagram) counterclockwise about point *A* until *C*, which now becomes *C''*, lies on a horizontal line through *C'*. When this is done, we can scale true deflections of the joints as *A''-A'*, *B''-B'*, *D''-D'*, and *C''-C'*, that is, from points on the rotated but undeflected truss (Mohr diagram) to corresponding points on the deflected truss (Williot diagram). This follows because the rotated and deflected trusses represented by the combined Williot-Mohr diagram are in the same relative positions as the unloaded and loaded trusses resting on the same abutment.

In retrospect, if a truss is symmetrical in shape and loading, it is not necessary to draw a Mohr rotation diagram if the Williot diagram is started using a nonrotating member of symmetry. If the structure or loading is unsymmetrical, it becomes necessary to use a Mohr rotation diagram.

The following rules of procedure can be used to advantage in drawing the Williot diagram with its Mohr correction. They should be considered in conjunction with Fig. 8.7c and e.

1. The left end of the truss A forms the starting point A' in Fig. 8.7c. Since the member AD is assumed to remain horizontal, D' moves to the right or left of A' depending upon whether the member lengthens or shortens (direction of actual movement with A' fixed) by the change in length of the member $(+\Delta AD)$. Draw ΔAD to the right from A', its reference point, and establish D'.

2. Each successive deflected joint is located from its movement relative to two points located previously. Thus, B' is located from the known positions of A' and D'. Consider members AB and DB disconnected at B. The movement of B relative to the point A' is downward and to the left parallel to AB by the amount ΔAB because of the shortening of member AB. Draw ΔAB from A', the point to which it is referred. Point B also moves upwards relative to D' because of the lengthening of member DB. The plotted deformation ΔDB is drawn parallel to member DB starting from D' in the Williot diagram, the point to which it is referred. Perpendiculars are drawn at the ends of these plotted deformations to represent rotation of the bars about their reference points. The intersection of these perpendiculars establishes the location of B', the new position of joint B.

3. All other joints are located in succession following the procedure established in step 2. The final deflections obtained (measured from A' to B', A' to D', and A' to C') are relative to A as a fixed point and to AD as a fixed direction. For example, for joint B, A' represents its original position in space, and B' represents its final position as determined by the Williot diagram.

When the Williot diagram does not give true deflections, a Mohr rotation diagram must be used. The following rules of procedure are helpful:

4. Place A'' at A' since the final movement A''-A' of point A is zero.
5. Place C'' on a vertical line through A' and on a horizontal line through C'. This fixes the scale of the Mohr diagram.
6. Construct the Mohr diagram on A''-C'' in the position the truss takes when it is rotated counterclockwise through 90° about the fixed end A.
7. True deflections can now be scaled for the various joints (A'' to A' for joint A, B'' to B' for joint B, etc.).

8.9 COMPOSITE STRUCTURES

These are structures in which some members are subjected chiefly to direct forces and others to bending forces. It is important to recognize which are two-force members and which are three-force members. A two-force member is

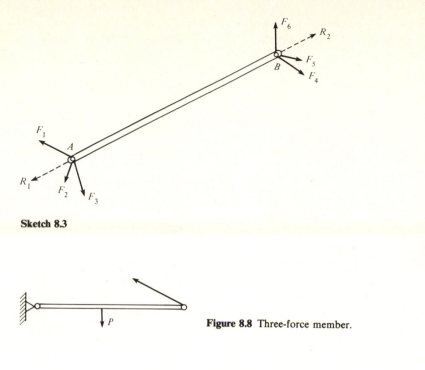

Sketch 8.3

Figure 8.8 Three-force member.

one which can be considered to be pin-connected at both ends and is not subjected to any loads between the end joints. A member that does not satisfy these requirements is a three-force member. A typical free-body diagram of a two-force member is given in Sketch 8.3.

Member AB is pin-connected and subjected to forces F_1, F_2, and F_3 at A and F_4, F_5, and F_6 at B. The forces are in equilibrium. R_1 is the resultant of forces F_1, F_2, and F_3, and R_2 is the resultant of forces F_4, F_5, and F_6. Since R_1 and R_2 are in equilibrium, they must be equal, opposite, and collinear. As a result member AB is subject to direct axial loads, either tension or compression.

In three-force members the forces acting at each end of the member can also be combined into resultants, but additional forces act on the member, in between the joints (Fig. 8.8). For problem solution, it is convenient to resolve the resultant forces at the ends of the members into vertical and horizontal components.

Example 8.5 Analyze the structure shown in Fig. E8.5a. The beam weighs 30 lb per lineal foot.

SOLUTION The external reaction components at A and B are indicated in Fig. E8.5a. The free-body diagrams of members AC and BC are given in Fig. E8.5b and c. Member AC is a two-force member. The resultant of H_A and V_A must be the tensile force in the steel cable. Member BC is a three-force member because of the weight of the member and the 800-lb load. Note that the force components at C in Fig. E8.5b are equal and opposite to those at C in Fig. E8.5c.

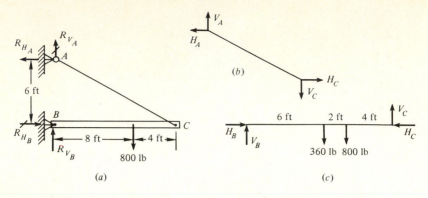

Figure E8.5

Consider the free-body diagram of member BC (Fig. E8.5c):

$$\overset{\curvearrowright}{\Sigma M_B} = 0: \qquad\qquad (360)(6) + (800)(8) - 12V_c = 0$$

$$V_c = 713.3 \text{ lb}$$

$$\uparrow \Sigma F_y = 0: \qquad\qquad 713.3 + V_B - 360 - 800 = 0$$

$$V_B = 446.7 \text{ lb}$$

Consider the free-body diagram of structure ABC (Fig. E8.5a)

$$\overset{\curvearrowright}{\Sigma M_B} = 0: \qquad\qquad (6)(360) + (8)(800) - 6R_{H_A} = 0$$

$$R_{H_A} = 1426.7 \text{ lb}$$

Consider the free-body diagram of member AC (Fig. E8.5b)

$$\overset{\leftharpoonup}{\Sigma F_x} = 0: \qquad\qquad 1426.7 - H_C = 0$$

$$H_C = 1426.7 \text{ lb}$$

Consider the free-body diagram of member BC (Fig. E8.5c)

$$\overset{\leftharpoonup}{\Sigma F_x} = 0: \qquad\qquad 1426.7 - H_B = 0$$

$$H_B = 1426.7 \text{ lb}$$

Example 8.6 Analyze completely the structure shown in Fig. E8.6a.

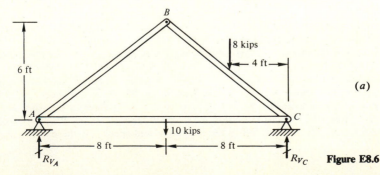

Figure E8.6

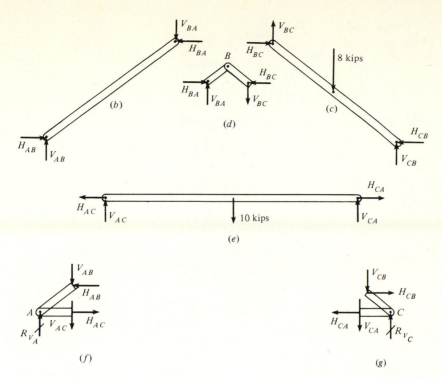

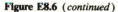

Figure E8.6 (*continued*)

SOLUTION AB is a two-force member, and BC and AC are three-force members. By treating the entire structure as a free body we get

$\Sigma M_A = 0$:

$$(8)(10) + (8)(12) - 16R_{V_C} = 0$$

$$R_{V_C} = \frac{80 + 96}{16} = 11 \text{ kips}$$

$\Sigma M_C = 0$:

$$(-8)(4) - (8)(10) + 16R_{V_A} = 0$$

$$R_{V_A} = \frac{80 + 32}{16} = 7 \text{ kips}$$

$\downarrow\Sigma F_y = 0$:

$$8 + 10 - 11 - 7 = 0 \quad \text{check}$$

Consider each member and joint as free-body diagrams. To avoid confusion the end reactions in free-body diagrams of joints are shown acting in the member a short distance from the joint. The solution involves selecting appropriate free-body diagrams in which unknowns can be determined by the use of equilibrium equations. Select member AC (Fig. E8.6e).

$\Sigma M_A = 0$:

$$(8)(10) - 16V_{CA} = 0$$

$$V_{CA} = 5 \text{ kips}$$

$\downarrow\Sigma F_y = 0$:

$$10 - 5 - V_A = 0$$

$$V_{AC} = 5 \text{ kips}$$

Consider joint C (Fig. E8.6g)

$\uparrow \Sigma F_y = 0$: $\qquad\qquad\qquad\qquad$ $11 - 5 - V_{CB} = 0$

$$V_{CB} = 6 \text{ kips}$$

Consider member BC (Fig. E8.6c)

$\curvearrowright \Sigma M_B = 0$: $\qquad\qquad\qquad$ $(8)(4) - (6)(8) + 6H_{CB} = 0$

$$H_{CB} = 2.67 \text{ kips}$$

$\downarrow \Sigma F_y = 0$: $\qquad\qquad\qquad\qquad$ $8 - 6 - V_{BC} = 0$

$$V_{BC} = 2 \text{ kips}$$

$\overrightarrow{\Sigma F_x} = 0$: $\qquad\qquad\qquad\qquad$ $H_{BC} - 2.67 = 0$

$$H_{BC} = 2.67 \text{ kips}$$

Consider joint B (Fig. E8.6d)

$\downarrow \Sigma F_y = 0$: $\qquad\qquad\qquad\qquad$ $2 - V_{BA} = 0$

$$V_{BA} = 2 \text{ kips}$$

$\overleftarrow{\Sigma F_x} = 0$: $\qquad\qquad\qquad\qquad$ $2.67 - H_{BA} = 0$

$$H_{BA} = 2.67 \text{ kips}$$

Consider member AB (Fig. E8.6b)

$\downarrow \Sigma F_y = 0$: $\qquad\qquad\qquad\qquad$ $2 - V_{AB} = 0$

$$V_{AB} = 2 \text{ kips}$$

$\overleftarrow{\Sigma F_x} = 0$: $\qquad\qquad\qquad\qquad$ $2.67 - H_{AB} = 0$

$$H_{AB} = 2.67 \text{ kips}$$

Consider joint A (Fig. E8.6f)

$\overleftarrow{\Sigma F_x} = 0$: $\qquad\qquad\qquad\qquad$ $2.67 - H_{AC} = 0$

$$H_{AC} = 2.67 \text{ kips}$$

$\downarrow \Sigma F_y = 0$: $\qquad\qquad\qquad\qquad$ $7 - 2 - V_{AC} = 0$

$$V_{AC} = 5 \text{ kips} \qquad \text{check}$$

Consider member AC (Fig. E8.6e)

$\uparrow \Sigma F_x = 0$: $\qquad\qquad\qquad\qquad$ $2.67 - H_{CA} = 0$

$$H_{CA} = 2.67 \text{ kips}$$

Consider joint C (Fig. E8.6g)

$\downarrow \Sigma F_y = 0$: $\qquad\qquad\qquad\qquad$ $5 - 11 + 6 = 0 \qquad \text{check}$

$\overrightarrow{\Sigma F_x} = 0$: $\qquad\qquad\qquad\qquad$ $2.67 - 2.67 = 0 \qquad \text{check}$

It is important to realize that if two members are connected by a pin at a joint, the forces exerted between the members are internal forces within the system but when they are disconnected these forces exist in equal and opposite pairs (Fig. E8.6d).

If more than two members are associated with a particular joint or two members are connected to a support, the forces exerted between the members will not consist of equal and opposite pairs but will be in equilibrium with the other forces at the joint (Fig. E8.6f).

8.10 SUMMARY

Trusses are composed of bars lying in one plane and connected at their ends to form a rigid framework. The basic repeating shape within a truss is the triangle.

When trusses are loaded at joints, the members become two-force members and develop only axial forces, for which two methods of analysis are available.

Method of joints. Each joint is isolated as a concurrent system of forces in equilibrium. This permits solving for two unknowns either graphically or algebraically.

Method of sections. This method isolates half of the truss by means of a cutting plane. The fact that the resulting nonconcurrent system of forces is in equilibrium allows for the solution of three unknown forces. They can be solved for either graphically or algebraically.

In the design of roof trusses it is important to consider combinations of loadings involving dead, snow, and wind loads.

The deflection of a truss is an important consideration in its design. The Williot-Mohr method is a graphical method for determining the deflection of structures made up of two-force members. It is based on changes in length of individual members due to applied loads, temperature changes, or errors in fabricating lengths.

Composite structures are pin-connected structures containing both two- and three-force members. Three-force members are members having loads acting at an angle to the axis of the member at a point other than the ends of the member. Such a member is subject to bending. Analysis of such structures involves isolating individual members with pin reactions resolved into vertical and horizontal components. A complete solution of pin reactions permits the analysis and design of three-force members.

PROBLEMS

8.1 (*a*) Determine the forces in all members of the cantilever truss shown in Fig. P8.1 by the method of joints.

(*b*) Determine the forces in members U_1U_2, U_1L_2, and L_1L_2 by the method of sections.

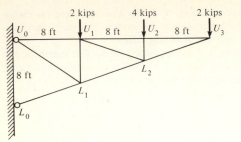

Figure P8.1

8.2 (*a*) Determine the forces in all members of the truss shown in Fig. P8.2 by the method of joints.
(*b*) Check the answers for members U_0U_1 and U_1L_1 by the method of sections.

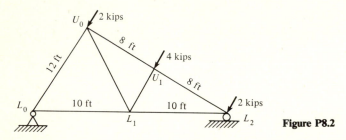

Figure P8.2

8.3 The truss in Fig. P8.3 supports a roof spanning a distance of 48 ft between two walls. This building is located in the Chicago area. Determine the design forces for each member considering possible combinations of dead loads, wind loads, and snow loads. Consider a bay size of 16 ft.

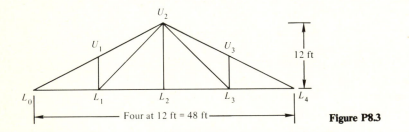

Figure P8.3

8.4 (*a*) Determine the forces in the members of the truss (Fig. P8.4).
(*b*) Design the truss completely (joints and member selection) using no. 1 Douglas fir timber.

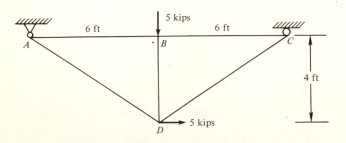

Figure P8.4

8.5 Determine the force in each member of the truss shown in Fig. P8.5. The load is applied at the upper panel points.

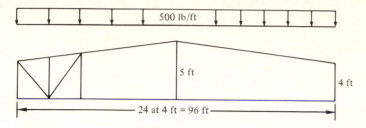

Figure P8.5

8.6 For the truss in Fig. P8.6 determine the reaction at supports A and F and the force in each of the members.

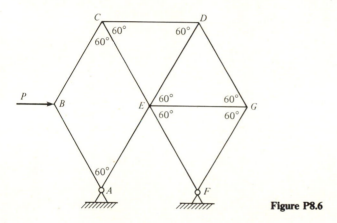

Figure P8.6

8.7 Determine the forces in all the bars of the trusses in Fig. P8.7.

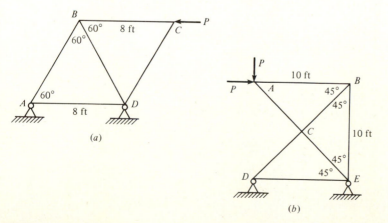

Figure P8.7

8.8 Analyze the structures in Fig. P8.8. Determine support and pin-reaction components.

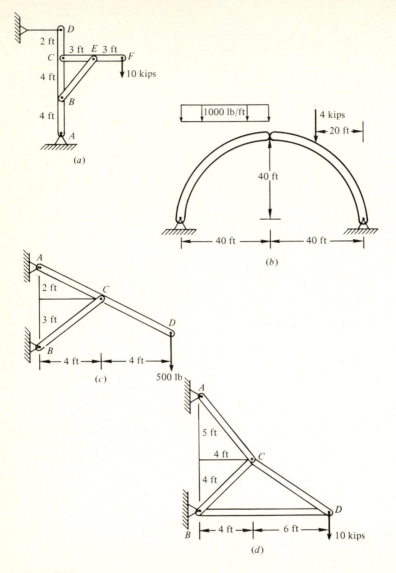

Figure P8.8

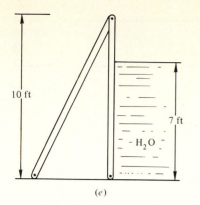

(e)

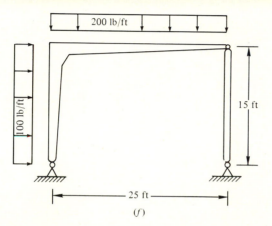

(f)

Figure P8.8 (continued)

CABLES

9.1 CABLE-SUPPORTED STRUCTURES

Steel cables, first used to support long-span bridges, are now being used to support large-span roof systems and other structures such as guyed towers. The increased interest in cabled structures lies in their high strength and relatively low dead load. Cables are flexible and develop only tension, a very important fact because it keeps cable structures statically determinate. When a cable is loaded, it spontaneously acquires a shape that will allow it to remain in tension. This shape is the familiar funicular polygon. A cable is unstable in the sense that it changes shape with changing loads. We are interested in determining the funicular shape of a cable, its reactions, the maximum tensile force in the cable, and its required cross section.

The design of cables in structures differs from the design of more conventional structural components such as beams and columns in both anchorage forces and dynamic behavior. Since cables are always in tension, this tension must be resisted. The resistance can be provided in a number of ways. The cable ends are frequently anchored into foundation abutments or continuous structural members such as rings (Figs. 2.2 to 2.4).

Individual suspended cables or a grid of cables are susceptible to motion, usually referred to as *flutter* when caused by such exterior dynamic forces as wind, moving loads, and seismic loads. Cables are made in a wide assortment of types and sizes (see Sec. 9.6).

9.2 LOAD SYSTEMS

Loads on cables can be considered in three categories, as shown in Fig. 9.1. Cables are normally identified by the horizontal span l between supports, the vertical distance h between supports, and sag S, the vertical distance between a chord through the supports and the lowest point on the cable. Figure 9.1a shows a cable supporting a concentrated load. If the weight of the cable is neglected, the cable forms straight lines between the load and the supports. Figure 9.1b shows a uniform load per unit length of span. The shape of the cable becomes that of a parabola. Figure 9.1c shows a loading that is uniform per unit length of cable. Under these circumstances the shape of the cable is that of a catenary, which lies outside the parabola of equal span and sag. For all practical purposes the catenary may be taken to be a parabola for shallow cables with sag-span ratios equal to or less than one-tenth. Under these circumstances, a cable carrying only its dead weight can be treated approximately on the assumption that the dead weight is uniform per horizontal foot of span.

In many cases, the weight of the cable can be neglected. Uniform and other distributed loads are often supported by cables by means of vertical hangers and constitute a series of concentrated loads.

9.3 SINGLE-CONCENTRATED-LOAD SYSTEMS

Figure 9.2 shows a cable with supports at different levels and a single concentrated load. The support reactions and load P constitute a system of forces in equilibrium. The location of at least a single point on the cable, other than the

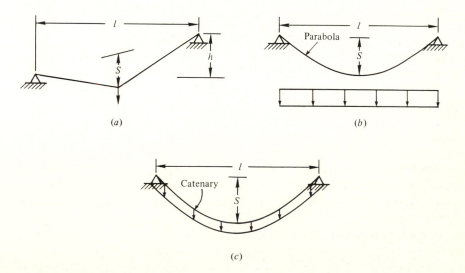

Figure 9.1 Types of loads on cables.

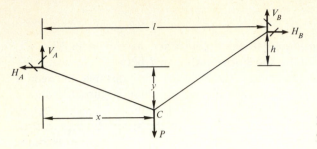

Figure 9.2 Cable carrying a concentrated load.

supports, must be known, and it must be realized that the cable cannot develop a resistance to bending (internal bending moment). This last assumption provides the additional equation required to solve for the four reaction components. The tensile force in a cable is different in different parts of the cable. The maximum tensile force is of prime interest and is the value used in the design of the cable.

From the free-body diagram (Fig. 9.2) the four equations of statics available for solving the reaction components are

$$\Sigma M_A = 0: \qquad\qquad V_B L - Px - H_B h = 0 \qquad\qquad (9.1)$$

$$\Sigma M_C = 0 \text{ (left portion of structure)}: \quad V_A x - H_A y = 0 \qquad\qquad (9.2)$$

$$\uparrow \Sigma F_Y = 0: \qquad\qquad V_A + V_B - P = 0 \qquad\qquad (9.3)$$

$$\overleftarrow{\Sigma F_X} = 0: \qquad\qquad H_A - H_B = 0 \qquad\qquad (9.4)$$

Example 9.1 illustrates the use of these equations.

Example 9.1 A cable is supported at points 100 ft apart. The right support is at a level 5 ft higher than the left support. The cable carries a concentrated load of 10 kips at midspan with a final sag of 5 ft relative to the left support. Determine the maximum tensile force in the cable. See Fig. E9.1*a*.

SOLUTION Consider the entire cable as a free body

$$\Sigma M_1 = 0: \qquad\qquad 5H_2 - 100V_2 + (10)(50) = 0 \qquad\qquad (1)$$

Consider the cable cut at the concentrated load and the right-hand portion as a free body (Fig. E9.1*b*)

$$\Sigma M_0 = 0: \qquad\qquad -50V_2 + 10H_2 = 0 \qquad\qquad (2)$$

Substituting V_2 in terms of H_2 from (2) into (1) gives

$$(5)\left(\tfrac{50}{10}V_2\right) - 100V_2 + (10)(50) = 0 \qquad V_2 = \frac{(10)(50)}{75} = 6.67 \text{ kips}$$

$$\uparrow \Sigma F_y = 0: \qquad\qquad V_1 + 6.67 - 10 = 0 \qquad V_1 = 3.33 \text{ kips}$$

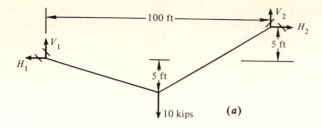

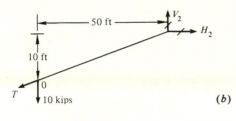

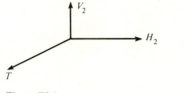

Figure E9.1a to c

From Eq. (2)

$$- (50)(6.67) + 10H_2 = 0 \qquad H_2 = 33.35 \text{ kips}$$

$\overrightarrow{\Sigma F_x} = 0$: $\qquad\qquad H_2 - H_1 = 0 \qquad H_1 = 33.5 \text{ kips}$

$\stackrel{\curvearrowright}{\Sigma M_2} = 0$: $\qquad (3.33)(100) - (10)(50) + (33.35)(5) = 0 \qquad$ check

The maximum tensile force in the cable will occur at the support with the maximum vertical reaction component ($V_2 = 6.67$ kips) since the horizontal component of the force in the cable is a constant. The maximum force in the cable becomes (Fig. E9.1c)

$$T = \sqrt{V_2^2 + H_2^2} = \sqrt{(6.67)^2 + (33.35)^2}$$
$$= 34.0 \text{ kips}$$

9.4 MULTIPLE-CONCENTRATED-LOAD SYSTEMS

When several concentrated loads are supported by a cable, it is necessary, as for a single concentrated load, to know the elevation of a point on the cable relative to the supports and to assume that the moment in the cable is always zero. Consequently the solution of a problem involving several concentrated loads is similar to the solution for a cable carrying a single load, as Example 9.2 illustrates.

Example 9.2 A cable is supported at points 200 ft apart and at the same elevation. The cable supports several concentrated loads as shown in Fig. E9.2a. Determine the maximum force in the cable and the force in each of the cable components. Determine the sag of the cable at each load point.

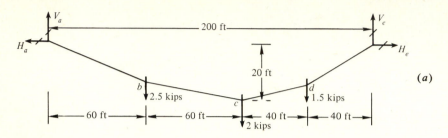

(a)

Figure E9.2

SOLUTION The vertical component of the reactions can be determined by taking moments about the supports

$\Sigma M_a = 0$: $\qquad (2.5)(60) + (2)(120) + (1.5)(160) - 200V_e = 0$

$$V_e = \frac{630}{200} = 3.15 \text{ kips}$$

$\Sigma M_e = 0$: $\qquad (-1.5)(40) - (2)(80) - (2.5)(140) + 200V_a = 0$

$$V_a = \frac{570}{200} = 2.85 \text{ kips}$$

$\uparrow \Sigma F_y = 0$: $\qquad 3.15 + 2.85 - 2.5 - 1.5 - 2.0 = 0 \qquad$ check

The value of H_e can be found by taking moments about c using the free-body diagram of the right portion of the cable (Fig. E9.2b)

$\Sigma M_c = 0$: $\qquad (1.5)(40) - (3.15)(80) + 20H_e = 0 \qquad H_e = 9.6 \text{ kips}$

$\Sigma F_x = 0$: $\qquad H_a - H_e = 0 \qquad H_a = 9.6 \text{ kips}$

The distance from the cable cord to the points of load on the cable can be determined by taking moments about the load points using free-body diagrams of portions of the cable. From Fig. E9.2c

$\Sigma M_b = 0$: $\qquad (2.85)(60) - 9.6y_b = 0 \qquad y_b = 17.8 \text{ ft}$

and from Fig. E9.2d

$\Sigma M_d = 0$: $\qquad (-3.15)(40) + 9.6y_d = 0 \qquad y_d = 13.1 \text{ ft}$

The maximum force in the cable occurs at the support with the largest vertical reaction component ($V_e = 3.15$ kips) because the horizontal component of the cable force is constant (Fig. E9.2e)

$$T_{\text{max}} = \sqrt{V_e^2 + H_e^2} = \sqrt{(3.15)^2 + (9.6)^2} = 10.1 \text{ kips}$$

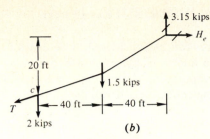

(b)

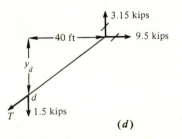

(c)

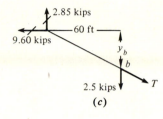

(d)

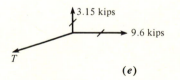

(e)

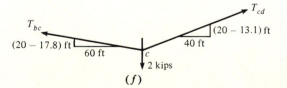

(f)

Figure E9.2 (*continued*)

Isolation of the 2-kip force at c as a free body enables us to determine the cable forces T_{cb} and T_{cd} (Fig. E9.2f)

$\uparrow \Sigma F_y = 0$: $\qquad\qquad 2 - \dfrac{2.2}{60.04} T_{cb} - \dfrac{6.9}{40.59} T_{cd} = 0$

$\overrightarrow{\Sigma F_x} = 0$: $\qquad\qquad \dfrac{40}{40.59} T_{cd} - \dfrac{60}{60.04} T_{cb} = 0$

Solving above equations simultaneously gives $T_{cd} = 9.71$ kips and $T_{cb} = 9.57$ kips.

9.5 CABLE WITH HORIZONTALLY DISTRIBUTED LOAD

As indicated previously, most cable loadings can be considered to be distributed horizontally. In suspension bridges and parallel roof cables this loading can be considered to be uniform. Usually the supports will be at the same elevation (Fig. 9.3). The fact that this type of loading results in a parabolic configuration for the cable can be proved by writing the equation for static moment equilibrium for a portion of the cable (Sketch 9.1). Because the cable does not develop an internal bending moment, we have

$$\overset{\curvearrowleft}{\Sigma M_0} = 0: \qquad V_A x - H_A y - \frac{wx^2}{2} = 0$$

$$y = \frac{wx^2}{2H_A} - \frac{V_A x}{H_A} \qquad (9.5)$$

Equation (9.5) is of the form of a second-degree parabola.

Consider the entire structure as a free body; then

$$\overset{\curvearrowleft}{\Sigma M_A} = 0: \qquad V_B l - \frac{wl^2}{2} = 0$$

$$V_B = \frac{wl}{2} \qquad (9.6)$$

$$\uparrow \Sigma F_y = 0: \qquad V_A + \frac{wl}{2} - wl = 0$$

$$V_A = \frac{wl}{2} \qquad (9.7)$$

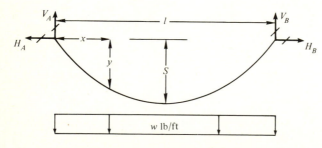

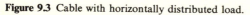

Figure 9.3 Cable with horizontally distributed load.

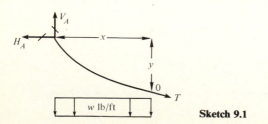

Sketch 9.1

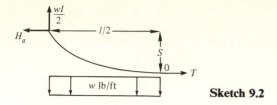

Sketch 9.2

The evaluation of H is possible by taking moments about the point where the sag is known. In this case, at the center of the span (Sketch 9.2)

$$\overset{\curvearrowleft}{\Sigma M_0} = 0: \qquad \frac{wl^2}{2}\frac{l}{2} - \frac{wl}{2}\frac{l}{4} - H_A S = 0$$

$$H_A = \frac{wl^2}{8S} \qquad (9.8)$$

Note that the horizontal reaction component H is directly proportional to the square of the span and inversely proportional to the sag.

The tension at any point in the cable can be found by determining the horizontal and vertical components of T at that point and applying the Pythagorean theorem.

The value of T will be a maximum at a support (in this case, at both supports). The value of T will be a minimum at the point where the sag is a maximum. Since the tangent to the cable will be horizontal, at this point the tension in the cable will be equal to the values of the horizontal reaction components

$$T^2_{\text{max}} = H^2 + V^2 = \frac{w^2 l^4}{64S^2} + \frac{w^2 l^2}{4}$$

$$T_{\text{max}} = \frac{wl}{2}\sqrt{1 + \frac{l^2}{16S^2}} \qquad (9.9)$$

A cable with supports at different elevations can be treated in a similar manner (Example 9.3).

Example 9.3 Determine the maximum force developed in a cable supporting a uniform load of 1.2 kips/ft over a span of 150 ft. The supports differ in elevation by 10 ft, and the sag is 30 ft (Fig. E9.3a).

SOLUTION Consider both the left and right portions of the cable relative to the point of maximum sag (Fig. E9.3b).

$$\overrightarrow{\Sigma F_x} = 0: \qquad T - H = 0 \qquad T = H$$

$$\overset{\curvearrowleft}{\Sigma M_A} = 0: \qquad 20T - \frac{1.2x^2}{2} = 0 \qquad T = \frac{1.2x^2}{40} = H$$

$$\overset{\curvearrowleft}{\Sigma M_B} = 0: \qquad 30T - \frac{(1.2)(150-x)^2}{2} = 0 \qquad T = \frac{(1.2)(150-x)^2}{60} = H$$

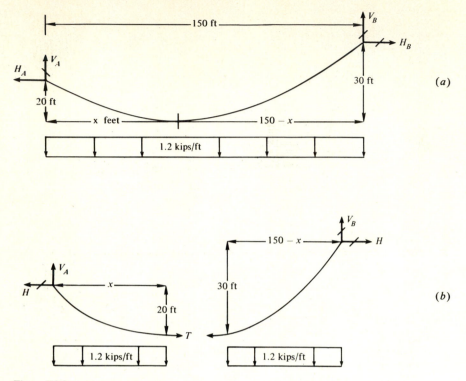

Figure E9.3

Since H is the same in both equations,

$$x^2 = \frac{40}{60}(150 - x)^2 \qquad x = 67.5 \text{ ft}$$

$$H = \frac{(1.2)(67.5)^2}{40} = 136.7 \text{ kips}$$

Using the left free-body diagram gives

$$\Sigma M_T = 0: \qquad 20H - 67.5V_A + 1.2\frac{(67.5)^2}{2} = 0$$

$$V_A = 80.9 \text{ kips}$$

Consider the free-body diagram of the entire structure:

$$\uparrow \Sigma F_y = 0: \qquad 80.9 - (1.2)(150) + V_B = 0 \qquad V_B = 99.1 \text{ kips}$$

Maximum force in the cable occurs at support B:

$$T_{max} = \sqrt{(136.7)^2 + (99.1)^2} = 168.8 \text{ kips}$$

9.6 CABLE DESIGN

A *cable* is normally defined as a flexible tension member consisting of one or more strands or wire ropes. A *strand* is an arrangement of wires helically laid

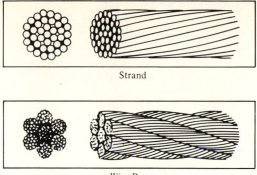

Strand

Wire Rope

Figure 9.4 Wire rope and strand.
(*Bethlehem Steel Corp.*)

Table 9.1 Mechanical properties for strand and wire rope [2]

Zinc coating class	Nominal diameter		Minimum stress at 0.7% extension under load		Minimum ultimate tensile strength		Total elongation in 10 in or 250 mm %
	in	mm	lb/in^2	MPa	lb/in^2	MPa	
A	0.040 to 0.110	1.016 to 2.794	150,000	1030	220,000	1520	2.0
	0.111 and up†	2.820 and up†	160,000	1100	220,000	1520	4.0
B	0.090 and up†	2.286 and up†	150,000	1030	210,000	1450	4.0
C	0.090 and up†	2.286 and up†	140,000	970	200,000	1380	4.0

† This is not to imply that larger wire will be manufactured to any unlimited diameter. It only implies that the wire sizes chosen by the strand manufacturer must meet the requirements of this specification.

Table 9.2 Minimum weight of coating per uncoated wire surface [2]

Nominal diameter of coated wire†		Coating class					
in	mm	A oz/ft^2	B oz/ft^2	C oz/ft^2	A g/m^2	B g/m^2	C g/m^2
0.040–0.061	1.016–1.549	0.40	0.80	1.20	122.00	244.00	366.00
0.062–0.079	1.575–2.007	0.50	1.00	1.50	152.50	305.00	457.50
0.080–0.092	2.032–2.337	0.60	1.20	1.80	183.00	366.00	549.00
0.093–0.103	2.362–2.616	0.70	1.40	2.10	213.50	427.00	640.50
0.104–0.119	2.642–3.023	0.80	1.60	2.40	244.00	488.00	732.00
0.120–0.142	3.048–3.607	0.85	1.70	2.55	259.25	518.50	777.75
0.143–0.187	3.632–4.750	0.90	1.80	2.70	274.50	549.00	823.50
0.188 and up	4.775 and up	1.00	2.00	3.00	305.00	610.00	915.00

† This is not to imply that larger wire will be manufactured to any unlimited diameter. It only implies that the wire sizes chosen by the strand manufacturer must meet the requirements of this specification.

Table 9.3 Properties of zinc-coated steel structural wire rope [2]

Nominal diameter, in	Minimum breaking strength, tons Coating class throughout			Approximate gross metallic area, in^2	Approximate weight, lb/ft
	A	B	C		
$\frac{3}{8}$	6.5	6.2	5.9	0.065	0.24
$\frac{7}{16}$	8.8	8.4	8.0	0.091	0.32
$\frac{1}{2}$	11.5	11.0	10.5	0.119	0.42
$\frac{9}{16}$	14.5	13.8	13.2	0.147	0.53
$\frac{5}{8}$	18.0	17.2	16.4	0.182	0.65
$\frac{11}{16}$	21.5	20.5	19.5	0.221	0.79
$\frac{3}{4}$	26.0	24.8	23.6	0.268	0.95
$\frac{13}{16}$	30.0	28.6	27.3	0.311	1.10
$\frac{7}{8}$	35.0	33.4	31.8	0.361	1.28
$\frac{15}{16}$	40.0	38.2	36.4	0.414	1.47
1	45.7	43.6	41.5	0.471	1.67
$1\frac{1}{8}$	57.8	55.1	52.5	0.596	2.11
$1\frac{1}{4}$	72.2	68.9	65.6	0.745	2.64
$1\frac{3}{8}$	87.8	83.8	79.8	0.906	3.21
$1\frac{1}{2}$	104.0	99.2	94.5	1.076	3.82
$1\frac{5}{8}$	123.0	117.0	112.0	1.270	4.51
$1\frac{3}{4}$	143.0	136.0	130.0	1.470	5.24
$1\frac{7}{8}$	164.0	156.0	149.0	1.690	6.03
2	186.0	177.0	169.0	1.920	6.85
$2\frac{1}{8}$	210.0	200.0	191.0	2.170	7.73
$2\frac{1}{4}$	235.0	224.0	214.0	2.420	8.66
$2\frac{3}{8}$	261.0	249.0	237.0	2.690	9.61
$2\frac{1}{2}$	288.0	275.0	262.0	2.970	10.60
$2\frac{5}{8}$	317.0	302.0	288.0	3.270	11.62
$2\frac{3}{4}$	347.0	331.0	315.0	3.580	12.74
$2\frac{7}{8}$	379.0	362.0	344.0	3.910	13.90
3	412.0	393.0	374.0	4.250	15.11
$3\frac{1}{4}$	475.0	453.0	432.0	5.040	18.00
$3\frac{1}{2}$	555.0	529.0	504.0	5.830	21.00
$3\frac{3}{4}$	640.0	611.0	582.0	6.670	24.00
4	730.0	696.0	664.0	7.590	27.00

Table 9.4 Properties of zinc-coated steel structural strand [2]

| Nominal diameter, in | Minimum breaking strength, tons | | | Approx gross metallic area, in^2 | Approx weight, lb/ft |
| | Class A coating, throughout | Class A coating inner wires | | | |
		Class B coating, outer wires	Class C coating, outer wires		
$\frac{1}{2}$	15.0	14.5	14.2	0.150	0.52
$\frac{9}{16}$	19.0	18.4	18.0	0.190	0.66
$\frac{5}{8}$	24.0	23.3	22.8	0.234	0.82
$\frac{11}{16}$	29.0	28.1	27.5	0.284	0.99
$\frac{3}{4}$	34.0	33.0	32.3	0.338	1.18
$\frac{13}{16}$	40.0	38.8	38.0	0.396	1.39
$\frac{7}{8}$	46.0	44.6	43.7	0.459	1.61
$\frac{15}{16}$	54.0	52.4	51.3	0.527	1.85
1	61.0	59.2	57.9	0.600	2.10
$1\frac{1}{16}$	69.0	66.9	65.5	0.677	2.37
$1\frac{1}{8}$	78.0	75.7	74.1	0.759	2.66
$1\frac{3}{16}$	86.0	83.4	81.7	0.846	2.96
$1\frac{1}{4}$	96.0	94.1	92.2	0.938	3.28
$1\frac{5}{16}$	106.0	104.0	102.0	1.03	3.62
$1\frac{3}{8}$	116.0	114.0	111.0	1.13	3.97
$1\frac{7}{16}$	126.0	123.0	121.0	1.24	4.34
$1\frac{1}{2}$	138.0	135.0	132.0	1.35	4.73
$1\frac{9}{16}$	150.0	147.0	144.0	1.47	5.13
$1\frac{5}{8}$	162.0	159.0	155.0	1.59	5.55
$1\frac{11}{16}$	176.0	172.0	169.0	1.71	5.98
$1\frac{3}{4}$	188.0	184.0	180.0	1.84	6.43
$1\frac{13}{16}$	202.0	198.0	194.0	1.97	6.90
$1\frac{7}{8}$	216.0	212.0	207.0	2.11	7.39
$1\frac{15}{16}$	230.0	226.0	221.0	2.25	7.89
2	245.0	241.0	238.0	2.40	8.40
$2\frac{1}{16}$	261.0	257.0	253.0	2.55	8.94
$2\frac{1}{8}$	277.0	273.0	269.0	2.71	9.49
$2\frac{3}{16}$	293.0	289.0	284.0	2.87	10.05
$2\frac{1}{4}$	310.0	305.0	301.0	3.04	10.64
$2\frac{5}{16}$	327.0	322.0	317.0	3.21	11.24
$2\frac{3}{8}$	344.0	339.0	334.0	3.38	11.85
$2\frac{7}{16}$	360.0	355.0	349.0	3.57	12.48
$2\frac{1}{2}$	376.0	370.0	365.0	3.75	13.13
$2\frac{9}{16}$	392.0	386.0	380.0	3.94	13.80
$2\frac{5}{8}$	417.0	411.0	404.0	4.13	14.47

Table 9.4 (*Continued*)

Nominal diameter, in	Class A coating, throughout	Minimum breaking strength, tons		Approx gross metallic area, in^2	Approx weight, lb/ft
		Class A coating, inner wires			
		Class B coating, outer wires	Class C coating, outer wires		
$2\frac{11}{16}$	432.0	425.0	419.0	4.33	15.16
$2\frac{3}{4}$	452.0	445.0	438.0	4.54	15.88
$2\frac{7}{8}$	494.0	486.0	479.0	4.96	17.36
3	538.0	530.0	522.0	5.40	18.90
$3\frac{1}{8}$	584.0	575.0	566.0	5.86	20.51
$3\frac{1}{4}$	625.0	616.0	606.0	6.34	22.18
$3\frac{3}{8}$	673.0	663.0	653.0	6.83	23.92
$3\frac{1}{2}$	724.0	714.0	702.0	7.35	25.73
$3\frac{5}{8}$	768.0	757.0	745.0	7.88	27.60
$3\frac{3}{4}$	822.0	810.0	797.0	8.43	29.50
$3\frac{7}{8}$	878.0	865.0	852.0	9.00	31.50
4	925.0	911.0	897.0	9.60	33.60

around a center wire, resulting in a symmetrical section (Fig. 9.4). *Wire rope* consists of several strands laid helically around a core (Fig. 9.4), which may be fiber or steel. Steel cores can consist of an independent wire rope (IWRC) or, for small-diameter rope, a wire-strand core (WSC). Both strand and wire rope are normally zinc-coated for protection against corrosion. This coating is available in different thicknesses (see Table 9.2). Strand is designated by the number of wires and wire rope by the number of wires in a strand and the number of strands in the rope. For example, a 6 by 7 class wire rope has 6 strands with 7 wires per strand. The number and size of wires in a strand and the number of strands in a wire rope are normally determined by the manufacturer.

Tables 9.3 and 9.4 present technical data for zinc-coated structural wire rope and strand according to ASTM specifications A603 and A586. When strands and wire rope are tested under tension, they fail under an ultimate load, which can be used for design purposes. Table 9.3 gives representative minimum breaking strengths in tons (2000 lb) per strand and wire rope. Appropriate factors of safety have to be used with these breaking-strength data. A factor of safety of 2 is permissible for a small static structure; a value of 11 is required by code for elevators.

Example 9.4 Consider Example 9.1. Using a factor of safety of 2, select a wire rope size from Tables 9.1 to 9.4.

SOLUTION The maximum T is 34 kips. For a factor of safety of 2, the minimum breaking strength required is 2(34) = 68 kips = 34 tons. Use a $\frac{7}{8}$-in steel structural wire rope with a class A coating.

9.7 SUMMARY

Cable-supported structures involve different loading systems. The most important systems are concentrated loads and a uniform load per unit length of span.

Cable structures are considered statically determinate because they are flexible and develop only tensile forces.

Cables consist of one or more strands or wire rope. Strands and wire rope are normally zinc-coated for protection against corrosion. Design data are given as minimum breaking strength in tons. The factor of safety used in design depends upon the type of structure and its use.

PROBLEMS

9.1 Determine the tension in the cable of the suspended systems in Fig. P9.1.

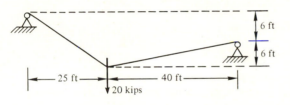

(a)

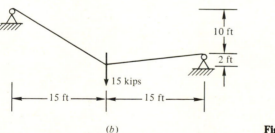

(b)   **Figure P9.1**

9.2 Find the cable tension between each force and determine the required length of the cable for the system in Fig. P9.2.

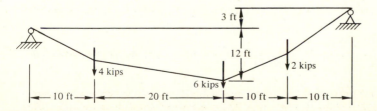

Figure P9.2

9.3 Find the required size of cable with a factor of safety of 2 for the system in Fig. P9.3.

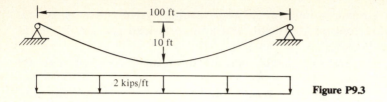

Figure P9.3

9.4 The distributed load on a cable supporting a circular roof of radius 150 ft is shown in Fig. P9.4. Determine the thrust and maximum tension in the cable for $l = 300$ ft and $S = 40$ ft. Select a cable size using a factor of safety of 3.

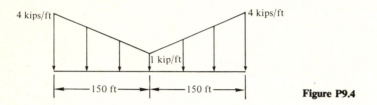

Figure P9.4

9.5 The shallow cable roof system shown in Fig. P9.5 is supported by radial cables connecting an outer compression ring and an inner tension ring set 50 ft below the outer ring. The cables are spaced 30 ft on center on the outer ring and 5 ft on center on the inner ring. The dead and live loads acting on the system add up to 120 lb/ft^2, which is carried to the cables by joists evenly spaced at 30 ft. Determine the size of cable required using a factor of safety of 3. Estimate the size of the compression ring using thin-walled-cylinder theory if the average allowable compressive strength of concrete used is 3000 lb/in^2.

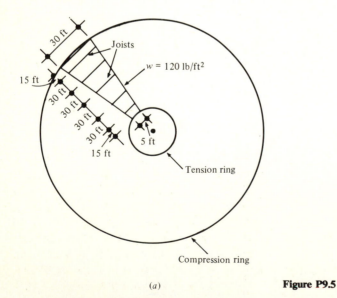

(a)

Figure P9.5

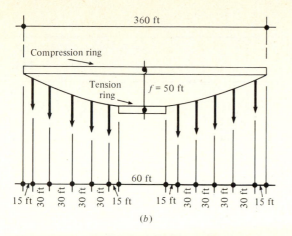

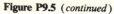

Figure P9.5 (*continued*)

REFERENCES

1. American Iron and Steel Institute, "Manual for Structural Applications of Steel Cables for Buildings," New York, 1973.
2. American Society for Testing and Materials, "Annual Book of ASTM Standards," Philadelphia, 1979.

PROPERTIES OF SECTIONS

10.1 INTRODUCTION

In subsequent chapters it will be necessary to locate the centroid of areas and to calculate the second moment of these areas about various axes. The second moment is more commonly referred to as the *moment of inertia* of a section about some axis, normally a centroidal axis. The *radius of gyration*, which is important relative to the buckling behavior of a member, is directly related to the moment of inertia and will also be considered in this chapter.

10.2 CENTROIDS

The *center of gravity*, used in reference to bodies having weight, refers to the position of the resultant of the earth's pull on the body. Since the name center of gravity is inappropriate for lines and areas, *centroid* is used. The concept of the centroid of an area is most easily understood if we consider a thin plate of uniform thickness made of a homogeneous material. The plate can be divided into a number of small elements of equal size, as shown in Fig. 10.1a. If the weight of each element is w, acting at its center, a system of parallel forces is formed whose resultant W, the weight of the entire plate, passes through the center of gravity of the plate.

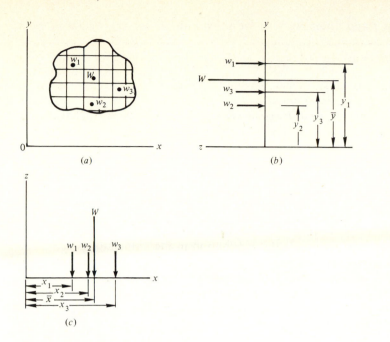

Figure 10.1 Center of gravity of a flat plate.

Let the weights of these elements be equal to w_1, w_2, etc., and let their respective coordinates be (x_1, y_1) and (x_2, y_2), etc. Let the coordinates of the resultant weight W be $\bar{x}$ and $\bar{y}$. Now by the principle of moments, the moment of the resultant W must equal the algebraic sum of the moments of the separate components w_1, w_2, etc., about a common axis. As a result, by moments about O in the xz plane

$$W\bar{x} = w_1 x_1 + w_2 x_2 + w_3 x_3 + \cdots = \Sigma wx$$

and by moments about O in the yz plane

$$W\bar{y} = w_1 y_1 + w_2 y_2 + w_3 y_3 + = \Sigma wy$$

or
$$\bar{x} = \frac{\Sigma wx}{W} \quad \text{and} \quad \bar{y} = \frac{\Sigma wy}{W} \tag{10.1}$$

thus locating the center of gravity.

If we now imagine that the plate thickness is gradually reduced until it reaches zero, the plate no longer has weight but only surface area. The point in the plate which formerly was the center of gravity now becomes the centroid of the area. As a result, Eq. (10.1) becomes

$$\bar{x} = \frac{\Sigma ax}{A} \quad \text{and} \quad \bar{y} = \frac{\Sigma ay}{A} \tag{10.2}$$

This can be proved mathematically by considering each element of area a to have a thickness t and a density γ. If we substitute γta for the weight of each element, we get the moment equation

$$\gamma tA\bar{x} = \gamma ta_1x_1 + \gamma ta_2x_2 + \gamma ta_3x_3 + \cdots = \gamma t\Sigma ax$$

from which the γt can be cancelled, leaving us with Eq. (10.2).

On the basis of Eq. (10.2) about any axis, *the moment of an area equals the algebraic sum of the moments of its component areas.* The moment of an area is defined as the product of the area and the perpendicular distance from the moment axis to the centroid of the area.

10.3 CENTROIDS OF SIMPLE GEOMETRIC AREAS

From Eq. (10.2) we can conclude that the first moment with respect to a centroidal axis must be zero since the summation in the numerator is zero. The moments of the areas on each side of the centroidal axis must be equal and opposite. As a result an axis of symmetry must go through the centroid.

Consider a rectangle. Divide it into narrow strips running parallel to the two opposite sides. The centroid of each strip lies at its midpoint, the centroids of all the parallel strips will lie on a line bisecting the two opposite sides, and the centroid of the entire area will be at the intersection of the two bisecting lines, which is at their midpoints.

In a similar manner, the centroid of a triangle lies at the intersection of its medians. If a triangle is divided into narrow strips running parallel to the base, the centroid of each strip is at its midpoint, the centroid of all the parallel strips will lie on the median of the base, and the centroid of the triangle lies at the intersection of the three medians.

The centroids of more complicated figures such as semicircles can be best determined by integral calculus. The location of centroids of the more common geometric shapes is given in Table 10.1.

10.4 CENTROIDS OF COMPOSITE AREAS

The principle of moments, which is used in locating the centroid of an area, can also be used directly in locating the centroid of a composite area, i.e., an assemblage of areas the location of whose centroids are known.

Equation (10.2) can readily be modified to give

$$\bar{x} = \frac{a_1x_1 + a_2x_2}{a_1 + a_2} \qquad \bar{x} = \frac{\Sigma ax}{A} \qquad (10.3)$$

and

$$\bar{y} = \frac{a_1y_1 + a_2y_2}{a_1 + a_2} \qquad \bar{y} = \frac{\Sigma ay}{A} \qquad (10.4)$$

where a_1, a_2, etc., are component areas with coordinates x_1, y_1, and x_2, y_2, etc. The total area A is the sum of the component areas, and the coordinates of this

Table 10.1 Moment of inertia and radius of gyration of simple areas

Area	Moment of inertia	Radius of gyration
Rectangle	$I_{x_0} = \dfrac{bh^3}{12}$ $I_x = \dfrac{bh^3}{3}$	$r_{x_0} = \dfrac{h}{\sqrt{12}} = 0.289h$ $r_x = \dfrac{h}{\sqrt{3}} = 0.577h$
Triangle	$I_{x_0} = \dfrac{bh^3}{36}$ $I_x = \dfrac{bh^3}{12}$	$r_{x_0} = \dfrac{h}{\sqrt{18}} = 0.236h$ $r_x = \dfrac{h}{\sqrt{6}} = 0.408h$
Circle	$I_{x_0} = \dfrac{\pi R^4}{4}$ $= 0.7854 R^4$ $I_{x_0} = \dfrac{\pi D^4}{64}$ $= 0.0491 D^4$	$r_{x_0} = \dfrac{R}{2} = 0.5R$
Semicircle	$I_{x_0} = \left(\dfrac{\pi}{8} - \dfrac{8}{9\pi}\right) R^4$ $= 0.110 R^4$ $I_x = I_{y_0} = \dfrac{\pi R^4}{8}$ $= 0.3927 R^4$	$r_{x_0} = \dfrac{\sqrt{9\pi^2 - 64}}{6\pi} R$ $= 0.264 R$ $r_x = r_{y_0} = \dfrac{R}{2} = 0.5R$
Quarter circle	$I_{x_0} = 0.055 R^4$ $I_x = \dfrac{\pi R^4}{16}$	$r_{x_0} = 0.264 R$ $r_x = \dfrac{R}{2}$

composite area are $\bar{x}$ and $\bar{y}$. This approach is illustrated in the following examples.

Example 10.1 A beam has the cross section shown in Fig. E10.1a. Locate the centroid of the cross section.

SOLUTION Figure E10.1b shows the cross section of the beam divided into two rectangular parts, 1 and 2. Since the cross section is symmetrical, the centroid is located on the axis of symmetry and only $\bar{y}$ has to be determined.

$$a_1 = (4)(2) \text{ in} = 8 \text{ in}^2 \qquad y_1 = 2 \text{ in}$$

$$a_2 = (4)(2) \text{ in} = 8 \text{ in}^2 \qquad y_2 = 4 \text{ in} + 1 \text{ in} = 5 \text{ in}$$

$$A = a_1 + a_2 = 16 \text{ in}^2$$

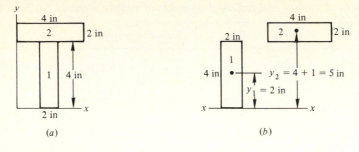

Figure E10.1

Applying Eq. (10.4) gives

$$\bar{y} = \frac{a_1 y_1 + a_2 y_2}{a_1 + a_2} = \frac{(8)(2) + (8)(5)}{16} = 3.5 \text{ in}$$

Areas with a portion cut out can also be considered in this way. The figure can be thought of as an entire area a_1 minus an area a_2. For this case the area a_2 is considered to be a negative area having a negative moment about the moment axis.

Example 10.2 Locate the centroid of a rectangle with the cutout shown in Fig. E10.2.

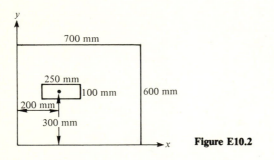

Figure E10.2

SOLUTION

$$a_2 = (700)(600) = 4.2 \times 10^5 \text{ mm}^2 \qquad x_2 = 350 \text{ mm} \qquad y_2 = 300 \text{ mm}$$

$$a_1 = (250)(100) = -2.5 \times 10^4 \text{ mm}^2 \qquad x_1 = 200 \text{ mm} \qquad y_1 = 300 \text{ mm}$$

Applying Eqs. (10.3) and (10.4), we get

$$\bar{y} = \frac{\Sigma ay}{A} = \frac{(4.2 \times 10^5)(300) - (2.5 \times 10^4)(300)}{(4.2 \times 10^5) - (2.5 \times 10^4)} = 300 \text{ mm}$$

$$\bar{x} = \frac{\Sigma ax}{A} = \frac{(4.2 \times 10^5)(350) - (2.5 \times 10^4)(200)}{(4.2 \times 10^5) - (2.5 \times 10^4)} = 359.5 \text{ mm}$$

It should be pointed out that a choice of reference axis for moments can simplify the solution.

Using Table 10.1 enables us to use the above concept for more complex areas. The procedure consists of dividing the figure into the simple areas given in Table 10.1 and then selecting a convenient set of reference axes.

Example 10.3 Locate the centroid of Fig. E10.3.

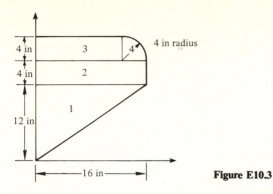

Figure E10.3

SOLUTION The figure is divided into four parts, the location of the centroids being known. The solution is best handled by a tabular format (Table E10.3).

Table E10.3

Section	Area	x	Ax	y	Ay
	$\dfrac{(12)(16)}{2}$	$\dfrac{16}{3}$	$\dfrac{(96)(16)}{3}$	8	(8)(96)
	(4)(16)	8	(8)(64)	14	(14)(64)
	(4)(12)	6	(6)(48)	18	(18)(48)
	$\dfrac{(4)^2\pi}{4}$	$12 + (0.42)(4)$	171.8	$16 + (0.42)(4)$	222.1
	$\Sigma = 220.6$		$\Sigma = 1483.8$		$\Sigma = 2750.1$

Using Eqs. (10.3) and (10.4), we get

$$\bar{x} = \frac{\Sigma Ax}{A} = \frac{1483.8}{220.6} = 6.73 \text{ in}$$

$$\bar{y} = \frac{\Sigma Ay}{A} = \frac{2750.1}{220.6} = 12.46 \text{ in}$$

10.5 THE SECOND MOMENT OF AREA: MOMENT OF INERTIA

When Newton's second law of motion is applied to the angular motion of a rigid body, one of the terms developed is a second moment of the mass about the axis of rotation. Known as the *moment of inertia*, it reflects the resistance of a mass to rotation about a particular axis. The larger the moment of inertia the larger the torque or moment required to bring about a certain angular rotation.

An analogous term is developed relative to an area in the consideration of stresses and deflections of a beam. Although the term *second moment of area* is more suitable in this case, moment of inertia is well established and a physical analogy is possible. With an increase in moment of inertia (second moment of area) a beam has a greater resistance to an applied moment and consequently develops smaller internal stresses.

The moment of inertia, or second moment of area, is defined mathematically as

$$I = \int \rho^2 \, dA \tag{10.5}$$

which indicates that an area is divided into small parts such as dA. Each area is multiplied by the square of its moment arm about the reference axis, and these products are then summed up for the entire area.

If, as shown in Fig. 10.2, the coordinates of the center of the differential area dA are x and y, the moment of inertia relative to the x axis will be

$$I_x = \int y^2 \, dA \tag{10.6}$$

and relative to the y axis will be

$$I_y = \int x^2 \, dA \tag{10.7}$$

referred to as *rectangular moments of inertia*. If the reference axis is the z axis (perpendicular to the xy plane), the moment of inertia is known as the *polar moment of inertia*. Since $r^2 = x^2 + y^2$, the polar moment of inertia is

$$J = I_x + I_y \tag{10.8}$$

In words, this equation states that the polar moment of inertia for an area with respect to an axis perpendicular to its plane is equal to the sum of the moments of inertia about any two mutually perpendicular axes in that plane that intersect on the polar axis.

Examination of the integral $\int \rho^2 \, dA$ indicates a distance squared multiplied by an area. Thus if L is the unit of distance, the unit of I will be L^4. Convenient units for L are feet and inches or centimeters and millimeters. The sign of I will be independent of the sign of the moment arm because it is squared. It depends on the sign of the area. A positive area is one which adds to the area of a figure; a negative area reduces it. For a net area the moment of inertia must be positive.

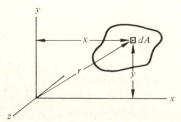

Figure 10.2 Moment of inertia of element dA.

In determining the moment of inertia by integration, it is desirable to select the differential area with care. It is usually convenient to have all parts of the differential area at the same distance from the reference axis. The following example illustrates the technique used in the integration method.

Example 10.4 Determine the moment of inertia for a rectangle of base b and depth h with respect to (a) an axis coinciding with the base, (b) a centroidal axis parallel to the base, and (c) the polar moment of inertia about a centroidal axis perpendicular to the rectangle.

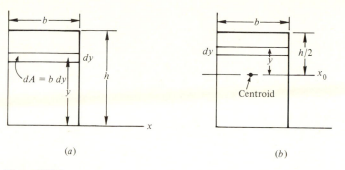

(a) (b)

Figure E10.4

SOLUTION (a) The differential element is selected (Fig. E10.4a) so that the element is parallel to the reference axis. Then by applying Eq. (10.6)

$$I_x = \int y^2 \, dA = \int_0^h y^2 b \, dy = \left[\frac{by^3}{3} \right]_0^h$$

$$I_x = \frac{bh^3}{3}$$

(b) The differential element is selected as shown in Fig. E10.4b and by applying Eq. (10.6) we get

$$I_{x_0} = \int y^2 \, dA = \int_{-h/2}^{h/2} y^2 b \, dy = b \left[\frac{y^3}{3} \right]_{-h/2}^{h/2} \quad \text{and} \quad I_{x_0} = \frac{b}{3} \left(\frac{h^3}{8} + \frac{h^3}{8} \right) = \frac{bh^3}{12}$$

(c) The polar moment of inertia J_0 can be obtained by application of Eq. (10.8)

$$J_0 = I_{x_0} + I_{y_0} = \frac{bh^3}{12} + \frac{hb^3}{12} = \frac{bh}{12}(h^2 + b^2)$$

Using similar approaches, the moment of inertia of various geometric shapes has been determined about various axes. The more common and useful values are given in Table 10.1.

10.6 PARALLEL-AXIS TRANSFER THEOREM

The most useful reference axis for a moment of inertia is an axis through the centroid of an area. The symbol I_{x_0} is used to denote the moment of inertia of an area about an x axis through the centroid of the area. It is possible to obtain the moment of inertia about any axis parallel to that passing through the

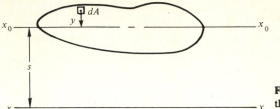

Figure 10.3 Moment-of-inertia transfer theorem.

centroid. Consider Fig. 10.3. The moment of inertia about a centroidal x_0 axis is known to be $I_{x_0} = \int y^2 \, dA$. We are interested in the moment of inertia about an x axis parallel to the centroidal axis x_0 and a distance s from it

$$I_x = \int (y + s)^2 \, dA$$

Expanding gives

$$I_x = \int y^2 \, dA + \int 2sy \, dA + \int s^2 \, dA = \int y^2 \, dA + 2s \int y \, dA + s^2 \int dA$$

Considering each term on the right, we find that $\int y^2 \, dA$ is the definition of I_{x_0}; the second term contains the first moment of the area about its centroidal axis, which by definition must be zero; and the last term yields $s^2 A$. As a result we have

$$I_x = I_{x_0} + As^2 \tag{10.9}$$

and by analogous reasoning

$$J_0 = \bar{J}_0 + As^2 \tag{10.10}$$

Equation (10.9), known as the parallel-axis theorem, states that *the moment of inertia of an area with respect to any rectangular axis is equal to the moment of inertia of the area about a parallel centroidal axis plus the product of the area times the square of the distance separating the axes.* It should be noted that I_x always exceeds I_{x_0}.

10.7 THE MOMENT OF INERTIA OF COMPOSITE AREAS

Most irregular areas of interest to us can be divided into squares, rectangles, circles, etc., for which the rectangular moments of inertia about the centroidal axis and also other convenient parallel axes are known (Table 10.1). As a result, the moment of inertia of a composite area can readily be obtained by using this information and the transfer theorem [Eq. (10.9)].

Example 10.5 Determine the moment of inertia of the I section in Fig. E10.5 with respect to its centroidal axes x_0 and y_0.

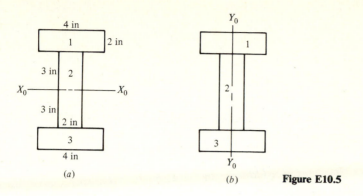

Figure E10.5

SOLUTION The centroidal axis is an axis of symmetry 5 in above the baseline of the section (Fig. E10.5a). Use a tabular format.

Area	I_{x_0}	s	As^2	$I_{X_0} = I_{x_0} + As^2$
1	$\dfrac{(4)(2)^3}{12} = 2.67$	4	$(8)(4)^2 = 128$	130.7
2	$\dfrac{(2)(6)^3}{12} = 36$	0	0	36
3	$\dfrac{(4)(2)^3}{12} = 2.67$	4	$(8)(4)^2 = 128$	130.7

giving

$$I_{X_0} = \Sigma I_{x_0} = 297.4 \text{ in}^4$$

The centroidal axis y_0 is not only an axis of symmetry for the cross section but passes through the centroid of each component area. Consequently the moment of inertia I_{Y_0} is equal to the sum of the moments of inertia of the three components about their centroidal axis y_0.

Area	I_{y_0}
1	$\dfrac{bh^3}{12} = \dfrac{(2)(4)^3}{12} = 10.67$
2	$\dfrac{bh^3}{12} = \dfrac{(6)(2)^3}{12} = 4$
3	$\dfrac{bh^3}{12} = \dfrac{(2)(4)^3}{12} = 10.67$

$$I_{Y_0} = \Sigma I_{y_0} = 25.34 \text{ in}^4$$

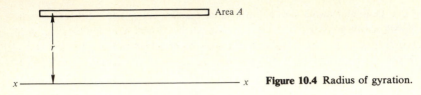

Figure 10.4 Radius of gyration.

10.8 RADIUS OF GYRATION

The *radius of gyration* is a mathematical expression that appears frequently in column formulas and is defined by

$$r = \sqrt{\frac{I}{A}} \qquad \text{or} \qquad I = Ar^2 \tag{10.11}$$

where r = radius of gyration
$\quad I$ = moment of inertia of cross section
$\quad A$ = cross-sectional area

This concept can be visualized with the help of Fig. 10.4. Assume the area A squeezed into a long thin strip parallel to the x axis at a distance r from the axis. As a result each differential element of area dA is at a distance r from the axis. The moment of inertia is given by

$$I = \int r^2 \, dA = r^2 \int dA = Ar^2$$

The strip may be placed on either side of the axis since the term r^2 will always be positive.

Example 10.6 Determine the radius of gyration of the cross section of Example 10.5 about both the X_0 and Y_0 axes.

SOLUTION From Example 10.5

$$I_{X_0} = 297.4 \text{ in}^4 \qquad \text{and} \qquad I_{Y_0} = 25.3 \text{ in}^4$$

Since $r = \sqrt{I/A}$ [Eq. (10.11)],

$$r_X = \sqrt{\frac{297.4}{28}} = 3.26 \text{ in} \qquad \text{and} \qquad r_Y = \sqrt{\frac{25.3}{28}} = 0.95 \text{ in}$$

10.9 SUMMARY

This chapter develops techniques for locating the centroid of an area and evaluating expressions useful in later chapters.

An extension of the theorem of moments is used to locate the centroid of simple and composite areas: the moment of an area about a given axis is equal to the algebraic sum of the moments of its component areas about the same axis.

The moment of inertia, or second moment of area, is defined mathematically as

$$I = \int \rho^2 \, dA$$

and can be evaluated for rectangular coordinates as well as polar axes. Values for most geometric areas have been evaluated and tabulated.

A parallel-axis transfer theorem is available which enables the moment of inertia to be determined for composite areas

$$I_x = I_{x_0} + As^2$$

The radius of gyration, used in column formulas, is defined by

$$r = \sqrt{\frac{I}{A}}$$

PROBLEMS

10.1 Locate the centroid of the triangle in Fig. P10.1 with respect to the y axis.

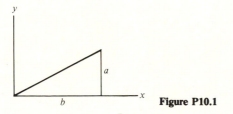

Figure P10.1

10.2 Locate the centroid with respect to the x axis of the parabolic segment in Fig. P10.2.

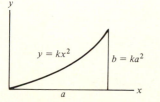

Figure P10.2

10.3 Locate the centroid of the equal leg angle in Fig. P10.3.

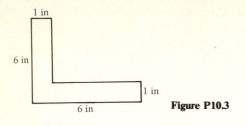

Figure P10.3

10.4 Locate the centroid of the tee section in Fig. P10.4.

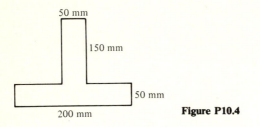

Figure P10.4

10.5 Locate the centroid of the section in Fig. P10.5.

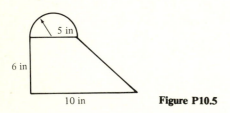

Figure P10.5

10.6 Determine the moment of inertia about the x centroidal axis of the triangle in Prob. 10.1.

10.7 Determine the moment of inertia about the y centroidal axis of the parabolic segment in Prob. 10.2.

10.8 Determine the moment of inertia about the x and y centroidal axes of the angle in Prob. 10.3.

10.9 Determine the polar moment of inertia about the centroid of the section in Prob. 10.4.

10.10 Determine the moment of inertia about the centroidal x axis of Prob. 10.5.

10.11 The cross section of a beam is composed of elements as shown in Fig. P10.11. Determine the moment of inertia with respect to the centroidal x axis.

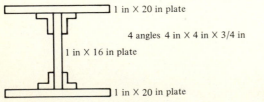

Figure P10.11

10.12 The built-up section in Fig. P10.12 is composed of two angles 6- by 6- by $\frac{3}{4}$-in welded to a 1-by 12-in plate. Determine the moment of inertia with respect to the centroidal $\bar{x}$ axis.

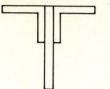

Figure P10.12

ANALYSIS AND DESIGN OF BEAMS

11.1 INTRODUCTION

Beams are among the most commonly used structural elements. Since most loads are vertical (gravity) and most usable surfaces are horizontal, beams are used to transfer vertical loads horizontally. A *beam* can be defined as a member that is subjected to bending by loads having a component transverse to their longitudinal axis.

Some of the beams used in buildings are joists, lintels, spandrels, stringers, and girders (Fig. 11.1). *Joists* are closely spaced beams used to support floors and roofs; *lintels* span openings in masonry walls such as windows and doors. A *spandrel* beam supports the exterior walls of buildings. The term *girder* is used rather loosely to designate a larger beam, one into which smaller ones are framed.

Beams are sometimes classified on the basis of their supports and whether they are statically determinate or not (Fig. 11.2). A *simple beam* is supported by a hinged support at one end and a roller support at the other and is not otherwise restrained. A *cantilever beam* is supported at one end only, with sufficient restraint to prevent rotation. An *overhanging beam* is a simple beam with one or both ends extending beyond the supports. These beams are all statically determinate since their reactions can be determined by the equations of static equilibrium.

Other methods of supporting beams are shown in Fig. 11.3. The *propped beam*, the *fixed-ended beam*, and the *continuous beam* have at least one more

Figure 11.1 Structural skeleton of building. Six-story expansion. Maryland National Bank Operations Center, Baltimore, Maryland. (*Bethlehem Steel Company*.)

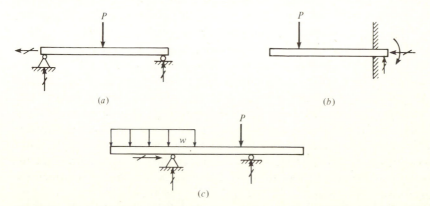

Figure 11.2 Statically determinate beams: (*a*) simple beam, (*b*) cantilever beam, (*c*) overhanging beam.

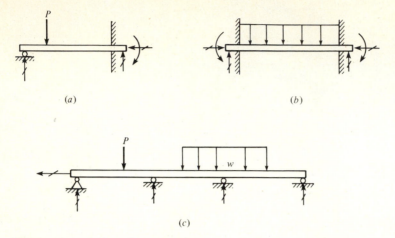

Figure 11.3 Statically indeterminate beams: (*a*) propped beam, (*b*) fixed-ended beam, (*c*) continuous beam.

reaction than is necessary for support. Such beams are statically indeterminate and require additional equations for the evaluation of their reaction components.

11.2 INTERNAL FORCES IN BEAMS

The ultimate aim of analysis and design is to enable one to determine the adequacy of a structure to carry the loads for which it is being designed. The relevant criteria involve comparing the stresses developed by the applied loads with the allowable stresses of the material, and is best done by considering the internal forces acting on the transverse section of a beam. The following discussions are limited to straight beams and beams on which the loads and reactions lie in a single plane containing the centroidal axis of the beam. The additional problem of twisting will not be covered.

Consider the beam in Fig. 11.4, which is cut into two portions by a transverse section *mn*. The external forces acting on each portion of the beam are not in equilibrium. Since the beam as a whole was in equilibrium, internal forces must be acting on the exposed imaginary surface *mn*. These forces must be of such magnitude and direction that their resultant effect balances that of the external forces on the isolated portions and therefore maintains the portion in a state of static equilibrium. In the free-body diagrams of Fig. 11.4 the internal forces are represented by *V* and *N* acting through the centroid of the cross section and a resisting moment *M*. It is important to note that the resultant effects *V*, *N*, and *M*, acting on the section *mn* of the right-hand portion of the beam, are equal but opposite those acting on the left-hand portion. That this is true becomes apparent in the following discussion.

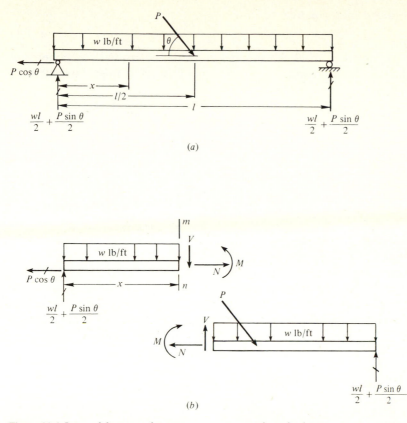

(a)

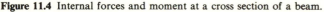

(b)

Figure 11.4 Internal forces and moment at a cross section of a beam.

The resultant of the external forces acting on the beam to the left of section *mn* can be calculated, and the resultant of the external forces to the right of *mn* can also be calculated. Since the beam as a whole is in equilibrium, the two resultants previously calculated must be equal, opposite in sense, and collinear. Similar reasoning also applies to the resultant effects V, N, and M acting on the cross section *mn* of the left- and right-hand portions of the beam.

11.2.1 Development of Shear, Bending-Moment, and Axial-Force Equations

The particular values of V, N, and M can be calculated by using the equations of static equilibrium. The beam illustrated in Fig. 11.4a shows a simply sup- ported beam carrying a uniform load plus a concentrated load at the center of the beam acting at an angle θ from the horizontal. The weight of the beam is considered part of the uniform load. Assume that a cutting plane *mn* is located at a distance x from the left reaction. Consider the left portion of the beam (Fig. 11.4b). To satisfy the condition $\Sigma X = 0$, the magnitude of N must be $P\cos\theta$. In a more generalized mathematical form

$$N = (\Sigma H)_L \tag{11.1}$$

where the subscript L restricts the summation of the horizontal components of external forces to the left portion of the beam. This force is an internal axial force.

To satisfy $\Sigma Y = 0$, the vertical unbalance caused by the left reaction and the uniform load requires a resisting force

$$V = \frac{wl}{2} + \frac{P\sin\theta}{2} - wx$$

The generalized equation is

$$V = (\Sigma Y)_L \tag{11.2}$$

where the subscript L restricts the summation of the vertical components of the external forces to the left portion of the beam. The force V is an internal shearing force because it is parallel to the section mn.

For complete equilibrium the summation of moments of all the forces acting on the left portion of the beam must be equal to zero. For convenience the summation is taken around the centroid of the cross section of plane mn. As a result

$$M = \left(\frac{wl}{2} + \frac{P\sin\theta}{2}\right)x - \frac{wx^2}{2}$$

This can be stated in a more generalized form as

$$M = (\Sigma M)_L \tag{11.3}$$

where the subscript L limits the summation of moments to the forces on the left portion of the beam about the centroidal axis of section mn.

Equations (11.1) to (11.3) are mathematical expressions for the axial force, shear force, and bending moment developed at a given cross section.

11.2.2 Definitions and Sign Conventions

The most widely used sign convention is based on the assumption that positive bending moment produces bending of the beam concave upward, as shown in Sketch 11.1a. This can be attained by considering forces acting up as being *positive* and downward forces as *negative*. These forces will develop correspondingly *positive* and *negative* bending moments regardless of which portion of the beam is being considered (Sketch 11.1b). Considering only the left-hand portion of a beam, positive forces (acting up) induce *positive shear* at the cross section (shear force acting downward) (Sketch 11.1c). Axial forces causing tension are normally considered positive (Sketch 11.1d).

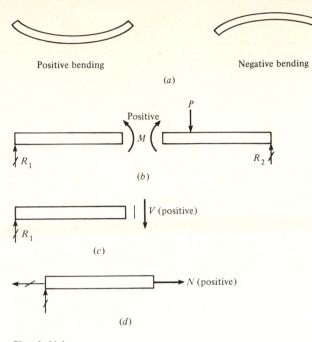

Positive bending   Negative bending

(a)

(b)

(c)

(d)

Sketch 11.1

Example 11.1 Write shear-force and bending-moment equations for the beam loaded as shown in Fig. E11.1a.

SOLUTION The initial step is to calculate the reactions from the equations of statics. $\Sigma M_{R_L} = 0$ gives $R_R = 850$ lb, and $\Sigma M_{R_R} = 0$ gives $R_L = 550$ lb. A check can be made by $\Sigma F_y = 0$.

The sections in the beam at which the loading conditions change are called *change-of-load points*, designated A, B, C, and D. Shear-force and bending-moment equations can be written for loading conditions between these points.

If a section aa is taken through the beam anywhere between A and B, the free-body diagram of the left portion of the beam is as shown in Fig. E11.1b. The shearing force and internal bending moment exposed are designated V and M. There is no axial force N since there are no horizontal components of external loads.

From Eqs. (11.2) and (11.3) the shear-force and bending-moment equations can be developed for the portion of the beam between points A and B:

$$V = (\Sigma Y)_L: \qquad\qquad V_{AB} = 550 - 200x \qquad\qquad 0 \leq x \leq 5$$

$$M = (\Sigma M)_L: \qquad\quad M_{AB} = 550x - 200x\frac{x}{2} = 550x^2 - 100x^2 \qquad 0 \leq x \leq 5$$

To obtain shear-force and bending-moment equations for the beam between points B and C, assume another exploratory section bb taken anywhere between B and C. The free-body diagram is given in Fig. E11.1c. The location of bb is still given in terms of x measured from the left support. The range of x is now between the limits of 5 and 10 ft:

$$V = (\Sigma Y)_L: \qquad\qquad V_{BC} = 550 - (200)(5) = -450 \text{ lb} \qquad 5 \leq x \leq 10$$

$$M = (\Sigma M)_L: \qquad\qquad M_{BC} = 550x - (200)(5)(x - 2.5)$$

$$= -450x + 2500 \qquad\qquad 5 \leq x \leq 10$$

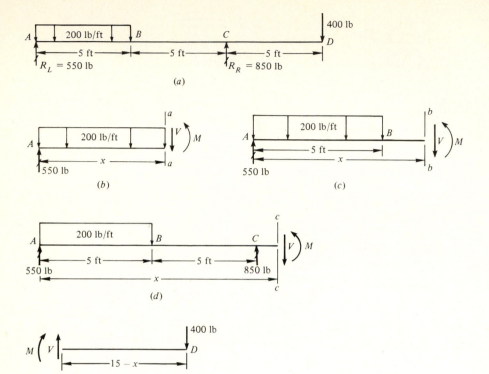

Figure E11.1

The shear-force and bending-moment equations are similarly found for segment CD by using a section cc located between C and D (the free-body diagram is given in Fig. E11.1d):

$$V = (\Sigma Y)_L: \qquad V_{CD} = 550 - 1000 + 850 = -400 \qquad\qquad 10 \leq x \leq 15$$

$$M = (\Sigma M)_L: \qquad M_{CD} = 550x - (1000)(x - 2.5) + (850)(x - 10)$$

$$= 400x - 6000 \qquad\qquad 10 \leq x \leq 15$$

It is important to note that the same moment equation could have been obtained more easily by considering the free-body diagram of the right portion of the beam (Fig. E11.1e):

$$M = (\Sigma M)_R: \qquad M_{CD} = (-400)(15 - x) = 400x - 6000 \qquad\qquad 10 \leq x \leq 15$$

To summarize, shear-force and bending-moment equations were developed for consecutive portions of the beam within which the loading remained constant. Equations (11.2) and (11.3) were used by considering free-body diagrams of appropriate portions of the beam.

11.2.3 Shear-Force and Bending-Moment Diagrams

These diagrams are the graphical representation of the shear-force and bending-moment equations plotted on V-vs.-x and M-vs.-x axes, usually plotted below the loading diagram. These diagrams for Example 11.1 are shown in Fig. 11.5.

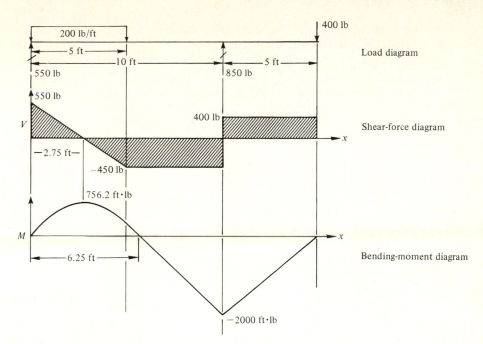

Figure 11.5 Shear-force and bending-moment diagrams for loaded beam.

11.3 RELATIONSHIPS BETWEEN LOAD, SHEAR FORCE, AND BENDING MOMENT

Important relationships existing between load, shear force, and bending moment are extremely useful in constructing shear-force and bending-moment diagrams. They are not independent of the basic definitions and can be used in conjunction with them.

The beam in Fig. 11.6a is assumed to carry any general loading. The free-body diagram of a segment dx of the beam is isolated in Fig. 11.6b. The effects of the loads to the left of the segment are reflected in the shear force V

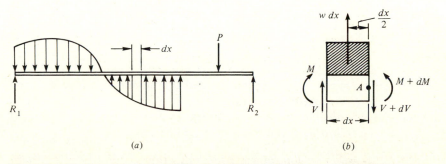

(a)   (b)

Figure 11.6 Development of relationships between load, shear force, and bending moment.

and the bending moment M. Although the loading is variable, it can be considered uniform over the differential length dx. As a result, the shear force and bending moment on the right face of the segment will have increased by differential amounts to values of $V + dV$ and $M + dM$, respectively.

Since the free-body diagram of the segment (Fig. 11.6b) is in equilibrium, a summation of forces must equal zero

$$\uparrow \Sigma Y = 0: \qquad\qquad V + w\,dx - (V + dV) = 0$$

and
$$dV = w\,dx \qquad\qquad (11.4)$$

and similarly a summation of moments must equal zero

$$\Sigma \overset{\frown}{M_A} = 0: \qquad\qquad M + V\,dx + w\,dx\frac{dx}{2} - (M + dM) = 0$$

The third term of a moment summation is the square of a differential, which makes it negligible compared with the other terms. As a result

$$dM = V\,dx \qquad\qquad (11.5)$$

Integrating, Eq. (11.4) gives

$$\int_{V_1}^{V_2} dV = \int_{x_1}^{x_2} w\,dx$$

in which the limits are the shear force V_1 at position x_1 and the shear force V_2 at position x_2. As a result, the left-hand term reduces to $V_2 - V_1$ and represents the change in shearing force between sections x_2 and x_1. In the right-hand term, the product $w\,dx$ represents the area of the load diagram acting on the segment. The definite integral $\int_{x_1}^{x_2} w\,dx$ represents the summation of such terms and represents the area under the load diagram between positions x_1 and x_2. As a result the integration of Eq. (11.4) yields

$$V_2 - V_1 = \text{area under load diagram between sections } x_2 \text{ and } x_1 \quad (11.6)$$

Similarly the integration of Eq. (11.5) gives

$$\int_{M_1}^{M_2} dM = \int_{x_1}^{x_2} V\,dx$$

which yields

$$M_2 - M_1 = \text{area under shear diagram between sections } x_2 \text{ and } x_1 \quad (11.7)$$

since the product $V\,dx$ between sections x_2 and x_1 represents the area under the shear diagram.

Equations (11.6) and (11.7) provide a convenient means of computing changes in shear force and bending moment and the numerical values at given sections. The areas associated with various curves in Fig. 11.7 can be used to advantage in these calculations (see Examples 11.2 through 11.5). Of equal importance are the following variations of Eqs. (11.4) and (11.5):

$$w = \frac{dV}{dx} = \text{slope of shear-force diagram} \qquad\qquad (11.8)$$

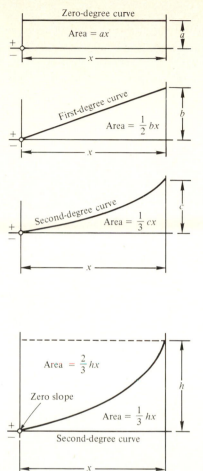

Figure 11.7 Area under curves.

and

$$V = \frac{dM}{dx} = \text{slope of bending-moment diagram} \qquad (11.9)$$

which can be used to sketch shear-force and bending-moment diagrams.

For an application of these principles, consider Fig. 11.8. According to our sign convention, forces acting up are considered positive. Normal gravity forces acting downward are negative (Fig. 11.8a). The area of such downward-acting loads is therefore considered negative and represents a decrease in shearing force. Positive shearing forces are plotted upward from the x axis, and those lying above the x axis represent increases in bending moment (Fig. 11.8b and c).

If we consider the load diagram with a section just to the right of R_1, the shearing force at that section will be equal to R_1 [Eq. (11.2) and Fig. 11.8b]. Since the loads are negative (downward-acting), the value of the shearing force decreases as we consider sections to the right of R_1. Since the load becomes larger (increasingly negative) as we approach section x_1, the slope of the

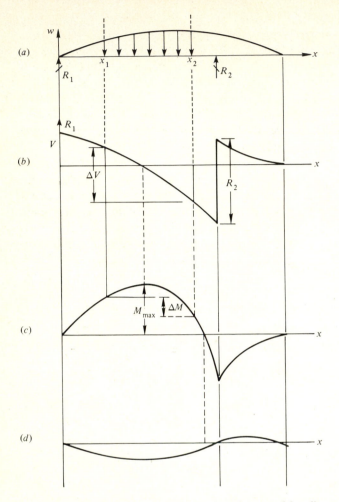

Figure 11.8 Relationship between load, shear force, and bending moment: (*a*) load diagram, (*b*) shear-force diagram, (*c*) bending-moment diagram, (*d*) elastic curve.

shear-force diagram becomes increasingly negative [Eq. (11.8) and Fig. 11.8*a* and *b*]. A similar application of Eq. (11.9) indicates that while there is positive shear force as section x_1 is approached, it is decreasingly positive; consequently the slope of the bending-moment diagram is positive but becomes decreasingly positive. The bending-moment diagram reaches a maximum positive value when the shear force reaches zero from a positive value. As the shear force becomes more negative, so does the slope of the moment diagram. According to Eq. (11.6), as we move from section x_1 to section x_2, the change in shear force ΔV (Fig. 11.8*b*) represents the area under the load diagram between these two sections. Since it is a negative loading, we have a decrease in shearing force.

In a similar manner on the basis of Eq. (11.7), the change in bending moment ΔM between sections x_1 and x_2 will be the area under the shear-force diagram between the two sections. It must be remembered that shear-force areas above the line are positive and below the line are negative. Because R_1 represents a hinge support, the moment at this support will be equal to zero.

Another point of interest is the abrupt change in shear force (from a negative value to a positive value) as the section crosses from one side of reaction R_2 to the other side. As a result the slope of the bending-moment curve changes from negative to positive and forms a cusp.

Sketching the elastic curve (the shape taken by the centerline of the beam under loading) can be of value in describing the behavior of a structure. An exaggerated elastic curve is given in Fig. 11.8d. Initially the beam is curved concave up because of the positive bending moment. Because of the hinge support the beam has an initial slope at the reaction. Point A is of interest because it represents a point of zero bending moment, a point of inflection where the curvature of the beam changes from concave up to concave down. The bending moment also changes from positive to negative. The elastic curve must go through the support R_2.

The steps used in constructing shear-force and bending-moment diagrams can be summarized as follows:

1. Calculate the reactions.
2. Calculate values of shear force at the change-of-load points using either Eq. (11.2) or (11.6).
3. Sketch the shear-force diagram, determining the shape from Eq. (11.8); the magnitude of the load ordinate equals the slope at the corresponding ordinate of the shear-force diagram.
4. Locate points of zero shear force.
5. Calculate values of bending moment at the change-of-load points and at the points of zero shear force, using Eq. (11.3) or (11.7).
6. Sketch the bending-moment diagram through the ordinates of the bending moments computed in step 5. The shape of the diagram is determined from Eq. (11.9); the magnitude of the shear-force ordinate equals the slope at the corresponding ordinate of the bending-moment diagram.

Example 11.2 The beam shown in Fig. E11.2a carries a single concentrated load of 18 kips at point B. Draw the shear-force and bending-moment diagrams for the beam (neglect the weight of the beam).

SOLUTION The reactions are determined by the summation of the moments of all forces, first about reaction R_1 and then R_2, and equating the sum to zero in each case

$$R_1 = 6 \text{ kips} \qquad R_2 = 12 \text{ kips}$$

The next step involves determining the shear-force values at change-of-load positions (see Fig. E11.2b). To the left of the left reaction the shear force is zero. Immediately to the right of reaction R_1 the shear force becomes the value of the reaction ($+6$ kips). Since no new loads are

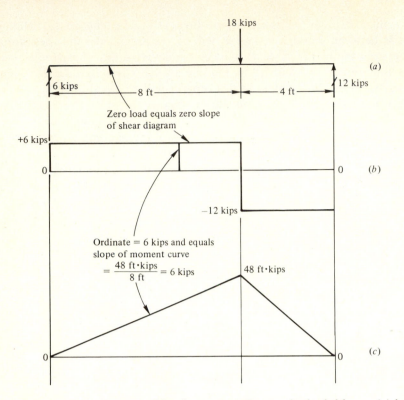

18 kips

(*a*)

6 kips

8 ft

4 ft

12 kips

Zero load equals zero slope
of shear diagram

+6 kips

0

0

(*b*)

−12 kips

Ordinate = 6 kips and equals
slope of moment curve

$= \dfrac{48 \text{ ft} \cdot \text{kips}}{8 \text{ ft}} = 6 \text{ kips}$

48 ft·kips

0

0

(*c*)

Figure E11.2 Shear force and bending-moment diagrams for loaded beam: (*a*) load diagram, (*b*) shear-force diagram, (*c*) bending-moment diagram.

applied between the left reaction and the 18 kips applied load, the slope of the shear-force diagram is zero and remains constant. Passing from the left side of the 18-kip load to the right side of the load (passing through a concentrated negative load) results in a change in shear force (reduction) by the amount of the load. As a result, the shear-force diagram changes from + 6 to − 12 kips and remains constant until the right reaction R_2 is reached. In passing through this reaction (+ 12 kips) the shear-force diagram returns to zero.

The bending moment is zero at the supports of a simply supported beam (see Fig. E11.2*c*). Since the shear force is constant at 6 kips between *A* and *B*, the slope of the bending-moment curve is constant at 6 kips [(6 kips/ft) ft]. The change in value of the bending moment from the left support *A* (where it is zero) to *B* (applied load) is 48 ft·kips, which is equal to the area under the shear-force diagram between these two points [(6 kips) (8 ft) = 48 ft·kips].

In a similar manner, the slope of the bending-moment curve from *B* to *C* is 12 kips, the value of the shear force from *B* to *C*. The area under the shear diagram from *B* to *C* is −48 ft·kips [(− 12 kips)(4 ft)], which is the change in value of bending moment from 48 ft·kips at *B* to zero at *C*.

Example 11.3 Construct the load, shear-force, and bending-moment diagrams for the beam shown in Fig. E11.3*a*. Evaluate all principal ordinates and determine the maximum shear force and bending moment.

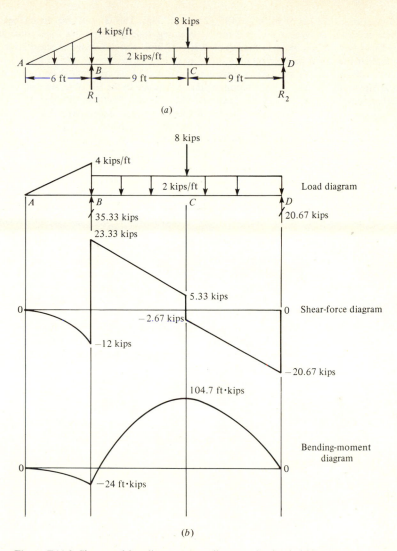

Figure E11.3 Shear and bending-moment diagrams for loaded beam.

SOLUTION Solving for reactions, we get

$M_{R_1}:$

$$(8)(9) + (2)(18)(9) - \frac{(4)(6)(2)}{2} - 18R_2 = 0$$

$$R_1 = 20.67$$

$M_{R_2}:$

$$(-8)(9) - (2)(18)(9) - \frac{(6)(4)(20)}{2} + 18R_1 = 0$$

$$R_2 = 35.33$$

$\uparrow \Sigma F = 0:$

$$35.33 + 20.67 - 8 - (2)(18) - \frac{(4)(6)}{2} = 0 \qquad \text{check}$$

The values of vertical shear force are next calculated at the change-of-load positions. At A the shear force is zero. At the left of B applying $V = (\Sigma Y)_L$ gives the shear force as the area of the load diagram to the left of B, or $[-(4)(6)/2] = -12$ kips. This is of course the change in shear force occurring between A and B. Since loads acting down give negative shear force, the resulting shear force in the beam to the left of B is -12 kips. The difference in shear force in passing from the left of B to the right of B is the magnitude of reaction R_1 (35.33 kips). Since the reaction acts up, the change in shear force is positive (35.33 kips), giving a shear force value of $+23.33$ kips to the right of B. In determining the shear force to the left of C, we sum up the forces acting to the left of C $[V = (\Sigma Y)_L]$ and get $-12 + 35.33 - (2)(9) = 5.33$ kips. The value of the shearing force to the right of C (concentrated load of 8 kips) differs from that to the left of C by -8 kips (the magnitude of the concentrated load) and equals -2.67 kips. The shearing force in the beam just to the left of the reaction R_2 at D can be best obtained by considering the forces on the right-hand portion of the beam (the reaction of 20.67 kips acting up). Since the shearing force on the left portion of the beam is considered positive when acting down, it will be positive when acting up on the right-hand portion of the beam (forces acting on opposite faces of a section must be equal and opposite). Consequently the shearing force on the section just to the left of D is -20.67 kips. On passing through the concentrated load representing the reaction the shearing force on a section just to the right of the reaction becomes zero.

The bending-moment diagram can best be developed by using Eqs. (11.5) and (11.7). Equation (11.5) can be used for sketching purposes and Eq. (11.7) for determining critical values of bending moment. From Eq. (11.7) we realize that the change in bending moment from point A of the beam, where it is zero, to point B is equal to the area under the shear-force curve between the points. Using the information provided in Fig. 11.7, we can calculate the area under the second-degree shear-force curve to be $-(\frac{1}{3})(6)(12) = -24$ ft·kips. On the basis of Eq. (11.5) we realize that since the shearing force between the two sections becomes increasingly negative according to a first-degree curve, the slope of the bending-moment curve will become increasingly negative, giving a second-degree curve. At B, since the shearing force changes from negative to positive, the slope of the bending-moment curve must also change from negative to positive. As we proceed from section B to C, the shearing force is positive but decreasingly positive, giving a moment diagram with a slope that is positive but decreasingly positive. The change in bending moment from B to C is equal to the area under the shear-force curve from B to C

$$\frac{23.33 + 5.33}{2}(9) = 128.7 \text{ ft·kips}$$

Since the bending moment at B was -24 ft·kips, the bending moment at C will be $128.7 - 24 = 104.7$ ft·kips. From section C to D the shearing force is negative and increasingly negative. As a result, the slope of the bending-moment curve is negative and increasingly negative. The change in bending moment between C and D is equal to the area under the shear-force diagram between C and D

$$\frac{-2.67 - 20.67}{2}(9) = -104.7 \text{ ft·kips}$$

The bending moment at D, the end of the beam, must be zero. This is verified by subtracting the negative area of 104.7 ft·kips from the bending moment at C, namely 104.7 ft·kips, which equals zero.

The shear-force and bending-moment diagrams are shown in Fig. E11.3b. The maximum shear force of 23.33 kips occurs at reaction R_1 at B. The maximum bending moment of 104.7 ft·kips occurs at point C in the beam.

Example 11.4 Construct the load, shear-force, and bending-moment diagrams for the cantilever beam shown in Fig. E11.4. Evaluate all principal ordinates.

SOLUTION See Fig. E11.4.

Example 11.5 Construct the load, shear-force, and bending-moment diagrams for the beam shown in Fig. E11.5. Evaluate all principal ordinates.

SOLUTION See Fig. E11.5.

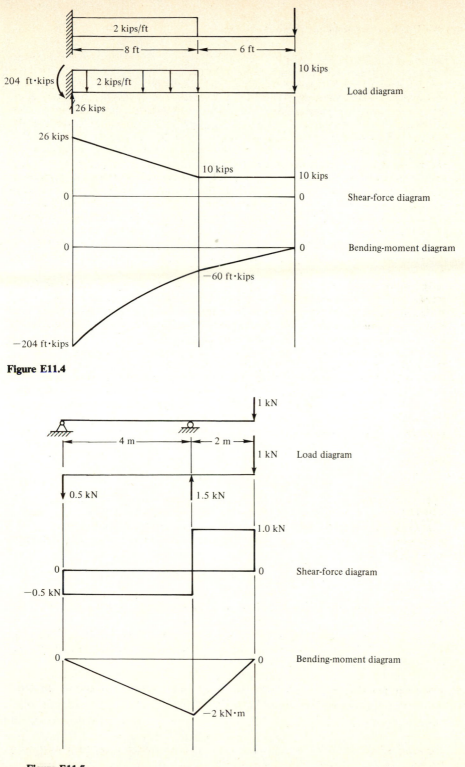

Figure E11.4

Figure E11.5

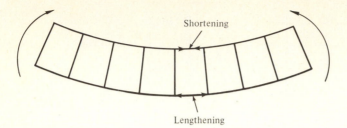

Figure 11.9 The shortening and lengthening of fibers in a bent beam.

11.4 FLEXURAL STRESSES

11.4.1 Introduction

As observed previously, beams are bent by transverse loads. Figure 11.9 shows the bending of a rubber model beam under transverse loads. Its bending is exaggerated compared with that of most steel and timber beams. It is important to realize that plane sections through the beam remain plane and that because of the bending the fibers near the top of the beam shorten and those near the bottom of the beam lengthen. *Fibers* are imaginary long thin elements lying parallel to the longitudinal axis of a beam. The shortening and lengthening of fibers, when considered in connection with Hooke's law, indicate the development of so-called *fiber stresses*, often referred to as *bending* or *flexural stresses*.

11.4.2 Derivation of Flexural Formula

In deriving the relationship between bending moment and flexural stresses several assumptions are made:

1. Plane sections of the beam remain plane after bending.
2. The beam is made out of a homogeneous material which obeys Hooke's law.
3. The moduli of elasticity for tension and compression are equal.
4. The beam is initially straight and of constant cross section.

Some of these assumptions can be proved experimentally, and others are reasonable in terms of normal construction materials.

Figure 11.10a shows two adjacent sections *ab* and *cd* separated by the distance *dx*. Because of the bending caused by loads *P*, sections *ab* and *cd* rotate relative to each other by the angle *dθ* (Fig. 11.10b). These planes remain plane according to assumption 1. Fiber *ac* at the top of the beam is shortened, and fiber *bd* at the bottom is lengthened. Somewhere between these two fibers is a fiber *ef* whose length is unchanged. The plane containing fiber *ef* is called the *neutral surface*. Drawing the line *c'd'* through *f* parallel to *ab* shows that fiber *ac*

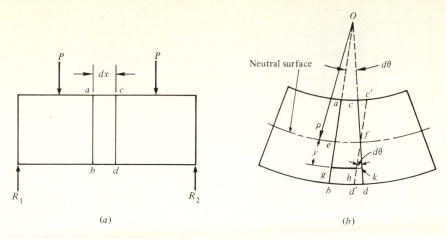

Figure 11.10 Deformations in beam fibers.

is shortened by an amount cc' and is in compression and that fiber bd is lengthened by an amount $d'd$ and is in tension.

Consider the deformation of a typical fiber gh located y units below the neutral surface. Its elongation hk is the arc of a circle of radius y subtended by an angle $d\theta$ and is given by

$$\delta = hk = y\,d\theta$$

The strain is found by dividing the deformation by the original length ef of the fiber

$$\varepsilon = \frac{hk}{gh} = \frac{\delta}{L} = \frac{y\,d\theta}{ef}$$

If the radius of curvature of the neutral surface is ρ, the curved length ef is equal to $\rho\,d\theta$. Consequently the strain is

$$\varepsilon = \frac{y\,d\theta}{\rho\,d\theta} = \frac{y}{\rho}$$

On the basis of Hooke's law the stress in the fiber gh is given by

$$f_b = \varepsilon E = \frac{E}{\rho}y \qquad (11.10)$$

indicating that the stress in any fiber in a beam varies directly with its distance y from the neutral surface.

To complete the derivation of the flexural formula we consider the equations of equilibrium. Consider a portion of the beam to the left of section ab as a free body (Fig. 11.11a). If we consider section ab in Figure 11.11b, the forces acting on a typical element dA are the shearing force $S\,dA$ and a flexural force $f_b\,dA$. The line of intersection between the neutral surface and a transverse section of a beam is called the *neutral axis*.

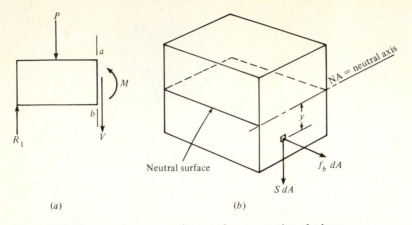

(a)   *(b)*

Figure 11.11 Forces acting on any element of a cross section of a beam.

Since there are no horizontal components of forces acting on the free body, to satisfy the condition that a horizontal summation of forces is equal to zero ($\Sigma X = 0$) we must have

$$\int f_b \, dA = 0$$

on the cross section *ab*.

By replacing f_b with the value given in Eq. (11.10) this becomes

$$\frac{E}{\rho} \int y \, dA = 0$$

The constant E/ρ can be placed outside the integral sign. Since $y \, dA$ is the moment of the differential area dA about the neutral axis, the integral $\int y \, dA$ is the total moment of area and

$$\frac{EA\bar{y}}{\rho} = 0$$

where A is the total area of the cross section and $\bar{y}$ is the distance from the centroid of the area to the neutral axis.

In this expression only $\bar{y}$ can be equal to zero. Consequently the distance from the neutral axis to the centroid of the cross section must be zero. As a result *the neutral axis must pass through the centroid of the section.*

The condition that a vertical summation of forces acting on the free body must be equal to zero ($\Sigma y = 0$) will be used later in deriving the shearing-stress formula.

Finally, to satisfy the condition of equilibrium relative to moments ($\Sigma M = 0$), the external bending moment acting at a cross section of a beam must be balanced by the internal resisting moment at that section. The resisting moment about the neutral axis of a typical element is $y f_b \, dA$ (Fig. 11.11*b*). As a result, for all the elements of cross sections

$$M = \int y(f_b \, dA)$$

which, by replacing f_b by Ey/ρ from Eq. (11.10), becomes

$$M = \frac{E}{\rho} \int y^2 \, dA$$

$\int y^2 \, dA$ is defined as I, the moment of inertia of the area about a reference axis [Sec. 10.5, Eq. (10.5)]. In this case the reference axis is the neutral axis, which coincides with the centroidal axis. Consequently,

$$M = \frac{EI}{\rho}$$

The importance of summing the moments of external forces about the centroidal axis (Sec. 11.2.1) is now apparent because the internal moment is also obtained by summation about the same axis.

A more useful form of the above equation is

$$\frac{1}{\rho} = \frac{M}{EI} \qquad (11.11)$$

Because curvature is equal to the reciprocal of the radius of curvature, Eq. (11.11) indicates that curvature is directly proportional to bending moment. This is the basis of the sign convention recommended in Sec. 11.2.2.

By equating the ratio E/ρ from Eq. (11.10) with its value from Eq. (11.11) we have

$$\frac{E}{\rho} = \frac{M}{I} = \frac{f_b}{y}$$

which gives us the flexural formula

$$f_b = \frac{M}{I} y \qquad (11.12)$$

where f_b = flexural stress, lb/in^2
  M = bending moment, $in \cdot lb$
  I = moment of inertia, in^4
  y = distance of fiber being considered, in

This expression indicates that flexural stress is proportional to the distance the fiber is from the neutral axis. By replacing y with the distance to the outermost fiber c, we obtain an expression for the maximum flexural stress

$$f_{b\,max} = \frac{Mc}{I} \qquad (11.13)$$

A better visualization of the above derivation is possible by considering a rectangular cross section. The stress distribution is such that the upper portion

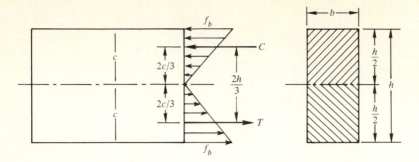

Figure 11.12 Resisting moment in a beam due to internal stresses.

of the beam is in compression and the lower portion in tension. The stresses vary linearly with distance from the neutral axis, which goes through the center of the beam (Fig. 11.12). The resultant of the compressive stresses must be equal and opposite in direction to the resultant of the tensile stresses. These resultants are parallel and equal to

$$\frac{f_b}{2}\frac{bh}{2} = \frac{f_b bh}{4}$$

and act at the centers of pressure, which, for a triangle, are two-thirds the distance from the apex to the base. As a result, the forces which constitute a couple are a distance $2h/3$ apart and generate an internal resisting moment of magnitude

$$f_b \frac{bh}{4}\frac{2h}{3} = \frac{f_b bh^2}{6}$$

$$M = \frac{f_b bh^2}{6}$$

For a rectangle, $I = bh^3/12$ and $c = h/2$. We know that

$$M = f_b \frac{I}{c} = \frac{f_b bh^3}{12h/2} = \frac{f_b bh^2}{6}$$

which is equal to the value obtained above.

11.4.3 Modulus of Rupture

Equation (11.13) can be used to determine the flexural stress in a beam loaded to failure in a testing machine. Since the proportional limit of the material is exceeded, the stress determined in this manner is not a true stress. It is called the *modulus of rupture*. It is useful in comparing beams of various sizes and materials.

11.4.4 Section Modulus

The moment of inertia I and the distance c are properties of the cross section of the beam. The ratio I/c is called *section modulus*, often denoted by S. This property is part of the structural information made available for rolled steel and timber sections (Tables A.2, A.3, and A.5).

11.4.5 Economic Sections

A beam with a rectangular or circular cross section has a lot of material near the neutral axis which is understressed. This makes the cross section inefficient in bending. The flexural formula written as $M = f_b I/c$ indicates that if the area of a rectangular beam can be rearranged to have the same depth but a larger moment of inertia, the cross section will be more efficient because it can develop a greater resistance moment. This has been accomplished by the development of a wide-flange section (Sketch 11.2).

Physically, the resisting moment is increased because more fibers are located at a greater distance from the neutral axis. Such fibers carry a greater stress and have a larger moment arm relative to the neutral axis. Wide-flange beams are normally specified by depth and weight per lineal foot. For example, W24 × 76 indicates a wide-flange beam 24 in deep weighting 76 lb/ft (Table A.5).

Sketch 11.2

11.4.6 Unsymmetrical Sections

All the beams discussed so far have been symmetrical with respect to the neutral axis. Since flexural stresses are proportional to the distance from the neutral or centroidal axis, such beam sections are desirable when the material used is equally strong in tension and compression. For a material weaker in tension, such as cast iron, it is desirable to use beams that are unsymmetrical with respect to the neutral axis. An ideal example would be one in which the ratio of distances to outermost fibers for the unsymmetrical section is equal to the ratio of allowable stresses in tension and compression.

Example 11.6 A timber beam 6 by 10 in supports the loads shown. Determine the maximum flexural stress.

SOLUTION See Fig. E11.6. The maximum bending moment is 9600 ft·lb, and

$$f_b = \frac{Mc}{I} = \frac{M}{S}$$

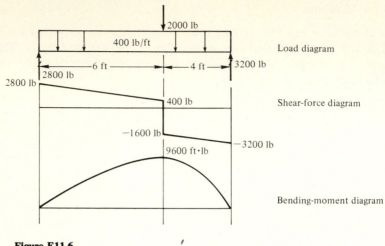

Figure E11.6

For 6 by 10 in, $S = 82.73$ in^3 from Table A.2. Then

$$f_b = \frac{(9600 \text{ ft} \cdot \text{lb})(12 \text{ in/ft})}{82.73 \text{ in}^3} = 1392 \text{ lb/in}^2$$

Example 11.7 A structural-steel wide-flange section W8 × 28 used as a cantilever beam 10 ft long supports a uniform load of 500 lb/ft (Fig. E11.7). Calculate the maximum bending stress.

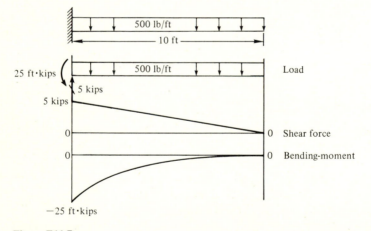

Figure E11.7

SOLUTION

$$\text{Maximum bending moment} = 25{,}000 \text{ ft} \cdot \text{lb}$$

$$F_b = \frac{M}{S}$$

$$S = 17.0 \text{ in}^3 \text{ for W8} \times 28 \qquad \text{Table A.5}$$

$$f_b = \frac{(25{,}000)(12)}{17} = 17{,}650 \text{ lb/in}^2$$

Example 11.8 Compute the maximum tensile and compressive stresses developed in the loaded beam shown in Fig. E11.8.

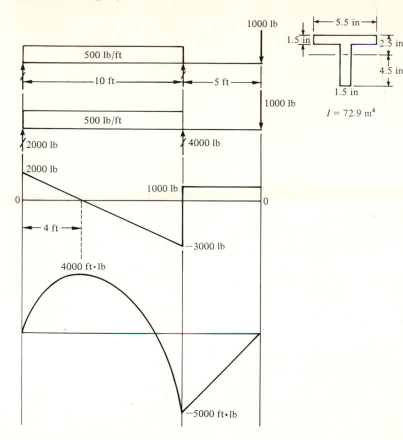

Figure E11.8

SOLUTION Sections of zero shear are at $x = 4$ ft and $x = 10$ ft. The bending moments at these points are $+4000$ ft·lb and -5000 ft·lb. The positive moment at $x = 4$ ft indicates curvature concave up. As a result, the upper fibers are in compression and the lower fibers are in tension.

$$f_c = \frac{Mc}{I} = \frac{(4000)(12)(2.5)}{72.9} = 1646 \text{ lb/in}^2$$

$$f_t = \frac{(4000)(12)(4.5)}{72.9} = 2963 \text{ lb/in}^2$$

At $x = 10$, the negative bending moment is associated with a curvature concave downward. This means that the upper fibers are in tension and the lower fibers are in compression.

$$f_t = \frac{(5000)(12)(2.5)}{72.9} = 2058 \text{ lb/in}^2$$

$$f_c = \frac{(5000)(12)(4.5)}{72.9} = 3704 \text{ lb/in}^2$$

As a result, the maximum tensile stress of 2963 lb/in^2 occurs at $x = 4$ ft, and the maximum compressive stress of 3704 lb/in^2 occurs at $x = 10$ ft. For unsymmetrical sections with curvature reversal it becomes necessary to investigate stresses at all sections of zero shear.

Example 11.9 A timber beam 9 cm by 18 cm by 4 m long supports a concentrated load of 6 k N at midspan. Determine the maximum flexural stress in the beam (Fig. E11.9a).

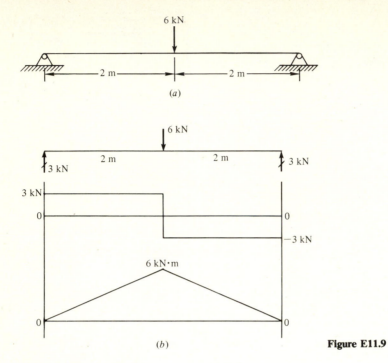

(a)

(b)

Figure E11.9

SOLUTION See Fig. E11.9b.

$$\text{Maximum bending moment} = 6 \text{ kN} \cdot \text{m}$$

The moment of inertia of the cross section is

$$\frac{bd^3}{12} = \frac{(0.09)(0.18)^3}{12} = 4.374 \times 10^{-7} \text{ m}^4$$

$$f_b = \frac{Mc}{I} = \frac{(6 \text{ kN} \cdot \text{m})(4.5 \times 10^{-2} \text{ m})}{4.374 \times 10^{-7} \text{ m}^4} = 6.17 \text{ kN/m}^2 = 6.17 \text{ MPa}$$

11.5 SHEAR STRESSES IN BEAMS

11.5.1 Introduction

On bending, a beam made up of layers would produce the effects shown in Sketch 11.3a. The separate layers would slide past each other, and the strength of the beam would be the sum of the layers. If the sliding of the layers relative to each other is prevented, the beam is much stiffer and stronger. Prevention of this sliding between layers in a solid beam involves the development of shear stresses between the horizontal layers (Sketch 11.3b). In Sec. 11.2.1 it became apparent that one of the internal forces acting on a cross section of a beam was a shearing

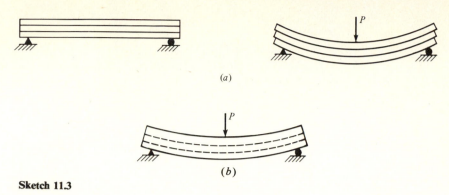

(a)

(b)

Sketch 11.3

force. This shearing force is responsible for the development of shear stresses on the section.

11.5.2 Relationship between Vertical and Horizontal Shearing Stresses in Beams

Consider the loaded beam shown in Fig. 11.13a. As developed in Sec. 11.2.1, the shearing force V at section aa (Fig. 11.13b) is equal to the sum of the shear stresses on the cut section (Fig. 11.13c).

If the small darkened rectangular prism of this beam is isolated, the free-body diagram shown in Fig. 11.14 can be drawn. If the dimensions of the prism a and c are infinitesimally small, the normal stresses on the right and left faces of the prism are equal and opposite and consequently form a system of forces in equilibrium. As a result, these forces need not be considered.

Let S_v and S_h be the vertical and horizontal shearing stresses on the right and top sides of the prism. The areas of these faces are ac and bc, respectively. Since the forces acting on the prism must be in equilibrium, it follows that in taking moments about O we have

$$S_v acb = S_h bca$$

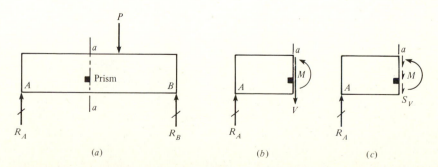

Figure 11.13 Vertical or transverse shearing stresses in beams.

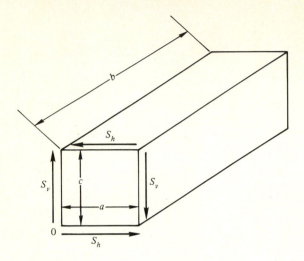

Figure 11.14 Relationship between vertical and horizontal shearing stresses.

since the moments of the shearing forces on the left and bottom sides are equal to zero. This gives

$$S_v = S_h$$

Thus, at any point in a stressed beam, the vertical and horizontal shear stresses are equal. As a result, the symbol S_s can be used to denote all shear stresses in a beam.

11.5.3 Derivation of Formula for Horizontal Shearing Stress

Consider two adjacent sections 1 and 2 in a beam separated by a distance dx (Fig. 11.15a). Isolate as a free body the shaded portion shown in Fig. 11.15b. The bending moment at section 2 is larger than at section 1, resulting in larger flexural stresses at section 2. The resultant H_2 of the flexural stresses acting at section 2 of the free body will be larger than H_1 acting at section 1. This difference between H_2 and H_1 is balanced by the shear force $S_s b\,dx$ developed on the bottom fibers of the free body. There are no external forces acting on the top or faces of the free body.

A horizontal summation of forces gives

$\Sigma x = 0$:
$$S_s b\,dx = H_2 - H_1$$

H_2 is the summation of thrusts on all elements dA in the shaded area in Fig. 11.15c

$$H_2 = \int_{y_1}^{c} f_2\,dA$$

In a similar manner

$$H_1 = \int_{y_1}^{c} f_1\,dA$$

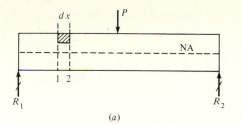

(a)

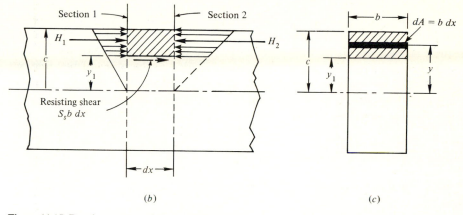

(b)   (c)

Figure 11.15 Development of shear-stress formula.

Replacing f with its equivalent, My/I, gives

$$S_s b\,dx = \frac{M_2}{I}\int_{y_1}^{c} y\,dA - \frac{M_1}{I}\int_{y_1}^{c} y\,dA = \frac{M_2 - M_1}{I}\int_{y_1}^{c} y\,dA$$

Since $M_2 - M_1$ represents the differential change in bending moment dM in the distance dx,

$$S_s = \frac{dM}{Ib\,dx}\int_{y_1}^{c} y\,dA$$

From Eq. (11.9) we know that $dM/dx = V$ is the vertical shear at the section in question, and consequently, we obtain

$$S_s = \frac{V}{Ib}\int_{y_1}^{c} y\,dA$$

where $\int_{y_1}^{c} y\,dA$ is the statical moment of the area dA between fiber level y_1 and the top of the beam about the neutral axis. This moment of area is often identified as Q.

As a result, we obtain the *general shear formula*

$$S_s = \frac{VQ}{Ib} \tag{11.14}$$

where S_s = shear stress (horizontal or vertical) at given level in cross section of beam

V = total transverse shear at section

Q = statical moment about neutral axis of that part of cross-sectional area lying on either side of horizontal level at which shear stress is being calculated

I = moment of inertia of entire cross-sectional area of beam about its neutral axis

b = width of beam at plane where S_s is being calculated

Without saying so explicitly, it has been assumed that the shearing stress is uniform across the width of the section. This is a realistic assumption when the flexural stresses are distributed evenly over the width of a section. This condition is present in rectangular and W sections loaded in the plane of bending but does not exist in triangular sections or circular sections, for example. Even in these cases the shear-stress equation (11.14) is considered to give the average shearing stress across the layer.

11.5.4 Variation of Shearing Stresses in a Beam

The design of beams involves resisting the maximum stresses developed in a beam, both flexural and shear. As a result it becomes important to be able to determine not only the cross section with the maximum shearing force but also the location and magnitude of the maximum shearing stress in the critical cross section. From studying the general shear formula, it is apparent that the shear stress will vary with location in a given section.

Consider a rectangular beam as shown in Fig. 11.16a. If the shearing stresses are calculated on equally spaced horizontal planes in the cross section, as caused by a shearing force V, the shear stresses increase from zero at the uppermost fibers to a maximum at the neutral axis and then decrease to zero at the bottom fibers (Fig. 11.16b).

In the case of an I beam or a W (wide-flange) section we have the variation shown in Fig. 11.17. The dashed curve indicates the stress variation in a rectangular beam of width b. The sudden increase in shear stress at the bottom of the upper flange comes from the change in section width from b to t. Similar

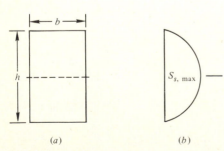

(a)   (b)

Figure 11.16 Variation of shearing stress in rectangular beam.

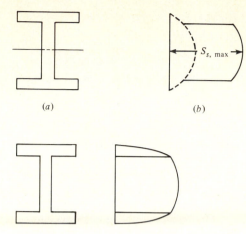

(a)   (b)

Figure 11.17 Variation of shearing stress in a W beam.

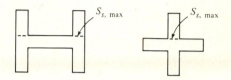

Figure 11.18 Variation of shearing force in a W beam.

Figure 11.19 Examples where S_s is not a maximum at the centroidal axis.

changes occur for the lower flange. If the shearing stress is multiplied by the width of the beam at that level, we obtain Fig. 11.18, which indicates the relative amount of the shear force taken up by the flange and the web. It becomes apparent immediately that, in the case of a W section, the web resists most of the shearing force on a given section.

On the basis of the above examples, it can be shown that the *maximum shearing stress normally occurs at the neutral axis*. The exceptions are cross sections having an increased width b at the neutral axis, e.g., H beams and a cross (Fig. 11.19).

11.5.5 Special Shear Formulas

Since the design of beams requires determining the maximum shear stress, it is convenient to develop special formulas for common sections.

Rectangular-beam shear formula Since the maximum shear stress occurs at the neutral axis of a rectangular beam, the shear stress can readily be calculated by evaluating both I and Q in terms of the dimensions of the beam (Sketch 11.4).

$$I = \frac{bh^3}{12} \qquad Q = \frac{bh}{2}\frac{h}{4}$$

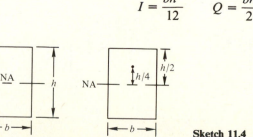

Sketch 11.4

Substituting these values into the general shear formula (11.14) gives

$$S_{s,\max} = \frac{Vbh^2}{8bh^3b/12} = \frac{3V}{2bh} \tag{11.15}$$

It is interesting to note that this value of maximum shear stress is 50 percent greater than the average shear stress $S_{s,\max} = V/bh$.

Web shear formula As indicated in Sec. 11.5.4, structural sections like I beams, channels, and W sections develop most of their shear resistance in their relatively thin webs. Consequently, an expression based on the average shear stress in the web, neglecting the area of the flanges, gives an average web shear stress very close to the maximum value obtained with the general shear formula (this type of relationship is more convenient than the general shear formula):

$$S_{s,\text{web}} = \frac{V}{t_w(d - 2t_f)} \tag{11.16}$$

where
S_s = average shear stress in web, lb/in^2
V = shear force, lb
t_w = thickness of web, in
d = total depth of section, in
t_f = average thickness of flange, in

Example 11.10 A simply supported beam 4 in wide, 8 in deep, and 16 ft long carries a uniformly distributed load of 400 lb/ft. (*a*) Compute the shearing stress developed at horizontal layers 2 in apart from top to bottom of beam at a section 4 ft from the beam end. (*b*) Compute the maximum shearing stress developed in the beam.

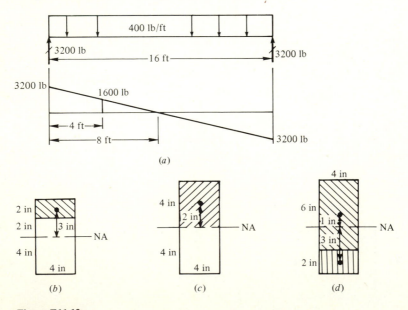

(*a*)

(*b*)   (*c*)   (*d*)

Figure E11.10

SOLUTION See Fig. E11.10a. The maximum shearing force is 3200 lb, and the shear at the section 4 ft from beam end is 1600 lb.

(a) The moment of inertia of the cross section about the neutral axis is

$$I = \frac{bh^3}{12} = \frac{(4)(8)^3}{12} = 170.7 \text{ in}^4$$

For the section at a level 2 in down from the top of the beam (Fig. E11.10b)

$$Q = (2)(4)(3) = 24 \text{ in}^3$$

$$S_s = \frac{VQ}{Ib} = \frac{(1600)(24)}{(170.6)(4)} = 56.3 \text{ lb/in}^2$$

For the section at a level 4 in down from the top of beam (to neutral axis) (Fig. E11.10c)

$$Q = (4)(4)(2) = 32 \text{ in}^3$$

$$S_s = \frac{VQ}{IB} = \frac{(1600)(32)}{(170.6)(4)} = 75.0 \text{ lb/in}^2$$

For the section at a level 6 in down from the top of beam or 2 in up from the bottom of the beam (Fig. E11.10d)

$$Q = (6)(4)(1) = 24 \text{ in}^3$$

or considering area below fiber level,

$$Q = (2)(4)(3) = 24 \text{ in}^3$$

$$S_s = \frac{VQ}{Ib} = \frac{(1600)(24)}{(170.6)(4)} = 79.71 \text{ lb/in}^2$$

This confirms the symmetrical distribution of shear stress relative to the neutral axis. It also illustrates that Q can be the statical moment about the neutral axis of the area either above or below the plane in which the shear stress is being determined.

(b) The maximum shearing force in the beam occurs at the supports and is equal to 3200 lb. The maximum shear stress will occur at the neutral axis

$$S_s = \frac{VQ}{Ib} = \frac{(3200)(32)}{(170.6)(4)} = 150.0 \text{ lb/in}^2$$

Example 11.11 The beam in Fig. E11.11a is made up of three 2- by 8-in pieces of southern pine and is to be used as a 12-ft beam carrying 300 lb/ft. Determine the spacing of lag bolts near the supports and the maximum shear stress in the beam.

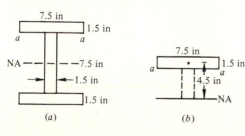

(a)

(b)

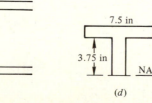

(c)

(d)

Figure E11.11

SOLUTION The maximum shearing force in the simply supported uniformly loaded beam is $[(12)(300)/2] = 1800$ lb. The moment of inertia of the cross section about the neutral axis is

$$\frac{(7.5)(10.5)^3}{12} - \frac{(6)(7.5)^3}{12} = 512.6 \text{ in}^4$$

Lag bolts will be used to develop shear at the interface between the web and flanges (plane aa) (Fig. E11.11b).

$$Q = (7.5)(1.5)(4.5) = 50.6 \text{ in}^3$$

$$S_s = \frac{VQ}{Ib} = \frac{(1800)(50.6)}{(512.6)(1.5)} = 118.4 \text{ lb/in}^2$$

The shearing force per inch of beam length at the support equals $(118.4)(1.5)(1.0) = 177.6$ lb.

Use a $\frac{3}{8}$-in-diameter lag screw $4\frac{1}{2}$ in long. This provides 11 diameters of penetration of lag screw (Sec. 7.8).

The lateral resistance of one lag screw is

$$KD^2 = (4280)\left(\tfrac{3}{8}\right)^2 = 601 \text{ lb}$$

$$\text{Spacing of lag screws} = \frac{\text{lateral resistance of lag screw}}{\text{shear force required per inch of beam}}$$

$$= \frac{601}{177.6} = 3.38 \text{ in} \qquad \text{Fig. E11.11}c$$

If desired, the spacing of lag screws can be increased toward the center of the beam because the shearing force in the beam reduces linearly, becoming zero at the center of the beam.

The maximum shear stress will occur at the neutral axis in the cross section having a maximum shear (Fig. E11.11d).

$$V = 1800 \text{ lb}$$

$$Q = (7.5)(1.5)(4.5) + (1.5)\frac{(3.75)^2}{2} = 50.6 + 10.5 \text{ in}^3 = 61.1 \text{ in}^3$$

$$S_s = \frac{(1800)(61.1)}{(512.6)(1.5)} = 143 \text{ lb/in}^2$$

Example 11.12 Consider the problem of Example 11.7 in which a W8 × 28 is used as a cantilever beam 10 ft long supporting a uniform load of 500 lb/ft. Determine the maximum shear stress in the beam.

SOLUTION The maximum shear force occurs at the support and is equal to the reaction, $(10)(528) = 5280$ lb. The maximum shear stress in the beam is approximated by Eq. (11.16):

$$S_s = \frac{V}{t_w(d - 2t_f)}$$

From Table A.5 for a W8 × 28

$$t_w = 0.285 \text{ in} \qquad d = 8.00 \text{ in} \qquad t_f = 0.465 \text{ in}$$

Consequently,

$$S_s = \frac{5280}{(0.285)[8.0 - (2)(0.465)]} = 2620 \text{ lb/in}^2$$

11.6 LOCAL BUCKLING

11.6.1 General Considerations

Whenever thin plates, such as webs and flanges, are subjected to bending stresses, there is a tendency for local buckling. Compression-flange buckling will be discussed in Section 11.8; this section deals with web buckling. The principal functions of the web are to separate the flanges as much as possible for beam efficiency and to resist the shear stresses. Since the shear stresses are normally small, webs are made as thin as possible. As a result, webs may be weak in the vicinity of concentrated loads.

11.6.2 Local Web Buckling

Figure 11.20a shows four types of vertical web buckling. Which one occurs depends on the nature of the support provided by the construction associated with the top flange. This type of buckling could develop only in the vicinity of concentrated loads such as reactions. Another type of web failure, shown in Fig. 11.20b is known as *web crippling*. Since experience shows a beam safe from web crippling to be safe from vertical buckling as well, beams are normally designed against web crippling. Plate girders and built-up members should be checked for both web crippling and vertical buckling.

AISC specifications assume that the web acts as a column with a cross-sectional area equal to $(N + k)t$ at reactions and $(N + 2k)t$ at interior loads. The distance N is equal to the length of bearing (Fig. 11.21). The design procedure permits a value greater than N, that is, k or $2k$ to allow for the distribution of load over the distance k before reaching the section of web of thickness t. The distance k is equivalent to the distance from the top of the flange to the web toe of the fillet (Fig. 11.21). Values of k can be found in AISC Steel Manual tables [3] and Table A.5.

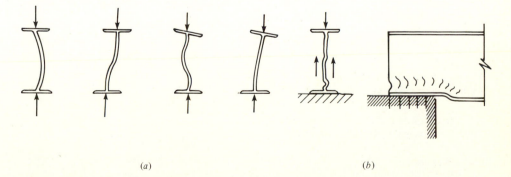

(a) (b)

Figure 11.20 Local web buckling: (a) web buckling, (b) web crippling.

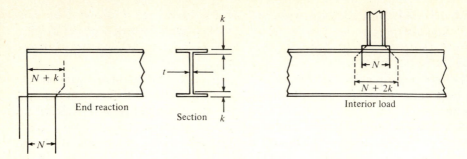

Figure 11.21 Design considerations for web buckling.

The compressive stress developed in a cross section at a reaction equals

$$f_c = \frac{R}{t(N+k)}$$

AISC specifications permit an allowable compressive stress in the web of $0.75F_y$. Therefore, the required minimum length of bearing for a given reaction is

$$N_{\min} = \frac{R}{0.75F_y t} - k \qquad (11.17)$$

At an interior point of load application to the top flange, the minimum length of bearing becomes

$$N_{\min} = \frac{P}{0.75F_y t} - 2k \qquad (11.18)$$

Where these required lengths cannot be obtained, a beam with a thicker web must be used or the thinner web must be reinforced by vertical stiffeners.

Example 11.13 The end reaction of a W12 × 50 of A36 steel is 40 kips. Determine the minimum required length of bearing to prevent web crippling.

SOLUTION For W12 × 50

$$t_w = 0.370 \text{ in} \qquad k = 1\tfrac{3}{8} \text{ in} = 1.375 \text{ in}$$

$$N = \frac{R}{0.75F_y t} - k = \frac{40}{(0.75)(36)(0.370)} - 1.375 = 4.00 - 1.375 = 2.63 \text{ in}$$

11.7 DESIGN OF BEAMS

11.7.1 General Considerations

Beams must be designed to withstand maximum shear and flexural stresses resulting from imposed loads. In the design of a beam the bending moment is known, and the allowable stresses F_b are based on the selected structural material. As a result, the size and shape of a beam are the unknown quantities which can be most readily determined by use of the flexural formula (11.13). The section modulus, $M/F_b = I/c = S$, depends only on the dimensional properties of the cross section.

Since it is too cumbersome to incorporate an appropriate shear formula into this design procedure, it is customary to design the beam on the basis of bending stresses and then to compare the shear stresses developed with the allowable value. Shear stresses are more likely to become critical in short, heavily loaded beams and timber beams because of the relatively low shear strength of wood.

The *weight of a beam* must be considered in its design; i.e., the beam must be capable of supporting not only the applied loads but also its own weight. In some instances, the weight of a beam may be insignificant, but in every case a check should be made to see whether the section modulus selected is large enough to take the weight of the beam into account.

The *most economical beam* is the one with the least cross-sectional area and consequently the least weight per foot of length. In general, for a given area, a deeper beam is stronger than a more shallow one. It is taken for granted in general practice that the beam selected is the most economical. Other factors, such as connection details, uniformity in details, and availability, may dictate other selections.

The compression flanges of beams have a tendency to buckle laterally between supports, particularly if the beam is relatively narrow and deep. In buildings most of the beams have lateral support which prevents this form of buckling, e.g., floor beams by the floor they are supporting. Exceptions, i.e, laterally unsupported beams, will be treated in Sec. 11.8.

11.7.2 Design of Timber Beams

Timber is not a manufactured material; it not only has natural defects, such as knots and slope of grain, but also is anisotropic in terms of structure and properties. It is also sensitive to duration of loading. These factors become important when using wood as a structural material.

The *span of timber beams* is considered to be from center to center of girders they are resting on. If beams are supported by hangers on the sides of girders, the span is from face to face of the girders. If floor girders rest on column caps, their span is considered to be from center to center of columns. Some of these details are illustrated in Fig. 11.22.

Although narrow, deep timbers are more economical than shallow ones, the width-depth ratio b/h usually varies from $\frac{1}{3}$ to $\frac{3}{4}$, often being about $\frac{1}{2}$. Exceptions to this rule are floor joists, which may have a width-depth ratio which is less, but special provisions have to be made for lateral bracing. Shear stresses can be critical in these joists. In terms of span, a rule of thumb allows about 1 in of depth for each foot of span for uniformly loaded beams. For girders carrying concentrated loads the depth is normally greater, about 1.25 in per foot of span.

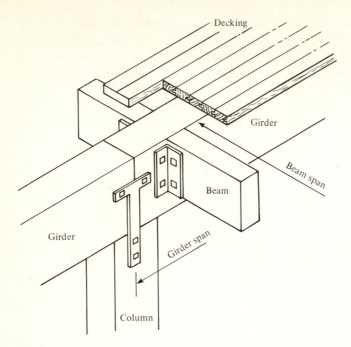

Figure 11.22 Structural timber frame.

The actual dimensions of rough timbers are presumed to equal their nominal dimensions. Actual dimensions and other properties of dressed timbers (S4S) are given in Table A.2.

Experiments have indicated that strains in wood beams are not directly proportional to their distance from the neutral axis. They deviate increasingly downward from the compression side, reaching a maximum at approximately one-third the beam depth. This strain pattern results in a shift of the neutral axis toward the tension side of the beam.

Another phenomenon observed in the behavior of wood beams is a variation in the value of the modulus of rupture with depth of beam. All other factors being equal, a constant value would be expected. As the depth of beam increases, the value of the modulus of rupture decreases. In 1924 Newlin and Trayer at the U.S. Forest Products Laboratory explained this anomaly by suggesting that fibers in a beam are capable of reinforcing each other because of the lignin cement between them. In deep beams the fibers adjacent to the ones in compression near the surface of the beam can reinforce the outermost ones only slightly because they are also highly stressed. In shallow beams the adjacent fibers are less stressed and consequently can reinforce the outermost fibers to a larger extent.

This variation in the modulus of rupture for beams of different depths has resulted in the introduction of a *depth factor* into the flexural formula. In its

Table 11.1 Values for three conditions of loading [2]

Depth d in	Uniformly distributed load	Single concentrated load	Third-point loading
< 12	1.00	1.00	1.00
12	1.00	1.08	0.97
19	0.95	1.02	0.92
31	0.90	0.97	0.87
52	0.85	0.92	0.82
90	0.80	0.86	0.77

modified form the formula is

$$f_b = \frac{Mc}{C_F I} \qquad \text{where } C_F = \left(\frac{12}{h}\right)^{1/9} \tag{11.19}$$

The values obtained from this formula are based on a uniformly loaded beam with a length-to-diameter ratio of 21. Tabulated values for three conditions of loading are given in Table 11.1. Straight-line interpolation can be used for intermediate depths. It is important to note that depth factors apply only for depths greater than 12 in and have use primarily with laminated timbers (Table A.3).

Example 11.14 Design a rectangular beam 20 ft long simply supporting 850 lb/ft. Use Douglas fir select structural-grade timber.

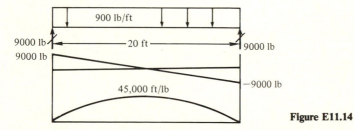

Figure E11.14

SOLUTION See Fig. E11.14. For select structural-grade Douglas fir (Table A.1),

$$F_b = 2100 \text{ lb/in}^2 \qquad F_v = 95 \text{ lb/in}^2$$

In the special case of simple-span timber beams carrying uniformly distributed loads, it is convenient, for design purposes, to add an estimated beam weight of 5 percent of the applied load.

Estimated weight of beam = 50 lb/ft   Total beam load = 900 lb/ft

Maximum bending moment = 45,000 ft·lb

$$\text{Section modulus} = S = \frac{I}{c} = \frac{M}{F_b}$$

$$S = \frac{(45,000)(12)}{1600} = 337.5 \text{ in}^3$$

Consider the use of rough timbers. S for a rectangular cross section is

$$\frac{I}{c} = \frac{bh^3}{12h/2} = \frac{bh^2}{6} \qquad S = 337.5 \text{ in}^3 = \frac{bh^2}{6} \qquad bh^2 = 2025 \text{ in}^3$$

Assume $b = h/2$; then

$$h^3 = 4050$$
and
$$h = 16 \text{ in}$$

Use 8- by 16-in timber. Check shear stress in beam

$$S_s = \frac{3}{2}\frac{V}{A} = \frac{3}{2}\frac{9000}{(8)(16)} = 105 \text{ lb/in}^2$$

This shear stress exceeds the allowable shear stress of 85 lb/in². A different, wider section is required; try a 10 by 16:

$$S = \frac{bh^2}{6} = \frac{(10)(16)^2}{6} = 426 > 337.5 \text{ in}^3 \qquad \text{OK}$$

$$S_s = \frac{3}{2}\frac{V}{A} = \frac{(3)(9000)}{(2)(10)(16)} = 84 \text{ lb/in}^2$$

The weight of a 10 by 16 is

$$\frac{(10)(16)}{(12)(12)}(32) = 35.5 \text{ lb/ft}$$

This is less than the estimated weight of 50 lb/ft. Use a 10 by 16.
Consider the use of dressed timber (S4S). As before,

$$S_{\text{req}} = 337.5 \text{ in}^3$$

Check Table A.2 for sections with adequate section modulus.

Nominal beam size, in	S
8 × 16	300.3
10 × 14	288.6
12 × 12	253.5
12 × 14	349.3
14 × 12	297.6

Previous calculations have shown that a minimum width of 10 in is required in shear. The most economical beam is the 12 × 14:

$$S = 349.3 > 337.5 \text{ in}^3 \qquad \text{OK}$$
$$S_s = \frac{3}{2}\frac{9000}{(12)(14)} = 80.4 < 85 \text{ lb/in}^2 \qquad \text{OK}$$

The weight of the beam is

$$\frac{(12)(14)}{(12)(12)}(32) = 37.3 < 50 \text{ lb/ft} \qquad \text{OK}$$

The depth-correction factor for this beam would be insignificant.

Example 11.15 Design a glued-laminated roof beam to meet the following requirements:

Span length 40 ft
Spacing 15 ft
Roof deck applied directly to beams
Dead load 10 lb/ft^2
Snow load 30 lb/ft^2 (load-duration factor of 1.15, Fig. 6.20)

$$F_b = (2400)(1.15) = 2760 \text{ lb/in}^2$$

$$F_v = (165)(1.15) = 190 \text{ lb/in}^2$$

SOLUTION Beam loading is

$$W_{DL} = (15)(10) = 150 \text{ lb/ft}$$

$$W_{SL} = (30)(15) = 450 \text{ lb/ft}$$

$$\text{Estimated weight of beam} = 30 \text{ lb/ft}$$

$$W_{TL} = \overline{630 \text{ lb/ft}}$$

The maximum bending moment for a uniformly loaded simply supported beam is

$$\frac{wl^2}{8} = \frac{(630)(40)^2}{8} = 126,000 \text{ ft·lb}$$

Assume a C_F (depth factor) of 0.90; then

$$S_{req} = \frac{(126,000)(12)}{(0.90)(2760)} = 608 \text{ in}^3$$

From Table A.3 the following sections are adequate:

Section	S in^3	A in^2
$3\frac{1}{8} \times 34\frac{1}{2}$	620	108
$5\frac{1}{8} \times 27$	623	139
$6\frac{3}{4} \times 23\frac{1}{4}$†	608	157
$8\frac{3}{4} \times 21$	643	184
$10\frac{3}{4} \times 18\frac{3}{4}$†	630	202

† Less desirable because unavailable in $1\frac{1}{2}$-in-thick laminations.

Try $5\frac{1}{8}$ by 27 laminated timber:

$$\text{Beam weight} = \frac{(139)(32)}{(12)(12)} = 31 \text{ lb/ft} \quad \text{OK}$$

Check beam for shear stresses:

$$V = (20)(631) = 12,620 \text{ lb}$$

$$S_s = \frac{3}{2} \frac{V}{bh} = \frac{3}{2} \frac{12,620}{139} = 136 < 190 \text{ lb/in}^2 \quad \text{OK}$$

$$\text{Depth factor for 27 in beam} = 0.92 \quad \text{OK}$$

Use a $5\frac{1}{8}$ by 27 laminated beam.

11.7.3 Design of Steel Beams

A complete treatment of the design of steel beams includes the design of built-up beams and plate girders. This section will be confined to the design (or more appropriately the selection) of rolled structural-steel beam sections and other beam sections of simple shape.

As indicated in Chap. 6, a number of structural steels are available not only with different strengths but also increased corrosion resistance and weldability. Codes usually relate the allowable bending stress of steel to its yield stress F_y.

The term *compact section* was developed to describe a cross section whose thin compression flanges and webs do not buckle before the development of yield stress in all fibers of the beam. Under these circumstances, some codes, including AISC, permit larger allowable bending stresses ($0.66F_y$ instead of $0.60F_y$). Most rolled sections are compact. Those which are not are appropriately identified in steel handbooks.

The *span* of a steel beam must be determined carefully. Figure 11.23 illustrates a beam-and-girder floor system. The beams (B_1, B_2, etc.) extend between webs of girders and have a span length from center to center of columns. The girders, however, span from flange to flange of columns. In this arrangement, the beams carry all the distributed load (see the crosshatched area associated with B_1), and the girders support only the concentrated loads from the beam ends and their own weight.

The *depth of steel floor beams*, according to a rule of thumb, should not be less than one-twenty-fourth of the span, or $\frac{1}{2}$ in per foot of span. Roof beams are normally not that deep (approximately one-thirtieth of the span).

Example 11.16 The floor of a factory building consists of a 6-in reinforced-concrete floor slab supported by steel beams 20 ft long and spaced at 6 ft on centers which are framed into webs of the steel girders and columns by means of standard connection angles. The girders are similarly

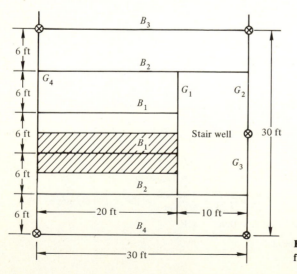

Figure 11.23 Beam-and-girder floor system.

connected to the flanges of the columns. The live load is 200 lb/ft^2. Figure E11.16a illustrates a typical interior panel. Design a typical beam and girder according to AISC specifications using A36 steel.

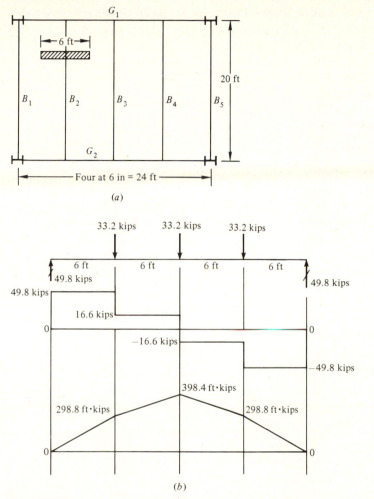

Figure E11.16

SOLUTION **Beam B_2** The beams and girders can be considered to be simply supported (standard connection angles). The uniform floor load carried by a typical beam (B_2), in pounds per foot of span, is indicated by the crosshatched area. The weight of a concrete slab 1 in thick can be assumed to be

$$(\tfrac{1}{12})(144) = 12 \text{ lb/ft}^2 \qquad \text{Table 3.1}$$

The beam loading is

$$\text{Live load} = 200 \text{ lb/ft}^2$$
$$\text{Concrete slab } (6)(12) = 72 \text{ lb/ft}^2$$
$$\overline{\text{Total load} = 272 \text{ lb/ft}^2}$$
$$w = (6)(272) = 1632 \text{ lb/ft} + \text{weight of beam}$$

The maximum bending moment for a uniformly loaded simply supported beam is $wl^2/8$; then

$$\frac{wl^2}{8} = \frac{(1632)(20^2)}{8} = 81,600 \text{ ft·lb}$$

According to AISC, $F_b = 0.66F_y$ for compact sections (Table 6.3), and for A36 steel

$$F_b = (0.66)(36) = 23.76 \text{ kips/in}^2 \quad \text{use 24 kips/in}^2$$

From Eq. (11.13)

$$\text{Section modulus} = S = \frac{I}{c} = \frac{M}{F_b}$$

$$S_{req} = \frac{(81.6)(12) \text{ in·kips}}{24 \text{ kips/in}^2} = 40.8 \text{ in}^3$$

Check Table A.5 for W sections with section moduli slightly larger than 40.8 in^3 (to take care of the weight of the beam):

W section	S
W16 × 31	47.2
W14 × 30	42.0
W12 × 35	45.6
W10 × 39	42.1
W 8 × 48	43.3

It is apparent that as the depth of the beam decreases, the weight of the beam becomes greater to provide a given section modulus. A W14 × 30 appears to be the most economical. A check must be made to see whether it is also capable of carrying both the applied loads and its own weight.

$$S \text{ required for own weight} = \frac{(30)(20)^2(12)}{(8)(24,000)} = 0.75 \text{ in}^3$$

$$\text{Required } S = 40.8 + 0.75 = 41.6 < 42 \text{ in}^3 \quad \text{OK}$$

Check the shear in the beam according to Eq. (11.16):

$$S_s = \frac{V}{t_w(d - 2t_f)}$$

V = maximum shear force in beam = reaction = (1662)(10)

From Table A.5 for W14 × 30

$$d = 13.84 \text{ in} \qquad t_f = 0.385 \text{ in} \qquad t_w = 0.270 \text{ in}$$

$$S_s = \frac{16.620}{0.270(13.84 - 0.77)} = 4725 \text{ lb/in}^2$$

The allowable shear stress F_v for A36 steel is

$$0.40F_y = (0.40)(36) = 14.4 > 4.7 \text{ kips/in}^2 \quad \text{OK}$$

Use W14 × 30.

Girder G_1 Girder G_1 (Fig. E11.16b) carries concentrated loads at quarter points. These concentrated loads are considered to be the reactions of two beams: (2)(10)(1662) = 33,240 lb. From Eq. (11.13)

$$S = \frac{I}{c} = \frac{M}{F_b} = \frac{(398.4)(12)}{24} = 199.2 \text{ in}^3$$

Try a W27 × 94 with $S = 243$ in³, $t_f = 0.745$ in, $d = 26.92$ in, $t_w = 0.490$ in (see Table A.5). The added section modulus required is

$$\frac{(94)(24)^2(12)}{24,000} = 27.0 \text{ in}^3$$

The required S is

$$199.2 + 27.0 = 226.2 < 243 \text{ in}^3 \text{ for W27} \times 94 \qquad \text{OK}$$

Check shear stress according to Eq. (11.16):

$$S_s = \frac{48,900}{(0.49)[26.92 - (2)(0.745)]} = 3924 \text{ lb/in}^2 < 0.40 F_y (14,000 \text{ lb/in}^2) \qquad \text{OK}$$

Use W27 × 94.

11.8 LATERAL BUCKLING

11.8.1 General Considerations

As observed in Sec. 11.4, a beam subjected to bending develops tension in one flange and compression in the other. In addition to the possibility of failure due to excessive bending stresses, there is the possibility of a form of instability termed *lateral buckling*, which causes the compression flange to flip sideways under load and the beam to collapse (Fig. 11.24). The vertical deflection is normal deflection and not due to buckling. The tendency to buckle increases as the ratio I_x / I_y increases. The deeper and narrower the section the more susceptible it is to buckling. Surrounding construction can prevent lateral buckling. As such it is classified as a beam with lateral support. Figure 11.25 illustrates conditions where lateral support is available.

It is important to differentiate between complete lateral support and situations where the support is intermittent and the compression flange can buckle between points of lateral support.

The rational analysis of the phenomenon has involved classical methods of elastic stability. The expressions obtained have been simplified for use in design. An approximate conservative approach is recommended by AISC.

Plan view of top flange

Section at
mid span

Figure 11.24 Lateral buckling.

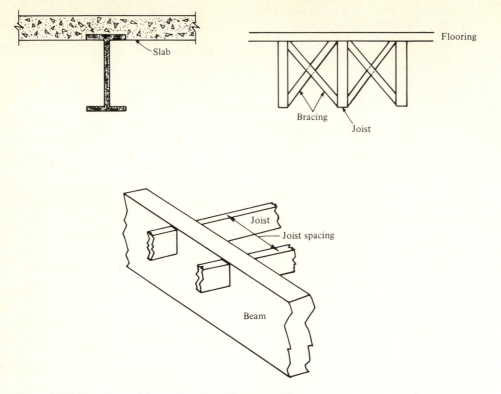

Figure 11.25 Complete and intermittent lateral support of beams.

11.8.2 Steel Beams

Beams without complete lateral support may have the allowable bending stress F_b reduced, depending upon the effective distance between lateral supports of the compression flange. Even though not continuous, if the spacing of lateral supports L_b is sufficiently small, no reduction in F_b is necessary.

Normally a beam will have adequate lateral support if the effective length between points of bracing L_b is such that

$$L_b \le L_c = \frac{76b_f}{\sqrt{F_y}} \tag{11.20}$$

This formula is based on the lateral bending strength of the compression flange.

Occasionally when torsional twisting becomes critical, the permissible length becomes

$$L_u = \frac{20,000}{(d/A_f)F_y} \tag{11.21}$$

As a result, since each rolled beam has a specific beam width, depth, and flange area, for a given grade of steel it will have a unique length which constitutes a limit for bracing points if it is to be classified as laterally supported. For example, a W10 × 33 beam of A36 steel can be considered to have adequate lateral support if

$$L_b \le L_c = \frac{76 b_f}{\sqrt{F_y}} = \frac{(76)(7.96)}{\sqrt{36}} = 100.8 \text{ in} = 8.4 \text{ ft}$$

$$L_b \le L_u = \frac{20{,}000}{(d/A_f)F_y} = \frac{20{,}000}{(2.83)(36)} = 196 \text{ in} = 16.4 \text{ ft}$$

Values of b_f and d/A_f are available from Table A.5 and the AISC Manual [3]. For the same beam section made of A242 steel ($F_y = 50$ kips/in^2), the allowable lengths become $L_c = 7.1$ ft and $L_u = 11.8$ ft. These lengths can also be obtained directly from appropriate tables in the AISC Manual [3]. It is generally conservative and usually acceptable to assume that the effective length is the span between lateral supports.

Unequal end moments on a span increase the lateral buckling capacity because of the beneficial effects of the moment gradient along the beam. This correction can be applied only if the beam is loaded in the plane of its web and is bent about its major axis. AISC evaluates this influence by means of a moment-gradient multiplier C_b ranging from 1.0 to 2.3, which can be evaluated by

$$C_b = 1.75 + 1.05 \frac{M_1}{M_2} + 0.3\left(\frac{M_1}{M_2}\right)^3 \qquad (11.22)$$

where M_1 is the smaller and M_2 the larger bending moment at the ends of the unbraced lengths. M_1/M_2 is positive for reverse-curvature bending and negative for single-curvature bending. When the bending moment at any point within an unbraced length is larger than at both ends, the value of C_b is taken as unity.

Once the value of C_b has been established, it is multiplied by L_u for comparison with L_b. *All charts and tables in the AISC Manual* [3] *assume* $C_b = 1.0$.

For further explanation of C_b, consider the beam in Sketch 11.5. Bracing

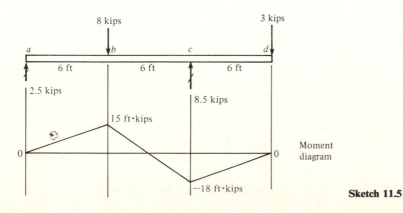

Sketch 11.5

points are located at reactions and each applied load. The C_b value for the length ab is

$$C_b = 1.75 + 1.05 \left(\tfrac{-0}{15}\right) + 0.3\left(\tfrac{-0}{15}\right)^3 = 1.75$$

For section bc

$$C_b = 1.75 + 1.05\left(\tfrac{15}{18}\right) + 0.3\left(\tfrac{15}{18}\right)^3 = 2.80 > 2.3 \qquad \text{use } 2.3$$

For section cd (cantilever)

$$C_d = 1.75 + 1.05\left(\tfrac{-0}{18}\right) + 0.3\left(\tfrac{-0}{18}\right)^3 = 1.75$$

Once it has been established that a beam has adequate lateral support, the allowable bending stress F_b will range between $0.66F_y$ and $0.60F_y$ and must be determined. The exact value depends upon the degree of compactness of the section being used.

A *compact section* is one that will develop its yield stress over the section depth without local buckling. Such a section inherently has a larger factor of safety against failure than one that fails by local buckling.

As a result, compact sections that are braced so that $L_b \leq L_c$ are permitted a higher allowable bending stress; that is, $F_b = 0.66F_y$. This compactness cannot be used with F_y greater than 65 kips/in² (A572 steel). Most rolled beam sections meet the requirements for compactness. According to AISC specifications, compact beams must meet the following requirement relative to the width-to-thickness ratio of unstiffened projecting elements of the compression flange:

$$\frac{b_f/2}{t_f} = \frac{65}{F_y} \qquad \text{and} \qquad \frac{d}{t_w} \leq \frac{640}{\sqrt{F_y}} \tag{11.23}$$

Beams which are braced so that $L_b \leq L_c$ but which are not fully compact because they exceed the above requirement can be designed using a value of F_b between $0.66F_y$ and $0.60F_y$; it is determined by the so-called *blending formula*

$$F_b = F_y\left(0.79 - \frac{0.002\,b_f}{2t_f}\sqrt{F_y}\right) \tag{11.24}$$

and is valid for $b_f/2t_f$ values up to $95/\sqrt{F_y}$, when the term in the parentheses becomes 0.60.

Beams that are braced so that $L_c \leq L_b \leq L_u$ are permitted an allowable stress of $F_b = 0.60F_y$. Compactness is not a factor under these circumstances.

When the spacing of lateral supports L_b exceeds L_u, the beam is not adequately supported laterally and the allowable bending stress F_b will be less than $0.60F_y$. Again compactness is not a factor in this case.

According to AISC, the susceptibility to buckling is related to the ratio of buckling length to effective radius of gyration or L_b/r_T. The radius of gyration r can be determined for any shape by the formula $r = I/A$, where r is always referenced to the same axis as I. AISC defines the effective section for r as the compression flange plus one-third of the compression-web area. Values of r_T are listed in Table A.5 under properties for designing. Units of L_b and r_T should be consistent.

If L_b/r_T is less than $\sqrt{1.02 \times 10^5 C_b/F_y}$, the lateral support is adequate and $F_b = 0.60F_y$ is permissible. If, however,

$$\sqrt{\frac{1.02 \times 10^5 C_b}{F_y}} \leq \frac{L_b}{r_T} < \sqrt{\frac{5.10 \times 10^5 C_b}{F_y}}$$

the allowable bending stress F_b becomes

$$F_{b_2} = \left[\frac{2}{3} - \frac{F_y(L_b/r_T)^2}{1.53 \times 10^6 C_b} \right] F_y \qquad (11.25)$$

If

$$\frac{L_b}{r_T} > \sqrt{\frac{5.10 \times 10^5 C_b}{F_y}}$$

F_b becomes

$$F_{b_3} = \frac{1.7 \times 10^5}{(L_b/r_T)^2} \qquad (11.26)$$

Both these formulas are based on the lateral bending stiffness of the compression flange and a small part of the web. They do not account for elastic torsional buckling strength and tend to be conservative.

In cases where the compression flange is solid, approximately rectangular in cross section, and having an area not less than that of the tension flange, AISC provides a simplified twisting formula for determining F_b

$$F_{b_1} = \frac{1.2 \times 10^4 C_b}{L_b(d/A_f)} \qquad (11.27)$$

This value must be compared with F_{b_2} or F_{b_3}, whichever applies, and the largest value is used. These interrelationships are illustrated in Fig. 11.26 and summarized below. For laterally supported beams

$$F_b = \begin{cases} 0.66F_y & L_c \geq L_b \leq C_b L_u & \text{beam compact} \\ 0.60F_y & & \text{beam not compact} \\ 0.60F_y & L_c < L_b \leq C_b L_u & \text{compactness not a consideration} \end{cases}$$

For laterally unsupported beams

$$F_b < 0.60F_y \qquad \text{where } L_c < L_b > C_b L_u$$

and

$$F_{b_1} = \frac{1.2 \times 10^4 C_b}{L_b(d/A_f)}$$

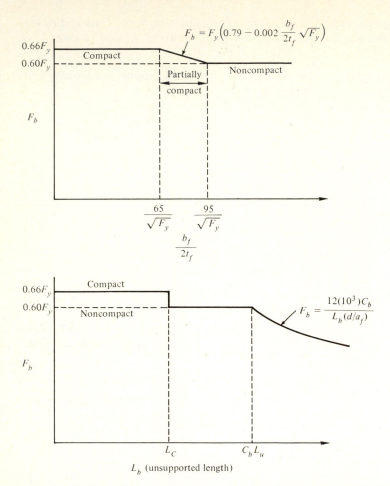

Figure 11.26 Effect of beam compactness and unsupported length on allowable bending stress.

for section symmetrical relative to the x axis

$$F_{b_2} = \left[\frac{2}{3} - \frac{F_y(L_b/r_T)^2}{1.53 \times 10^6 C_b} \right] F_y \quad \text{if} \quad \sqrt{\frac{1.02 \times 10^5 C_b}{F_y}} < \sqrt{\frac{L_b}{r_T} \frac{5.10 \times 10^5 C_b}{F_y}}$$

and
$$F_{b_3} = \frac{1.7 \times 10^5 C_b}{(L_b/r_T)^2} \quad \text{if} \quad \frac{L_b}{r_T} > \frac{5.10 \times 10^5 C_b}{F_y}$$

The largest value (F_{b_1}, F_{b_2}, or F_{b_3}) is used. If F_{b_1} is not a valid consideration, the allowable bending stress will be either F_{b_2} or F_{b_3}, depending upon L_b/r_T.

> **Example 11.17** A W14 × 30 of A36 steel has a simple span of 24 ft. Lateral support is supplied only at the reactions. Calculate the largest uniform load (including its own weight) it can support.

SOLUTION Determine the various critical limits of lateral support for the beam. From Table A.5 $d = 13.84$ in, $b_f = 6.73$ in, $t_f = 0.385$ in, $r_T = 1.74$ in, $S = 42$ in^3, and $C_b = 1.0$. Then

$$L_c = \frac{76b_f}{\sqrt{F_y}} = \frac{(76)(6.73)}{\sqrt{36}} = 85.2 \text{ in} = 7.1 \text{ ft}$$

$$L_u = \frac{20{,}000}{(d/A_f)F_y} = \frac{20{,}000}{[13.84/(6.73)(0.385)](36)} = 104 \text{ in} = 8.7 \text{ ft}$$

$$\sqrt{\frac{1.02 \times 10^5 C_b}{F_y}} = \sqrt{\frac{(102{,}000)(1)}{36}} = 53.2$$

$$\sqrt{\frac{5.10 \times 10^5 C_b}{F_y}} = \sqrt{\frac{(510{,}000)(1)}{36}} = 119$$

$$L_b = 24 \text{ ft} > L_c, L_u$$

$$\frac{L_b}{r_T} = \frac{(24)(12)}{1.74} = 165.5 > 119$$

Consequently the controlling equation is

$$F_{b_3} = \frac{1.7 \times 10^5 C_b}{(L_b/r_T)^2} = \frac{(170{,}000)(1)}{(165.5)^2} = 6.20 \text{ kips/in}^2$$

Since the beam is a rolled section, the twisting formula is also valid:

$$F_{b_1} = \frac{1.2 \times 10^4 C_b}{L_b(d/A_f)} = \frac{(12{,}000)(1)}{(24)(12)(13.84)/(6.73)(0.385)} = 7.80 \text{ kips/in}^2$$

Since $7.80 > 6.20$, the allowable bending stress is 7.80 kips/in^2. The maximum resisting moment equals

$$M = SF_b = \frac{(42)(7.80)}{12} = 27.3 \text{ ft·kips}$$

and since $M = wl^2/8$,

$$w = \frac{8M}{l^2} = \frac{(27.3)(8)}{(24)^2} = 0.36 \text{ kips/ft}$$

Example 11.18 Select the lightest available steel section for the beam shown in Fig. E11.18. Use AISC specifications and A36 steel. Lateral support is provided at the ends only.

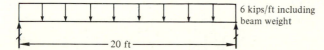

Figure E11.18

SOLUTION The maximum bending moment is

$$M = \frac{wl^2}{8} = \frac{(6)(20)^2}{8} = 300 \text{ ft·kips}$$

A 20-ft unsupported length is quite large and probably exceeds L_u, giving $F_b < 0.60F_y$. Assume

$F_b = 17$ kips/in^2; then

$$S_{req} = \frac{(300)(12)}{17} = 211.8 \text{ in}^3$$

Try W24 × 94, width $d = 24.31$ in, $b_f = 9.065$ in, $t_f = 0.875$ in, $S = 222$ in^3, $r_T = 2.33$, and $C_b = 1.0$. Then

$$\frac{L_b}{r_T} = \frac{(20)(12)}{2.33} = 103 < \sqrt{\frac{5.10 \times 10^5 C_b}{F_y}} = \sqrt{\frac{(510,000)(1)}{36}} = 119$$

$$> \sqrt{\frac{1.02 \times 10^5 C_b}{F_y}} = \sqrt{\frac{(102,000)(1)}{36}} = 53.2$$

$$F_{b_2} = \left[\frac{2}{3} - \frac{F_y (L_b/r_T)^2}{1.53 \times 10^6 C_b} \right] F_y = \left[\frac{2}{3} - \frac{(36)(103)^2}{(1,530,000)(1)} \right](36)$$

$$= (0.42)(36) = 15.1 \text{ kips/in}^2$$

Since it is a rolled section, the twisting formula also applies:

$$F_{b_1} = \frac{1.2 \times 10^4 C_b}{L_b(d/A_f)} = \frac{(12,000)(1)}{(20)(12)(24.31)/(9.065)(0.875)}$$

$$= 16.3 \text{ kips/in}^2$$

$F_{b_1} = 16.3$ kips/in^2 controls. Check required section modulus

$$S = \frac{(300)(12)}{16.3} = 222 > 213 \text{ in}^3 \qquad \text{OK}$$

Slightly undersigned. A W21 × 93 is adequate. (The beam must be checked for shear stress.)

11.8.3 Wood Beams

Economy in timber-beam design, particularly laminated beams, favors a deep and narrow section. As indicated previously, such sections increase the likelihood of lateral buckling of the compression flange. While many building codes have rules governing depth-to-breadth ratios to minimize such tendencies, theoretical calculations and testing verify the importance of the length of the compression flange as a governing parameter. A formula similar to that recommended by the AISC for the buckling of steel beams, where $L_b = f(d/A_f)$ is used, has been derived and checked for use with laminated beams.

The slenderness factor of a beam equals

$$C_S = \sqrt{\frac{l_e d}{b^2}} \tag{11.28}$$

where l_e = effective length of beam = function of unsupported length l_u, in (see Table 11.2)
        b = width of beam, in
        d = depth of beam, in

Table 11.2 Effective length of compression edge [2]

Type of span	Type of load	l_e/l_u
Single	Concentrated at midspan	1.61
	Uniformly distributed	1.92
	Equal end moments	1.84
Cantilever	Concentrated at end	1.69
	Uniformly distributed	1.06
Either	Any	1.92

When the compression edge of a beam is supported throughout its length, preventing its lateral displacement, and the ends at points of bearing have lateral support to prevent rotation, the unsupported length may be taken as zero. When lateral support is provided to prevent rotation at the points of end bearing but no other support to prevent rotation or lateral displacement is provided throughout the length of a beam, the unsupported length is the distance between such points.

The effective length of the compression edge depends on the unsupported length, type of span, and type of load (Table 11.2). Three possible ranges in behavior have been defined on the basis of the slenderness factor C_S. The allowable bending stress F_b' depends upon the range:

Short beams $C_S < 10$

$$F_b' = F_b \tag{11.29}$$

Intermediate beams $10 < C_S < C_K$

where

$$C_K = \sqrt{\frac{3E}{5F_b}}$$

and

$$F_b' = F_b\left[1 - \frac{1}{3}\left(\frac{C_S}{C_K}\right)^4\right] \tag{11.30}$$

Long beams $C_K < C_S < 50$

$$F_b' = \frac{0.40E}{C_S^2} \tag{11.31}$$

These ranges in behavior with their allowable bending stresses are illustrated in Fig. 11.27.

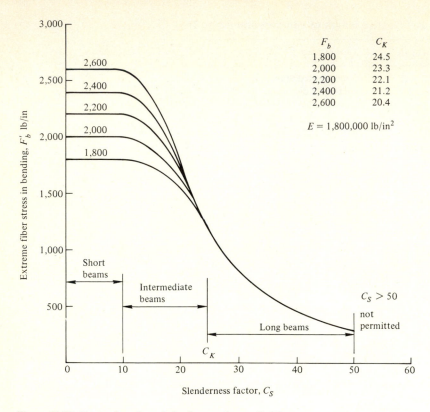

F_b	C_K
1,800	24.5
2,000	23.3
2,200	22.1
2,400	21.2
2,600	20.4

$E = 1,800,000 \text{ lb/in}^2$

Figure 11.27 F_b as a function of slenderness factor C_S [1].

Example 11.19 Design a glued-laminated roof beam to meet the following requirements:

Length 40 ft
Spacing 20 ft
Joists 24 in on centers
Snow load 30 lb/ft^2
Dead load 10 lb/ft^2

Use 22F structural glued-laminated southern pine.

SOLUTION From Table A.4, using 1.15 as the load-duration factor for snow, we get

$$F_b = (2200)(1.15) = 2530 \text{ lb/in}^2 \qquad F_v = (200)(1.15) = 230 \text{ lb/in}^2$$

$$E = 1.7 \times 10^6 \text{ lb/in}^2$$

$$w_{DL} = (20)(10) = 200 \text{ lb/ft}$$

$$w_{SL} = (20)(30) = \underline{600 \text{ lb/ft}}$$

$$w = 800 \text{ lb/ft}$$

The minimum bending moment is

$$\frac{wl^2}{8} = \frac{(800)(40)^2}{8} = 160,000 \text{ ft·lb}$$

Assume a depth factor D_F of 0.90. The section modulus required is

$$S = \frac{M}{F_b C_f} = \frac{(160,000)(12)}{(2530)(0.90)} = 843.2 \text{ in}^3$$

Try $6\frac{3}{4}$ by $28\frac{1}{2}$ in:

$$S = 914 \text{ in}^3 \qquad C_F = 0.908 \qquad \text{Table A.3}$$

This beam can now be checked for lateral support conditions.

Consider the beam braced laterally at the location of the joists (every 24 in). This case can be considered to be similar to equal end moments (Table 11.2), giving an effective length

$$l_e = (1.84)(24) = 44.2 \text{ in}$$

$$C_S = \sqrt{\frac{l_e d}{b^2}} = \sqrt{\frac{(44.2)(28.5)}{(6.75)^2}} = 5.25$$

Since $C_S < 10$, there is no reduction in bending stress because of slenderness and the section is adequate.

If the joists are not fastened sufficiently to be considered points of lateral support, the beam could buckle over the unsupported distance of 40 ft.

$$l_e = (1.92)(40)(12) = 921.6 \text{ in} \qquad \text{Table 11.2}$$

$$C_S = \sqrt{\frac{(921.6)(28.5)}{(6.75)^2}} = 24 \qquad C_K = \sqrt{\frac{3E}{5F_b}} = \sqrt{\frac{(3)(1.7 \times 10^6)}{(5)(2200)}} = 21.5$$

Since $C_K < C_S < 50$, the beam is a long beam and the allowable stress is

$$F_b' = \frac{0.40E}{C_S^2} = \frac{(0.40)(1.7 \times 10^6)}{(24)^2} = 1180 \text{ lb/in}^2$$

The section is not quite adequate since

$$(1180)(1.15) < (2220)(1.15)(0.91)$$

The depth factor C_F is not used in conjuction with the slenderness factor [2].

Try a new section; assume $F_b' = 950$ lb/in^2:

$$S_{req} = \frac{(160,000)(12)}{950} = 2021 \text{ in}^3$$

Try a $6\frac{3}{4}$- by $43\frac{1}{2}$-in section:

$$S = 2129 \text{ in}^3 \qquad C_S = \sqrt{\frac{(921.6)(43.5)}{(6.75)^2}} = 29.66$$

$$F_b' = \frac{0.40E}{C_S^2} = \frac{(0.40)(1.7 \times 10^6)}{(29.66)^2} = 773 \text{ lb/in}^2$$

$$F_b = (773)(1.15) = 889 \text{ lb/in}^2$$

The required section modulus is

$$\frac{(1.6 \times 10^6)(12)}{889} = 2159 > 2129 \text{ in}^3 \text{ by } 1.47\% \qquad \text{OK}$$

A better solution might be to ensure that the joists will provide lateral support.

11.9 BENDING ABOUT TWO AXES

Occasionally a beam is required to support loads inclined to its axes or applied perpendicular to both axes. The former may occur when the beam is tilted and the loads are vertical (Fig. 11.28a). An example of the latter is girts used on the outside of buildings to support materials enclosing the structure. They span from column to column and generally are channels or angles. Such members are subject to gravity and wind loads (Fig. 11.28b).

Beam theory presented earlier is based on the assumption that the plane of loading passes through the centroid of the beam section. Since this is still true for both beams shown in Fig. 11.28, their design can be kept reasonably simple. For a tilted beam, it is desirable to resolve the load into components parallel to the x and y axes. At the centroid of the beam, this produces a loading similar to that in Fig. 11.28b. For any point along the beam, the fiber stress becomes the algebraic sum of the stresses resulting from bending about each axis

$$f_b = \frac{M_x c_x}{I_x} \pm \frac{M_y c_y}{I_y} \tag{11.32}$$

where
M_x = moment causing bending about x axis
I_x = moment of inertia of section about x axis
c_x = distance from x axis to fiber in question measured perpendicular to x axis
M_y = moment causing bending about y axis
I_y = moment of inertia of section about y axis
c_y = distance from y axis to fiber in question measured perpendicular to y axis

Appropriate section moduli can be used instead of I_x/c_x and I_y/c_y.

In practice, the load rarely passes through the centroid of a tilted beam. Instead the load is applied at the center of the top of the flange. This loading

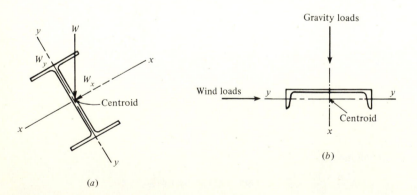

Figure 11.28 Biaxial bending: (a) tilted beam, (b) wall girt.

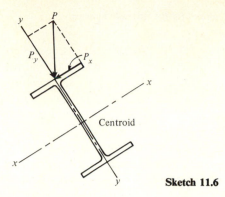

Sketch 11.6

causes twisting of the beam as well as bending (Sketch 11.6). A complete analysis of this torsion problem is beyond the scope of this text, but an approximate analysis can be used. It assumes that the resistance to bending, resulting from the component of load parallel to the top of the flange, is provided by the top flange. The moment of inertia of the flange is used in computing the bending stress from the parallel component P_x. This is approximately equal to $I_y/2$ since the web contributes little to I_y. As a result, the flexure formula becomes

$$f_b = \frac{M_x c_x}{I_x} \pm \frac{M_y c_y}{I_y/2} \tag{11.33}$$

The trial-and-error method of design is appropriate for situations involving bending about both axes.

Example 11.20 A beam is estimated to have a vertical bending moment of 100 ft·kips and a lateral bending moment of 25 ft·kips. These moments are assumed to include the weight of the beam. The loads can be assumed to pass through the centroid of the beam. Select a W section to resist these loads. Use a A36 steel and AISC specifications. Assume complete lateral support of the compression flange.

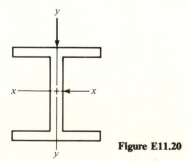

Figure E11.20

SOLUTION (Fig. E11.20)

$$F_b = 0.60 F_y = 22,000 \text{ lb/in}^2$$

A trial section will have to be larger than the S required for the vertical moments,

$$S_x = \frac{(100)(12)}{22} = 54.5 \text{ in}^3$$

Try W14 × 68:

$$S_x = 103 \text{ in}^3 \qquad S_y = 24.2 \text{ in}^3$$

$$f_b = \frac{(100)(12)}{103} + \frac{(25)(12)}{24.2} = 11.65 + 12.39$$

$$= 24.04 > 22 \text{ kips/in}^2 \qquad \text{NG}$$

Try W12 × 72:

$$S_x = 97.4 \text{ in}^3 \qquad S_y = 32.4 \text{ in}^3$$

$$f_b = \frac{(100)(12)}{97.4} + \frac{(25)(12)}{32.4} = 12.32 + 9.25 = 21.57 < 22 \text{ kips/in}^2 \qquad \text{OK}$$

Other selections are possible, some of which might be more economical.

Example 11.21 Redesign the beam of Example 11.20 if the lateral loads are assumed to be parallel to the top flange (Fig. E11.21). Equation (11.32) can be used as a basis for this design. The effective section modulus about the y axis becomes 50 percent of the tabulated value.

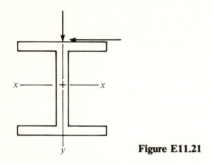

Figure E11.21

SOLUTION Try W14 × 90:

$$S_x = 143 \text{ in}^3 \qquad S_y = 49.9 \text{ in}^3$$

$$f_b = \frac{(100)(12)}{143} + \frac{(25)(12)}{49.9/2} = 8.39 + 12.02 = 20.41 < 22 \text{ kips/in}^2 \qquad \text{OK}$$

11.10 COMPOSITE BEAMS

11.10.1 Introduction

A *composite beam* is a beam made of more than one material but still acting and deflecting as a single unit. When such a system acts compositely, there is no relative slippage at the interface between the materials.

The two most common examples are so-called *flitch beams* (wood reinforced by steel) and steel beams supporting a cast-in-place reinforced-concrete slab. Typical examples are illustrated in Fig. 11.29. The design of the steel beam supporting a slab will be considered in Chap. 14.

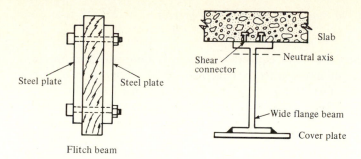

Figure 11.29 Composite-beam sections.

11.10.2 Composite Flexure Theory

The theory of flexure does not apply to composite beams because it is based on the assumption that the beam is homogeneous and that plane sections remain plane during bending. A basic approach to nonhomogeneous beams is to transform the section into an equivalent homogeneous section and apply the flexure theory developed earlier.

Consider the timber beam in Fig. 11.30. It is reinforced by a steel plate securely fastened to the timber so that no slip occurs between them as the beam is bent. To obtain an equivalent section of one or the other material, consider adjacent fibers of steel and wood at point A. Since the wood and steel are bolted together and do not slip relative to each other, the strains of the steel and wood fibers at A must be equal

$$\varepsilon_s = \varepsilon_w$$

Expressing this relationship in terms of stresses and moduli of elasticity gives

$$\frac{f_s}{E_s} = \frac{f_w}{E_w} \tag{1}$$

In order to be equivalent, the loads carried by the steel fiber and the equivalent wood fiber must be equal. As a result,

$$P_s = P_w \quad \text{or} \quad f_s A_s = f_w A_w \tag{2}$$

Combining Eqs. (1) and (2) gives

$$A_s \frac{E_s}{E_w} f_w = A_w f_w$$

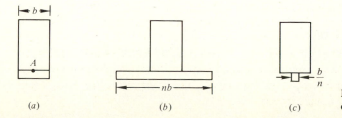

(a) (b) (c)

Figure 11.30 Development of composite-beam theory.

and denoting the ratio of moduli of elasticity by n gives

$$A_s n = A_w \tag{11.34}$$

This indicates that the area of equivalent wood is n times the area of steel. The location of the equivalent area is governed by the condition that the equivalent wood fibers must be located at the same distance from the neutral axis as the steel fibers they replace in order to satisfy the criterion of equal deformations [Eq. (1)]. This means that the equivalent wood area is n times as wide as the steel it replaces but remains as thick (Fig. 11.30b). An equivalent steel section would have the wood replaced by steel $1/n$ as wide as the wood (Fig. 11.30c).

The flexure formula (11.12) can now be applied directly to either equivalent section. With the equivalent wood section, the actual steel stress is n times the stress in the equivalent wood. With the equivalent steel section, the actual wood stress is $1/n$ times the stress in the equivalent steel.

The formula for horizontal shear stress [Eq. (11.14)] developed for homogeneous beams can be applied to the equivalent section of the composite beam. This is true because the shear stress is based on the difference between normal forces acting on two adjacent sections. The actions of the forces on the equivalent section are the same as those acting on the original section.

Deflections of composite beams can be calculated, as in homogeneous beams, by using the flexural rigidity EI of the equivalent section (see Sec. 11.11). This is true because deflection is the result of strain in fibers and the strain in fibers of the equivalent section must be equal to the strain in corresponding fibers of the original section.

Example 11.22 A flitch beam is made of two 2- by 12-in no. 1 Douglas fir with a 1/4-in A36 steel plate. Determine the allowable bending moment for a load of 2 months duration. Which material controls the design? What are the actual stresses in the wood and steel at the allowable moment?

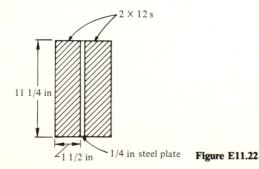

Figure E11.22

SOLUTION See Fig. E11.22. For Douglas fir no. 1 (Table A.1),

$$F_b = (1350)(1.15) = 1552 \text{ lb/in}^2 \quad \text{and} \quad E_w = 1.6 \times 10^6 \text{ lb/in}^2$$

(the load-duration factor does not apply to steel stresses)

$$F_v = (85)(1.15) = 98 \text{ lb/in}^2$$

Section properties of the 2 by 12 (from Table A.2) are $1\frac{1}{2}$ by $11\frac{1}{4}$ in and $I = 178$ in⁴. For A36 steel

$$F_b = 0.60F_y = 22,000 \text{ lb/in}^2 \qquad E = 2.95 \times 10^7 \text{ lb/in}^2$$

$$n = \frac{2.95 \times 10^7}{1.6 \times 10^6} = 18.43$$

Since $22,000/18.43 = 1194 < 1552$, steel stresses control. The moment of inertia of the equivalent wood section is

$$I_w = \frac{bh^3}{12} = \frac{\left[(2)(1.5) + \left(\frac{1}{4}\right)(18.43)\right](11.25)^3}{12} = 902.6 \text{ in}^4$$

The allowable moment is

$$\frac{I}{cF_b} = \frac{902.6}{11.25/2} \frac{1194}{12} = 15,966 \text{ ft} \cdot \text{lb}$$

For the allowable bending moment of 15,966 ft·lb the maximum wood stress is

$$\frac{Mc}{I} = \frac{(15,966)(12)}{902.6} \frac{11.25}{2} = 1194 \text{ lb/in}^2$$

and the maximum steel stress is

$$(1194)(18.43) = 22,000 \text{ lb/in}^2$$

Example 11.23 A uniformly distributed load of 400 lb/ft (including the weight of the beam) is simply supported on a 20-ft span. The depth of the beam cannot exceed 10 in. Determine the thickness of required 8-in side plates. Use no. 1 dense southern pine.

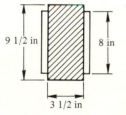

9 1/2 in   8 in

3 1/2 in   **Figure E11.23**

SOLUTION See Fig. E11.23. For no. 1 dense southern pine (Table A.1)

$$F_b = 1400 \text{ lb/in}^2 \qquad E_w = 1.6 \times 10^6 \text{ lb/in}^2$$

For A36 steel

$$F_b = 0.60F_y = 22,000 \text{ lb/in}^2 \qquad E_S = 2.95 \times 10^7 \text{ lb/in}^2$$

Try a 4 by 10:

$$n = \frac{2.95 \times 10^7}{1.6 \times 10^6} = 18.43$$

The maximum moment is

$$\frac{wl^2}{8} = \frac{(400)(20)^2(12)}{8} = 240,000 \text{ in} \cdot \text{lb}$$

Since

$$\frac{22,000}{18.43} \frac{4}{4.75} = 1005 < 1400 \text{ lb/in}^2$$

steel is critical. The required I of the wood section is

$$\frac{Mc}{F_b} = \frac{240,000}{1300}\frac{9.5}{2} = 876.9 \text{ in}^4$$

The moment of inertia of a 4 by 10 is 230.8 in^4. The moment of inertia contributed by the plates is

$$I = 876.9 - 230.8 = 646.1 \text{ in}^4$$

Then from

$$\frac{n(2th^3)}{12} = \frac{(18.43)\left[2t(8)^3\right]}{12}$$

we get

$$t = \frac{(646.1)(12)}{(18.43)(2)(8)^3} = 0.411 \text{ in}$$

Use two $\frac{7}{16}$-in plates.

11.11 BEAM DEFLECTION

11.11.1 Introduction

So far the design of beams has been based on resistance to failure in bending and shear. Another important aspect is excessive deflection. The most common results of excessive beam deflection are damage to associated materials, a curvature that is noticeable, and functional deficiencies such as an excessively springy floor. In order to avoid these conditions, it is customary to restrict the deflection. Over the years building codes and design practice have limited deflections of floors and roof framing to a maximum of $\frac{1}{360}$ times the span. This deflection would not cause cracking of an attached plaster ceiling. Additional criteria are necessary with changes in materials and activities.

A study of deflection phenomena is also helpful in the analysis of statically indeterminate structures by providing additional equations of geometric compatibility.

This treatment will be restricted to the deflection of statically determinate beams due to bending moments. Small amounts of deflection also develop in beams due to shear. Except in the case of very short deep beams, the contribution due to shear can be neglected.

Deflection equations can be derived in a number of ways. The most basic is the elastic-curve method, based on the solution of the differential equation for the elastic curve.

11.11.2 Derivation of the Elastic-Curve Equation

When a straight beam is loaded and the action is elastic, the longitudinal centroidal axis of the beam becomes a curve defined as the *elastic curve*. In Sec. 11.4.1 it was shown that the magnitude of the bending stress varies directly with the distance from the neutral axis

$$\frac{f}{y} = \frac{E}{\rho}$$

where ρ is the radius of curvature of the elastic curve. Using the general flexure expression $f = My/I$ and substituting the value of f in the above equation, we get

$$\frac{1}{\rho} = \frac{M}{EI} \qquad (11.35)$$

an expression that is the basis for calculation of deflections due to bending.

Equation (11.35) is applicable only when the bending moment is constant for the interval of the beam being considered. For most beams the bending moment is a function of the position along the beam. A more general expression can be developed on the basis of the geometry of the bent beam shown in Sketch 11.7.

$$\frac{dy}{dx} = \tan \theta$$

The elastic curve is normally very flat for beams and its slope very small at any point. The value of the slope, $\tan \theta = dy/dx$, can with little error be set equal to θ. As a result,

$$\theta = \frac{dy}{dx} \qquad \text{and} \qquad \frac{d\theta}{dx} = \frac{d^2y}{dx^2}$$

It is also evident from Sketch 11.7 that

$$dL = \rho \, d\theta$$

Again, because the elastic curve is flat, dL is practically equivalent to dx. Consequently

$$\frac{1}{\rho} = \frac{d\theta}{dL} = \frac{d\theta}{dx}$$

or

$$\frac{1}{\rho} = \frac{d^2y}{dx^2} \qquad (11.36)$$

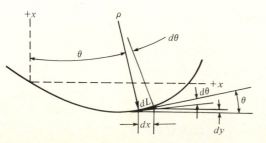

Sketch 11.7

Equating values of $1/\rho$ from Eqs. (11.11) and (11.36), we have

$$EI\frac{d^2y}{dx^2} = M \qquad (11.37)$$

This is known as the *differential equation of the elastic curve of a beam*. The product EI is termed the *flexural rigidity* of the beam and is usually constant for a given beam.

The sign convention established for bending moments in Sec. 11.2.2 is still applicable. Both E and I are always positive. As a result the sign of the second derivative must be the same as that of the bending moment, as illustrated in Sketch 11.8. For the interval A to B, the slope changes from positive to negative and gives a negative second derivative. For the interval B to C, d^2y/dx^2 and M are both positive. Unfortunately with this convention *y upward becomes positive*.

It is necessary to correlate the succession derivatives of the deflection y of the elastic curve with the physical quantities they represent in beam action:

$$\text{Deflection} = y \qquad \text{Slope} = \frac{dy}{dx} \qquad \text{Moment } M = \frac{EI\,d^2y}{dx^2}$$

$$\text{Shear } V = \frac{dM}{dx} = \frac{EI\,d^3y}{dx^3} \qquad \text{Load } w = \frac{dV}{dx} = \frac{EI\,d^4y}{dx^4}$$

Before proceeding with specific examples of calculating beam deflections, the assumptions used in the derivation of Eq. (11.37) should be emphasized. All the limitations that apply to the flexure formula apply to deflection calculations. In addition the slope of the beam was assumed to be small. Beam deflections due to shearing stresses are neglected with the assumption that plane sections remain plane during bending. The values of E and I are assumed to remain constant over the length of the beam.

Whenever the above assumptions are essentially correct, and if the bending moment equation is in integratable form, Eq. (11.37) can be used to solve for the deflection y of the elastic curve of a beam at any point x on the beam. Since the bending-moment equation requires double integration to give the equation of the elastic curve, two constants of integration have to be evaluated. This is done by using applicable *boundary conditions*, i.e., a known set of values of x and y or x and dy/dx for a given location on the beam. When a beam is subjected to abrupt changes in loading, it becomes impossible to write a single equation for the bending moment for the entire beam length. This can be overcome by

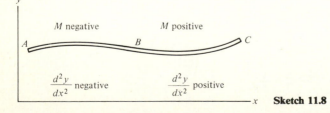

Sketch 11.8

developing appropriate equations for each interval of the beam. Since the beam is continuous at these points of abrupt changes in loading, the slope and deflections at the juncture of adjacent intervals must match. These matching conditions are used to help evaluate the constants of integration in a given interval.

The following examples illustrate the use of the double-integration method for calculating beam deflections. Other methods, often simpler in application, are available. Some depend on the bending-moment equation, e.g., the area-moment and conjugate-beam methods. The importance of the double-integration method is that it helps relate such physical attributes as deflection, slope, moment, shear, and load of a beam. The prime purpose of this section is to illustrate a method for deriving slope and deflection equations that can be used in structural design. In many instances these formulas will be obtained from tables in design manuals. A selection of these formulas is presented in Table 11.3. It is important to choose the formula which corresponds to the actual design conditions.

Table 11.3 Beam loadings, slope, and deflection formulas†

Type of load	Maximum moment	Slope at end‡	Maximum deflection¶
(cantilever, end load P)	$M = -PL$	$\theta = \dfrac{PL^2}{2EI}$	$\delta = \dfrac{PL^3}{3EI}$
(cantilever, load P at distance a)	$M = -Pa$	$\theta = \dfrac{Pa^2}{2EI}$	$\delta = \dfrac{Pa^2}{6EI}(3L - a)$
(cantilever, uniform load w lb/ft)	$M = -\dfrac{wL^2}{2}$	$\theta = \dfrac{wL^3}{6EI}$	$\delta = \dfrac{wL^4}{8EI}$
(cantilever, triangular load w lb/ft)	$M = -\dfrac{wL^2}{6}$	$\theta = \dfrac{wL^3}{24EI}$	$\delta = \dfrac{wL^4}{30EI}$
(cantilever, end moment M)	$M = -M$	$\theta = \dfrac{ML}{EI}$	$\delta = \dfrac{ML^2}{2EI}$
(simple beam, center load P)	$M = \dfrac{PL}{4}$	$\theta_L = \theta_R = \dfrac{PL^2}{16EI}$	$\delta = \dfrac{PL^3}{48EI}$

Table 11.3 (Cont'd.)

Type of load	Maximum moment	Slope at end‡	Maximum deflection¶
	$M = \dfrac{Pab}{L}$ at $x = a$	$\theta_L = \dfrac{Pb(L^2 - b^2)}{6EIL}$ $\theta_R = \dfrac{Pa(L^2 - a^2)}{6EIL}$	$\delta = \dfrac{Pb(L^2 - b^2)^{3/2}}{9\sqrt{3}\ EIL}$ at $x = \sqrt{\dfrac{L^2 - b^2}{3}}$ At center (not maximum) $\delta = \dfrac{Pb}{48EI}(3L^2 - 4b^2)$ when $a > b$
	$M = \dfrac{wL^2}{8}$	$\theta_L = \theta_R = \dfrac{wL^3}{24EI}$	$\delta = \dfrac{5wL^4}{384EI}$
	$M = \dfrac{wL^2}{9\sqrt{3}}$	$\theta_L = \dfrac{7wL^3}{360EI}$ $\theta_R = \dfrac{8wL^3}{360EI}$	$\delta = \dfrac{2.5wL^4}{384EI}$ at $x = 0.519L$
	$M = \dfrac{wL^2}{12}$	$\theta_L = \theta_R = \dfrac{5wL^3}{192EI}$	$\delta = \dfrac{wL^4}{120EI}$
	$M = M$	$\theta_L = \dfrac{ML}{6EI}$ $\theta_R = \dfrac{ML}{3EI}$	$\delta = \dfrac{ML^2}{9\sqrt{3}\ EI}$ at $x = \dfrac{L}{\sqrt{3}}$ At center (not maximum) $\delta = \dfrac{ML^2}{16EI}$
	$M = M$	$\theta_L = \dfrac{ML}{3EI}$ $\theta_R = \dfrac{ML}{6EI}$	$\delta = \dfrac{ML^2}{9\sqrt{3}\ EI}$ at $x = \left(L - \dfrac{L}{\sqrt{3}}\right)$ At center (not maximum) $\delta = \dfrac{ML^2}{16EI}$

† P in pounds, w in pounds per foot, L in feet, E in pounds per square inch, I in inches[4], and M in foot-pounds.

‡ To compute slope in radians multiply by $(12)^2 = 144$.

¶ To compute deflection in inches multiply by $(12)^3 = 1728$.

A *method of superposition* can be used wherein the results of a few simple loadings can be used to obtain those for more complicated loadings. This method determines the slope or deflection at any point in a beam as the resultant of the slopes and deflections at that point caused by each of the loadings acting separately. The only restriction on this method is that the effect produced by each load must be independent of that produced by the other loads. This technique of superposition is especially useful for loadings that combine the types tabulated in Table 11.3. Example 11.26 illustrates the use of the superposition method.

Example 11.24 Determine the maximum deflection of a simply supported beam of span L carrying a uniformly distributed load of w lb/ft (Fig. E11.24).

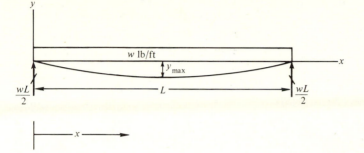

Figure E11.24

SOLUTION The bending-moment equation is

$$M = \frac{wLx}{2} - \frac{wx^2}{2}$$

Substituting into Eq. (11.37) gives

$$EI \frac{d^2 y}{dx^2} = \frac{wLx}{2} - \frac{wx^2}{2}$$

The slope equation is obtained by integrating the moment equation

$$EI \frac{dy}{dx} = \frac{wLx^2}{4} - \frac{wx^3}{6} + C$$

The constant of integration C is determined by considering the boundary condition $dy/dx = 0$ at $x = L/2$. This is true because of the symmetrical loading

$$EI \frac{dy}{dx} = \frac{wL}{4} \frac{(L)^2}{2} - \frac{w}{6} \frac{(L)^3}{2} + C = 0$$

$$C = - \frac{wL^3}{24}$$

The equation for the slope curve becomes

$$EI \frac{dy}{dx} = \frac{wLx^2}{4} - \frac{wx^3}{6} - \frac{wL^3}{24}$$

The equation for the elastic curve is obtained by integrating the slope equation

$$EIy = \frac{wLx^3}{(4)(3)} - \frac{wx^4}{(6)(4)} - \frac{wL^3x}{24} + C_1$$

The constant of integration C_1 is determined by using the boundary condition $y = 0$ at $x = 0$

$$EIy = 0 + C_1 \qquad C_1 = 0$$

This gives the equation of the elastic curve as

$$EIy = \frac{wLx^3}{12} - \frac{wx^4}{24} - \frac{wL^3x}{24}$$

The maximum deflection occurs at midspan because of the symmetrical loading

$$EIy = \frac{wL}{12}\left(\frac{L}{2}\right)^3 - \frac{w}{24}\left(\frac{L}{2}\right)^4 - \frac{wL^3}{24}\frac{L}{2}$$

$$y_{max} = -\frac{5wL^4}{384EI} = \frac{5wL^4}{384EI} \qquad \text{downward}$$

Example 11.25 Determine the deflection equation of a cantilever beam of length L carrying a uniformly distributed load of w lb/ft (Fig. E11.25).

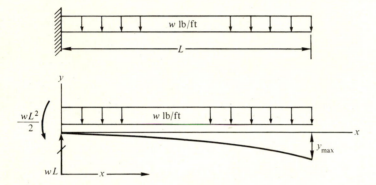

Figure E11.25

SOLUTION The bending-moment equation is

$$M = wLx - \frac{wL^2}{2} - \frac{wx^2}{2}$$

Substituting into Eq. (11.37) gives

$$EI\frac{d^2y}{dx^2} = wLx^2 - \frac{wL^2}{2} - \frac{wx^2}{2}$$

The slope equation is obtained by integrating the moment equation

$$EI\frac{dy}{dx} = \frac{wLx^2}{2} - \frac{wL^2x}{2} - \frac{wx^3}{(2)(3)} + C$$

The constant of integration C is determined by using the boundary condition $dy/dx = 0$ at $x = 0$. As a result,

$$EI\frac{dy}{dx} = C = 0$$

The equation for the slope curve becomes

$$EI\frac{dy}{dx} = \frac{wLx^2}{2} - \frac{wL^2x}{2} - \frac{wx^3}{6}$$

The equation for the elastic curve is obtained by integrating the slope equation

$$EIy = \frac{wLx^3}{(2)(3)} - \frac{wL^2x^2}{(2)(2)} - \frac{wx^4}{(6)(4)} + C_1$$

The constant of integration C_1 is evaluated by using the boundary condition $y = 0$ at $x = 0$:

$$EIy = C_1 = 0$$

This gives the equation of the elastic curve,

$$EIy = \frac{wLx^3}{6} - \frac{wL^2x^2}{4} - \frac{wx^4}{24}$$

The maximum deflection occurs at $x = L$,

$$EIy = \frac{wL(L^3)}{6} - \frac{wL^2(L^2)}{4} - \frac{w(L)^4}{24}$$

$$y_{max} = -\frac{wL^4}{8EI} = \frac{wL^4}{8EI} \qquad \text{downward}$$

Example 11.26 Determine the maximum deflection of the loaded cantilever beam shown in Fig. E11.26.

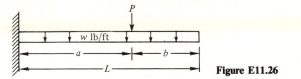

Figure E11.26

SOLUTION

$$y_{P,\,max} = \frac{Pa^2}{6EI}(3L - a) \qquad \text{Table 11.3}$$

$$y_{w,\,max} = \frac{wL^4}{8EI} \qquad \text{Table 11.3}$$

$$y_{max} = y_{P,\,max} + y_{w,\,max} = \frac{Pa^2}{6EI}(3L - a) + \frac{wL^4}{8EI} = \frac{4Pa^2(3L - a) + 3wL^4}{24EI}$$

11.11.3 Deflection of Wood Beams

Under service-load conditions wood beams behave like linear elastic members. As a result, the deflection formulas given in Table 11.3 can be used for the determination of deflections of wood beams.

As pointed out previously, these deflection equations represent deflections due to bending strains and neglect those due to shear. This component tends to be more significant for wood beams because the shear modulus G is much smaller than the modulus of elasticity E in the case of wood than for steel. There is, however, a built-in compensation: the listed modulus of elasticity E for any given species of wood is smaller than the actual value because it has not been corrected for the l/d ratio normally encountered in practice. There is no load-duration adjustment for the modulus of elasticity, but under permanently applied loads, a timber beam could be expected to creep to a deflection twice the initial elastic value. As a result, where there are specific constraints on the maximum allowable deflection, it is customary to design for one-half the deflection using the listed value of elastic modulus. With laminated timber it is not unusual to build in a camber equal to the dead-load deflection. This becomes very important in providing drainage of flat roofs and avoiding unsightly sagging.

Table 11.4 Recommended deflection limitations for beams [1][a]

Use classification	Applied load only[b]	Dead + applied load	$\dfrac{\text{Minimum camber}[c]}{\text{Dead-load deflection}}$
Roof beams, industrial[d]	$L/180$	$L/120$	1.5
Commercial and institutional:[e]			
Without plaster ceiling	$L/240$	$L/180$	1.5
With plaster ceiling	$L/360$	$L/240$	1.5
Floor beams (ordinary use[f])	$L/360$	$L/240$	1.5

[a] For special uses, such as beams supporting vibrating machines, beams over glass windows and doors, etc., more severe limitations may be required.

[b] Applied load is live load, wind load, snow load, etc., or any combination thereof.

[c] Curvature built into a glued-laminated beam opposite to that due to bending. If drainage is to be provided for by camber, add to above ratios.

[d] For sawn members, this classification applies to construction for which appearance is not of prime importance and for which adequate drainage is provided to avoid ponding.

[e] Applies to churches, schools, residences, etc., for which appearance, absence of visible deflection, and minimizing of plaster cracking are of prime importance.

[f] Construction in which walking comfort, minimized plaster cracking, and the elimination of objectionable springiness are of prime importance.

Recommended deflection limitations for solid sawn and glued-laminated beams are given in Table 11.4. In addition, recommended values of the ratio of minimum camber to dead-load deflection for glued-laminated beams are given.

Example 11.27 Design a simply supported roof beam given the following:

$$26\text{F Douglas fir} \qquad L = 50 \text{ ft} \qquad \text{Spacing} = 20 \text{ ft}$$

Dead load of joists, decking, and roof is 10 lb/ft^2, snow load is 30 lb/ft^2, and deflection limit is $L/180$.

SOLUTION Allowable stresses from Table A.4 are

$$F_b = 2600 \text{ lb/ft}^2 \qquad F_v = 165 \text{ lb/in}^2 \qquad E = 1.8 \times 10^6 \text{ lb/in}^2$$

Assume that the top of the beam is laterally supported. The applied loads are

$$\text{Dead loads} = 10 \text{ lb/ft}^2$$

$$\text{Snow loads} = \frac{30 \text{ lb/ft}^2}{40 \text{ lb/ft}^2}$$

$$(40 \text{ lb/ft}^2)(20 \text{ ft}) = 800 \text{ lb/ft}$$

$$\text{Beam weight} = \underline{50 \text{ lb/ft}}$$

$$\text{Total load} = \overline{850 \text{ lb/ft}}$$

The maximum bending moment is

$$M = \frac{wL^2}{8} = \frac{(850)(50)^2}{8} = 266{,}000 \text{ ft·lb}$$

The required section modulus is

$$C_F S = \frac{M}{F_b} = \frac{(266{,}000)(12)}{(2600)(1.15)} = 1069 \text{ in}^3$$

where 1.15 is the load-duration factor for snow. If $1\frac{1}{2}$-in laminations are used, an $8\frac{3}{4}$- by $28\frac{1}{2}$-in beam is adequate (Table A.3) and

$$C_F = 0.908 \qquad S = 1185 \text{ in}^3 \qquad C_F S = 1075 \text{ in}^3$$

$$I = 16{,}880 \text{ in}^4 \qquad A = 249 \text{ in}^2 \qquad \text{Weight per foot} = \frac{294}{4} = 62 \text{ lb} \qquad \text{OK}$$

Check the shear stress:

$$f_v = \frac{VQ}{Ib} = \frac{3V}{2A} = \frac{(25)(862)}{249} \frac{3}{2} = 130 < 165 \text{ lb/in}^2 \qquad \text{OK}$$

This calculation does not neglect the load within a distance of the depth of beam from each support.

Check deflection:

$$y = \frac{5}{384} \frac{wL^3}{EI} \qquad \text{Table 11.3}$$

$$= \frac{5}{384} \frac{(862)(50)^3(12)^3}{(1.8 \times 10^6)(16{,}880)} = 0.80 \text{ in}$$

$$y_{\text{all}} = \frac{(50)(12)}{180} = 3.33 > 0.80 \text{ in} \qquad \text{OK}$$

11.11.4 Deflection of Steel Beams

It is difficult to differentiate between beams that have an excessive deflection under a given load and beams that do not, but there are instances where excessive deflections will not only cause aesthetic and psychological problems but will also inhibit the performance of materials and equipment. For example, excessive floor deflections can be objectionable not only because of springiness but because of undesirable vibration characteristics and the effect on ceiling materials such as plaster. They can also cause misalignment of equipment. Excessive deflection of roof purlins can cause damage to roofing materials. The retention of water due to the deflection of flat roofs (*ponding*) can cause collapse and has consequently received special consideration by the AISC.

In most cases, deflection limitations are given as a proportion of the span, but there are occasions when absolute values are of importance. For example, if a lintel has $\frac{1}{2}$-in clearance over a glass panel the permissible deflection will be less than $\frac{1}{2}$ in, regardless of the span of the lintel.

The maximum deflection of a simply supported uniformly loaded beam is given by

$$y = \frac{5}{384} \frac{WL^3}{EI} \qquad \text{Table 11.3}$$

where W denotes the total load on the span. Since the maximum bending moment is

$$M = \frac{WL}{8}$$

we have by substitution

$$y = \frac{5}{48} \frac{ML^2}{EI}$$

Again since $M = f_b I/c$ and by substitution

$$y = \frac{5}{48} \frac{f_b}{E} \frac{L^2}{c}$$

Substituting $c = d/2$ and $E = 30{,}000$ kips/in^2, we have, with f_b expressed in kips per square inch,

$$y = \frac{f_b L^2}{144{,}000 d}$$

The ratio L/d of beam span to beam depth can be readily related to the ratio y/L of deflection to span

$$\frac{L}{d} = \frac{144{,}000}{f_b} \frac{y}{L}$$

If deflections are limited to $1/300$ times the span for steel beams designed for $f_b = 0.60 F_y$, we have

$$\frac{L}{d} = \frac{144{,}000}{0.6 F_y} \frac{1}{300} = \frac{800}{F_y}$$

The commentary to the AISC specification suggests this as a guide to deflection control of steel-beam floors. For purlins used in other than flat roofs, the value $1000/F_y$ is recommended.

Example 11.28 Select the lightest-weight section for a floor beam that will support a uniform load of 760 lb/ft on a simple span of 20 ft. Use A36 steel and AISC specifications.

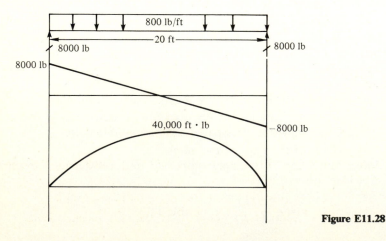

Figure E11.28

SOLUTION Assume that the beam has lateral support. Assuming a beam weight of 40 lb/ft gives a total beam load of 800 lb/ft (Fig. E11.28). The maximum shearing force is

$$R_L = R_R = V = 8000 \text{ lb}$$

and the maximum bending moment is

$$M = \frac{wl^2}{8} = \frac{(800)(20)^2}{8}(12) = 480,000 \text{ in·lb}$$

The required section modulus using $F_b = 0.66 F_y = 24 \text{ kips/in}^2$ is

$$S = \frac{M}{F_b} = \frac{480,000}{24,000} = 20 \text{ in}^3$$

Select the lightest-weight beam furnishing the required section modulus (Table A.5). Use $F_y/800$ as a depth guide.

$$d = \left(\tfrac{36}{800}\right)(20)(12) = 11 \text{ in}$$

Try a W12 × 19:

$$S = 21.3 \text{ in}^3 \quad d = 12.16 \text{ in} \quad t_w = 0.235 \text{ in} \quad t_f = 0.350 \text{ in} \quad I = 130.0 \text{ in}^4$$

Since the section modulus $S = 21.3$ is greater than the 20.0 in^3 required, $f_b < 24,000 \text{ lb/in}^2$ allowed. Check the shear stress:

$$f_v = \frac{V}{t_x(d-2t_f)} = \frac{8000}{(0.235)[12.16-(2)(0.350)]} = 2970 < (0.40F_y = 14,400) \text{ lb/in}^2 \quad \text{OK}$$

Check the maximum deflection at midspan:

$$y = \frac{5wl^4}{384EI} = \frac{(5)(780)(20)^4(12)^3}{(384)(2.95 \times 10^7)(129.5)} = 0.735 \text{ in}$$

$$y_{\text{all}} = \frac{L}{300} \frac{(20)(12)}{300} = 0.80 > 0.735 \text{ in} \quad \text{OK}$$

11.12 SUMMARY

The basic definitions of shear force and bending moment are developed and expressed by

$$V = (\Sigma Y)_L \quad \text{and} \quad M = (\Sigma M)_L = (\Sigma M)_R$$

in which upward-acting forces or loads cause positive effects.

Relationships between load, shear force, and bending moment are given by

$$w = \frac{dV}{dx} = \text{slope of shear-force diagram}$$

and

$$V = \frac{dM}{dx} = \text{slope of bending-moment diagram}$$

These relationships are developed to provide supplementary methods of computing shear forces and bending moments:

$$V_2 - V_1 = \Delta V = \text{area under load diagram between sections 1 and 2}$$

$$M_2 - M_1 = \Delta M = \text{area under shear-force diagram between sections 1 and 2}$$

For straight homogeneous beams carrying transverse loads in the plane of symmetry, the bending moment develops flexural or bending stresses given by

$$f_b = \frac{My}{I}$$

As a result, maximum flexural stresses occur in the extreme fibers of the section with the maximum bending moment. With the distance from the neutral axis (NA) to the extreme fibers denoted by c, the flexural formula can be revised for design purposes so that

$$S = \frac{I}{c} = \frac{M}{F_b}$$

where F_b is the allowable bending stress of the material and S is the section modulus, which relates only to the dimensions of the cross section. Values of S are tabulated for rolled steel and timber sections (Tables A.2, A.3, and A.5).

The vertical shear force on the cross section of a beam develops numerically equal horizontal and vertical shear stresses at a given location in the section which can be calculated by the general shear formula

$$S_s = \frac{VQ}{Ib}$$

Maximum shear stresses occur at the section of maximum shearing force V and usually at the NA.

For rectangular beams, the maximum shear stress is

$$S_{s,\,\mathrm{max}} = \frac{3}{2} \frac{V}{bh}$$

and for I and W sections a very close approximation is given by

$$S_{s,\,\mathrm{web}} = \frac{V}{t_w(d - 2t_f)}$$

Beams are normally designed on the basis of allowable bending stresses, after which maximum shear stresses are compared to allowable shear stresses.

The allowable bending stress F_b depends on the compactness of the section and on the lateral support provided to the compression flange.

$$F_b = \begin{cases} 0.66F_y & & \text{beam compact} \\ 0.60F_y & L_c \geq L_b \leq C_b L_u & \text{beam not compact} \\ 0.60F_y & L_c < L_b \leq C_b L_u & \text{compactness not a consideration} \end{cases}$$

For $L_c < L_b > C_b L_u$

$$F_b = \frac{1.2 \times 10^4 C_b}{L_b(d/A_f)} \qquad \text{for rolled sections}$$

but the values must be compared with values based on ranges of L_b/r_T.

Wood beams have three possible ranges of behavior defined on the basis of a slenderness factor $C_S = \sqrt{l_e d/b^2}$. The allowable bending stress F_b' depends upon the range

Short beams $\qquad C_S < 10$

$$F_b' = F_b$$

Intermediate beams $\qquad 10 < C_S < C_K$

where $\qquad\qquad\qquad C_K = \sqrt{\dfrac{3E}{5F_b}}$

and

$$F_b' = F_b\left[1 - \frac{1}{3}\left(\frac{C_S}{C_K}\right)^4\right]$$

Long beams $\qquad C_K < C_S < 50$

$$F_b' = \frac{0.40E}{C_S^2}$$

Occasionally beams are required to support loads inclined to the axis or applied parallel to both axes. In the former case, the load is resolved into components parallel to the x and y axes at the centroid of the beam. Under these circumstances the bending stress in a fiber can be determined by

$$f_b = \frac{M_x c_x}{I_x} \pm \frac{M_y c_y}{I_y}$$

Composite beam action is best handled by transforming the composite section into an equivalent section of one of the materials. This is done by using

$$nA_1 = A_2 \qquad \text{where } n = \frac{E_1}{E_2}$$

This transformation requires the equivalent fibers to be located the same distance from the neutral axis as the fibers replaced.

Limitations relative to the deflection of beams can be critical in terms of their design.

The differential equation of the elastic curve of a beam is

$$EI\frac{d^2y}{dx^2} = M$$

It is used to illustrate the development of deflection formulas for beams under different applied loads. A table of such formulas is presented.

PROBLEMS

11.1 Write shear and moment equations for the loaded beams in Fig. P11.1. Draw shear-force and bending-moment diagrams, specifying all critical values (change-of-load positions and points of zero shear). Neglect the weight of the beam itself.

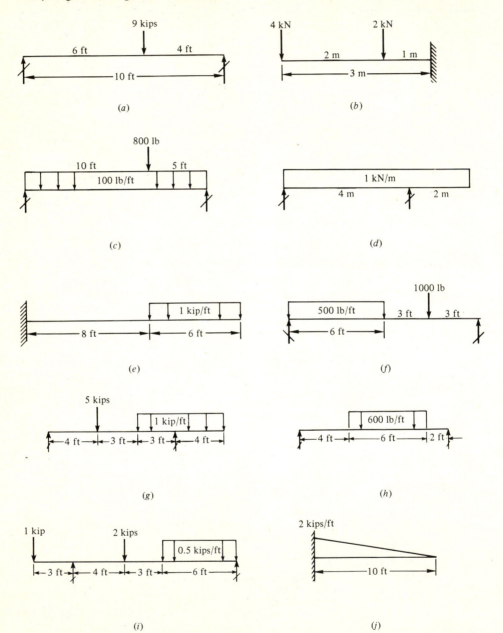

(a)

(b)

(c)

(d)

(e)

(f)

(g)

(h)

(i)

(j)

Figure P11.1

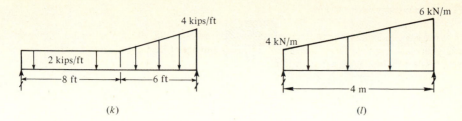

(k)   (l)

Figure P11.1 (*Cont'd.*)

11.2 Draw shear-force and bending-moment diagrams for the beams in Fig. P11.2. Calculate numerical values at all critical points.

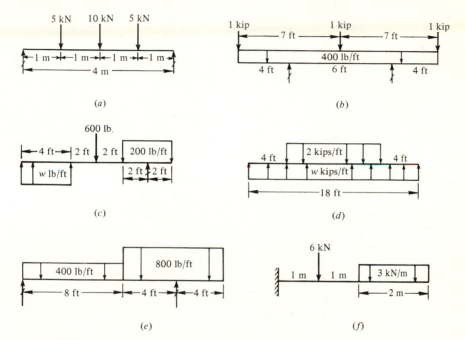

(a)   (b)

(c)   (d)

(e)   (f)

Figure P11.2

11.3 For the shear-force diagrams in Fig. P11.3 draw the corresponding load and bending-moment diagrams. Specify all critical values.

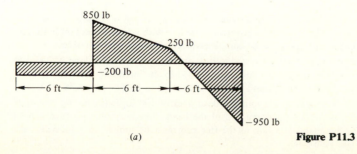

(a)   **Figure P11.3**

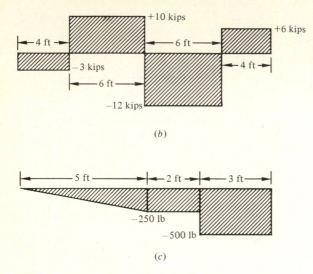

(b)

(c)

Figure P11.3 (*Cont'd.*)

11.4 A 6- by 10-in timber beam (S4S) supports the loads shown in Fig. P11.4. Determine the maximum flexural and shear stresses.

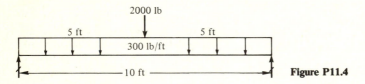

Figure P11.4

11.5 A 6- by 12-in timber beam (S4S) is subjected to a bending moment of 16 ft·kips. Calculate the maximum flexural stress. What percentage decrease in stress occurs if the beam has rough dimensions of 6 by 12 in?

11.6 What is the maximum bending moment that can be carried by a C10 × 15.3 with a $\frac{1}{4}$- by 7-in plate welded to its bottom flange? The plate and channel are A36 steel. Such a beam is often used as a lintel over openings in masonry construction.

11.7 A structural-steel W14 × 38 is used as a simply supported beam 24 ft long. It supports a uniform load of 800 lb/ft including its own weight. Calculate the maximum flexural stress in the beam. What is the average web shear stress?

11.8 A wood plank (2 by 12 in S4S) 15 ft long is supported at its ends. What is the maximum flexural stress produced in the plank by a 200-lb man standing at the center of the span if the plank is (*a*) on edge and (*b*) laid flat?

11.9 A beam is simply supported on a 24-ft span with a concentrated load of 18 kips at midspan. Neglect the weight of the beam. Select the most economical section with a load-deflection limitation of $L/240$. Consider the compression flange to be completely supported and use A36 steel.

11.10 A beam is simply supported and has a span of 26 ft. It carries a concentrated load of 2 kips at the center of the span and a uniformly distributed load of 450 lb/ft (including the weight of the beam). Select the most economical section. Limit the load deflection to $L/240$. Use A36 steel.

11.11 A cantilever beam 10 ft long carries a uniform dead load of 100 lb/ft (including weight of beam) and a concentrated load of 2 kips at the end of the beam. The concentrated load can be considered a live load. The live-load deflection at the free end must be limited to $\frac{3}{8}$ in. Select the most economical section. Use A36 steel.

11.12 Select a beam with a 12-ft span to carry a uniform load of 500 lb/ft (including the weight of the beam). Use Douglas fir no. 1.

11.13 What is the maximum fiber stress in bending in a 6- by 12-in wood beam 12 ft long loaded with a uniform load of 700 lb/ft? What is the maximum shear stress in the beam?

11.14 A W10 × 19 beam of A36 steel is 28 ft long with a 18-ft span and 10-ft overhang. It carries a concentrated load of 4 kips at the end of the overhang and a 8-kip concentrated load at midpoint of the supports. Determine the required length of bearing to prevent web crippling.

11.15 Design a glued-laminated roof beam to meet the following requirements. Beams have a 40-ft span and are spaced at 20-ft intervals. The loads are applied by joists with a spacing of 24 in. The snow load can be considered to be 30 lb/ft^2. The dead load equals 10 lb/ft^2. Deflection limits include $L/240$ for snow load and $L/180$ for total load. Use F24 timber.

11.16 Design an A36 steel beam to span a distance of 24 ft (simple supports). It carries a uniform load of 400 lb/ft (including its own weight) and three concentrated loads of 2 kips each at quarter points. Consider lateral supports to be provided at reactions and quarter points. Reconsider the problem if lateral support is provided only at the supports. What bearing length is required at the supports to prevent web crippling?

11.17 A W16 × 36 beam of A36 steel is simply supported on a 28-ft span. There are no deflection limitations.

   (*a*) Determine the maximum uniformly distributed load the beam can carry if it has complete lateral support.

   (*b*) Determine the maximum uniformly distributed load if lateral support is provided only at the supports.

11.18 The floor framing system for a typical interior bay is illustrated in Fig. P11.18. All connections are of the flexible type, providing simple supports. A one-way reinforced-concrete slab provides continuous support for the beams. Lateral support for the girders is provided only at the supports and load points. The floor and associated ceiling weigh 60 lb/ft^2. Limit the deflection resulting from live load to $L/360$. Design the beams and girders using A36 steel. Consider the load limit to be 100 lb/ft^2.

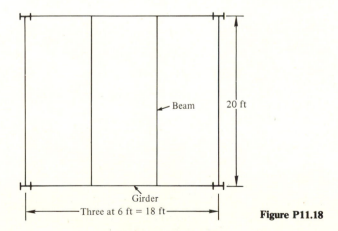

Figure P11.18

11.19 A channel girt of A36 steel spans between columns spaced at 18 ft. The flanges of the channel point down. The horizontal wind load is determined to be 250 lb/ft, and the vertical load resulting from the exterior surface can be considered to be equal to 70 lb/ft. Assume that the loads act through the centroid of the beam. Design the girt according to AISC specifications.

11.20 Determine the maximum deflection of a W12 × 16 that spans 22 ft and carries a distributed load of 500 lb/ft and a 5-kip concentrated load at midspan. Use deflection formulas and superposition.

11.21 Determine the maximum deflection of a 4- by 10-in timber beam that cantilevers 12 ft from a wall and carries a distributed load of 200 lb/ft. Use $E = 1.8 \times 10^6$ lb/in^2.

11.22 Find the shearing stress on the adhesive joint between 2 by 8s used to form an I if $V = 1000$ lb. What is the maximum shearing stress in the beam?

11.23 A church roof with a pitch of 8:12 is supported by 6- by 12-in no. 1 dense-stress-rated southern pine beams with a span of 18 ft. The roof decking and roofing weigh 6 lb/ft^2; the 2-month live load is 25 lb/ft^2 on a horizontal projection (Fig. P11.23). Determine whether the beams are adequate.

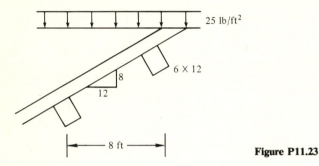

Figure P11.23

11.24 A floor beam is to carry 300 lb/ft (including beam weight) with a span of 18 ft. If the two side members are 2 by 10 no. 1 southern pine, what is the required thickness of the A36 plate if the plate is 8 in deep?

11.25 An 8- by 12-in no. 1 Douglas fir is reinforced by 6- by $\frac{1}{2}$-in plates top and bottom. Determine the maximum uniformly distributed load (including beam weight) that can be simply supported by the beam on an 18-ft span. Use AISC specifications and NDS.

11.26 A pair of C10 $\times$ 20 channels are securely bolted to an 8 by 12 and centered about the x axis of the timber beam. Determine the safe bending moment for the beam if bending occurs about the x axis.

REFERENCES

1. American Institute of Timber Construction, AITC 102-78 Standards for the Design of Structural Timber Framing. Englewood, Col., 1978.
2. National Forest Products Association, Design Specifications for Wood Construction, Washington, 1977.
3. American Institute of Steel Construction, "Steel Construction Manual," 8th ed., Chicago, 1978.

CHAPTER 12

COMPRESSION MEMBERS

12.1 INTRODUCTION

Straight members that are compressed by axial forces are known as *compression members* or *columns*. Their strength is often limited by buckling, a tendency to deflect laterally, at loads considerably less than those required to cause failure by crushing. It is assumed here that local buckling, the instability of an individual element of a member, does not occur before column buckling. When the action line of the load coincides with the axis of the member, the member is said to be *axially* or *concentrically loaded*; when they do not coincide, the member is considered to be *eccentrically loaded*.

Columns are usually subdivided into long, intermediate, and short, the distinction being based on their behavior. Long columns fail by buckling, intermediate ones by a combination of crushing and buckling, and short ones by crushing. Several types of structural members are subjected to this kind of action, but the column is the best known. Other types are the top chords of trusses, bracing members, the compression flanges of rolled beams and built-up sections, and members which are subjected simultaneously to bending and compressive loads. Columns are often thought of as being straight slender vertical members. Short vertical members subjected to compressive loads are often referred to as *struts* or compression members.

12.2 STRENGTH OF A PERFECT COLUMN

An ideal column is assumed to be a homogeneous member of constant cross section that is initially straight and subject to axial compressive loads. Such a column will shorten uniformly due to the compressive strains on transverse cross sections. As the load is increased, a point is reached where the column is on the verge of deflecting laterally in a bending mode. This load is the *buckling load* and, for practical purposes, is the maximum load the member can carry.

If the perfect column is made of a material that has a linear stress-strain curve, e.g., the idealized curve of a low-carbon structural steel shown in Fig. 12.1, the buckling stress is given within the elastic range by the *Euler equation*, derived in 1757 by Leonard Euler, a Swiss mathematician. It gives the critical or buckling load for ideal long columns with ends free to rotate. Since the mathematics involved is fairly complex and beyond the scope of this book, the formula is given without the derivation which is readily available in strength-of-materials textbooks. The Euler formula is

$$P = \frac{\pi^2 EI}{l^2} \tag{12.1}$$

which is usually written in a slightly different form involving a buckling stress and the slenderness ratio. Since $I = r^2 A$, where I is the moment of inertia, r the radius of gyration, and A the area of the cross section, the Euler formula can be written as

$$f_{cr} = \frac{P}{A} = \frac{\pi^2 E}{\left(\dfrac{l}{r}\right)^2} \tag{12.2}$$

where l/r is the *slenderness ratio* and f_{cr} the *buckling stress*.

It is of particular interest to note that the buckling stress, on the basis of the Euler equation, is independent of the strength of the material. The critical stress for columns with other end conditions can be expressed in terms of the hinged column by the use of a coefficient K, an *effective-length coefficient*. Kl/r becomes the effective-slenderness ratio. The relationship between the buckling stress and effective-slenderness ratio is given in Fig. 12.2 with an upper limit equal to the yield stress of the material. As indicated in Fig. 12.2, when Kl/r is less than $\sqrt{\pi^2 E/F_y}$, the column will buckle inelastically at a stress equal to F_y.

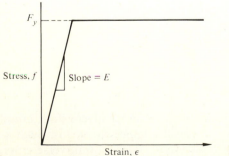

Figure 12.1 Idealized stress-strain curve for steel.

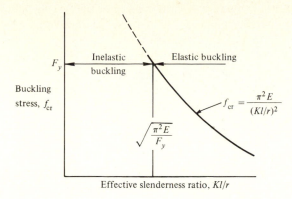

Figure 12.2 Column curve for perfect column made of steel with idealized stress-strain curve.

At lower slenderness ratios, general yielding rather than buckling occurs. In actual columns, at small slenderness ratios, strain hardening may allow structural-steel columns to reach a stress greater than F_y.

12.3 STRENGTH OF AN ACTUAL COLUMN

Because of the presence of residual stresses, initial column curvature, and accidental eccentricity of the applied load, the strength of an actual column is generally less than that of a perfect column. Stresses that remain in structural members after rolling or fabrication are known as *residual stresses*. In hot-rolled structural shapes and welded sections, the residual stresses result from uneven cooling of portions of the sections after rolling or welding. A typical residual-stress distribution in hot-rolled shapes is given in Fig. 12.3

As a result, practical column design is often based on formulas which have been developed to fit test data with reasonable accuracy. The testing of columns with various Kl/r values results in a scattered set of values, as shown in Fig.

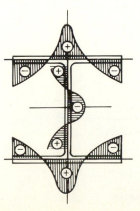

Figure 12.3 Residual stresses in W shapes.

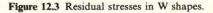

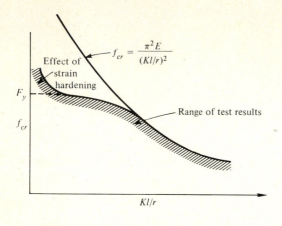

Figure 12.4 Compression-member test-result curve.

12.4. Usually formulas are developed to give design values which represent test values with an appropriate factor of safety (Fig. 12.5).

The allowable stresses given in AISC specifications [1] are applicable for the design of all types of structural-steel columns for buildings. These expressions were developed to incorporate the latest research information available concerning the behavior of steel columns and take into account the effect of residual stresses, the actual end-restraint conditions of the columns, and the varying strengths of different steels.

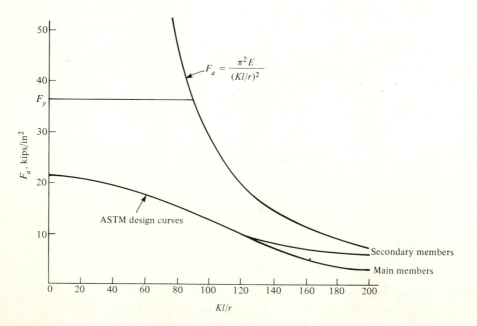

Figure 12.5 Comparison of ideal-column strengths and AISC design curves.

The AISC formulas assume that the upper limit of elastic buckling is defined by an average stress equal to $F_y/2$, or one-half of the yield stress. The corresponding slenderness ratio based on Euler's expression is C_c, where

$$\frac{F_y}{2} = \frac{\pi^2 E}{(Kl/r)^2} = \frac{\pi^2 E}{C_c^2}$$

$$C_c = \sqrt{\frac{2\pi E}{F_y}}$$

For slenderness ratios less than C_c, AISC uses a parabolic formula recommended by the Column Research Council. It gives the ultimate strength of a centrally loaded column with a factor of safety (FS)

$$F_a = \frac{\left[1 - (Kl/r)^2/2C_c^2\right]F_y}{\text{FS}} \tag{12.3}$$

For values of Kl/r greater than C_c a modified Euler formula is used. With a factor of safety of $23/12 = 1.92$ the expression becomes

$$F_a = \frac{12\pi^2 E}{23 Kl/r} \tag{12.4}$$

Figure 12.5 shows the relationship between the AISC design formulas and the theoretical curves for ideal columns (Fig. 12.2).

The factor of safety to be applied to Eq. (12.3) is determined from

$$\text{FS} = \frac{5}{3} + \frac{3Kl/r}{8C_c} - \frac{(Kl/r)^3}{8C_c^3} \tag{12.5}$$

This expression is a quarter sine wave, which equals 1.67 when Kl/r is equal to zero and increases up to 1.92 when Kl/r equals C_c. This variation in the factor of safety is in response to tests which have shown that short columns are not greatly affected by small eccentricities and that as columns become more slender, they become more susceptible to imperfections. Columns with a Kl/r value greater than C_c are considered long columns.

Lighter compression members, e.g., bracing and compression members of roof trusses, are often referred to as struts and classified as secondary members. AISC specifications allow slightly higher stresses for these members when the slenderness ratio is greater than 120. K is always considered 1 in these cases.

The increased allowable stress F_{as} is determined by dividing the applicable average F_a [Eqs. (12.3) and (12.4)] by $1.6 - l/200r$, giving

$$F_{as} = \frac{F_a}{1.6 - l/200r} \tag{12.6}$$

The upper limit of slenderness ratios permitted by AISC is 200.

Allowable stresses calculated according to AISC formulas [Eqs. (12.3), (12.4), and (12.6)] are given in Table 12.1.

Table 12.1 Allowable stress for compression members of 36 kips/in² specified yield stress steel [1]

$\dfrac{Kl}{r}$	F_a kips/in²	$\dfrac{Kl}{r}$	F_a kips/in²	$\dfrac{Kl}{r}$	F_a kips/in²	$\dfrac{Kl}{r}$	F_a kips/in²
Main and secondary members, $Kl/r \leq 120$							
1	21.56	31	19.87	61	17.33	91	14.09
2	21.52	32	19.80	62	17.24	92	13.97
3	21.48	33	19.73	63	17.14	93	13.84
4	21.44	34	19.65	64	17.04	94	13.72
5	21.39	35	19.58	65	16.94	95	13.60
6	21.35	36	19.50	66	16.84	96	13.48
7	21.30	37	19.42	67	16.74	97	13.35
8	21.25	38	19.35	68	16.64	98	13.23
9	21.21	39	19.27	69	16.53	99	13.10
10	21.16	40	19.19	70	16.43	100	12.98
11	21.10	41	19.11	71	16.33	101	12.85
12	21.05	42	19.03	72	16.22	102	12.72
13	21.00	43	18.95	73	16.12	103	12.59
14	20.95	44	18.86	74	16.01	104	12.47
15	20.89	45	18.78	75	15.90	105	12.33
16	20.83	46	18.70	76	15.79	106	12.20
17	20.78	47	18.61	77	15.69	107	12.07
18	20.72	48	18.53	78	15.58	108	11.94
19	20.66	49	18.44	79	15.47	109	11.81
20	20.60	50	18.35	80	15.36	110	11.67
21	20.54	51	18.26	81	15.24	111	11.54
22	20.48	52	18.17	82	15.13	112	11.40
23	20.41	53	18.08	83	15.02	113	11.26
24	20.35	54	17.99	84	14.90	114	11.13
25	20.28	55	17.90	85	14.79	115	10.99
26	20.22	56	17.81	86	14.67	116	10.85
27	20.15	57	17.71	87	14.56	117	10.71
28	20.08	58	17.62	88	14.44	118	10.57
29	20.01	59	17.53	89	14.32	119	10.43
30	19.94	60	17.43	90	14.20	120	10.28
Main members Kl/r 121–200							
121	10.14	133	8.44	145	7.10	157	6.06
122	9.99	134	8.32	146	7.01	158	5.98
123	9.85	135	8.19	147	6.91	159	5.91
124	9.70	136	8.07	148	6.82	160	5.83
125	9.55	137	7.96	149	6.73	161	5.76
126	9.41	138	7.84	150	6.64	162	5.69
127	9.26	139	7.73	151	6.55	163	5.62
128	9.11	140	7.62	152	6.46	164	5.55
129	8.97	141	7.51	153	6.38	165	5.49
130	8.84	142	7.41	154	6.30	166	5.42
131	8.70	143	7.30	155	6.22	167	5.35
132	8.57	144	7.20	156	6.14	168	5.29

Table 12.1 (*Cont'd.*)

$\frac{Kl}{r}$	F_a kips/in^2	$\frac{Kl}{r}$	F_a kips/in^2	$\frac{Kl}{r}$	F_a kips/in^2	$\frac{Kl}{r}$	F_a kips/in^2

Main members Kl/r 121–120 (*Cont'd.*)

$\frac{Kl}{r}$	F_a	$\frac{Kl}{r}$	F_a	$\frac{Kl}{r}$	F_a	$\frac{Kl}{r}$	F_a
169	5.23	177	4.77	185	4.36	193	4.01
170	5.17	178	4.71	186	4.32	194	3.97
171	5.11	179	4.66	187	4.27	195	3.93
172	5.05	180	4.61	188	4.23	196	3.89
173	4.99	181	4.56	189	4.18	197	3.85
174	4.93	182	4.51	190	4.14	198	3.81
175	4.88	183	4.46	191	4.09	199	3.77
176	4.82	184	4.41	192	4.05	200	3.73

Secondary members†, l/r 121–200

$\frac{Kl}{r}$	F_a	$\frac{Kl}{r}$	F_a	$\frac{Kl}{r}$	F_a	$\frac{Kl}{r}$	F_a
121	10.19	141	8.39	161	7.25	181	6.56
122	10.09	142	8.32	162	7.20	182	6.53
123	10.00	143	8.25	163	7.16	183	6.51
124	9.90	144	8.18	164	7.12	184	6.49
125	9.80	145	8.12	165	7.08	185	6.46
126	9.70	146	8.05	166	7.04	186	6.44
127	9.59	147	7.99	167	7.00	187	6.42
128	9.49	148	7.93	168	6.96	188	6.40
129	9.40	149	7.87	169	6.93	189	6.38
130	9.30	150	7.81	170	6.89	190	6.36
131	9.21	151	7.75	171	6.85	191	6.35
132	9.12	152	7.69	172	6.82	192	6.33
133	9.03	153	7.64	173	6.79	193	6.31
134	8.94	154	7.59	174	6.76	194	6.30
135	8.86	155	7.53	175	6.73	195	6.28
136	8.78	156	7.48	176	6.70	196	6.27
137	8.70	157	7.43	177	6.67	197	6.26
138	8.62	158	7.39	178	6.64	198	6.24
139	8.54	159	7.34	179	6.61	199	6.23
140	8.47	160	7.29	180	6.58	200	6.22

†K taken as 1.0 for secondary members.
Note: C_c = 126.1 for A36 steel.

12.4 EFFECTIVE COLUMN LENGTHS

The value of Kl used in the AISC formulas is the effective length of the column; K is an effective-length coefficient, and l is the column length between supports. For pin-ended or hinged columns, K equals 1.0. The effective length of a column can be defined as the distance between the inflection points of the column. This distance was found to vary for different columns according to the types of end restraint against both rotation and translation. Theoretical K values for several

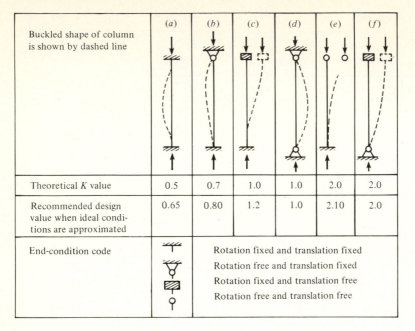

Buckled shape of column is shown by dashed line	(a)	(b)	(c)	(d)	(e)	(f)
Theoretical K value	0.5	0.7	1.0	1.0	2.0	2.0
Recommended design value when ideal conditions are approximated	0.65	0.80	1.2	1.0	2.10	2.0
End-condition code		Rotation fixed and translation fixed				
		Rotation free and translation fixed				
		Rotation fixed and translation free				
		Rotation free and translation free				

Figure 12.6 Effective-length factors for columns.

combinations of idealized end conditions are given in Fig. 12.6. Should a column be perfectly fixed at each end, inflection points would occur at the quarter points, giving an effective length of $l/2$ or a K value of 0.5. If the column ends are not completely fixed against rotation, the K value is generally larger than the theoretical value. Since complete end fixity is not usually obtained, AISC has recommended design values that are generally greater than the theoretical values, also given in Fig. 12.6. A base fully fixed against rotation can be obtained only when a column is rigidly anchored to a footing that does not rotate significantly under an overturning moment. A column top fully fixed against rotation can be approached when the top of the column is rigidly connected by welding or bolting (friction joint) to a heavy girder many times stiffer than the column. The translation of a column top relative to the base is termed *sidesway*. It can be prevented by the lateral support provided by rigid walls or diagonal bracing. If the amount of restraint against either rotation or translation is uncertain, it may be conservatively assumed to be nonexistent.

12.5 LOADS ON COLUMNS

The design load for a given column in a building is usually made up of (1) the live and dead loads on the floor area supported by the column being designed, (2) the load transmitted by the column above it if one exists, and (3) the weight

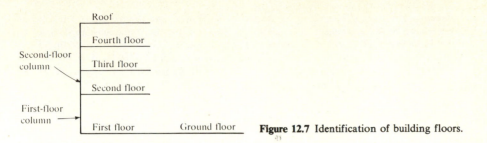

Figure 12.7 Identification of building floors.

of the column itself. The portion of the column load contributed by the floor can be found by adding the reactions of the beams and girders framing into it or by multiplying the contributing area of the supported floor by the live and dead loads imposed per square foot.

It is recognized by some codes that under certain conditions and types of occupancy, not all floors in a building will necessarily be subjected simultaneously to the full live load and reductions are permitted (see Sec. 3.5.1).

Columns are normally designated by the story through which they run (Fig. 12.7). For example, columns between the third and fourth floors of a building, supporting the fourth floor, are known as *third-floor columns*.

12.6 DESIGN AND ANALYSIS OF STEEL COLUMNS

The design of columns involves trial and error, a technique not unusual in structural design. The procedure involves using an estimated allowable stress f_a', from 10 to 20 kips/in^2 depending on the nature of the problem. If the load is large and the column short, a larger stress is allowable than for a smaller load and a longer column. This estimated allowable stress f_a' is used to calculate the probable required area $A = P/f_a'$.

The area A is used to select a trial section having approximately the required area and a large least radius of gyration (r or k depending on the handbook used). A most economical section will normally be one of the lightest of the square groups, 8 by 8, 10 by 10, 12 by 12, or 14 by 14.

On the basis of the trial section, calculate Kl/r and determine the actual allowable stress f_a from Eq. (12.3) or (12.4) or from Table 12.1.

Calculate allowable load $P' = f_a A$ on a trial section and the percentage to which it is stressed.

$$\text{Percentage stressed} = \frac{\text{actual load}}{\text{allowable load}}(100) = \frac{P}{P'}(100)$$

These steps are repeated if the section is stressed less than 95 percent or more than 102 percent. Remember that because a column is 100 percent stressed does not mean it is the most economical section.

Example 12.1 A six-story building has a structural-steel beam-and-column frame. The columns are spaced 20 ft on centers. The building code specifies that the frame must be designed to withstand the actual dead load of the structure plus a possible roof snow load of 30 lb/ft^2 and a live load on each floor of 125 lb/ft^2. The dead load is estimated to be 75 lb/ft^2 on the roof and 100 lb/ft^2 on each floor. The unsupported length of the ground-floor column is 20 ft; the others are 16 ft. Design an interior first-floor column using a W section.

SOLUTION The first-floor column supports the five floors above it and the roof:

$$
\begin{array}{rl}
\text{Roof dead load: } (400)(75) = & 30{,}000 \text{ lb} \\
\text{Snow load: } (400)(30) = & 12{,}000 \text{ lb} \\
\text{Floor dead loads: } (400)(100)(5) = & 200{,}000 \text{ lb} \\
\text{Floor live loads: } (400)(125)(5) = & \underline{250{,}000 \text{ lb}} \\
\text{Design load, first floor column} = & 492{,}000 \text{ lb}
\end{array}
$$

Assume an allowable stress of 15 kips/in^2. An estimate of the required area of the column cross section is $A = 492/15 = 32.8$ in^2. From Table A.5, a W12 × 106 gives the required area.

Properties of W12 × 106

$$\text{Area} = 31.2 \text{ in}^2 \qquad r_x = 5.47 \text{ in} \qquad r_y = 3.11 \text{ in}$$

The slenderness ratio for the column assuming $K = 1.0$ is $(1.0)(20)(12)/3.11 = 77.2$. Since this value is less than $C_c = 126.3$,

$$
F_a = \frac{\left[1 - (Kl/r)^2/2C_c\right]F_y}{5/3 + 3(Kl/r)/8C_c - (Kl/r)^3/8C_c^3}
$$

$$
= \frac{\left\{1 - (77.2)^2/\left[(2)(126.1)^2\right]\right\}(36)}{5/3 + \left[(3)(77.2)/(8)(126.1)\right] - (77.2)^3/(8)(126.1)^3} = 15.67 \text{ kips/in}^2
$$

This value can also be obtained from Table 12.1.

The allowable load for the selected column is

$$F_a A = (15.67)(31.2) = 488.9 \text{ kips}$$

The percentage of stress for the section is $492/488.9 = 100.6$. This is within permissible overdesign limits of about 2 percent.

It is not unusual to repeat the process because the initial section is either under- or overstressed.

Example 12.2 A W10 × 110 is used as a column with a column length of 16 ft. Using A242 steel, determine the allowable axial load permitted by AISC specifications; $K_x = 2.0$ and $K_y = 1.0$.

SOLUTION From Table A.5 for a W10 × 110

$$r_x = 4.66 \qquad r_y = 2.68 \qquad A = 32.9 \text{ in}^2$$

From Table 6.1, $F_y = 50$ kips/in^2 for A242 steel. Then

$$\frac{K_x L_x}{r_x} = \frac{(2)(16)(12)}{4.66} = 82.4 \qquad \text{controls}$$

$$\frac{K_y L_y}{r_y} = \frac{(16)(12)}{2.68} = 71.6$$

$$
F_a = \frac{\left[1 - (Kl/r^2)/2C_c^2\right]F_y}{5/3 + 3(KL/r)/8C_c - (KL/r)^3/8C_c^3}
$$

$$
C_c = \sqrt{\frac{2\pi^2 E}{F_y}} = \sqrt{\frac{2\pi^2(29{,}500)}{50}} = 108 > 82.4
$$

Use Eq. (12.3)

$$F_a = \frac{\left\{1 - (82.4)^2 / \left[(2)(108)^2\right]\right\}(50)}{5/3 + (3)(82.4)/(8)(108) - (82.4)^3/(8)(108)^3} = \frac{35.45}{1.90} = 18.6 \text{ lb/in}^2$$

The allowable axial load on the column, including its own weight is

$$P' = F_a A = (32.9)(18.6) = 612 \text{ kips}$$

Example 12.3 A 10-ft strut of A36 steel is made of two 5- by 3- by $\frac{1}{2}$-in angles with the 5-in legs back to back and separated by a $\frac{1}{2}$-in gusset plate (Fig. E12.3). Determine the load-carrying capacity of the strut in compression.

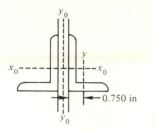

Figure E12.3

SOLUTION The radius of gyration of the pair of angles can be obtained directly from AISC tables or calculated by using information from Table A.5 and principles set forth in Chap. 10. The properties of 5- by 3- by $\frac{1}{2}$-in angles are

$$A = 3.75 \text{ in}^2 \qquad r_{x0} = 1.59 \text{ in} \qquad I_y = 2.58 \text{ in}^2 \qquad x = 0.75 \text{ in}$$

The calculated properties of a pair of angles are

$$r_{x0} = 1.59 \text{ in} \qquad I_{y0} = I_y + Ad^2 = (2.58) + (3.75)(0.75 + 0.25)^2 = 6.33 \text{ in}^4$$

$$r_{y0} = \sqrt{\frac{I_{y0}}{A}} = \sqrt{\frac{6.33}{(2)(3.75)}} = 0.92 \text{ in}$$

The slenderness ratio is

$$\frac{Kl}{r} = \frac{(1)(10)(12)}{0.92} = 130.4$$

The allowable average stress is

$$F_a = \frac{\left\{1 - (130.4)^2 / \left[(2)(126.1)^2\right]\right\}(36)}{5/3 + (3)(130.4)/(8)(126.1) - (130.4)^3/(8)(126.1)^3} = 8.80 \text{ kips/in}^2$$

The allowable load on the strut is equal to

$$P' = F_a A = (2)(3.75)(8.80) = 66.0 \text{ kips}$$

Note that in designing a single-angle strut, the least radius of gyration is about a diagonal axis, often referred to as the z axis (Table A.5).

12.7 WOOD COLUMNS

Many experimental investigations of wood-column behavior have been made. An extensive investigation carried out by the U.S. Forest Products Laboratory [5] has served as the basis of many working formulas. These tests showed that, within the elastic limit of the material, the behavior of long columns is best interpreted by Euler's equation

$$f_{cr} = \frac{\pi^2 E}{(KL/r)^2} \tag{12.7}$$

This type of formula is recommended by the National Design Specifications [2]. Almost all wood columns are designed as pinned since it is very difficult to obtain fixed end columns. As a result, K normally equals 1.0. Since most columns are rectangular, it is easier to obtain r as a function of the dimensions

$$r^2 = \frac{I}{A} = \frac{bd^3}{12bd} = \frac{d^2}{12}$$

$$r = \frac{d}{\sqrt{12}}$$

Substituting this value into Eq. (12.7) and letting $K = 1$ gives

$$f_{cr} = \frac{\pi^2 E}{12(L/d)^2} = \frac{0.823E}{(L/d)^2} \tag{12.8}$$

The formula gives the critical or buckling stress as a function of E and L/d. It is interesting to note that b drops from the expression and that d is the dimension of the column in the direction of buckling. If the unbraced length is the same for both axes, *d is the minimum dimension*. Note that L is used to designate column length in place of the l used with steel columns.

The allowable stress F_c' is obtained by applying an appropriate factor of safety to f_{cr} of Eq. (12.8). The factor of safety recommended by NDS is 2.74, which gives

$$F_c' = \frac{f_{cr}}{2.74} = \frac{0.823E}{2.74(L/d)^2}$$

or

$$F_c' = \frac{0.30E}{(L/d)^2} \tag{12.9}$$

To use this formula NDS suggests that the L/d should not exceed 50, and F_c' cannot exceed F_c, which is the allowable compressive stress parallel to the grain for short columns. To avoid the sharp transition from the long-column relationship to that of short columns, NDS introduces a column equation which represents experimental data and covers an intermediate range of column lengths. Short columns have an L/d ratio of 11 or less. Intermediate columns have an L/d ratio greater than 11 but less than K, where

$$K = 0.671\sqrt{\frac{E}{F_c}} \tag{12.10}$$

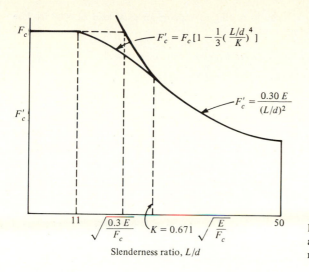

F_c

$F_c' = F_c [1 - \frac{1}{3}(\frac{L/d}{K})^4]$

$F_c' = \frac{0.30 E}{(L/d)^2}$

F_c'

11 $\sqrt{\dfrac{0.3 E}{F_c}}$ $K = 0.671 \sqrt{\dfrac{E}{F_c}}$ 50

Slenderness ratio, L/d

Figure 12.8 Relationship between allowable axial stress and slenderness ratio for wood columns.

In this range the allowable compressive stress F_c can be determined by

$$F_c' = F_c\left[1 = \frac{1}{3}\left(\frac{L/d}{K}\right)^4\right] \tag{12.11}$$

These relationships are illustrated in Fig. 12.8. It is important to remember that F_c' can be modified for duration of loading (Fig. 6.20).

NDS stipulates that the allowable load for a column of round cross section shall not exceed that for a square column of the same cross-sectional area. The equivalent design formula for a round column is

$$F_c' = \frac{3.619E}{(L/r)^2} \tag{12.12}$$

F_c' cannot exceed F_c, as previously indicated. The above formulas can be used in both the analysis and design of timber columns, but as in the case of steel columns the design procedure requires a trial-and-error method. Table 12.2 shows an interesting way of arriving at an initial trial setting; interpolation is permissible. Normally the two side dimensions should not differ by more than 2 in.

Table 12.2 Approximating F_c' on the basis of P/L [3]

$\frac{P}{L}$, lb/ft	Approximate F_c'	$\frac{P}{L}$, lb/ft	Approximate F_c'
600	375	3000	800
1000	475	4000	950
1500	575	5000	1150
2000	650	8000	1350

Example 12.4 Design a timber column 10 ft long to support an axial load of 6000 lb. Use Douglas fir no. 1, NDS, and normal loading.

SOLUTION From Table A.1 for Douglas fir no. 1 (posts and stringers)

$$F_c = 1000 \text{ lb/in}^2 \qquad E = 1.6 \times 10^6 \text{ lb/in}^2$$

From Table 12.2

$$\frac{P}{L} = \frac{6000}{10} = 600 \text{ lb/ft} \qquad F_c' = 375 \text{ lb/in}^2$$

The approximate area required is

$$\frac{P}{F_c'} = \frac{6000}{375} = 16.0 \text{ in}^2$$

Try 4 by 6 (area = 19.25 in^2):

$$\frac{L}{d} = \frac{(10)(12)}{3.5} = 34.3 \qquad K = 0.671\sqrt{\frac{E}{F_c}} = 0.671\sqrt{\frac{1.6 \times 10^6}{1000}} = 26.8$$

Since $L/d > K$, use the long-column formula [Eq. (12.9)]

$$F_c' = \frac{0.30E_2}{(L/d)^2} = \frac{(0.30)(1.6 \times 10^6)}{(34.3)^2} = 408 \text{ lb/in}^2$$

The load-carrying capacity of the 4 by 8 column is

$$P' = F_c'A = (408)(19.25) = 7854 > 6000 \text{ lb} \qquad \text{OK}$$

Try a 4 by 4 (area = 12.25 in^2):

$$P' = (408)(12.25) = 4998 < 6000 \text{ lb} \qquad \text{NG}$$

Use a 4 by 6.

Example 12.5 Design a timber column 15 ft long to support an axial load of 90,000 lb. Use Douglas fir no. 1 and NDS. Consider snow loading.

SOLUTION From Table A.1 for Douglas fir no. 1 (post and stringers)

$$F_c = 1000 \text{ lb/in}^2 \qquad E = 1.6 \times 10^6 \text{ lb/in}^2$$

From Table 12.2

$$\frac{P}{L} = \frac{90,000}{15} = 6000 \text{ lb/ft} \qquad F_c' = 1150 \text{ lb/in}^2$$

Since $1150 > F_c = 1000 \text{ lb/in}^2$, use $F_c' = 1000 \text{ lb/in}^2$. The area required is

$$\frac{90,000}{(1000)(1.15)} = 78.3 \text{ in}^2$$

Try a 10 by 10:

$$\text{Area} = 90.25 \text{ in}^2 \qquad \frac{L}{d} = \frac{(15)(12)}{9.5} = 18.9$$

$$K = 0.671\sqrt{\frac{E}{F_c}} = 0.671\sqrt{\frac{1.6 \times 10^6}{1000}} = 26.8 > 18.9$$

$$F_c' = F_c\left[1 - \frac{1}{3}\left(\frac{L/d}{K}\right)^4\right] = (1000)\left[1 - \frac{1}{3}\left(\frac{18.9}{26.8}\right)^4\right] = 918 \text{ lb/in}^2$$

Considering snow loading

$$F_c' = (918)(1.15) = 1056 \text{ lb/in}^2$$

The load-carrying capacity of a 10 by 10 is

$$P' = (1056)(90.25) = 95,300 > 90,000 \text{ lb} \qquad \text{OK}$$

Example 12.6 Determine the allowable normal axial load for an 8 by 8 southern pine no. 1 SR column 12 ft long.

SOLUTION From Table A.1

$$F_c = 775 \text{ lb/in}^2 \qquad E = 1.5 \times 10^6 \text{ lb/in}^2$$

$$\frac{L}{d} = \frac{(12)(12)}{7.5} = 19.2$$

$$K = 0.671\sqrt{\frac{E}{F_c}} = 0.671\sqrt{\frac{1.5 \times 10^6}{775}} = 29.5 > 19.2$$

The column is in the intermediate range; so

$$F_c' = F_c\left[1 - \frac{1}{3}\left(\frac{L/d}{K}\right)^4\right] = (775)\left[1 - \frac{1}{3}\left(\frac{19.2}{29.5}\right)^4\right] = 729 \text{ lb/in}^2$$

The load-carrying capacity of the column is

$$P' = F_c'A = (729)(56.25) = 41,000 \text{ lb}$$

12.8 COLUMNS WITH ECCENTRIC LOADS

Column loads considered so far have been assumed to be concentric. This assumption is valid when the load is applied uniformly to the top of the column or when the beam reactions are the same on opposing sides of a column (Sketch 12.1a), but if the beam reactions are not the same or if one is missing, as is the case with exterior columns, the column load becomes eccentric (Sketch 12.1b). The use of spandrel beams (Sketch 12.1c) will help to alleviate such loading.

In order to develop an expression that will take into account the eccentricity of a load, consider a short block with a load P eccentrically applied (Fig. 12.9a). This eccentric loading is equivalent to a concentric load P and a moment Pe (Fig. 12.9b), as developed in Sec. 4.11. The stress in any fiber on any cross section of the block such as AA may be considered to be the sum of the average axial stress P/A and the bending stress caused by the moment Pe (Fig. 12.10). As a result, a general expression applicable to symmetrical sections becomes

$$f = \frac{P}{A} \pm \frac{Pec}{I} \tag{12.13a}$$

where f is the stress at either edge of the section in the plane of bending, depending on whether the plus or minus sign is used, and I is the moment of inertia about a centroidal axis perpendicular to the direction of eccentricity.

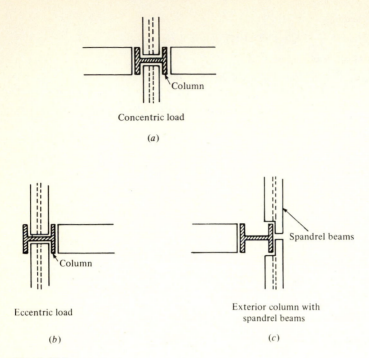

Concentric load

(a)

Eccentric load

(b)

Exterior column with
spandrel beams

(c)

Sketch 12.1

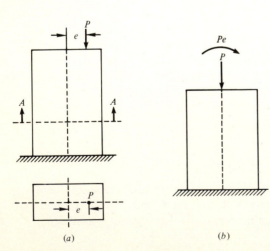

(a)

(b)

Figure 12.9 Converting an eccentric load
into an axial load plus a moment.

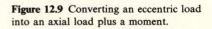

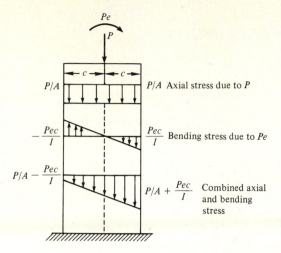

P/A Axial stress due to P

$\dfrac{Pec}{I}$ Bending stress due to Pe

$P/A + \dfrac{Pec}{I}$ Combined axial and bending stress

Figure 12.10 Superposition of axial and bending stresses.

In most buildings, a column will carry a direct axial load in addition to an eccentric load. Under these circumstances Eq. (12.13a) becomes

$$f = \frac{P}{A} \pm \frac{P''ec}{I} \qquad (12.13b)$$

where P is the total vertical load including the eccentric load P''.

Example 12.7 A 6 by 8 timber supports a 30-kip eccentric load as shown in Fig. E12.7. Determine the maximum and minimum stresses resulting from the load.

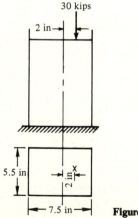

Figure E12.7

SOLUTION

$$A = 41.25 \text{ in}^2 \qquad I = 193.4 \text{ in}^4 \qquad C = \frac{7.5}{2} \text{ in}$$

$$f = \frac{P}{A} \pm \frac{Pec}{I} \qquad \text{Eq. (12.13a)}$$

$$= \frac{30,000}{41.25} \pm \frac{(30,000)(2)(7.5)}{(193.4)(2)} = 727 \pm 1163 \text{ lb/in}^2$$

The maximum compressive stress is

$$727 + 1163 = 1890 \, \text{lb/in}^2$$

and the minimum stress is

$$727 - 1163 = -436 \, \text{lb/in}^2 \qquad \text{tension}$$

Since the rigid frame construction provides continuity between beams and columns at a joint, the connection provides restraint to the rotation of the beam ends. Thus the beam not only supplies an eccentric reaction but also an induced bending moment to the column. The fiber stress in the column can be expressed as

$$f = \frac{P}{A} \pm \frac{P''ec}{I} \pm \frac{Mc}{I} \qquad (12.14)$$

This equation neglects the stresses resulting from lateral deflection of the column. These effects will be taken into account in Sec. 12.9.

12.9 DESIGN OF ECCENTRICALLY LOADED COLUMNS

In order to decide whether a column can carry a given set of loads safely, the computed stress must be compared with an allowable combined bending and axial stress. Early specifications compared the combined stress with the allowable axial stress. The goal of modern specifications is to isolate actual bending f_b and axial stresses f_a and compare them with allowable stresses F_a and F_b. One complication in this approach is the increased bending stresses resulting from the action of an axial load on a bent column.

A detailed explanation and derivation of applicable design formulas follows. Let

F_a = allowable axial stress

F_b = allowable bending stress

f_a = actual axial stress

f_b = actual bending stress

$I = Ar^2$ (Sec. 10.8), where r = radius of gyration

As a result, for an eccentrically loaded column,

$$f_a = \frac{P}{A} \qquad \text{and} \qquad f_b = \frac{Pec}{I} = \frac{Pec}{Ar^2}$$

In the design of the column, if it is subjected to axial loads only, the required cross-sectional area A_a becomes

$$A_a = \frac{P}{F_a}$$

and if subjected to bending action only, the required cross-sectional area becomes

$$A_b = \frac{P''ec}{F_b r^2}$$

For a column subjected to both axial and bending action, the total required area becomes

$$A = A_a + A_b = \frac{P}{F_a} + \frac{P''ec}{F_b r^2}$$

Dividing both sides of the equation by A gives

$$1 = \frac{P/A}{F_a} + \frac{P''ec/Ar^2}{F_b}$$

and by substitution

$$1 = \frac{f_a}{F_a} + \frac{f_b}{F_b} \tag{12.15}$$

This is known as the *fundamental interaction equation*. It is a straight line when plotted graphically, as shown in Fig. 12.11. A simple explanation of its meaning is that if the ratio $f_a/F_a = 0.4$, for example, the ratio f_b/F_b should not exceed 0.6.

This interaction equation takes into account the effect of column buckling by decreasing F_b as the value of Kl/r increases, but it does not take into account the additional stress resulting from the action of the axial load on a column bent by induced moments. This action results in an additional moment at midspan equal to $P \Delta_m$ where Δ_m is the maximum lateral deflection due to the initial bending moment M. This, in turn, causes more lateral deflection, causing more moment, etc. The final bending stress at midheight of the column due to this sequence of events will be the sum of the stresses caused by each action

$$f_b = \frac{Mc}{I} + \frac{\Delta_m Pc}{I}$$

The additional stresses caused by $P \Delta_m$ are difficult to evaluate mathematically, but it becomes apparent that a large slenderness ratio permits a large deflection and that for a given deflection the resulting bending stresses increase with the magnitude of P.

In order to keep the design procedure realistic, AISC recommends a method based on an interaction formula modified in the light of experimental test data.

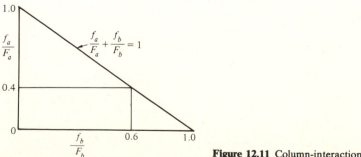

Figure 12.11 Column-interaction formula.

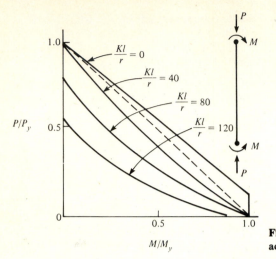

Figure 12.12 Test-data modification of interaction equation.

The curves shown in Fig. 12.12 are typical of such test data. Columns having equal end moments M but different slenderness ratios were tested to determine what additional axial load P could be applied before failure. These values of varying combinations of P and M are made dimensionless by dividing them by P_y and M_y, respectively. P_y is the axial load that causes yielding without any moment M, and M_y is the bending moment that causes yielding without any axial load. The interaction formula is shown as a dashed line. In all cases, except for columns with very small Kl/r ratios, the test data plot on the unsafe side of the interaction curve. The curves become more unsafe as Kl/r increases.

The formula that approximates these curves most closely is

$$\frac{P}{P_y} + \frac{1}{1 - P/P_e}\frac{M}{M_y} = 1 \tag{12.16}$$

where P_e is the elastic buckling load of the column when buckling occurs about the bending axis and is equal to

$$\frac{\pi^2 EI}{l^2} = F_e' A$$

The remaining terms can be interpreted as follows:

$$P = f_a A \qquad P_y = F_a A \qquad M = f_b\frac{I}{c} \qquad M_y = F_b\frac{I}{c}$$

When used in a design formula, F_e', F_a, and F_b must include a factor of safety instead of being the yield values of stress and bending stress.

Substituting these values into the basic curve formula (12.16) gives

$$f_a A + \left[\frac{1}{1 - f_a A/F_e' A}\right]\frac{f_b I/c}{f_b I/c} = 1.0$$

which reduces to

$$\frac{f_a}{F_a} + \left[\frac{1}{1 - f_a/F_e'} \right] \frac{f_b}{F_b} = 1.0 \qquad (12.17)$$

The term in the brackets is referred to as the *amplification factor*. F_e' is defined as the limiting Euler stress and can be evaluated by Eq. (12.4), which is Euler's equation with a factor of safety of $\frac{23}{12}$

$$F_e = \frac{12\pi^2 E}{23(Kl/r)^2}$$

The slenderness ratio is relative to the bending axis.

The basic column formula (12.16) was developed for equal end moments causing single-curvature deflection (Fig. 12.12). When F_e' is very large and/or the stress f_a is small, the amplification factor becomes negligible (approaches 1). When significant, it increases the ratio f_b/F_b and takes into account the cumulative effect of axial load and lateral deflection due to eccentric load or moment.

12.9.1 Steel Columns

According to AISC specifications, the basic interaction formula

$$\frac{f_a}{F_a} + \frac{f_b}{F_b} \leq 1.0$$

can be used if $f_a/F_a \leq 0.15$. For values larger than 0.15, the modified interaction equation (12.17) is further modified by a *reduction factor* C_m, which is applied to the bending-stress ratio and takes into account various end conditions

$$\frac{f_a}{F_a} + \frac{C_m}{1 - f_a/F_e'} \frac{f_b}{F_b} \leq 1 \qquad (12.18)$$

Not only must Eq. (12.18) be satisfied but the equation

$$\frac{f_a}{0.60F_y} + \frac{f_b}{F_b} \leq 1.0$$

must also be satisfied at points of lateral support in the plane of bending.

The value of C_m is evaluated on the basis of the relative size and direction of end moments. C_m is never larger than 1.0 and should never be considered less than 0.40. For braced columns

$$C_m = 0.6 - 0.4\frac{M_1}{M_2}$$

where M_1 is the smaller and M_2 the larger end moment.

As indicated in Fig. 12.13, the ratio M_1/M_2 is negative when the moments cause single curvature and is positive when they cause double curvature. AISC specifications also include procedures for columns subjected to lateral loads and sidesway conditions.

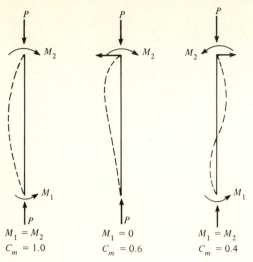

Figure 12.13 Reduction factors for end-braced columns.

$M_1 = M_2$
$C_m = 1.0$

$M_1 = 0$
$C_m = 0.6$

$M_1 = M_2$
$C_m = 0.4$

Example 12.8 A W14 × 53 column of A36 steel 15 ft long is loaded as shown in Fig. E12.8. The bending moments are acting relative to the major axes. The top and bottom of the column are braced in both directions. K values have been estimated as 1.0 about the y axis and 0.70 about the x axis. Check the safety of the column according to AISC specifications.

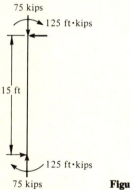

75 kips

125 ft·kips

15 ft

125 ft·kips

75 kips

Figure E12.8

SOLUTION The following properties of the W14 × 53 section are from Table A.5:

$$A = 15.6 \text{ in}^2 \qquad r_x = 5.89 \text{ in} \qquad r_y = 1.92 \text{ in} \qquad r_T = 2.15 \text{ in}$$

$$S = 77.8 \text{ in}^3 \qquad d = 13.92 \text{ in} \qquad b_f = 8.06 \text{ in} \qquad t_f = 0.660 \text{ in}$$

The axial stress is

$$f_a = \frac{P}{A} = \frac{75}{15.6} = 4.80 \text{ kips/in}^2$$

The bending stress is

$$f_b = \frac{M}{S} = \frac{(125)(12)}{77.8} = 19.28 \text{ kips/in}^2$$

The slenderness ratio is

$$\frac{K_x}{r_x} = \frac{(0.70)(15)(12)}{5.89} = 21.4 \qquad \frac{K_y}{r_y} = \frac{(1.0)(15)(12)}{1.92} = 93.8$$

On the basis of the larger slenderness ratio, the allowable axial stress F_b becomes

$$F_a = \frac{\left\{1 - \left[(K_y l)/r_y\right]^2/2C_c^2\right\}F_y}{5/3 + (3)(K_y l/r_y)/8C_c - \left[(K_y l)/r_y\right]^3/8C_c^3} = 13.74 \text{ kips/in}^2 \qquad \text{Table 12.1}$$

Then

$$\frac{f_a}{F_a} = \frac{4.80}{13.74} > 0.15$$

and consequently

$$\frac{f_a}{F_a} + \frac{C_m}{1 - f_a/F_e'}\frac{f_b}{F_b} \le 1$$

must be satisfied along with the expression

$$\frac{f_a}{0.60F_y} + \frac{f_b}{F_b} \le 1$$

$$F_e' = \frac{12\pi^2 E}{23(Kl/r)^2} = \frac{(12)(\pi^2)(29,500)}{(23)(21.4)^2} = 332 \text{ kips/in}^2$$

The amplification factor is

$$1 - \frac{f_a}{F_e'} = 1 - \frac{4.80}{332} = 0.985$$

The reduction factor is

$$C_m = 0.6 - 0.4\frac{M_1}{M_2} = 0.6 - (0.4)(75/125) = 0.36$$

Use 0.40:

$$L_c = \frac{76b_f}{F_y} = \frac{(76)(8.06)}{36} = 102 \text{ in}$$

$$L_u = \frac{20,000}{(d/A_f)F_y} = \frac{20,000}{(13.96)(36)/(8.06)(0.660)} = 212 \text{ in}$$

Since the column is supported at each end, $L_b = (15)(12) = 180$ in. Consequently since $102 < L_b < 212$,

$$F_b = 0.60F_y = 22 \text{ kips/in}^2$$

Substituting the above data in

$$\frac{f_a}{F_a} + \frac{C_m}{1 - f_a/F_e'}\frac{f_b}{F_b} \le 1.0$$

gives

$$\frac{4.80}{13.74} + \frac{0.40}{0.985}\frac{19.28}{22} = 0.35 + (0.41)(0.876) = 0.71 < 1.0$$

The column is considered safe. Continue to check with

$$\frac{f_a}{0.60F_y} + \frac{f_b}{F_b} \le 1.0$$

$$\frac{4.80}{22} + \frac{19.28}{22} = 0.218 + 0.876 = 1.094 > 1.0 \qquad \text{unsafe}$$

12.9.2 Wood Columns

Newlin [4] has adapted the experimental interaction equation (12.17) into the form

$$\frac{P/A}{F_c'} + \frac{M/S + \beta(P/A)(6e/d)}{F_b C_F - \gamma P/A} \le \bar{\alpha} \tag{12.19}$$

which can be used for both short and long columns. At the critical section of the column the bending moment is equal to M, the moment produced by the side loads, plus Pe, the moment produced by the eccentric longitudinal load, times a magnification factor β greater than 1. The magnification factor is taken as 1 for short columns and 1.25 for long columns, where it takes into account the additional moment created by secondary deflections. The net safe bending stress is represented by $F_b C_F - \gamma P/A$, in which γ is the amplification coefficient which depends on the slenderness ratio L/b. γ is equal to 1 for long columns and is zero for short columns. The coefficient C_F reflects the variation of bending strength with depth for rectangular sections (Sec. 11.7). Short columns have an $L/b \le \sqrt{0.3E/F_c}$. Long columns have $L/b > \sqrt{0.3E/F_c} < 50$. For most structural materials, especially wood, there is a tendency for strength to increase as the duration of loading is reduced (Sec. 6.5.4). This effect has been evaluated experimentally (Fig. 6.20). By definition the duration-of-load factor $\bar{\alpha}$ is 1.0 for normal loading (0.9 for permanent loads, 1.15 for snow, 1.33 for wind or earthquakes, and 2.0 for impact).

Equation (12.19) is limited by its derivation to the simple case of a prismatic wood column of rectangular section with both ends hinged. In addition, the ends are braced to prevent relative lateral displacement. In view of these limitations, the use of unbraced wood frames requires exact analysis, including the determination of secondary deformations.

NDS provisions for the design of members subjected to combined end loads, side loads, and eccentricity are essentially contained in Eq. (12.19).

Appendix H of the NDS gives the following general formula for members subjected to a combination of some or all of end loads, side loads, and eccentricity:

$$\frac{f_c}{F_c'} + \frac{f_b + f_c(6 + 1.5J)(e/d)}{F_b - Jf_c} \le 1 \tag{12.20}$$

The value of J is determined as

$$J = \frac{L/d - 11}{K - 11} \qquad 0 \le J \le 1$$

where f_c = stress in compression parallel to grain induced by axial load, lb/in^2
  f_b = flexural stress induced by side loads only, lb/in^2
  F_c' = allowable stress in compression parallel to grain that would be permitted if only axial stress existed, lb/in^2

$$F'_c = \begin{cases} F_c & \dfrac{L}{b} < 11 \\[2ex] F_c\left[1 - \dfrac{1}{3}\left(\dfrac{L/b}{K}\right)^4\right] & K \geq \dfrac{L}{b} \geq 11 \\[2ex] \dfrac{0.30E}{(L/b)^2} & K \leq \dfrac{L}{b} \leq 50 \end{cases} \quad \text{for } K = 0.671\sqrt{\dfrac{E}{F_c}}$$

where e = eccentricity, in
   L = length of columns, in
   d = side of rectangular column in direction of side loads, in
   b = minimum dimension of rectangular column, in

NDS also recommends a procedure for the design of columns with bracket loads which involves replacing the bracket load by an axial load of equal magnitude plus a side load $P = 3al'P/l^2$ at midheight (Sketch 12.2). The NDS general formula for column design (12.20) can be used for the new loading condition.

In the design of a column the values P, e, M, and L are known from the analysis of the structure. The load-duration factor $\bar{\alpha}$ can be determined from the nature of the loading. Allowable stresses F_c and F_b, to be used in the design, can be determined once the desired wood species and grading have been selected (Tables A.1 and A.4). The problem consists of determining an acceptable section for the column. The solution involves a trial-and-error process. A section with area A and section modulus S is assumed. Any section may be used for the initial trial. It is now possible to use Eq. (12.20). If the sum of the two terms is less than $\bar{\alpha}$, the assumed section is a possible solution. The closer the initial trial

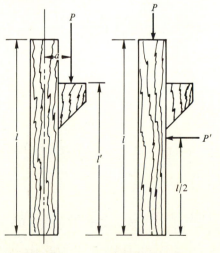

Actual loading.   Assumed loading.   **Sketch 12.2**

is to an acceptable solution, the fewer cycles required to get the most economical section. Example 12.9 illustrates this approach.

Example 12.9 Select a column 18 ft long of Douglas fir no. 1 to carry an eccentric load of 40 kips. The eccentricity is 4 in (Fig. E12.9). Use NDS and consider normal loading.

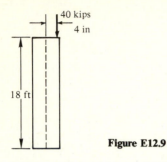

Figure E12.9

SOLUTION From Table A.1 for Douglas fir no. 1 (posts and timbers)

$$F_c = 1000 \text{ lb/in}^2 \qquad E = 1.6 \times 10^6 \text{ lb/in}^2 \qquad F_b = 1200 \text{ lb/in}^2$$

Estimate initial trial section by assuming that $f_c/F_c' = 0.5$ and F_c' is 800 lb/in². Then

$$A_{est} = \frac{P}{f_c} = \frac{40,000}{(0.5)(800)} = 100 \text{ in}^2$$

Try 10 by 12:

$$A = 109.25 \text{ in}^2 \qquad S = 209.4 \text{ in}^3$$

$$\frac{L}{b} = \frac{(18)(12)}{9.5} = 22.7 \qquad \frac{L}{d} = \frac{(18)(12)}{11.25} = 19.2$$

$$K = 0.671\sqrt{\frac{E}{F_c}} = 0.671\sqrt{\frac{1.6 \times 10^6}{1000}} = 26.8$$

$$J = \frac{L/d - 11}{K - 11} = \frac{19.2 - 11}{26.8 - 11} = 0.52$$

$$F_c' = F_c\left[1 - \frac{1}{3}\left(\frac{L/b}{K}\right)^4\right] = 1000\left[1 - \frac{1}{3}\left(\frac{22.7}{26.8}\right)^4\right]$$

$$= 830 \text{ lb/in}^2$$

Consider

$$\frac{f_c}{F_c'} + \frac{f_b + f_c(6 + 1.5J)(e/d)}{F_b - Jf_c} \le 1$$

$$f_b = 0 \qquad f_c = \frac{40,000}{109.25} = 366 \text{ lb/in}^2 \qquad e = 4 \text{ in} \qquad d = 11.25 \text{ in}$$

$$\frac{366}{830} + \frac{(366)[6 + (1.5)(0.52)](4)/11.25}{1200 - (0.52)(366)}$$

$$= 0.44 + 0.874 > 1 \qquad \text{NG}$$

Try a 10 by 14:

$$A = 128.25 \text{ in}^2 \qquad \frac{L}{b} = 22.7 \qquad K = 26.8 \qquad \frac{L}{d} = \frac{(18)(12)}{13.5} = 16 \qquad J = \frac{16 - 11}{26.8 - 11} = 0.316$$

$$F_c' = 830 \text{ lb/in}^2 \qquad f_c = \frac{40,000}{128.25} = 312 \text{ lb/in}^2 \qquad e = 4 \text{ in} \qquad d = 13.5 \text{ in}$$

Substitute into the interaction equation:

$$\frac{312}{830} + \frac{(312)[6 + (1.5)(0.316)](4)/13.5}{1200 - (0.316)(312)}$$

$$= 0.376 + 0.543 = 0.919 < 1 \qquad \text{OK}$$

12.9.3 Wood Spaced Columns

Spaced columns are formed by two or more individual members (leaves) with their longitudinal axes parallel. They are separated at the ends and at the middle of their length by blocking and are joined at the ends by connectors capable of developing the required shear between leaves and blocking. This shear is caused by the lateral deflection of the column. The spaces at the center of the column require only the use of bolts. Figure 12.14 shows a spaced column using connectors. Spaced columns are important in the design of timber structures using connectors as fasteners. Examples are compression chords and web members in trusses.

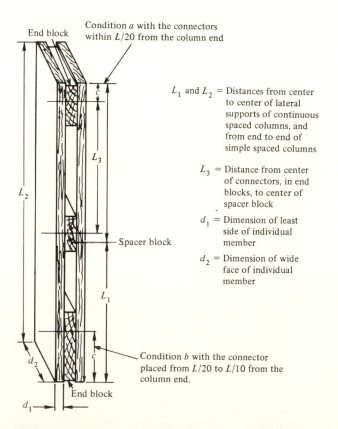

L_1 and L_2 = Distances from center to center of lateral supports of continuous spaced columns, and from end to end of simple spaced columns

L_3 = Distance from center of connectors, in end blocks, to center of spacer block

d_1 = Dimension of least side of individual member

d_2 = Dimension of wide face of individual member

End block

Condition *a* with the connectors within $L/20$ from the column end

Spacer block

Condition *b* with the connector placed from $L/20$ to $L/10$ from the column end.

End block

Figure 12.14 Details of a spaced column.

Spacer blocks must be at least as thick as the leaves; their grain direction should be parallel to the length of the column. The minimum length of end spacer blocks is determined by the end distance required by the connectors, and they should be as wide as the associated leaves.

NDS recognize two end conditions, which depend upon the location of the centroid of the end-block connector group relative to the end of the column (distance c). Condition a applies when $c \leq L/20$ and condition b when $L/20 < c \leq L/10$. For condition a the fixity factor is 2.5, giving a column formula

$$F'_c = \frac{0.30E(2.5)}{(L/d)^2} \qquad (12.21)$$

and for condition b, the fixity factor is 3.0, giving

$$F'_c = \frac{0.30E(3.0)}{(L/d)^2} \qquad (12.22)$$

with the additional requirement that $F'_c \leq F_c$.

End condition a is normally considered to exist in the compression chord of a truss, where web members and purlins prevent movement of the panel points. In this case, the unsupported length L is measured from center to center of panel points. End condition b is considered to exist where columns are anchored to a concrete footing. In single-story buildings the column length is considered to be from column end to column end. NDS also limit the slenderness ratio L/d to 80 and that of $L/2d$ to 40 for the leaves (see Fig. 12.14).

The total safe column load of a spaced column is given by the smaller of (1) the sum of the products of F'_c [Eq. (12.21) or (12.22)] and the cross-sectional area of each leaf and (2) the capacity of a solid column bending about axis x-x (see Fig. 12.14).

Table 12.3 End-spacer-block constants for connector-joined spaced columns [2]

L_1/d_1 ratio of individual member in spaced column†	Connector load group			
	A	B	C	D
0–11	0	0	0	0
15	38	33	27	21
20	86	73	61	48
25	134	114	94	75
30	181	155	128	101
35	229	195	162	128
40	277	236	195	154
45	325	277	229	181
50	372	318	263	208
55	420	358	296	234
60–80	468	399	330	261

†Constants for intermediate L/d ratios can be obtained by straight-line interpolation.

Each end block of the spaced column requires a connector or connectors adequate to resist the shear between leaves and blocking. This shearing force is considered to be equal to the cross-sectional area of the leaf multiplied by an end-spacer constant, which depends on the L/d ratio and the species of wood being used. Table 12.3 gives end-spacer-block constants. The load is considered to be applied in either direction parallel to the longitudinal axis of the individual leaves. Table 7.11 can be used to identify the species group. Example 12.10 illustrates the design of a spaced column.

Example 12.10 Evaluate the load-carrying capacity (live and wind) of the following spaced column:

Three 12-ft 3 by 10s
End condition b
4 in split rings
Douglas fir no. 1 (posts and timbers)

SOLUTION From Table A.1

$$F_c = 1000 \text{ lb/in}^2 \qquad F_b = 1200 \text{ lb/in}^2 \qquad E = 1.6 \times 10^6 \text{ lb/in}^2$$

$$\frac{L}{d} = \frac{(12)(12)}{2.5} = 57.6$$

$$F'_c = \frac{0.30 E(3)(1.33)}{(L/d)^2} = \frac{(0.30)(1.6 \times 10^6)(3)(1.33)}{(57.6)^2} = 577 \text{ lb/in}^2$$

where 1.33 is the load-duration factor for wind. Load-carrying capacity of the column is

$$F'_c A = (577)(3)(23.13) = 40,500 \text{ lb}$$

Check as a solid column using the greater d:

$$\frac{L}{d} = \frac{(12)(12)}{9.5} = 15.2$$

$$F'_c = \frac{0.30 E(1.33)}{(L/d)^2} = \frac{(0.30)(1.6 \times 10^6)(1.33)}{(15.2)^2}$$

$$= 2760 > (1000)(1.33) \qquad \text{use } 1333 \text{ lb/in}^2$$

$$P = (1333)(3)(23.13) = 92,497 > 40,500 \text{ lb} \qquad \text{OK}$$

Consider rings required for shear transfer:

$$\frac{L}{d} = 57.6 \qquad \text{group B}$$

From Table 12.3, the end-spacer-block constant is 379. The ring load is $(379)(23.13) = 8766$ lb. The allowable load per 4-in ring (Table 7.21) is $(5000)(1.33) = 6650$ lb. Use two rings.

12.10 SUMMARY

The behavior of compression members is complicated by a buckling phenomenon which involves a tendency to deflect laterally at loads considerably less than the crushing strength of the material.

Euler's relationship

$$P_{cr} = \frac{\pi^2 E}{(Kl/r)^2}$$

is used to predict the buckling stress over an elastic buckling range.

AISC specifications recommend the use of two formulas based on slenderness ratio Kl/r, where K is the effective length coefficient taking end conditions into account. For slenderness ratios less than $C_c = \sqrt{2\pi^2 E/F_y}$

$$F_a = \frac{\left[1 - (Kl/r)^2 / 2C_c^2\right] F_y}{FS} \qquad \text{where} \qquad FS = \frac{5}{3} + \frac{3Kl/r}{8C_c} - \frac{(Kl/r)^3}{8C_c^3}$$

For slenderness ratios greater than C_c

$$F_a = \frac{12\pi^2 E}{23(Kl/r)^2}$$

For wood compression members NDS recommends the use of three formulas, depending upon the slenderness ratio L/d:

$$F_c' = \begin{cases} F_c & \dfrac{L}{d} < 11 \text{ (short columns)} \\[2ex] F_c\left[1 - \dfrac{1}{3}\left(\dfrac{L/d}{K}\right)^4\right] & K \ge \dfrac{L}{d} \ge 11 \text{ (intermediate columns)} \\[3ex] & \hspace{3em} K = 0.671\sqrt{\dfrac{E}{F_c}} \\[3ex] \dfrac{0.030E}{(L/d)^2} & K \le \dfrac{L}{d} \le 50 \text{ (long columns)} \end{cases}$$

The stress in an eccentrically loaded compressive member can be considered to be the sum of the axial stress and the bending stress caused by the equivalent moment

$$f = \frac{P}{A} \pm \frac{Pec}{I}$$

The interaction equation

$$\frac{f_a}{F_a} + \frac{f_b}{F_b} \le 1$$

is the basis for expressions used to evaluate the safety of eccentrically loaded columns or beams with axial loads (beam columns). AISC specifications recommend

$$\frac{f_a}{F_a} + \frac{C_m}{1 - f_a/F_e'} \frac{f_b}{F_b} \le 1$$

where C_m is a *reduction factor* which takes end conditions into account and $1 - f_a/F_e'$ is an *amplification factor* which accounts for additional stresses resulting from the action of the axial load on a column bent by induced moments.

A similar approach can be used for timber beam columns. The expression

$$\frac{P/A}{F_c'} + \frac{M/S + (\beta P/A)(6e/d)}{F_b C_F - \gamma P/A} \leq \bar{\alpha}$$

where β = a magnification factor
  γ = amplification factor
  $\bar{\alpha}$ = load-duration factor

contains all the provisions of the NDS for the design of members subjected to some or all of end load, side load, and eccentricity.

The more unusual end conditions possible with spaced timber columns are recognized by NDS:

$$F_c' = \begin{cases} \dfrac{0.30E(2.5)}{(L/d)^2} & \text{condition } a \\[3mm] \dfrac{0.30E(3.0)}{(L/d)^2} & \text{condition } b \end{cases}$$

PROBLEMS

12.1 A W8 × 40 of A36 steel is used as a column with an unbraced length of 12 ft. $K = 1.0$. Determine the maximum axial load.

12.2 A W6 × 25 of A36 steel is used as a column with an unbraced length of 14 ft. $K = 1.0$. Determine the maximum axial load.

12.3 A column consists of a W8 × 24, 20 ft long, laterally supported in all directions at both ends but at midheight only in a direction perpendicular to the web. Determine the allowable load according to AISC specifications. Use A36 steel.

12.4 A W8 × 35 column of A242 steel has an unbraced length of 16 ft for its major axis and 12 ft for its minor axis. Determine the maximum load if K is 1 for both axes.

12.5 The top chord of a truss consists of two 4- by 3- by $\frac{3}{8}$-in angles with the short legs back to back but separated $\frac{3}{8}$ in by fillers and gusset plates. Lateral support is provided parallel to the short legs at 5-ft intervals and parallel to the long legs at 10-ft intervals. What is the maximum axial compressive load according to AISC specifications? Use A36 steel.

12.6 Determine the axial load-carrying capacity of the following columns under normal loading conditions. Use NDS:

    (*a*) 4 by 4 by 3 ft Douglas fir no. 1 (posts and timbers)
    (*b*) 6 by 6 by 9 ft southern pine no. 1 SR
    (*c*) 6 by 6 by 16 ft Douglas fir dense no. 1 (posts and timbers)

12.7 Determine the allowable axial snow load for a 6- by 6-in column 16 ft long. Use southern pine no. 1 SR and NDS.

12.8 Determine the wind-load capacity of the following spaced column:

Two 3 by 10s 12 ft long
No. 1 Douglas fir
3- by 10-in spacer blocks
4- in split rings
End condition *b*

12.9 Select the most economical steel W section capable of carrying an axial load of 70 kips. The length between supports is 16 ft. Use AISC specifications.

12.10 Select a steel angle 6 ft long to resist an axial compressive load of 18 kips. Use AISC specifications and A36 steel.

12.11 Select the most economical double-angle truss compression member to carry a load of 78 kips. Gusset plates are $\frac{3}{8}$ in thick. Lateral support is provided in the vertical plane at 5-ft centers and in the horizontal plane at 10-ft centers. Use AISC specifications and A36 steel.

12.12 What is the lightest W section permitted by AISC specifications for a column to carry an axial load of 76 kips when it is 30 ft long and laterally restrained at the ends and in only one plane at its midpoint? Use A36 steel.

12.13 A column of A36 steel has an unbraced length of 16 ft and must carry an axial load of 210 kips, including its own weight. Select the most economical rolled section. $K = 1.0$.

12.14 A column of A242 steel has an unbraced length of 12 ft and must carry a load of 60 kips. Select the lightest-weight rolled column section. Consider the ends to be fixed.

12.15 A steel-frame building has eight floors and a roof. A typical interior bay is 20 by 20 ft. The live load on a typical floor is 80 lb/ft^2 and dead load 60 lb/ft^2. Determine the design load for a typical interior ground-floor column. The combined dead and snow loads for the roof are 100 lb/ft^2. Select an appropriate W section using AISC specifications.

12.16 Select a W section for a column with a 30-kip axial load. The bottom of the column is fixed, but the top is laterally restrained only in one direction. The length of the column is 16 ft. Use AISC specifications.

12.17 Select a no. 1 SR southern pine column 14 ft long to carry 82 kips. Use NDS.

12.18 Select a no. 1 Douglas fir column 10 ft long to carry a 40-kip load. Use NDS.

12.19 Select a no. 1 SR southern pine column 12 ft long to carry a snow load of 32 kips. Use NDS.

12.20 A 7-ft-long compression diagonal in a truss consists of individual members separated by a 3 by 12 member at each end. Design the diagonal as a spaced column with end condition *a*. Use no. 1 Douglas fir timber. The load combination controlling the design is dead load plus wind load, equaling 32,500 lb. Member thickness is limited to 3 in. Use NDS.

12.21 Design a steel column of A36 steel 18 ft long to support an axial load of 150 kips. Moments of 100 ft·kips act at both ends of the column in such a way as to form a single curvature in deflection. K_x and K_y can be assumed to equal 1. Use AISC specifications.

12.22 A W14 × 53 column of A36 steel has an unbraced height of 12 ft for the *y-y* axis and 24 ft for the *x-x* axis. K_x and K_y can be assumed equal to 1.0. A 100-kip load is carried at the outside flange face of the column along the *y* centroidal axis of the section. Using AISC specifications, determine whether the column is safe.

12.23 An A36 steel column 16 ft long is braced top and bottom in both directions. $K_x = K_y = 1.0$. The column is to carry 80 kips with an eccentricity of 2 in relative to the *x* axis. Select an appropriate W section using AISC specifications.

12.24 Design a timber column to carry a concentric load of 30 kips and a lateral wind load of 400 lb/ft. The column is 14 ft long. Use no. 1 Douglas fir and NDS.

12.25 Design a timber column to carry an eccentric load of 20 kips ($e = 2$ in) and a lateral concentrated load at midheight of 2 kips. The column is 12 ft long. Lateral support is provided at the column ends. Use no. 1 Douglas fir and NDS.

REFERENCES

1. American Institute of Steel Construction, "Manual of Steel Construction," 8th ed., Chicago, 1978.
2. National Forest Products Association, National Design Specifications for Wood Construction, Washington, 1977.
3. A. Jensen and H. Chenoweth, "Applied Strength of Materials," 2d ed., McGraw-Hill, New York, 1975.
4. J. A. Newlin, Formulas for Columns with Side Loads and Eccentricity, *Build. Stand. Mon.*, December 1940.
5. J. A. Newlin and J. M. Gahagan, Tests of Large Timber Columns and Presentation of the Forest Products Laboratory Formula, *U.S. Dept. Agr. Tech. Bull.* **167**, 1930.

BUILDING CONNECTIONS

13.1 INTRODUCTION

Chapter 7 developed the failure mechanisms involved with different types of connectors (fasteners and welds) and used them as a basis for the design of shear connections for tension members. This chapter is concerned with beam-to-beam, beam-to-column, and bracket connections used in buildings.

The behavior of connections in general is very complex. Most connections are statically indeterminate, the forces and stresses depending upon the deformations of the component parts and the connectors. Stress concentrations are also usual. It is practically impossible to analyze most connections by a rigorous procedure. Thus analysis and design are approximate and are based on simplifying assumptions and prototype testing.

In building frames, connections may be subjected to shear and bending moments resulting from continuous structural action. The amount of continuity depends upon the ability of the connection to resist moment.

AISC specifications recognize three basic types of steel construction, according to how one member is attached to another.

Type 1, *rigid frame*, assumes that beam-column connections have sufficient rigidity to prevent the rotation of one member relative to the other. The main reason for using this type of connection in buildings is to accommodate lateral forces (Chap. 3).

Type 2, *simple framing*, assumes that the ends of the beams and girders are connected for shear only and are free to rotate under gravity loads.

Type 3, *semirigid framing*, assumes that the connection possesses a dependable and known moment capacity intermediate in degree between types 1 and 2.

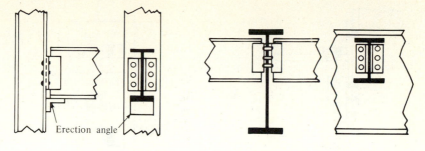

Erection angle

Beam to column

Beam to girder

(*a*)

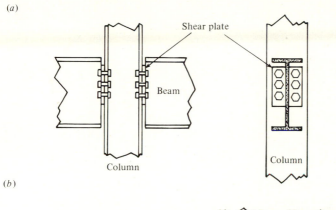

Shear plate

Beam

Column

Column

(*b*)

Clip angle

Clip angle

Erection bolts

Clip angle

Shelf angle

Welded plates or
Structural tee

Beam to girder

(*c*)

Beam to column

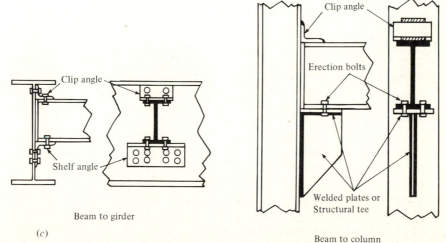

Figure 13.1 Simple framing connections: (*a*) framed connections, (*b*) shear plates, (*c*) beam seat connections.

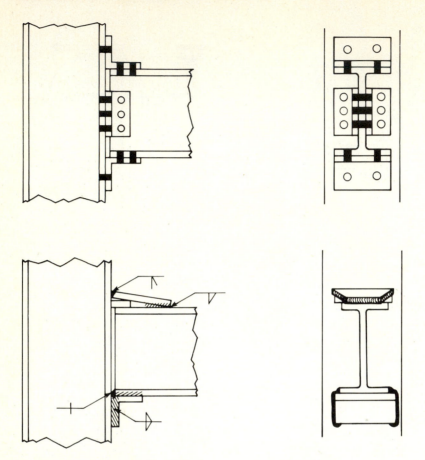

Figure 13.2 Semirigid framing connections.

From a practical point of view connections cannot be rigid or completely flexible. It is common practice to classify them on the basis of the percentage of rigidity developed, i.e., roughly, simple connections 0 to 20 percent, semirigid 20 to 90 percent, and rigid above 90 percent. Examples of these types of connections are illustrated in Figs. 13.1 to 13.3. As a result it is possible to make the design of members and connections compatible in terms of basic assumptions. The importance of this compatability can readily be illustrated by considering simply supported and fixed-ended beam designs. If the fixed-ended beam shown in Sketch 13.1 is designed to resist the given bending moments and the connections are then designed as simply supported, giving free rotation, the beam will not be strong enough to resist the bending moment developed.

Connections of the types indicated above involve a number of mechanisms not previously discussed in Chap. 7. They include tension loads on bolted joints, eccentric loads in the plane of the connector (which induce additional shear forces due to eccentricity), and eccentric loads outside the plane of connector

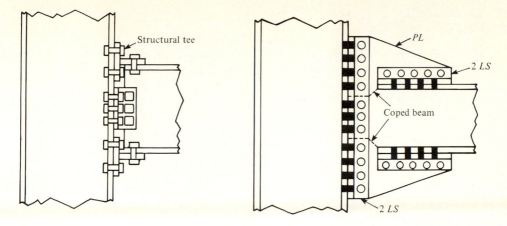

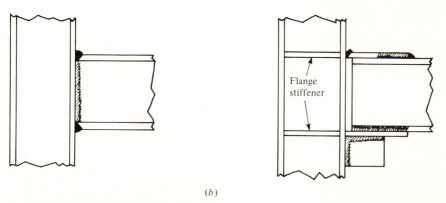

(b)

Figure 13.3 Rigid connections: (a) structural tee and bracket connections, (b) welded moment-resisting connection.

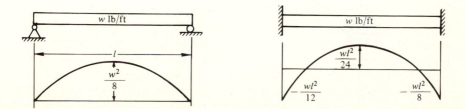

Sketch 13.1

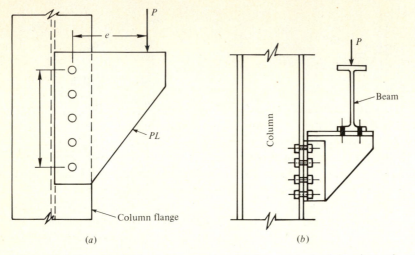

Figure 13.4 Eccentrically loaded connections: (*a*) eccentric load in plane of connectors, (*b*) eccentric load out of plane of connectors.

shears (which induce combined shear and tension in the connector). Connections involving these loading mechanisms are illustrated in Fig. 13.4.

13.2 TENSION LOADS ON BOLTED JOINTS

Bolted and riveted connections subjected to tensile loads tend to be avoided, but they are sometimes necessary and convenient in moment connections (type 1) and hanger connections for supporting pipe systems, etc. Sketch 13.2 shows a bolted connection where some of the bolts are subjected to tensile loads.

There has been a reluctance among designers to apply tensile loads to hot-driven rivets and tightened bolts for fear of increasing their existing tensile

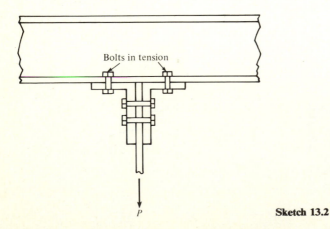

Sketch 13.2

stresses. Careful consideration indicates, however, that when external tensile loads are applied to connections of this type, the fasteners will experience little if any change in stress.

Consider the application of tensile load P at the contact surface between the flange and angles. It tends to reduce the thickness of the plates, but at the same time the contact pressure between the plates is reduced and the plates tend to expand by the same amount. The theoretical result is no change in angle thickness or fastener tension. This situation continues until P equals the initial tensile force in the fasteners, when an increase in P will result in separation of the plates; thereafter increases in P will increase the tensile force in the fastener.

If the load is applied to the outer surface of the angle, there will be an immediate strain increase in the fastener. This increase in fastener length will be accompanied by an expansion of the angle. The increase in the load of the fastener is relatively small because the load goes to the angle and the fasteners roughly in proportion to their stiffness and the angle is much stiffer than the fastener.

In most situations the transfer of load P falls between the cases discussed, thus keeping the stress condition in the fastener relatively constant even with the application of load P.

Another factor that requires consideration in this type of connection is the possibility of prying action. A tensile connection subject to prying action is illustrated in Sketch 13.3. Should the flanges of the connection be thick and stiff, the prying action will probably be negligible. On the other hand, if the flanges are thin and flexible, the prying action could be significant. It is usually desirable to limit the number of rows of fasteners because a large percentage of the load would be carried by the inner row even at ultimate load.

AISC specifications provide empirical equations that produce conservative designs for situations involving prying action. Figure 13.5 indicates the nature of the action of these forces. The prying force Q is assumed to be a line load acting along the edge of the flange over a distance L. The fasteners must be designed for the load F due to the moment and the load Q due to the prying action. Bending of the flange (Fig. 13.5b) results in the bending-moment diagram in Fig. 13.5a. Either M_1 or M_2 could be critical in the design of the flange. They are assumed to occur at the fastener line and at a point $\frac{1}{16}$ in from the near face of the outstanding leg of the angle or tee:

$$M_1 = Qa \quad \text{and} \quad M_2 = (F + Q)b - Q(a + b)$$

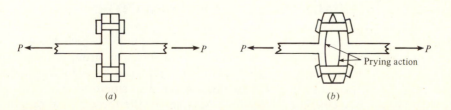

(a)   (b)

Sketch 13.3

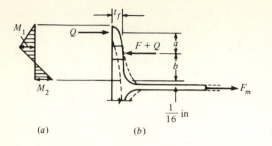

Figure 13.5 Prying forces: (*a*) bending moment in the flange, (*b*) tee-flange bending.

(*a*)   (*b*)

These moments generate bending stresses in the flange which must be kept below the allowable ($F_b = 0.75F_y$). These stresses can be determined by use of the flexure formula $f_b = M/S$, where S is the section modulus of the length of flange involved with a particular fastener w.

It is difficult to establish exact values for the prying forces Q. AISC empirical formulas based on test results provide conservative values:

$$Q = \frac{F\left[100b(d_b)^2 - 18w(t_f)^2\right]}{70a(d_b)^2 + 21w(t_f)^2} \qquad \text{connections using A325 bolts}$$

$$Q = \frac{F\left[100b(d_b)^2 - 14w(t_f)^2\right]}{62a(d_b)^2 + 21w(t_f)^2} \qquad \text{connections using A490 bolts}$$

where Q = prying force per bolt, kips
         F = external applied load per fastener, kips = F_m/n
      F_m = force in flange due to bending moment $M = M/d$
         d = depth of beam
        w = length of flange involved with each bolt, in
           = $L/2$ for two lines of bolts
       d_b = nominal diameter of bolt, in
         a = distance from bolt line to edge of flange $< 2t_f$, in
         b = distance from fastener line to near face of outstanding leg of angle
            or structural tee less $\frac{1}{16}$ in

13.3 ECCENTRIC LOADS IN PLANE OF CONNECTORS

The connectors are assumed to be bolts subjected to a load P which has an eccentricity e relative to the center of gravity of the fastener group (Sketch 13.4*a*). To analyze this situation the eccentric load is replaced by a concentric load plus a moment Pe (Sketch 13.4*b*) (see Sec. 4.11). The situation shown in Sketch 13.4*b* in no way changes the loading. As a result, the force acting on a

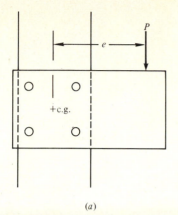

(a)

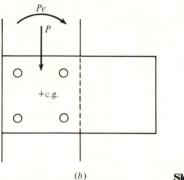

(b)   **Sketch 13.4**

particular fastener is the resultant of the force due to the concentric load and a force due to the moment. Consideration will now be given to determining the force in each connector due to the applied moment and to the direct load. The discussion includes the possibility that connectors will be of different sizes and strengths.

The moment Pe tends to rotate the connection in a clockwise manner about the center of gravity of the connectors and causes a shear load in each connector (fastener or unit length of weld). For this derivation, the gusset plates are considered rigid and the connector elastic. Rotation will be the greatest at the connector which is farthest from the center of rotation (center of gravity). The resulting shear force can be considered to act perpendicular to a line joining the center of rotation and the connector being considered. The magnitude of the shear load caused in each unit will depend on the amount of shearing and bearing deformation undergone as a result of the rotation of the joint and on its resistance to this deformation, which can be assumed to be proportional to the cross-sectional area, the bearing area, or the ultimate strength of the unit. While none of these assumptions is exact, the last is probably most accurate.

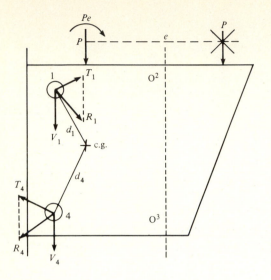

Figure 13.6 Connection with eccentric load in plane of connectors.

The notation to be used in the derivation is as follows (see Fig. 13.6):

S_1 = ultimate strength of connector 1
d_1 = distance of connector 1 from center of rotation (center of gravity of connection)
T_1 = load on connector 1 due to moment Pe
V_1 = load on connector 1 due to concentric load P
R_1 = total shearing load on connector 1 = resultant of T_1 and V_1

For other connectors the subscript 1 is replaced by the number of the connector involved.

From the above discussion we can assume that the load T is proportional to the distance d and to the strength S of the unit. In the above terms we have

$$\frac{T_1}{d_1 S_1} = \frac{T_2}{d_2 S_2} = \frac{T_3}{d_3 S_3} \cdots$$

Consequently,

$$T_2 = T_1 \frac{d_2 S_2}{d_1 S_1} \qquad T_3 = T_1 \frac{d_3 S_3}{d_1 S_1} \cdots$$

For equilibrium

$$Pe = M = T_1 d_1 + T_2 d_2 + T_3 d_3 + \cdots$$

or

$$M = \frac{T_1 d_1^2 S_1}{d_1 S_1} + \frac{T_1 d_2^2 S_2}{d_1 S_1} + \cdots$$

Collecting terms, we get

$$M = \frac{T_1 \Sigma S d^2}{d_1 S_1}$$

Solving for T_1 gives

$$T_1 = \frac{Md_1 S_1}{\Sigma Sd^2}$$

The term ΣSd^2 may be considered to be the polar moment of inertia of the strengths of the connectors about the center of gravity of the strengths. Since the load T for any unit depends on the ratio of its strength to that of another, that ratio or relative strength may be used instead of the actual strength S. Thus the load on a connector due to moment only is

$$T = \frac{Md \times \text{relative strength } S \text{ of unit}}{J \text{ (of relative strengths of all units)}} \tag{13.1}$$

If the connection is subject to direct load in addition to moment, the total shear load on any unit is the vector sum, or resultant, of the load due to moment and the load due to direct load. The applied load is assumed to be divided among the connectors on the basis of their relative strengths

$$V = \frac{P \times \text{relative strength of unit}}{\text{sum of relative strengths of all units}} \tag{13.2}$$

These equations are simpler in the usual case where the relative strengths of all units in the connection are equal and become

$$T = \frac{Md}{J} \tag{13.3}$$

and

$$V = \frac{P}{n} \tag{13.4}$$

where n is the number of connectors

It is often more convenient to break up the moment component of force into vertical and horizontal components (Sketch 13.5). On the basis of Sketch 13.5,

$$T = \frac{Md}{J} \quad \text{and} \quad T_v = \frac{Mh}{J} \quad \text{and} \quad T_h = \frac{Mv}{J}$$

These components can be combined algebraically or graphically with the force due to the applied load.

Since the design of a connector requires determining the number of fasteners or length of weld, it is usually necessary to use trial-and-error methods.

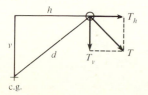

c.g.   **Sketch 13.5**

Example 13.1 Figure E13.1a shows a bracket connected to a column flange by A325 bolts. The load P applied to the bracket is the end reaction of the beam. If P is 20 kips, determine the bolt size required.

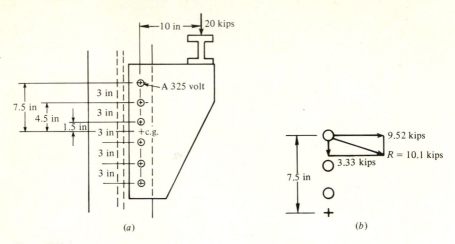

(a)   *(b)*

Figure E13.1

SOLUTION Since the bolts are the same size, the relative strength is unity. The most highly loaded bolt will be the top or bottom bolt (Fig. E13.1b)

$$M = 20(10) = 200 \text{ in·kips} \qquad J = (2)\left[(1.5)^2 + (4.5)^2 + (7.5)^2 \right] = 157.5 \text{ in}^2$$

$$T = \frac{Md}{J} = \frac{(200)(7.5)}{157.5} = 9.52 \text{ kips} \qquad V = \frac{20}{6} = 3.33 \text{ kips}$$

$$R = \sqrt{(9.52)^2 + (3.33)^2} = 10.1 \text{ kips}$$

A $\frac{3}{4}$-in A325 bolt is capable of carrying $(\pi/4)(3/4)^2(30) = 13.25$ kips in single shear, which is adequate for the above connection.

Example 13.2 For the bracket shown in Fig. E13.2a determine the fillet size required if E60 electrodes are used.

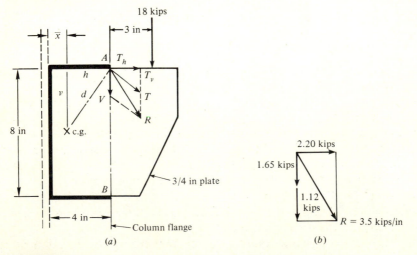

(a)   *(b)*

Figure E13.2

SOLUTION Since the weld is the same size and strength per inch, the relative strength can be considered unity. Assume a weld of unit width; then

$$A = 8 + 4 + 4 = 16 \text{ in}^2/\text{in of width} \qquad \bar{x} = \frac{(2)(4) + 2(4)}{16} = 1 \text{ in}$$

$$I_x = 1/12(1)(8)^3 + (2)(4)(4)^2 = 170.7 \text{ in}^4/\text{in of width}$$

The moments of inertia of the horizontal welds about their own centroidal x axis were neglected.

$$I_y = (2)(1/3)(3)^3 + (2)(1/3)(1.0)^3 + (8)(1.0)^2 = 18 + 0.67 + 8 = 26.7 \text{ in}^4/\text{in of width}$$

$$J = I_x + I_y = 196.7 \text{ in}^4/\text{in of width}$$

The most stressed portions of the welds will be farthest from the center of gravity of the weld (points A and B in Fig. E13.2a). In a problem of this kind it is advantageous to calculate the horizontal and vertical components of the force resulting from the moment and acting on the extreme inch of weld (Fig. E13.2b).

$$T_h = \frac{Mv}{J} = \frac{(18)(6)(4)}{196.7} = 2.20 \text{ kips/in}$$

$$T_v = \frac{Mh}{J} = \frac{(18)(6)(3)}{196.7} = 1.65 \text{ kips/in}$$

$$V = \frac{P}{A} = \frac{18}{16} = 1.12 \text{ kips/in}$$

$$R = \sqrt{(2.20)^2 + (1.65 + 1.12)^2} = 3.5 \text{ kips/in}$$

The allowable stress on 1-in fillet weld (E60 electrode) is

$$1(0.707)(0.30)(60) = 12.7 \text{ kips/in of weld length}$$

The weld size required is

$$\frac{3.5}{12.7} = 0.275 \text{ in}$$

Use 5/16-in weld.

13.4 ECCENTRIC LOADS OUTSIDE THE PLANE OF CONNECTOR SHEARS

Connections supporting loads falling outside the plane of shear are generally of the type shown in Fig. 13.4. They normally involve a bracket attached to a column flange or a stiffened seat of type 2 construction. The moment resulting from the eccentric load tends to separate the bracket from the column flange at the top and press the bracket against the flange at the bottom. The load also places the connectors in direct shear (see Fig. 13.7). As a result, design procedures normally consider a resulting state of combined stress. The effect of the moment is evaluated by the use of the flexure formula. Factors such as an uncertain stress distribution in the bearing of the bracket on the flange and the degree of tightness or prestress in the fasteners affect the location of the neutral axis associated with the stress distribution.

There are two acceptable approaches for determining the location of the neutral axis and defining areas for calculating moments of inertia. One method

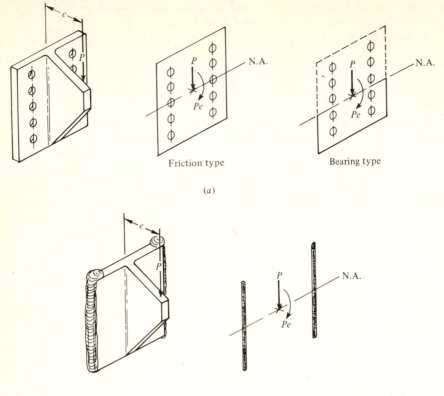

Friction type   Bearing type

(a)

(b)

Figure 13.7 Connections with eccentric load out of plane of connectors: (a) bracket connection with fasteners, (b) welded bracket connection.

applies to high-strength bolts in a friction type of connection (prestress in fastener, A325 and A490 bolts). The other method applies to fasteners in a bearing type of connection (no reliable prestress in fastener-rivets and A307 bolts). In a welded connection, the location of the center of gravity and the moment of inertia can readily be determined by considering the weld configuration to have a unit width (see Example 13.2).

In equation form, the tensile stress resulting from the moment (eccentric load) equals

$$f_t = \frac{Pe}{I/c} \tag{13.5}$$

where I is the moment of inertia of the appropriate area of stress at the contact surface between the bracket and column and c is the distance from the neutral axis to the stress center (fastener or unit length of weld). For shear stress

$$f_v = \frac{P}{A_S} \tag{13.6}$$

where A_s is equal to the area of the *total* number of bolts or the area of the weld configuration.

Tests on rivets and bolts subjected to combined shear and tension have shown an elliptical interaction curve of the form

$$\left(\frac{f_v}{kF_u}\right)^2 + \left(\frac{f_t}{F_u}\right)^2 = 1$$

where k is a constant (< 1.0) which depends on the type of fastener and location of the shear plane, f_v and f_t are simultaneous shearing stress and tensile stress at failure, and F_u is the tensile strength of fastener in tension only.

According to AISC specifications, the allowable tensile stress F_t in rivets and bolts subject to combined shear and tension in bearing connections should not exceed the values computed from the formulas in Table 13.1, where f_v, the shear stress produced by the same forces, should not exceed the values given in Table 7.5.

For A325 and A490 bolts used in friction-type connections, the maximum allowable shear stress (Table 7.5) should be multiplied by the *reduction factor*

$$1 - \frac{f_t A_b}{T_b}$$

where f_t = tensile stress due to applied load
$\quad T_b$ = specified pretension load on bolt
$\quad A_b$ = area of bolt

The specified pretension load T_b is based on the minimum tension values shown in Table 13.2.

When allowable stresses are increased for wind and seismic loads, the F_t values of Table 13.1 are adjusted by increasing the constants in the appropriate formulas by one-third. In friction connections the adjusted f_v value is increased by one-third.

In a welded connection the combined effect of the applied load and moment will be a vectorial sum of the shearing stress f_v and the tensile stress f_t.

Table 13.1 Allowable tension stress F_t for fasteners in bearing-type connections [1]

Description of fastener	Threads not excluded from shear planes	Threads excluded from shear planes
Threaded parts:		
A449 bolts over $1\frac{1}{2}$-in diameter	$0.43F_u - 1.8f_v \le 0.33F_u$	$0.43F_u - 1.4f_v \le 0.33F_u$
A325 bolts	$55 - 1.8f_v \le 44$	$55 - 1.4f_v \le 44$
A490 bolts	$68 - 1.8f_v \le 54$	$68 - 1.4f_v \le 54$
A502, grade 1 rivets	$30 - 1.3f_v \le 23$	
Grades 2 and 3 rivets	$38 - 1.3f_v \le 29$	
A307 bolts	$26 - 1.8f_v \le 20$	

Table 13.2 Minimum bolt tension, kips [1]†

Bolt size, in	$\frac{1}{2}$	$\frac{5}{8}$	$\frac{3}{4}$	$\frac{7}{8}$	1	$1\frac{1}{8}$	$1\frac{1}{4}$	$1\frac{3}{8}$	$1\frac{1}{2}$
A325	12	19	28	39	51	56	71	85	103
A490	15	24	35	49	64	80	102	121	148

† Equal to 0.70 times specified minimum tensile strengths of bolts, rounded off to nearest kip.

Since they act at right angles to each other,

$$f = \sqrt{(f_t)^2 + (f_v)^2}$$

For a safe design the maximum stress should not exceed the allowable stress $0.30F_u$, where F_u is the ultimate tensile strength of the electrode (Table 7.7).

13.4.1 Friction-Type Connection

In this type of connection the eccentric load on the bracket tends to bend the top of the bracket away from the column and push the bottom against the column. Unless the tendency to pull away from the column is greater than the compressive stresses at the interface due to the initial tension in the bolts, there will be almost no change in bolt tension (Sec. 13.2). As a result, the complete face of the bracket remains in contact with the column flange and the neutral axis of the cross section can be considered to pass through the centroid of the contact surface. The fastener holes in the bracket are normally neglected to simplify the calculations.

In the design of such a connection, if the tensile stress due to the eccentricity exceeds the initial compressive stress due to the bolt prestress, the trial connection will not function as a friction connection and a new connection must be selected.

Example 13.3 Determine the stress in the $\frac{3}{4}$-in A325 bolts used in the friction connection shown in Fig. E13.3a.

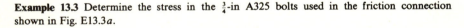

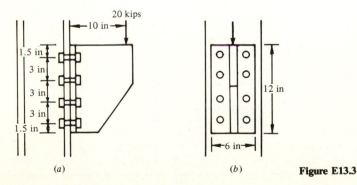

(a)   (b)   **Figure E13.3**

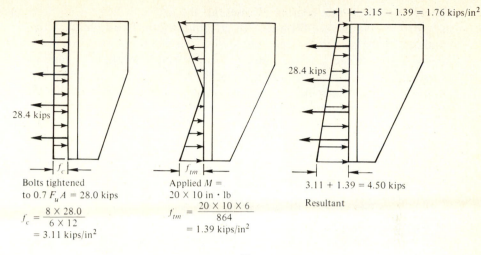

(b)

Figure E13.3 (*Cont'd.*)

SOLUTION Calculate the moment of inertia of contact surface

$$I = \frac{bh^3}{12} = \frac{(6)(12)^3}{12} = 864 \text{ in}^4$$

From Eq. (13.6) the average shear stress in the bolts is

$$f_v = \frac{20}{(8)(\pi/4)\left(\frac{3}{4}\right)^2} = 5.66 \text{ kips/in}^2$$

and from Eq. (13.7) the allowable shear stress is determined from

$$F_v = 17.5\left(1 - \frac{f_t A_b}{T_b}\right)$$

The portion of bolt tension caused by moment (top bolts)

$$f_t A_b = \frac{(20)(10)(4.5)}{864} \frac{(3)(6)}{2} = 9.37 \text{ kips}$$

$$f_v = (17.5)\left(1 - \frac{9.37}{28.40}\right) = 11.7 \text{ kips/in}^2$$

$$f_v = 5.66 < 11.7 \text{ kips/in}^2 \quad \text{OK}$$

Practically, the tension in the bolt does not change, although the pressure distribution has.

13.4.2 Bearing-Type Connection

Since in this case prestress in the fastener cannot be relied on, under the action of the moment the upper portion of the bracket may be considered to be separated from the column and the lower portion in contact. Only the areas of the fasteners above the neutral axis and the area in bearing below are considered in the calculations of the moment of inertia. The width of the compression area is sometimes taken as the width of the bracket. Because of deformation of the flanges, however, the bearing stresses at the edges will be less than those indicated by the flexure formula. An effective width, five-eighths of the entire width, has been used in conservative design. To determine the distance $\bar{y}$ to the neutral axis, it is necessary to assume its approximate location. A good estimate is one-sixth of the bracket length up from the bottom. After the tensile and shear loads have been determined, it is necessary to compare them with the allowable combined-stress condition.

Example 13.4 Reinvestigate Example 13.3 if the fasteners are 3/4-in A307 bolts. Assume no initial tension.

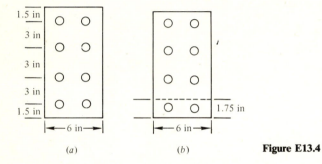

(a)   (b)   **Figure E13.4**

SOLUTION Assume the location of the neutral axis to be (Fig. E13.4a)

$$\bar{y} = \left(\tfrac{1}{6}\right)(12) = 2 \text{ in}$$

$$\Sigma M \text{ compression area about NA} = \Sigma M \text{ tension area about NA}$$

Check location:

$$\frac{5}{8}\frac{6\bar{y}^2}{2} = (6)(0.44)(7.5 - \bar{y})$$

$$\bar{y}^2 + 0.70\bar{y} - 5.3 = 0$$

$$\bar{y} = 1.97 \text{ in}$$

The holes in the bearing area can be neglected in computing $\bar{y}$ but should be considered in calculating the moment of inertia (Fig. E13.4b). The moment of inertia of the effective section is

$$I = \left(\tfrac{1}{3}\right)\left(\tfrac{5}{8}\right)(6)(1.97) - (0.88)(0.47)^2 + (2)(0.44)\left[(2.53)^2 + (5.53)^2 + (8.53)^2\right]$$

$$= 9.56 - 0.19 + (0.88)(6.40 + 30.6 + 72.8) = 106.0 \text{ in}^4$$

From Eq. (13.5) the tensile stress in the top bolt is

$$f_t = \frac{(20)(10)(8.53)}{106.0} = 16.1 \text{ kips/in}^2$$

and from Eq. (13.6) the shear stress in the top bolt is

$$f_v = \frac{20}{(8)(0.44)} = 5.68 \text{ kips/in}^2$$

According to AISC specifications, the allowable tensile stress is

$$F_t = 55 - (1.8)(5.68) = 44.8 \text{ kips/in}^2 \quad \text{use } 44.0 \text{ kips/in}^2 \text{ (Table 13.1)} \quad \text{OK}$$

13.4.3 Welded Connection

It is usually assumed that the stress due to moment is given by the flexure formula. If the flexure formula holds, the intensity of horizontal shear and consequently vertical shear should be given by the general shear formula. Since the abrupt change in section from the weld to column face causes stress concentrations at the extremities of the bracket, however, design procedures normally distribute the shear stress uniformly over the weld.

Example 13.5 A tee bracket made from a WT12 × 38 is welded to a column flange by two welds going the length of the bracket. Calculate the required weld size to resist the 60-kip eccentric load. Use A36 steel and E60 electrodes, (Fig. E13.5).

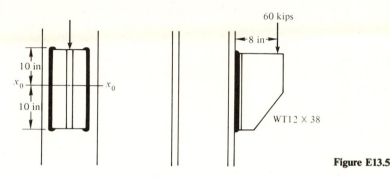

Figure E13.5

SOLUTION Consider a 1-in weld size. The average vertical shear force per unit length of weld equals

$$F_v = \frac{P}{L} = \frac{60}{40} = 1.50 \text{ kips/in}$$

The centroidal axis for the weld configuration is located 10 in up from the base. The moment of inertia of welds is

$$I_{x0} = \frac{(2)(1)(20)^3}{12} = 1333.3 \text{ in}^4/\text{in width}$$

The maximum force in the welds due to the moment is

$$F_t = \frac{Pe}{I_x/c} = \frac{(60)(8)}{1333.3/10} = 3.60 \text{ kips/in}$$

The vectorial sum of these two forces is

$$F = \sqrt{F_v^2 + F_t^2} = \sqrt{(1.50)^2 + (3.60)^2} = 3.9 \text{ kips/in}$$

The shear strength of a 1-in weld of E60 electrode from Eq. (7.5) is

$$(1)(0.707)(0.30)(60) = 12.7 \text{ kips/in}$$

The required weld size is

$$\frac{3.9}{12.7} = 0.307 \text{ in} \qquad \text{use 5/16-in weld}$$

This weld size is permissible for a WT12 × 38 with a flange thickness of 0.770 in.

13.5 SIMPLE CONNECTIONS

Simple connections are designed to be flexible. Although they have some moment resistance, it is assumed to be negligible and they are designed to resist shear only. The connections between beams and girders use framing angles or shear plates, while the connections between beams and columns also include seats (Fig. 13.1). It is not unusual to use two different types of connectors with the same connection. For example, a common practice is to shop-weld the web angles to the beam web and field-bolt them to the column or girder.

13.5.1 Framed Connections

Figure 13.1*a* and *b* shows typical framed connections. This type of connection consists of a pair of flexible web angles, probably shop-connected to the web of the supported beam and field-connected to the supporting beam or column. The legs of the angle attached to the supported beam are normally called *connected legs*. The other legs are called *outstanding legs*. The thickness of the angles is purposely kept small to provide flexibility. The AISC Manual [1] arbitrarily limits the thickness to $\frac{5}{8}$ in. It is often convenient to have an angle, called an *erection seat*, to support the beam during construction (Fig. 13.1). For this kind of connection the eccentricity is neglected unless welding is used as the connector.

The stresses that must be considered are shear in the connector; bearing on the beam webs, column flange, and framing angle (if bearing type); and gross shear on a vertical section through the beam web and framing angle.

The AISC Manual tabulates framed connections for various beams using bolts, rivets, and welds. Figure 13.8 gives details for $\frac{3}{4}$- and $\frac{7}{8}$-in fasteners.

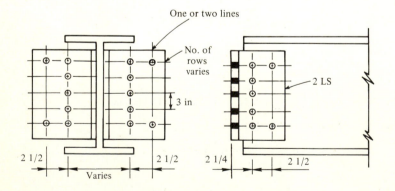

Figure 13.8 Details of framed connections using fasteners [1].

Example 13.6 Design a framing-angle connection for a W16 × 50 girder framing into the flange of a W10 × 39 column, both of A36 steel (Fig. E13.6a). The end reaction of the girder is 55 kips. Use $\frac{3}{4}$-in A325 bolts in a bearing-type connection with threads in the shear plane.

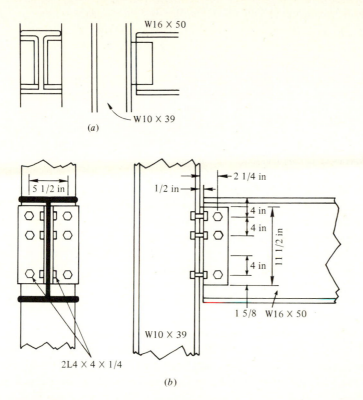

Figure E13.6

SOLUTION

W16 × 50:

$$T = 13\tfrac{3}{4} \text{ in} \qquad t_w = 0.380$$

For the connecting legs from Table 7.5 for bolts in double shear

$$(2)(21)\left[\frac{\pi}{4}\left(\frac{3}{4}\right)^2\right] = 18.6 \text{ kips/bolt}$$

For web of W16 × 50

$$\text{Bearing} = (1.50)(58)(0.75)(0.380) = 24.8 \text{ kips/bolt} \qquad \text{shear controls}$$

The number of bolts required is

$$n = \frac{55}{18.6} = 2.96 \qquad \text{use 3 bolts}$$

The minimum angle thickness is

$$\frac{55}{(1.50)(58)(0.75)(4)} = 0.21 \text{ in} \qquad \text{use } \tfrac{1}{4} \text{ in}$$

Consider the outstanding legs:

$$\text{Bolts in single shear} = (21)\frac{\pi}{4}\left(\frac{3}{4}\right)^2 = 9.3 \text{ kips/bolt}$$

$$\text{Bearing in angle} = (1.50)(58)(0.25)(0.75) = 16.3 \text{ kips} \qquad \text{shear controls}$$

The number of bolts required is

$$n = \frac{55}{9.3} = 5.9 \qquad \text{use 6}$$

Suitable details for the connection can be arranged with the help of Fig. 13.8. The T depth (see Table A.5, Dimensions for Detailing) for a W16 × 50 is 13 5/8 in which is adequate for a row of three bolts. A suitable arrangement is illustrated in Fig. E13.6b. Use two angles 4 by 4 by 1/4 in. Check shear on the gross area of the angles:

$$F_v = 0.4(36) = 14.4 \text{ kips/in}^2$$

There are two angles each 11 1/2 in long. The shear resistance that can be developed is

$$(2)(11.5)\left(\tfrac{1}{4}\right)(14.4) = 82.8 > 55 \text{ kips} \qquad \text{OK}$$

Welded framing-angle connections are shown in detail in Fig. 13.9. The framing angles are placed on each side of the web of the connecting beam. The angles extend out from the beam web by approximately $\frac{1}{2}$ in, as shown. This distance is referred to as the *setback*. Fillet welds are placed along the top, bottom, and sides of the connecting leg. These welds are considered to be eccentric loads within the plane of the welds. The angles usually have 3-in legs and the thickness necessary for the size of weld. The length of the angles depends on strength requirements but cannot exceed the T distance of the beam nor should it be less than one-half of T.

Frequently, the outstanding leg is the field connection and bolts are used instead of welds. To allow for bolting, the outstanding leg must have a minimum width of 4 in. When welded, the outstanding legs are welded to the supporting member only along the sides of the angle. This helps provide the flexibility required by type 2 construction. These vertical welds are normally *returned* about $\frac{1}{2}$ in on the top for added strength. These field welds are subject to

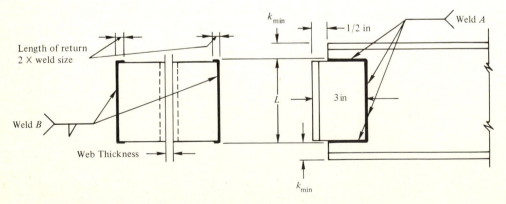

Figure 13.9 Details of welded framed connections [1].

eccentric loads, as shown in Fig. 13.9. This moment causes rotation of the web angles which forces them against the beam web at the top and pushes them apart at the bottom. Different design procedures proposed to take this into account differ primarily in the distance over which the pressure is distributed near the top of the angle.

A satisfactory approach is to assume that the pressure near the top of the angle occurs over a distance of one-sixth the length of the weld L. The stress is assumed to vary linearly from a maximum at the top to zero at $L/6$. The lower remaining distance ($\frac{5}{6} L$) develops a counterforce which is considered to have a linear distribution reaching a maximum at the bottom. The force-equilibrium diagram for the outstanding leg is presented in Sketch 13.6.

The internal couple developed by T and P opposes the applied external moment $Re/2$

$$\frac{T2L}{3} = \frac{Re}{2}$$

$$T = \frac{3Re}{4L}$$

The maximum horizontal force per unit length of weld at the bottom will be

$$F_2 = \frac{9}{5} \frac{Re}{L^2} \tag{13.7}$$

The vertical shear force per unit length of weld is

$$F_1 = \frac{R/2}{L} \tag{13.8}$$

These two forces must be added vectorially. As a result the maximum force per unit length of weld becomes

$$F = \sqrt{F_1^2 + F_2^2} \tag{13.9}$$

and it determines the size of the weld.

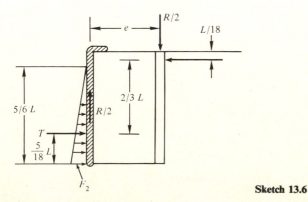

Sketch 13.6

Example 13.7 Design a welded framing-angle connection between a W18 × 70 beam and a W10 × 49 column. The end reaction of the beam is 55 kips. All steel is A36. Use E70 electrodes. The end of the beam is to be set back $\frac{1}{2}$ in from the column flange, (Fig. E13.7a).

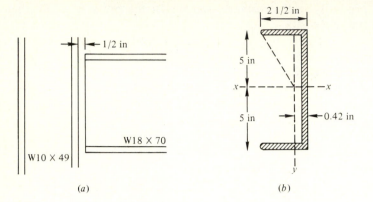

(a) (b)

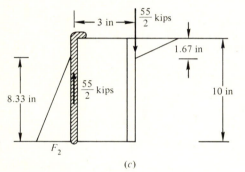

(c)

Figure E13.7

SOLUTION

W18 × 70: $\qquad\qquad\qquad\qquad T = 15\frac{1}{8}$ in $\qquad t_w = 0.438$ in

W10 × 49: $\qquad\qquad\qquad\qquad t_f = 0.558$ in

Assume a 3- by 3-in angle 10 in long. The T dimension of the beam is $15\frac{1}{8}$ in. The weld configuration for a connected leg is shown in Fig. E13.7b, and

$$\bar{x} = \frac{(2)(2.5)(1.25)}{15} = 0.42 \text{ in}$$

$$I_x = \left(\frac{10}{12}\right)^3 + (2)(2.5)(5)^2 = 83.3 + 125 = 208 \text{ in}^4/\text{in of width}$$

$$I_y = \frac{(2)(2.5)^3}{12} + (2)(2.5)(0.83)^2 + (10)(0.42)^2$$

$$= 2.60 + 3.44 + 1.76 = 7.8 \text{ in}^4/\text{in width}$$

$$J = I_x + I_y = 208 + 7.8 = 215.8 \text{ in}^4/\text{in width}$$

From Eq. (13.6) the shear force per unit length of weld is

$$F = \frac{55}{(2)(15)} = 1.83 \text{ kips/in}$$

From Eq. (13.5) the shear force in the outermost portion of weld configuration A due to moment is

$$F_2 = \frac{(27.5)(2.58)(5.42)}{215.8} = 1.78 \text{ kips/in}$$

The vertical component of F_2 is

$$F_{2v} = 1.78 \frac{2.08}{5.42} = 0.68 \text{ kips/in}$$

and the horizontal component of F_2 is

$$F_{2h} = 1.78 \frac{5}{5.42} = 1.64 \text{ kips/in}$$

The vectorial sum of the shear forces acting at point A equals

$$F = (1.83 + 0.68)^2 + (1.64)^2 = 3.0 \text{ kips/in}$$

The shear strength of a 1-in weld of E70 electrode is

$$(1)(0.707)(0.30)(70) = 14.84 \text{ kips/in}$$

The required size of weld is therefore

$$\frac{3.0}{14.84} = 0.20 \text{ in}$$

Use a $\frac{1}{4}$-in weld.

The shear strength of the web should be equal to or greater than the shear force developed in the welds

$$(2)(0.20)(0.707)(0.30)(70) = 0.438(0.4)(36)$$
$$= 5.94 < 6.30 \text{ kips/in} \quad \text{OK}$$

The required angle thickness is $\frac{1}{4} + \frac{1}{16} = \frac{5}{16}$ in. The weld configuration for the outstanding leg is shown in Fig. E13.7c.

From Eq. (13.7) the maximum shearing force in the bottom of the weld due to the eccentricity is

$$F_2 = \frac{9}{5} \frac{Re}{L^2} = \frac{9}{5} \frac{(55)(3)}{10^2} = 2.97 \text{ kips/in}$$

From Eq. (13.8) the vertical shear force in the weld is

$$F_1 = \frac{R}{2L} = \frac{55}{(2)(10)} = 2.75 \text{ kips/in}$$

and the vectorial sum of these shears is

$$\sqrt{(2.97)^2 + (2.75)^2} = 4.0 \text{ kips/in}$$

The required weld size is

$$a = \frac{4.0}{14.8} = 0.27 \text{ in} \quad \text{use } \tfrac{5}{16}\text{-in weld}$$

$$\text{Required angle size} = \tfrac{5}{16} + \tfrac{1}{16} = \tfrac{3}{8} \text{ in}$$

Summarizing, we have

$$\text{Angle size} = 3 \text{ by } 3 \text{ by } \tfrac{3}{8} \text{ in}$$

$$\text{Weld size of connected leg} = \tfrac{1}{4} \text{ in}$$

$$\text{Weld size of outstanding leg} = \tfrac{5}{16} \text{ in}$$

13.5.2 Shear Plates

A shear plate welded to the end of a beam may be used instead of the paired angles used as framing angles (Fig. 13.1). Bolts are then used to connect the plate to the supporting column flange or beam web. Shear-plate connections require closer fabricating tolerances for the beam. The design procedure for the bolts holding the plate to the supporting member was covered in Sec. 13.5.1 and Example 13.6.

The length of the plate is made less than the T depth of the beam so that all the welding will be confined to the web. This and the limitation of plate thickness to $\frac{1}{4}$ to $\frac{3}{8}$ in provide the flexibility and end rotation required for type 2 construction. No eccentricity is considered in the design of the welds or the bolts.

The AISC Code requires one fillet weld on each side of the beam. The code specifies that the welds should not be returned across the web. As a result, each weld has a length equal to the plate length minus twice the weld size. Since the weld is placed on both sides of the web, the theoretical limitation of weld size in terms of web thickness must be kept in mind: the shear strength of the web must be equal to that required of the welds.

Example 13.8 Design an end shear-plate connection for a W12 × 45 beam having a reaction of 34 kips. The supporting column is a W10 × 30. All steel is A36, and E70 electrodes are specified.

SOLUTION Try the $\frac{1}{4}$-in minimum shear-plate thickness. Use $\frac{3}{4}$-in A325 bolts in a bearing-type connection. The bolts are in a single shear with threads in the shear plane (Table 7.5). The number of bolts required is

$$\frac{34}{(21)(\pi/4)(3/4)^2} = 3.6 \qquad \text{use 4 bolts}$$

The maximum length of the shear plate is

$$d - 2k = T = 9\tfrac{1}{2} \text{ in}$$

Use an $8\frac{1}{2}$-in plate (standard). Assume $\frac{1}{4}$-in welds. The effective length becomes $8\frac{1}{2} - 2(\frac{1}{4}) = 8$ in. The weld size required is

$$(16)(0.707)(0.30)(70)a = 34$$

$$a = \frac{34}{(16)(0.707)(0.30)(70)} = 0.143 \text{ in}$$

Use $\frac{3}{16}$-in weld. Check the beam web thickness against the required weld size.

$$\text{Shear strength of web} = (0.4)(36)(0.335) = 4.82 \text{ kips/in}$$

$$\text{Shear strength of welds} = (0.3)(70)(2)(0.143)(0.707) = 4.24 \text{ kips/in}$$

The web thickness is adequate.

13.5.3 Beam Seats

These connections also allow the end rotation needed for type 2 construction (Fig. 13.1). One advantage of beam seats is that a surface is provided for resting the beam during construction, eliminating the need for temporary erection bolts. The beam transfers its entire load by direct bearing to the seat angle. The load which can be supported by this type of connection is limited by the bending strength of the horizontal leg of the seat angle (about 40 to 50 kips for A36 steel). For heavier loads it is necessary to use stiffened seats (Fig. 13.1). A top clip angle is used to provide lateral support to the beam. Should space limitations provide a problem above the beam, the clip angle can be placed on the side and attached to the beam web.

The connected leg of the angle is attached to the supporting member (column of girder web) in the shop. A 4-in outstanding leg (seat) with a minimum of two fasteners is usually sufficient since theoretically there is no load on the fastener. The beam must have at least $\frac{1}{2}$-in clearance from the column flange, but $\frac{3}{4}$ in is frequently used. This leaves a design bearing length of $3\frac{1}{2}$ in. If this is not sufficient to prevent web crippling, a longer outstanding leg must be used. The seat angle must be thick enough to avoid failure in bending since it acts as a cantilever beam. The critical section for bending is usually assumed to be at the edge of the fillet, which is approximately $t + 3/8$ in from the back of the vertical leg (Sketch 13.7). The beam reaction R is assumed to be concentrated at the center of the theoretical required bearing length (Sec. 11.6).

$$e_R = \frac{1}{2} + \frac{N}{2} - t - \frac{3}{8} \qquad \text{and} \qquad M = Re_R$$

It is not unusual to increase the allowable bending stress to $0.75F_y$ in order to increase the flexibility of the seat.

The connected leg of the beam seat must be large enough to accommodate the number of fasteners required to transfer the beam reaction to the supporting member. Fasteners are designed for direct shearing force only. Usually, one, two, or three lines of fasteners in two or three rows are used. Figure 3.10 illustrates several of the standard types shown in the AISC Manual [1], which also gives the corresponding safe loads. Types D, E, and F, with three rows, cannot be used on the flanges of columns because the center row would coincide with the web. Designs other than those shown are possible.

In welded seat connections the length of the connected leg (vertical) is determined by the required length of weld. A length can be assumed and the

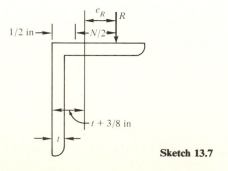

Sketch 13.7

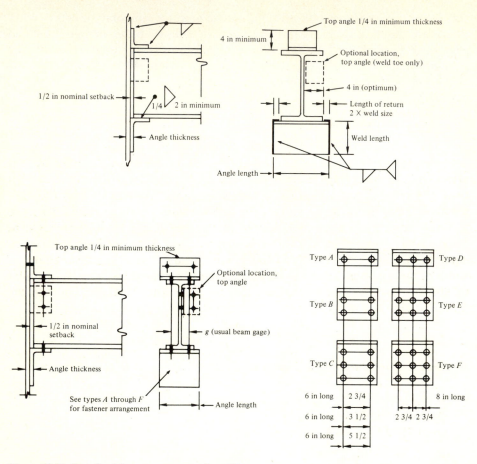

Figure 13.10 Details of beam seat connections [1].

necessary weld size for the depth determined. After one trial, another may have to be selected to give a reasonable weld size.

There are differing opinions concerning the stress variation in the vertical welds. A simple assumption is normally used, i.e., that the neutral axis occurs at middepth.

Stiffened beam seats often consist of structural tees or two plates welded into the shape of a tee. As in the case of unstiffened beam seats, a stiffened seat must provide sufficient bearing length to avoid web crippling. As the beam rotates on the stiffened seat, the location of the beam reaction moves out toward the end of the seat. A reasonable assumption has the reaction centered a distance from the outer edge of the seat equal to one-half of the theoretical bearing length required for web crippling (Sketch 13.8).

The stems of beam seats are in little danger of buckling. The usual practice is to make the stem thickness at least equal to the web thickness of the beam.

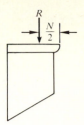

Sketch 13.8

The length of the stem is determined by the length of weld required for attachment to the supporting member. The width of the seat should be at least twice the weld size wider than the beam flange. Since the vertical welds are close together, they provide little resistance to lateral twisting. It is therefore desirable to weld horizontally along the bottom of the seat flange a distance equal to one-fourth to one-half the vertical weld length. Beam seats usually have a width equal to the 6- or 8-in length associated with standard beam seats or the flange width of the columns. The analysis and design of stiffened beam seats is similar to that of rigid brackets (Examples 13.3 to 13.5). The AISC Manual has the tables for selecting connections of these types.

Example 13.9 Design an unstiffened beam seat for a W12 × 40 beam with a reaction of 30 kips and a W8 × 35 column, both of A36 steel. Use $\frac{3}{4}$-in A325 bolts in a bearing-type connection with threads in the shear plane.

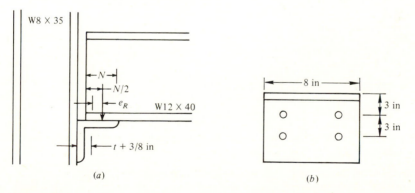

Figure E13.9

SOLUTION See Fig. E 13.9a.

W8 × 35: $t_f = 0.560$ in

W12 × 40: $t_w = 0.295$ in $\quad k = 1\frac{1}{4}$ in

From Eq. (11.18) the end bearing length n required to avoid web crippling equals

$$N = \frac{R}{(0.75\, F_y)t_w} - k = \frac{30}{(0.75)(36)(0.295)} - 1.25 = 3.77 - 1.25 = 2.52 \text{ in}$$

This is less than the $3\frac{1}{4}$-in bearing length available with a 4-in angle.

The eccentricity of the beam reaction relative to the critical section of the seat leg is

$$e_R = \frac{1}{2} + \frac{N}{2} - t - \frac{3}{8} = \frac{1}{2} + \frac{2.52}{2} - t - \frac{3}{8} = 1.38 - t$$

Assume a seat width of 8 in which is equal to the column flange width. The required thickness of the seat leg is determined on the basis of the flexural formula (11.13)

$$f_b = \frac{Mc}{I}$$

$$0.75 F_y = \frac{R e_R t/2}{b t^3/12}$$

$$(0.75)(36) = \frac{(1.38 - t)(30)}{8 t^2/6}$$

$$36 t^2 = 41.4 - 30 t$$

$$t^2 + 0.83 t = 1.15$$

Solve for t by completing the square

$$t^2 + 0.83 t + \left(\frac{0.83}{2}\right)^2 = 1.15 + \left(\frac{0.83}{2}\right)^2$$

$$(t + 0.41)^2 = 1.32$$

$$t + 0.41 = 1.15$$

$$t = 0.74 \text{ in} \qquad \text{use a } \tfrac{3}{4}\text{-in thick angle}$$

The bolts connecting the connected leg of the seat angle to the column act in single shear. The number of bolts required equals

$$n = \frac{30}{(21)(\pi/4)(3/4)^2} = 3.24 \qquad \text{use 4 bolts}$$

The bolts are not considered to be loaded eccentrically. Use an 8- by 4- by $\tfrac{3}{4}$-in seat angle with four $\tfrac{3}{4}$-in A325N bolts in a pattern similar to type B (Figs. 13.10 and E13.9*b*).

The design of a bolted stiffened beam seat was covered in Sec. 13.4 dealing with eccentric loads outside of the plane of shear (see Examples 13.3 and 13.4).

Example 13.10 Design a welded unstiffened seat for a W16 × 36 beam with a reaction of 20 kips on a W10 × 49 column. Both members are A36 steel. Use E60 electrode and AISC specification.

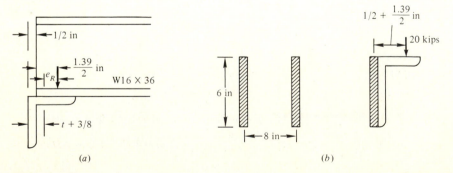

(a)   *(b)*

Figure E13.10

SOLUTION For W16 × 36

$$t_w = 0.295 \text{ in} \qquad b_f = 6.99 \text{ in} \qquad k = 1\tfrac{1}{8} \text{ in}$$

The required length of bearing is

$$N = \frac{R}{0.75 F_y t_w} - k = \frac{20}{(0.75)(36)(0.295)} - 1.125 = 2.51 - 1.12 = 1.39 \text{ in}$$

Use a 4-in horizontal leg (seat). Assume the width of seat angle to be equal to the width of supporting column flange, 8 in (Fig. E13.10a). Then

$$e_R = \frac{1.39}{2} + 0.50 - t - \frac{3}{8} = 0.82 - t$$

The required thickness of the seat leg is determined on the basis of the flexural formula (11.13)

$$f_b = \frac{Mc}{I}$$

$$0.75 F_y = \frac{R e_R t / 2}{b t^3 / 12}$$

$$(0.75)(36) = \frac{(20)(0.82 - t)}{8 t^2 / 6}$$

$$36 t^2 = 16.4 - 20t$$

$$t^2 + 0.55t = 0.456$$

Solve for t by completing the squares

$$t^2 + 0.55t + \left(\frac{0.55}{2}\right)^2 = 0.456 + \left(\frac{0.55}{2}\right)^2$$

$$(t + 0.275)^2 = 0.532$$

$$t + 0.275 = \pm 0.729$$

$$t = 0.729 - 0.275 = 0.454 \text{ in}$$

Use $\tfrac{1}{2}$-in angle thickness.

Assume that the welds are 6 in deep (6-in connected leg) (Fig. E13.10b); then the moments on each weld are

$$\frac{20}{2}\left(\frac{1}{2} + \frac{1.39}{2}\right) = 12 \text{ in} \cdot \text{kips}$$

Consider a 1-in weld. The moment of inertia of weld about the centroidal axis is

$$I = \frac{(1)(6)^3}{12} = 18 \text{ in}^4 / \text{in weld}$$

$$f_h = \frac{Mc}{I} = \frac{(12)(3)}{18} = 2.0 \text{ kips/in}$$

$$f_v = \frac{R}{L} = \frac{20}{12} = 1.67 \text{ kips/in}$$

$$f_r = \sqrt{(2.0)^2 + (1.67)^2} = 2.6 \text{ kips/in}$$

$$\text{Weld size required} = \frac{2.6}{\text{shear strength of 1-in fillet weld}}$$

$$a = \frac{2.6}{(1)(0.707)(0.3)(60)} = 0.204 \text{ in} \qquad \text{use } \tfrac{1}{4}\text{-in weld}$$

Use a 6- by 4- by $\tfrac{1}{2}$-in seat angle with 6 in of $\tfrac{1}{4}$-in E60 weld on both sides of 6-in leg.

The design of a welded stiffened beam seat was covered in Sec. 13.4 dealing with eccentric loads outside of the plane of shear (see Example 13.5).

13.6 MOMENT-RESISTING CONNECTIONS

The main reason for using moment-resisting connections in building is to accommodate lateral forces. When they are used, gravity loads will induce negative moments at the ends of beams. The behavior of moment-resisting connections is complex. Assumptions are made to facilitate their design. For example, the web angles are assumed to carry the shear force, and the upper and lower flange connections to carry the moment.

It is difficult to develop moment-rotation curves by analytical methods because of the highly indeterminate interaction of the components. Only tests will give realistic data. A typical set of data for the common connections using fasteners is given in Fig. 13.11.

By combining web angles with a seat and a top angle it is possible to develop a connection that has greater moment resistance than any of the previously discussed flexible connections. This type of connection is termed *semirigid*. Structural tees used in place of the top and bottom angles greatly increase the rigidity of the connection. Moment-resisting connections that provide practically 100 percent restraint are illustrated in Fig. 13.12. A relatively simple welded connection is to butt-weld the beam to the column flanges on one end and to connect the beam on the other end by a stiffened seat and top plate. Usually the width of the top plate is narrower than the width of the beam flange so that the weld can be placed on top of the flange. The bottom plate is usually wider than the flange to permit placing the weld on the top of the plate and to

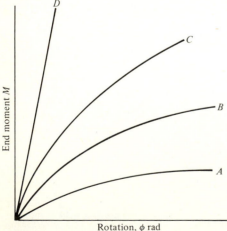

Figure 13.11 Moment-rotation curves.

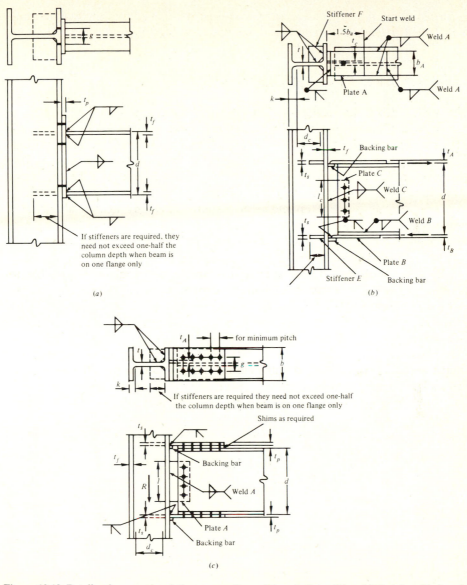

Figure 13.12 Details of moment-resisting connections: (*a*) end plate, (*b*) welded, (*c*) welded and bolted [1].

avoid overhead welding. AISC specifications may require stiffeners for the column flange (Fig. 13.12). It should be continuous if beams are connected to both flanges.

Example 13.11 Design a moment-resisting connection of the type shown in Fig. E13.11*a* to resist a shearing force of 50 kips and a bending moment of 80 ft·kips. Use A36 steel, $\frac{3}{4}$-in A325 bolts, and AISC specifications.

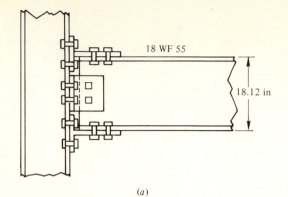

(a)

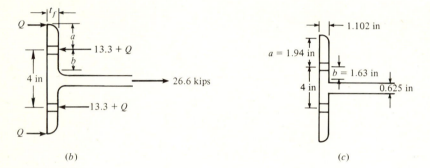

(b)   (c)

Figure E13.11

SOLUTION Assume two angles 4 by 4 by $\frac{1}{4}$ in are used in the framing connection for shear (see Example 13.6 for design). The force on the tee due to the moment is

$$F_T = \frac{M}{d} = \frac{(80)(12)}{18.12} = 53 \text{ kips}$$

The tensile load-carrying capacity of the bolt (see Table 13.1) is

$$F_t = 44\frac{\pi}{4}\left(\frac{3}{4}\right)^2 = 19.4 \text{ kips}$$

The number of bolts required per tee is

$$\frac{F_T}{F_t} = \frac{53.0}{19.4} = 2.7 \qquad \text{use 4 bolts}$$

The number of bolts required to transfer the load from the flange to the tee is

$$\frac{F_T}{\text{single-shear capacity of bolt (friction connected)}} = \frac{53}{(17.5)(\pi/4)(3/4)^2} = 6.9$$

Use 8 (2 rows of 4). Another possibility is to use two rows of two $1\frac{1}{8}$-in A325 bolts.

Selection of structural tee (Fig. E13.11b) The length of the tee would be 8 in, and

$$F = \frac{53}{4} = 13.3 \text{ kips}$$

Assume

$$Q = 0.5F \approx 6.7 \text{ kips}$$

$$b \approx \frac{4 - 0.5}{2} = 1.75$$

$$a \approx \frac{1.75}{2}$$

$$M_1 \approx 6.7 \frac{1.75}{2} = 5.9 \text{ in} \cdot \text{kips}$$

$$M_2 \approx (13.3 + 6.7)(1.75) - 6.7\left(\frac{1.75}{2} + 1.75\right) = 17.5 \text{ in} \cdot \text{kips}$$

$$F_b = \frac{M}{S} \qquad S = \frac{M}{F_b} = \frac{bt^2}{6}$$

$$t = \sqrt{\frac{6M}{4F_b}} = \sqrt{\frac{(6)(17.5)}{(4)(0.75)(36)}} = 0.99 \text{ in}$$

The web thickness for the tee is

$$F_t = (0.6)(36) = 22 \text{ kips/in}^2$$

$$8t_w(22) = 59.6 \text{ kips}$$

$$t_w = 0.34 \text{ in}$$

Select the tee section on basis of

$$t_f = 0.99 \text{ in} \qquad t_w = 0.34 \text{ in}$$

Try ST12 $\times$ 52.95 (Fig. E13.11c):

$$t_f = 1.102 \text{ in} \qquad b_f = 7.875 \text{ in} \qquad t_w = 0.625 \text{ in} \qquad d = 12.00 \text{ in}$$

$$w = 4 \text{ in}$$

$$b = 2 - \frac{0.625}{2} - \frac{1}{16} = 1.63 \text{ in}$$

$$a = \frac{7.875 - 4}{2} = 1.94 < 2t_f \text{ in}$$

$$F \text{ (one bolt)} = \frac{53.0}{4} = 13.3 \text{ kips}$$

Check the prying force Q;

$$Q = F\frac{100b(d_b)^2 - 18w(t_f)^2}{70a(d_b)^2 - 21w(t_f)^2} = 13.3\frac{(100)(1.63)(3/4)^2 - (18)(4)(1.102)^2}{(70)(1.94)(3/4)^2 - (21)(4)(1.102)^2}$$

$$= 13.3\frac{91.8 - 87.4}{76.5 - 101.6} = 0.3 \text{ kip}$$

The bolt is adequate in tension since

$$13.3 + 0.4 = 13.7 < 19.4 \text{ kips}$$

Check bending stresses in the flange:

$$M_1 = Qa = (0.3)(1.94)$$

$$M_2 = (F + Q)b - Q(a + b) = (13.3 + 0.3)(1.63) - (0.3)(1.63 + 1.94)$$

$$= 22.2 - 1.1 = 21.1 \text{ in} \cdot \text{kips} \qquad \text{controls}$$

$$f_b = \frac{Mc}{I} = \frac{(21.1)(0.55)}{(4)(1.102)^3/12} = 26.0 < 27.0 \text{ kips/in}^2 \qquad \text{OK}$$

Use an ST12 $\times$ 52.95 12 in long.

Example 13.12 Design a welded moment-resisting connection of the type shown in Fig. E13.12. The beam is a W18 × 40, and the column is a W10 × 54. The beam has an end reaction of 25 kips and an end moment of 80 ft·kips. All steel is A36. Use E60 electrodes.

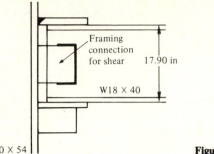

Figure E13.12

SOLUTION Assume that two 4- by 4- by $\frac{1}{4}$-in framing angles have been selected for shear (see Example 13.7 for design). The horizontal force that must be transmitted by the flange connection is

$$F = \frac{(80)(12)}{17.90} = 53.6 \text{ kips} \qquad F_t = 0.6F_y = 22 \text{ kips/in}^2$$

The required area of the top and bottom plates is

$$A_p = \frac{53.6}{22} = 2.43 \text{ in}^2$$

Since the beam flange is 6 in wide, select a plate width of $4\frac{1}{2}$ in. The required thickness of the plate is

$$\frac{2.43}{4.5} = 0.54 \text{ in} \qquad \text{use } \tfrac{5}{8}\text{-in plate}$$

$$L = \frac{53.6}{\text{shear strength of } \frac{1}{4}\text{-in weld}}$$

$$= \frac{53.6}{\left(\frac{1}{4}\right)(0.707)(0.30)(60)} = 16.8 \text{ in}$$

Distribute this distance as follows: 4.5 in on one end and 6.5 in along each side of the plate. Select a $7\frac{1}{2}$-in-wide plate for the bottom flange to permit welding between the edge of the flange and the plate. The plate thickness required is

$$t = \frac{A_p}{7.5} = \frac{2.43}{7.5} = 0.324 \text{ in} \qquad \text{use } \tfrac{3}{8} \text{ in}$$

A complete-penetration single-groove weld will be adequate between the plate and column flange

$$(7.5)\left(\tfrac{3}{8}\right)(22) = 61.9 > 53.6 \text{ kips}$$

With a $\frac{1}{4}$-in weld, the total length required is the same as for the top plate. Use $9\frac{1}{2}$-in length of weld on each side of plate. Column stiffeners would have to be selected according to AISC specifications.

13.7 CONNECTIONS FOR COMPRESSION MEMBERS

The column splice and baseplate involve important compression-member connections. The column splice is necessary because the total length of a column often runs the height of the building, while column lengths are usually only two or three stories. The baseplate connection transmits the column load to the foundation. The area of the baseplate must be such that the concrete footing is not overstressed. Both connections may have to transmit axial load, shear, and moment. Figures 13.13 and 13.14 show details of splices and baseplates for steel columns. Figure 13.15 illustrates baseplate connections for timber columns.

13.7.1 Column Splices

A column splice must be able to transfer the loads and moments that occur in a column at the location of the splice. Splice plates are attached to the column flanges by bolts or welds which act in shear to transmit the load from the column flange to the splice and back again. The splice plates can be designed to carry the entire load. In this case the ends of the columns do not have to be especially prepared. The other extreme would be to use the splice plates for alignment and to transfer the total load directly from one column to the other by milled surfaces. Design procedures normally assume that 50 percent of the total load is transmitted by bearing and that the remainder is carried by the splice plates. If bending is involved, as much as 75 percent of the total load may have to be carried by the splice plates.

Figure 13.13*a* shows splices with columns having the same nominal depth. W sections of the same series have the same inside distance between flanges although their total depth may vary greatly. When this is the case, filler plates are required (Fig. 13.13*b*). When column sizes differ so greatly that large portions of the column areas do not match, a bearing plate is required. The filler plates could be thick, requiring extra rows of bolts (Fig. 13.13*c*). Welded splices are similar to bolted ones (Fig. 13.13). Erection bolts are usually required.

13.7.2 Column Baseplates

Column baseplates can be classified as those transmitting axial load only and those transmitting bending moment and axial load (Fig. 13.14). For small columns the baseplates can be shop-welded to the column. In this case provision must be made to have uniform bearing under the plate in its final position. For large columns the baseplates are grouted into position before the column is erected. Anchor bolts required to secure the column are embedded in the concrete footing and project through the baseplate to engage lug angles shop-welded to the flange or web.

The design of a baseplate involves providing a plate large enough to transmit the column load to the footing without exceeding the allowable bearing stress of the concrete. The plate must also be thick enough to keep the bending stresses within allowable limits. The bending stress is determined by

$$f = \frac{Mc}{I} = \frac{M}{bt^2/6}$$

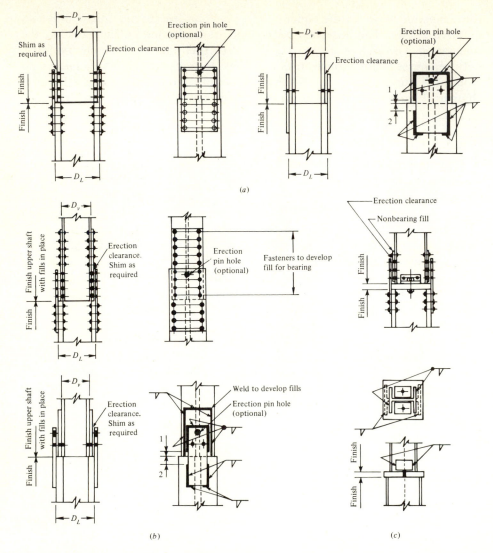

Figure 13.13 Steel-column splice plates: (a) D_U and D_L normally the same, (b) D_U nominally 2 in less than D_L, (c) use of butt plate [1].

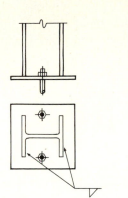

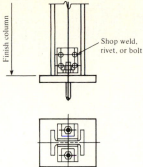

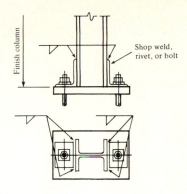

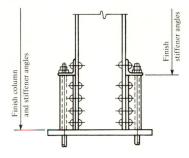

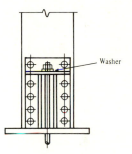

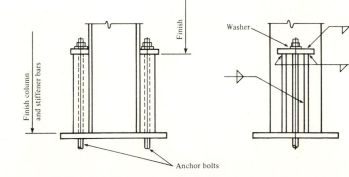

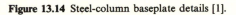

Figure 13.14 Steel-column baseplate details [1].

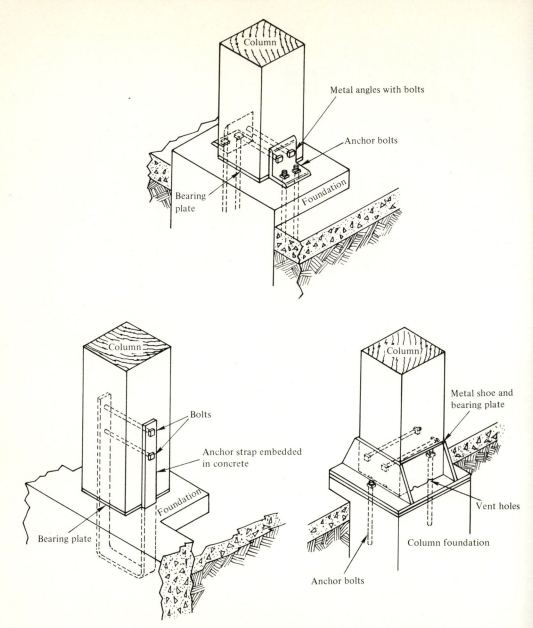

Figure 13.15 Timber-column baseplate details [1].

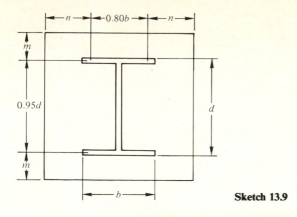

Sketch 13.9

where M is the bending moment in the plate. The actual value of M is indeterminate, primarily because the pressure distributions between the column and plate and footing are not uniform. Two approximate methods are in use. One assumes uniform pressure distribution on all contact surfaces. The other assumes uniform pressure distribution on an assumed reduced contact area between column and plate. AISC recommends that this area be defined by $b = 0.80b$ and $d = 0.95d$ and that the bending moments be distributed uniformly across the width of the plate (Sketch 13.9).

The pressure acting upon the bearing plate is

$$F_p = \frac{P}{A}$$

The following moment expressions apply to the critical sections for a 1-in width of plate

$$M = F_p n \frac{n}{2} = F_p \frac{n^2}{2} \qquad M = F_p m \frac{m}{2} = F_p \frac{m^2}{2}$$

According to AISC, F_b can equal $0.75F_y$.

For column bases transmitting moment, angles or pipes anchored in the footing may be used provided they can resist uplift. When moments become large, more complicated systems are required (Fig. 13.14).

Example 13.13 Design a column baseplate of A36 steel to support a W12 × 58 column with a load of 350 kips. The concrete in the footing has a compressive strength $f_c' = 3000 \text{ lb/in}^2$.

SOLUTION The allowable bearing stress of the concrete is

$$0.385f_c' = (0.385)(3000) = 1155 \text{ lb/in}^2$$

$$\text{Area required} = \frac{P}{1155} = \frac{350,000}{1155} = 303 \text{ in}^2$$

Use 17 × 18 plate (Fig. E13.13); then

$$\text{Area} = 306 \text{ in}^2 \qquad F_p = \frac{P}{A} = \frac{350,000}{306} = 1144 \text{ lb/in}^2$$

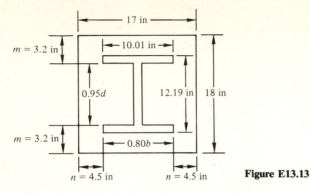

Figure E13.13

The distance $n = 4.5$ in is critical:

$$t = \frac{(3)(1144)(4.5)^2}{27,000} = 1.60 \text{ in}$$

Use 17- by 18- by $1\frac{5}{8}$-in plate.

13.8 SUMMARY

AISC specifications recognize three types of steel construction according to how members are connected to each other:

Type 1, rigid framing
Type 2, simple framing
Type 3, semirigid framing

Another means of classifying eccentric connections depends upon whether the load is inside or outside the plane of connector shears. If the loads are eccentric but inside the plane of shears, the resultant load on the connector is a combination of shear forces. When the load is outside the plane of shears, the resultant force on the connector involves a shear force and a tensile force. This type of connection uses bolts with and without prestress as well as welds. The simple connections involving framed connections, shear plates, and beam seats all come under the category of connections with loads outside the plane of shears.

Moment-resisting connections also come under the category of connections with the loads outside the plane of shears.

PROBLEMS

13.1 A member is connected to a gusset plate by three rivets (Fig. P13.1). Compute the load in each rivet when P is 15 kips.

13.2 Find the resultant load on each bolt of the connection in Fig. P13.2. Show the direction and magnitude of the resultant on a sketch.

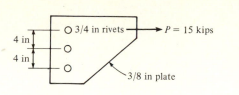

Figure P13.1

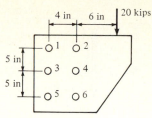

Figure P13.2

13.3 A 4- by 4- by $\frac{3}{8}$-in angle is bolted to the flange of a W14 × 87 column (Fig. P13.3). Both members are A36 steel. Determine which bolt is critical and what size is necessary for the 8.0-kip load. Use A325 bolts. Consider both friction- and bearing-type connections.

13.4 For the joint in Fig. P13.4, what is the shear per inch of weld at the most highly stressed point?

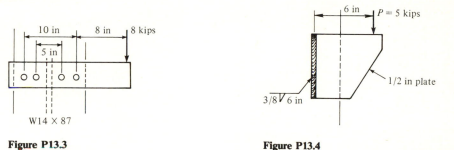

Figure P13.3

Figure P13.4

13.5 The beam in Fig. P13.5 is to be supported by a bracket welded to the near side of the column. Design the bracket and the welded connection required. Use AISC specifications and A36 steel.

13.6 Design an unstiffened seat of the type shown in Fig. P13.6 supporting a W14 × 34 having a reaction of 10 kips. Design both a welded connection and one using $\frac{3}{4}$-in A325F bolts. Use AISC specifications and A36 steel.

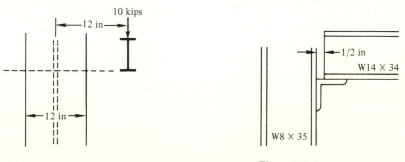

Figure P13.5

Figure P13.6

13.7 Design a welded connection for the beams in Fig. P13.7. The two beams are to be made continuous. The end moment of each 14-in beam is 35 ft·kips, and the end reaction is 15 kips. Use AISC specifications.

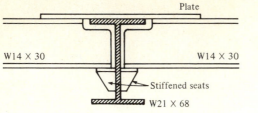

Plate

W14 × 30   W14 × 30

Stiffened seats

W21 × 68   **Figure P13.7**

13.8 A W16 × 36 beam of A36 steel has a reaction of 18 kips and an end moment of 25 ft·kips. Design a bolted connection of the type shown in Fig. P13.8. Use AISC specifications and assume that the moment and reaction act separately.

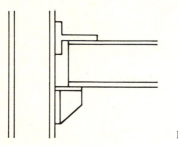

Figure P13.8

13.9 Design a framing angle connection between a beam (W12 × 40) and the flange of a column (W10 × 53), both of A36 steel. The beam reaction is 40 kips. Design the connection using:

    (*a*) $\frac{3}{4}$-in A325N bolts
    (*b*) $\frac{7}{8}$-in A325F bolts
    (*c*) E70 electrode in a welded connection

13.10 Design an end shear-plate connection to field-bolt a W12 × 45 beam to the flange of a W10 × 45 column, both of A36 steel. The beam reaction is 24 kips. Use A307 standard bolts for the field connection and E70 electrode for welding the plate to the beam.

13.11 A seat angle welded to a column uses a 5- by 3- by $\frac{1}{2}$-in angle with the long leg down. Assuming that a top angle does not transmit any load to the column, determine the size of weld required to support a beam reaction of 24 kips (AISC specifications).

13.12 A connection has two $\frac{1}{2}$-in and two $\frac{3}{4}$-in A502 grade 2 rivets in a bearing-type connection (Fig. P13.12). It transmits a load of 12 kips. Calculate the load carried by rivets *A* and *B*.

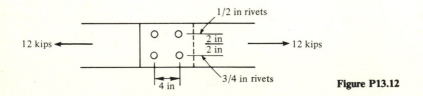

1/2 in rivets

12 kips ←   ○ ○   → 12 kips

2 in
2 in

4 in   3/4 in rivets   **Figure P13.12**

13.13 A framing-angle connection uses $3\frac{1}{2}$- by $2\frac{1}{2}$- by $\frac{3}{8}$-in angles (Fig. P13.13). The beam reaction is 24 kips. Determine the size and arrangement of welds required.

13.14 A $\frac{1}{2}$-in steel plate connected to a $\frac{3}{4}$-in plate with six A325 bolts of the same size carries a load of 15 kips (Fig. P13.14). Consider a bearing-type connection. Which bolt is stressed most? Determine its load and the size of the bolt.

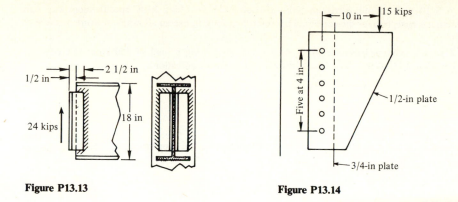

Figure P13.13

Figure P13.14

13.15 A plate is connected to a column with four A502 grade 2 rivets as shown in Fig. P13.15. It carries a load of 24 kips. Calculate the size of rivet required.

13.16 A W18 × 50 beam is connected to a column by two structural tees WT12 × 38 to transmit a moment M (Fig. P13.16). Compute the maximum allowable moment as controlled by (a) $\frac{7}{8}$-in A325 bolts and (b) the structural tees.

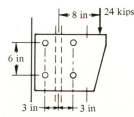

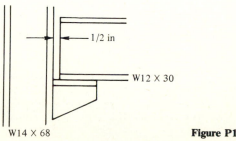

Figure P13.15

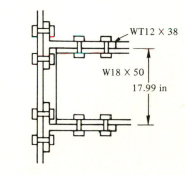

Figure P13.16

13.17 A W12 × 30 beam is connected to the flanges of a W14 × 68 column as shown in Fig. P13.17. The reaction from the beam on the seat is 28 kips. Design a stiffened seat using (a) welded connection and (b) $\frac{7}{8}$-in A325 bolts in bearing-type connection.

Figure P13.17

13.18 A W10 × 39 column carries an axial load of 150 kips. Using AISC specifications and A36 steel, design a baseplate for the column. The supporting footing has an allowable bearing pressure of 1000 lb/in^2.

13.19 Design a column baseplate for a W14 × 145 supporting a load of 750 kips. The concrete footing has an f_c' of 4000 lb/in^2. Use A36 steel and AISC specifications.

REFERENCES

1. American Institute of Steel Construction, "Manual of Steel Construction," 8th ed., Chicago, 1978.

CHAPTER 14

ELEMENTS OF STEEL BUILDING DESIGN

14.1 INTRODUCTION

The material in this chapter pertains to the design of steel buildings of varying heights. It has application to apartment buildings, warehouses, schools, and institutional and office buildings. The distinguishing characteristic of multistory buildings is the influence of wind loads. In general wind loads become significant when the building height becomes greater than twice its least lateral dimension. One way of classifying buildings is as single story, multistory, and special buildings using long-span structures. Another classification system involves their type of construction.

14.2 TYPES OF CONSTRUCTION

These types of construction include bearing-wall, steel-skeleton, and long-span construction. More than one type may be used in a given building.

14.2.1 Bearing-Wall Construction

This type of construction is very common in single-story commercial buildings. The ends of beams, joists, or trusses are supported by the walls, which transfer the loads to the foundations. Bearing plates are normally required to transfer the load from the beam or truss to the masonry wall. Reinforced-concrete masonry with increased load-carrying capacity is finding application in multistoried buildings. This type of construction eliminates the need for thick, massive wall sections, and clear spans of 35 or 40 ft are possible.

14.2.2 Skeleton Construction

In skeleton construction the loads are transferred to the foundations by a framework of steel beams and columns. The exterior walls, partitions, floor slabs, etc., are all supported by the frame. Lateral loads, such as wind and earthquakes, are withstood by lateral bracing and rigid connections between the beams and columns.

In this type of construction the frame usually consists of columns spaced 20, 25, or 30 ft apart (bay size), with beams and girders framed into them from both directions at each floor level. This arrangement of members is illustrated in Fig. 14.1. Other arrangements of beams and girders may be used with different floor systems. The exterior walls are generally referred to as *curtain walls* or *nonbearing walls* and are carried by exterior *spandrel beams*.

14.2.3 Long-Span Structures

In field houses, auditoriums, theaters, etc., very long spans are required between columns and skeleton construction may not be adequate. In some cases ordinary W sections can be replaced by built-up beams, trusses, arches, cables, or rigid frames. Limitations in depth will also influence the choice of structure. These types of structures, referred to as long-span structures, are illustrated in Fig. 14.2.

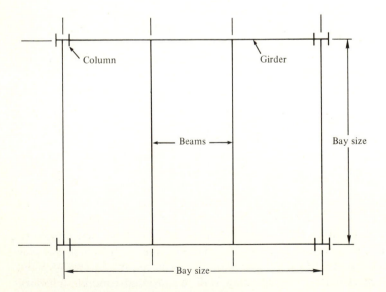

Figure 14.1 Arrangement of beams and columns.

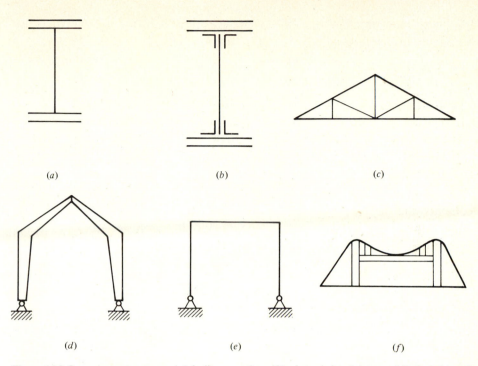

Figure 14.2 Long-span structures: (*a*) built-up section, (*b*) plate girder, (*c*) truss, (*d*) three-hinged arch, (*e*) rigid frame, and (*f*) cable system.

14.3 TYPES OF FLOOR CONSTRUCTION

Concrete floor slabs of one type or another are generally used with steel-framed buildings. Concrete slabs are strong and have excellent fire ratings and good acoustical properties, but they require the use of formwork and reinforcement and involve considerable dead weight. Other factors involved in their selection include live load, sound and heat transmission, needs for conduits, wiring, piping, etc., depth of floor space available, and time for construction. Some of the more common systems are described briefly below.

14.3.1 Concrete Slabs on Beams

Two basic systems are used, one- and two-way reinforced slabs. A *one-way slab* is shown in Sketch 14.1. The span is in the short direction, as indicated by the arrows. The span is normally from 6 to 8 ft with slab thicknesses of 4 in or more, which are heavier than some of the other systems available. One advantage of this system is that the formwork required can be supported on the beams, doing away with the need for vertical shoring. This system is still competitive for heavier loads supplying a rigid durable slab.

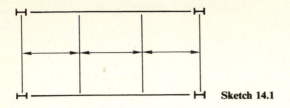

Sketch 14.1

It is possible to take advantage of composite action in the design of one-way systems. This requires bonding the slab to the beam so that they act as a unit in resisting the total load, which the beam would otherwise have to resist on its own. As a result the composite section uses the compressive strength of the concrete and the tensile strength of the steel beam, generally giving a lighter steel section. The bond between the slab and the steel section is normally improved by the use of shear connectors (Sketch 14.2).

The *two-way slab* system is feasible when the slabs are square or nearly so with supporting beams at all four edges. Reinforcement has to be provided in both directions (Sketch 14.3). Other characteristics are similar to those of the one-way reinforced slab.

Steel decking is available for composite construction with concrete slabs. It acts as formwork and becomes an integral part of the slab. Steel decking is available in a wide range of depths and gauges. A $2\frac{1}{2}$-in concrete topping combined with spray-on fire protection of the deck underside can provide the required fire rating. The small construction depth makes this system applicable to residential construction. This type of construction is shown in Sketch 14.4.

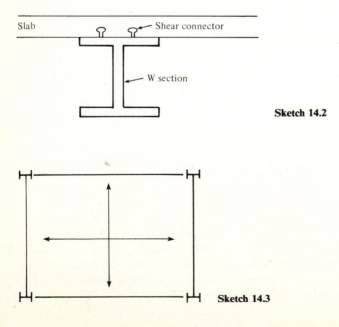

Slab — Shear connector — W section

Sketch 14.2

Sketch 14.3

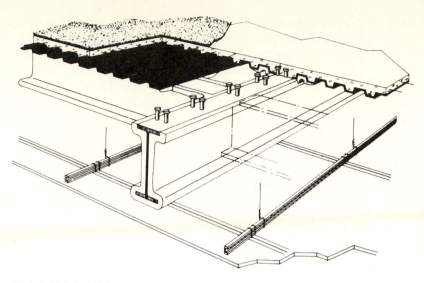

Sketch 14.4 (*Bethlehem Steel Company*.)

14.3.2 Concrete Slabs on Open-Web Steel Joists

This system is very common for small frame buildings. The joists are basically small chord trusses made from bars, angles, and rolled shapes. They are available in varying depths and lengths in regular carbon and high-strength steel. Their spacing normally ranges from 24 in for floors to 30 in for roofs. An example of such a system is given in Sketch 14.5.

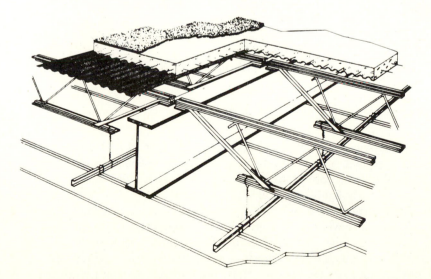

Sketch 14.5 (*Bethlehem Steel Company*.)

Joists must be braced laterally to keep them from buckling and twisting. This bracing, called *bridging*, involves horizontal and cross bracing made of rods. Floors may consist of cast-in-place or precast concrete. Slabs, which should be at least 2 in thick, can be formed not only on sheet decking but also on ribbed metal lath and paper-backed welded-wire fabric capable of supporting the slab at the designated joist spacing. The open spaces in the webs are used for conduits, piping, wiring, and ducts.

Open-web floor and roof systems are well suited for the lighter loads found in schools, apartment houses, office buildings, and other low-level buildings.

14.3.3 Precast Concrete Floors

Lightweight concrete precast units (planks) have had extensive use in roofs and are becoming increasingly popular in floor systems. They are erected quickly and eliminate the use of formwork. These hollow-core precast elements are available in a wide range of thicknesses and varying widths to satisfy fire-protection and long-span requirements. Common sections are illustrated in Sketch 14.6. Concrete grouting is always required, and concrete topping is sometimes required to provide fire rating, as well as a smooth level floor. Floor openings can be preengineered into the units.

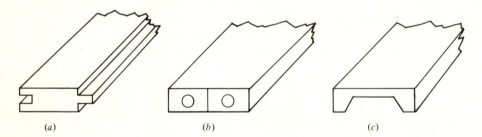

    (a)   (b)   (c)

Sketch 14.6 (*a*) Plank; (*b*) hollow-cored slab; (*c*) channel slab.

14.4 TYPES OF ROOF CONSTRUCTION

The factors normally considered in selecting a specific type of roof construction include strength, weight, span, insulation, acoustics, appearance, and covering required. Since the loads applied to roofs are normally smaller than those on floors, the use of weaker lightweight-aggregate concretes is possible. Insulation is of particular importance and if not inherent in the slab must be added through insulation boards and covering. Among the many lightweight aggregates used, expanded shale, vermiculite, gypsum, and foams are quite common.

Precast units made with these aggregates are light and easily erected and can be sawed and nailed. For cast-in-place slabs the concrete can be pumped into place on open deck joists and steel decking.

14.5 FRAMES

14.5.1 Introduction

Rigid frames are structures with moment-resisting joints which prevent relative rotation of the members when loaded. These structures are also called *bents*. Their advantages are economy and appearance. They accomplish with less waste space what can be done by columns and trusses. Rigid frames have found extensive use in churches, auditoriums, field houses, and other structures requiring large open spaces. They can be economical for spans of 25 to 200 ft. This section is confined to single-span single-story structures. Common frames are illustrated in Fig. 14.3.

14.5.2 Supports

As indicated in Fig. 14.3, theoretically frames can be either hinged or fixed. Practically, the hinge support is more common since it represents an anchor bolt

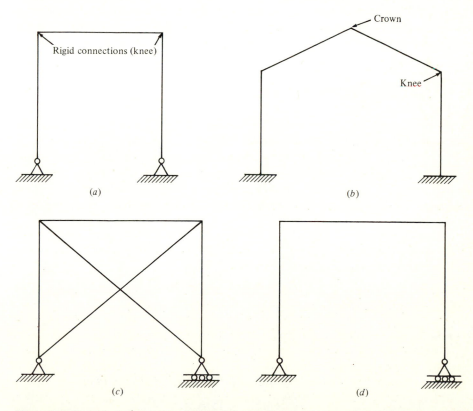

Figure 14.3 Statically indeterminate frames: (*a*) with hinged supports and (*b*) with fixed supports; statically determinate frames: (*c*) braced and (*d*) unbraced.

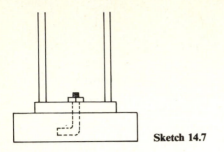

Sketch 14.7

passing through a steel baseplate into a concrete footing. To facilitate hinge action the anchor bolts are put along the neutral axis of the columns (Sketch 14.7). As the spans become longer or as the height-to-span ratio decreases, the horizontal thrust acting on footings become so large that horizontal ties between footings become necessary if they cannot be bedded in rock. These tie rods normally have turnbuckles to permit proper adjustment before concreting into the floor. The tie rods may be prestressed to counter the effects of dead loads.

Fixed supports are difficult to attain in practice. Since the only possibilities involve anchorage in bedrock or extremely large concrete footings, the design of rigid frames with fixed supports is unusual.

14.6 ANALYSIS OF FRAMES

As indicated previously, frames can be divided into a number of categories. The prime distinguishing feature is whether the frame is statically determinate or indeterminate. The braced bent is a special case of the statically determinate frame if the diagonals are only tension-resistant members. The frame becomes a vertical truss. The diagonals come into play when the bent is subject to horizontal loads, such as wind or earthquake forces, or from unsymmetrical vertical loads.

14.6.1 Statically Determinate Frames

The reactions for the statically determinate bent shown in Fig. 14.4 are found by first taking moments about the pinned column which is not on rollers.

$$\overset{\curvearrowright}{\Sigma M_A} = 0: \qquad (10)(12) + (15)(1)\left(\tfrac{15}{2}\right) - 15V_D \qquad \text{Fig. 14.4}b$$

$$V_D = \frac{(10)(12)}{15} + \frac{(15)(1)}{15}\frac{15}{2} = 8 + 7.5 = 15.5 \text{ kips}$$

$$\uparrow \Sigma F_y = 0: \qquad V_A + 15.5 - 15 = 0$$

$$V_A = -0.5 \text{ kips}$$

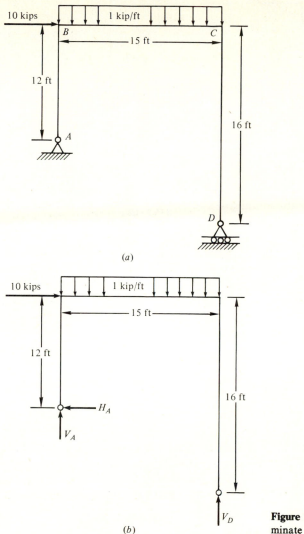

(a)

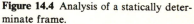

(b)

Figure 14.4 Analysis of a statically determinate frame.

The negative sign means that the force acts in a direction opposite that assumed.

$$\sum \vec{F}_x = 0: \qquad\qquad 10 \text{ kips} - H_A = 0$$
$$H_A = 10 \text{ kips}$$

The shear and bending-moment diagrams can be determined once the reactions are known. These diagrams can be readily developed when the members have been isolated as free-body diagrams, as illustrated in Fig. 14.5a. The shear and bending-moment diagrams are developed in Fig. 14.5b and c.

The value of shear at any point in the members is determined in the same manner as beams. With the horizontal reaction at A acting to the left, the shear

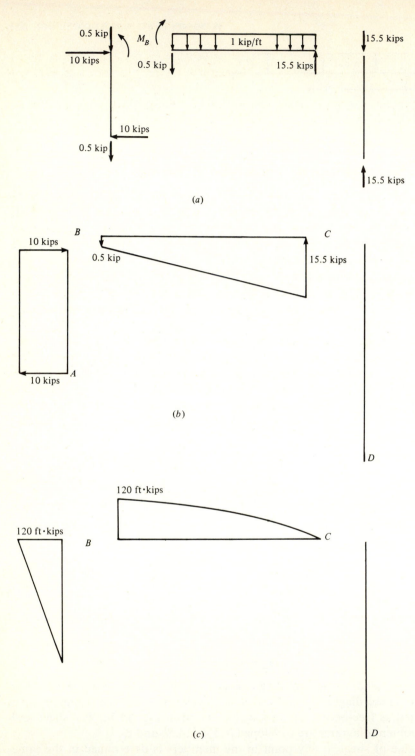

(a)

(b)

(c)

Figure 14.5 (a) Load, (b) shear, and (c) bending-moment diagrams for a determinate frame.

diagram is drawn on the left side of the member. The horizontal member *BC* is treated as a normal horizontal beam. The vertical member *CD* is theoretically a column with axial loads only.

The moment diagrams, determined by using the same method as developed for beams, are plotted on the compression side of the beam.

14.6.2 Single-Story Statically Indeterminate Frames

The bents shown in Fig. 14.3*b* are statically indeterminate. Additional equations beyond those supplied by static equilibrium are necessary to develop the shear and bending-moment diagrams of the members. These equations are based on slope and deflection relationships for exact solutions. It is important to realize that the shear and bending-moment diagrams in statically determinate bents are independent of the cross-sectional properties of the members and the material of which they are made. For statically indeterminate frames, the additional equations required (slope and deflection) involve *EI*, the stiffness of the member (Sec. 11.11).

The shear and moments in a single-story, single-bay bent can be approximated with considerable accuracy for lateral loads by simple assumptions. This is particularly true for bents that have columns of equal length and equal moments of inertia and are pin-ended at their bases. In this case only one assumption is necessary since the bent is statically indeterminate to the first degree. The assumption for this case is that the horizontal shear is equally divided between the column bases.

The analysis of such a bent follows (see Sketch 14.8). The lateral force of 10

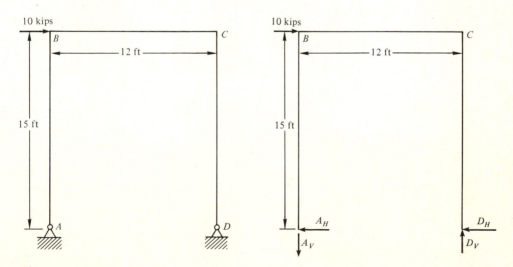

Sketch 14.8

kips is divided between the reactions at A and D

$$A_H = D_H = \frac{10}{2} = 5 \text{ kips}$$

The vertical reaction components can be calculated by summing moments about one of the supports

$\Sigma M_A = 0:$ $\qquad\qquad (10)(15) - 12D_V = 0$

$$D_V = \frac{(10)(15)}{12} = 12.5 \text{ kips}$$

$\uparrow \Sigma V = 0:$ $\qquad\qquad 12.5 - A_V = 0$

$$A_V = 12.5 \text{ kips}$$

The shear and bending-moment diagrams can be developed from the free-body diagrams of the individual members (Fig. 14.6).

With a fixed-end condition at the base of the columns, the frame is statically indeterminate to the third degree: the frame has three more unknown reactions than equations of equilibrium. The first assumption is the same as that used in a pin-ended frame; i.e., the horizontal shear is divided equally between the two column bases. The second and third assumptions are similar because a point of contraflexure† is assumed to occur at midheight of each column. This would be the case for an infinitely stiff beam. Some designers recommend assumed points of contraflexure slightly different from midpoint. If the footing is only partially fixed against rotation, the point of contraflexure moves down.

The fixed-ended bent analyzed in Sketch 14.9 has the same loading and dimensions as the bent in Fig. 14.6. Assuming a point of contraflexure at midheight of the columns provides a condition equivalent to pins at these points. The frame is cut at these points, and the upper portion is considered as a free-body diagram (Sketch 14.10).

† A point where the slope, and consequently the bending moment, changes sign.

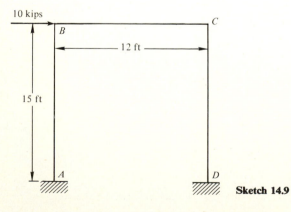

Sketch 14.9

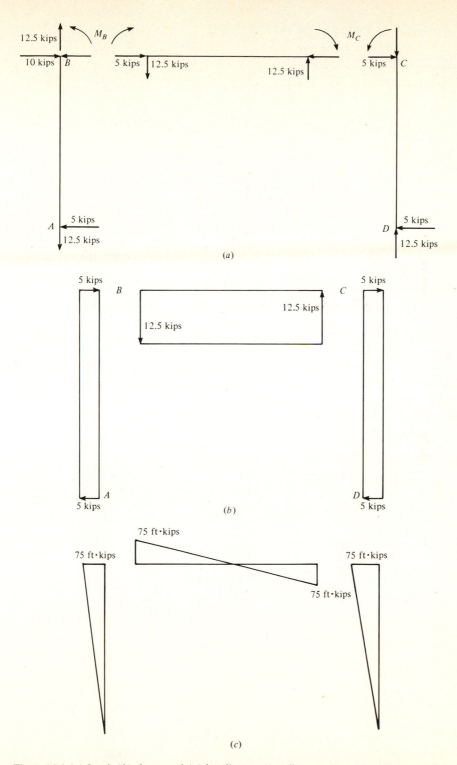

Figure 14.6 (*a*) Load, (*b*) shear, and (*c*) bending-moment diagrams for a pin-ended bent.

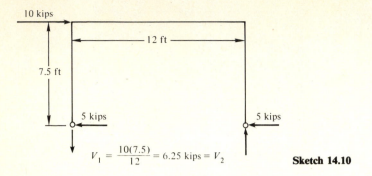

$$V_1 = \frac{10(7.5)}{12} = 6.25 \text{ kips} = V_2$$

Sketch 14.10

The horizontal load of 10 kips is divided equally between the horizontal reaction components at the points of contraflexure. The vertical reaction components (axial load in the columns) can be obtained by taking moments about one of the points and then summing forces in the vertical direction

$$\overset{\frown}{\Sigma M} = 0: \qquad (10)(7.5) - 12V_2 = 0$$

$$V_2 = \frac{(10)(7.5)}{12} = 6.25 \text{ kips}$$

$$\uparrow \Sigma V = 0: \qquad 6.25 - V_1 = 0$$

$$V_1 = 6.25 \text{ kips}$$

The reactions at the base of the columns can be determined by considering the lower half of each column as a free body. The shear and axial loads at the point of contraflexure in the upper portion of the frame act on the lower portion of the column in the opposite direction (Sketch 14.11). By summing moments about the base and then summing forces in the vertical and horizontal directions

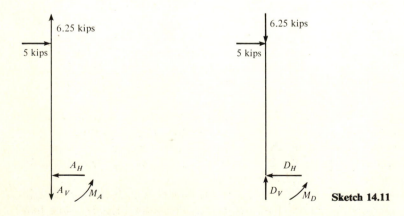

Sketch 14.11

the reactions at A and D are obtained

$$\overset{\curvearrowright}{\Sigma M_A} = 0: \qquad\qquad (5)(7.5) - M_A = 0$$
$$M_A = 37.5 \text{ ft·kips}$$

$$\uparrow \Sigma F_y = 0: \qquad\qquad 6.25 - A_V = 0$$
$$A_V = 6.25 \text{ kips}$$

$$\overset{\rightarrow}{\Sigma F_H} = 0: \qquad\qquad 5 - A_H = 0$$
$$A_H = 5 \text{ kips}$$

In a similar manner the reactions at D can be obtained as

$$D_H = 5 \text{ kips} \qquad D_V = 6.25 \text{ kips} \qquad \text{and} \qquad M_D = 37.5 \text{ ft·kips}$$

The load, shear, and bending-moment diagrams are shown in Fig. 14.7. Once the shear and bending-moment diagrams are available, the members can be designed by the principles developed for beam columns in Sec. 12.9.

14.7 MULTISTORY BUILDINGS

14.7.1 Introduction

Multistory, or high-rise, buildings date back to early historical times. Buildings 10 stories high were in use in early Roman cities. The principle of masonry bearing walls was fundamental to this type of construction. It appeared to reach its limit in the 16-story Monadnock Building in Chicago (1891), which required walls 6 ft thick at their base.

The subsequent development of lightweight frame systems out of steel allowed for greater heights and more and larger openings in the walls. The nine-story second Rand McNally Building in Chicago (1889) used an all-steel frame for the first time. Improved design methods allowed buildings to be made higher. In 1905 the 50-story Metropolitan Tower Building became a reality in New York City, followed by the 102-story Empire State Building in 1931.

In the meantime concrete started to establish itself as a common structural material. Shortly after the turn of the century, the 16-story Ingall Building in Cincinnati became the first reinforced-concrete frame building. The evolvement of new construction techniques and high-quality materials after World War II resulted in such structures as the 65-story Marina City Towers in Chicago (1963).

It was also in this post-World War II era that there was a resurgence in the use of masonry bearing-wall construction. The inherent disadvantage of wall thickness had been overcome by the use of reinforced-masonry construction. Multistoried structures such as the 20-story Park Lane Towers, Denver (1970), with a wall thickness of 10 in, became possible.

The loads acting on a structure can be classified as artificial or natural (geophysical). The former include live loads while the latter include loads that

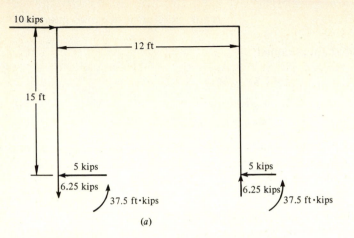

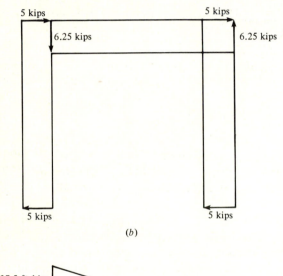

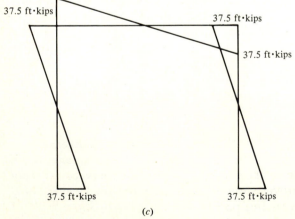

Figure 14.7 (*a*) Load, (*b*) shear, and (*c*) bending-moment diagrams for a fixed-ended bent.

are gravitational (dead), meteorological (wind and snow), and seismological (earthquake). The nature and means of calculating the magnitude of these forces was covered in considerable detail in Chap. 3.

14.7.2 Common Multistory Building Structures

The basic structural elements of buildings are beams and columns, walls, and floor systems. These elements can be combined into numerous building types. A few common types are illustrated in Fig. 14.8 and discussed below.

Parallel bearing walls (Fig. 14.8a) This structure consists of planar vertical elements of reinforced masonry prestressed by their own weight, which facilitates handling lateral loads. The parallel-wall system is particularly advantageous for apartment construction, where large free spaces are not needed and mechanical systems do not require core structures.

Cores and bearing walls (Fig. 14.8b) This structure involves vertical planar elements in the exterior walls around an interior core, which houses the mechanical and vertical transportation systems. The size of the open interior spaces depends upon the span capabilities of the floor system.

Flat slab (Fig. 14.8c) The horizontal planar system consists of concrete floor slabs supported on columns. If it is a flat-slab system, it does not use drop panels at the column-floor interface. This system minimizes story height.

Rigid frame (Fig. 14.8d) Rigid joints are used to connect the beams and columns. Both vertical and horizontal systems of columns and beams are normally on a rectangular grid. The rigidity of the resulting skeleton is dependent upon the strength and rigidity of the members and their connections. As a result, story height and column spacing become primary design considerations.

Trussed frame (Fig. 14.8e) This structural system combines a rigid or hinged frame with vertical trusses. The vertical truss may be used to provide the resistance to lateral loads acting on the hinged frame in one direction with a rigid frame performing similarly in the other direction. The frames would also normally resist the gravitational loads. Various combinations of vertical or horizontal trusses can be used in this way.

Tube in tube (Fig. 14.8f) The exterior tube is made up of columns and beams forming a rigid frame which provides only enough space for windows. The entire building acts as a hollow tube cantilevered out of the ground. The interior tube provides a core for mechanical and vertical transportation systems as well as sharing loads with the exterior skeleton.

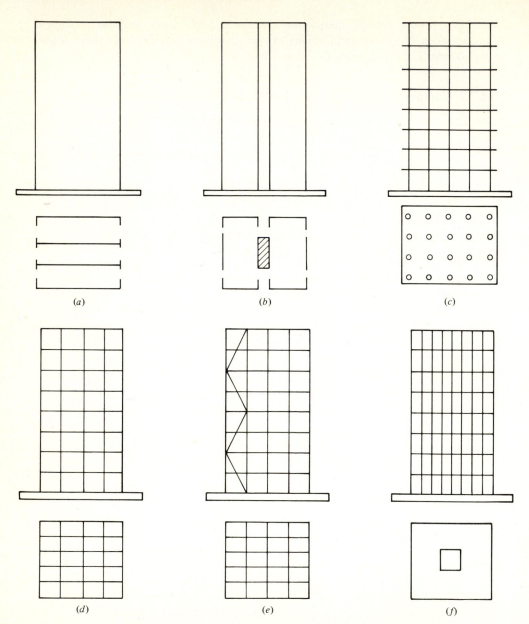

Figure 14.8 Common high-rise structures: (*a*) parallel bearing walls, (*b*) core and bearing walls, (*c*) flat slab, (*d*) rigid frame, (*e*) trussed frame, (*f*) tube in tube.

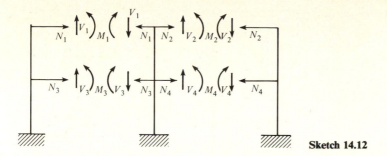

Sketch 14.12

14.8 APPROXIMATE ANALYSIS AND DESIGN OF RIGID FRAMES

A rigid-frame structure is made up of columns and beams rigidly connected together. This continuity of members is necessary for the resistance of lateral forces and asymmetrical vertical forces. Such a frame is highly indeterminate. The degree to which it is indeterminate can be realized by considering Sketch 14.12. Each beam is cut at center span. The resulting structure becomes statically determinate if the unknown values of bending moment, shear, and normal force at the cut section are removed. Each column would be a cantilever. As a result, if there are n girders in the structure, it is necessary to remove or solve for $3n$ redundants to make the structure statically determinate. Such a structure is statically indeterminate to the $3n$th degree.

Even if the approximate method of analysis is not used in the final design, it can be very helpful for a first approximation of member sizes in the early stages of design.

The approximate analysis and design of a rigid frame structure can be considered to involve a number of steps:

1. The overall stability of the building against overturning and sliding
2. The analysis for vertical loads (gravity)
3. The analysis for lateral loads (wind and earthquake)
4. The design of the structure for the critical load combination

14.8.1 Overall Stability

Codes normally require that the overturning moment due to the wind load not exceed 75 percent of the moment of stability resulting from the dead load of the building unless the building is anchored to resist the excess (BOCA building code [2]). These codes also require that the sliding force due to lateral forces not exceed soil friction or soil friction plus lateral resistance provided by anchors.

14.8.2 Analysis for Vertical Loading

A rigid frame deforms under vertical loads as indicated in Sketch 14.13. The deflected shape of a typical girder AB shows two points of contraflexure where

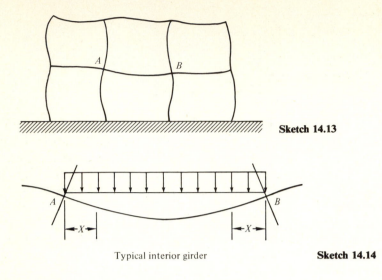

Sketch 14.13

Typical interior girder

Sketch 14.14

moments are zero (Sketch 14.14). If these points are known, the structure becomes statically determinate. An approximate analysis depends upon the selection of these inflection points. An insight into the location of these points can be gained by considering extreme conditions. With fixed-ended beams (full restraint) the inflection points are a distance of $0.21l$ from each support. In a simply supported beam the inflection point moves to the support (Sketch 14.15). In a rigid frame there is partial restraint at the ends of the beams, and the inflection points will be located somewhere between $0.21l$ and $0.00l$. The unpredictability of live loads and their location made it reasonable to assume that the inflection points (zero moment) of the partially restrained girders will be at a distance of $0.10l$ from the supports (Sketch 14.16). The beam can be visualized as two cantilevered beams supporting a simply supported beam (Sketch 14.17).

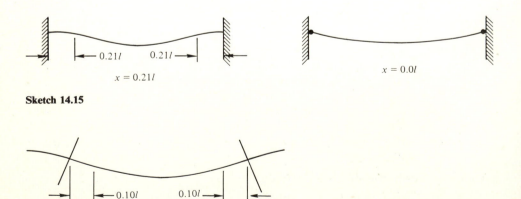

$x = 0.21l$

$x = 0.0l$

Sketch 14.15

Assumed

Sketch 14.16

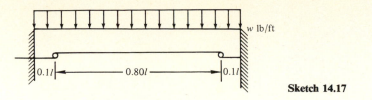

Sketch 14.17

In addition to the assumption relative to the location of the inflection points, the axial force in the girder is considered zero because of the relatively small magnitude. These assumptions provide the three additional equations required to make the beam statically determinate. The maximum moment for the simply supported beam is

$$M = \frac{wl^2}{8} = \frac{w(0.8l)^2}{8} = 0.08wl^2 \tag{14.1}$$

The reaction R of this beam is

$$R = \frac{0.8lw}{2} = 0.4wl$$

The support moment of the cantilevered beam at its connection to the column

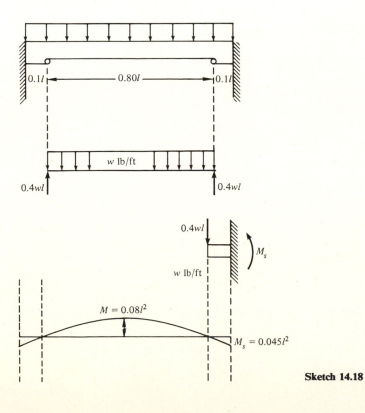

Sketch 14.18

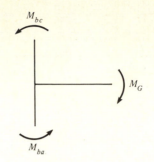

Sketch 14.19   **Sketch 14.20**

becomes (Sketch 14.18)

$$M_s = -\left[0.4wl(0.1l) + \frac{w(0.1l)^2}{2}\right] = 0.045wl^2 \tag{14.2}$$

The support moments must be distributed to the associated columns. Maximum loads for interior columns are obtained by loading the bays on both sides of the column. In general, moments in interior columns may be neglected since the end moments of the girders tend to cancel each other, as shown in Sketch 14.19.

The end column has to carry the girder moment, which is divided between the columns as shown in Sketch 14.20. Rotational equilibrium about joint b yields

$$M_{ba} + M_{bc} = M_G$$

According to slope-deflection equations, the distribution of the girder moment between the columns will be on the basis of their stiffness I/l

$$M_{bc} = M_G\frac{I_{bc}/l_{bc}}{I_{bc}/l_{bc} + I_{ba}/l_{ba}} \tag{14.3}$$

An assumption of relative stiffness values becomes necessary for an initial solution. If necessary, these values can be adjusted on the basis of initial design values.

Example 14.1 Determine the moment and axial forces in the two-story rigid frame shown in Fig. E14.1a. Assume that the column moments of inertia at ground level are 50 percent larger than those of the second floor.

SOLUTION

Axial forces in columns It is assumed that the columns carry one-half of the respective beam loads. At column line A

$$N_{GD} = (1.5)(\tfrac{24}{2}) = 18.0 \text{ kips} \qquad \text{compression}$$

$$N_{DA} = (1.5)(\tfrac{24}{2}) + (2)(\tfrac{24}{2}) = 42.0 \text{ kips} \qquad \text{(comp)}$$

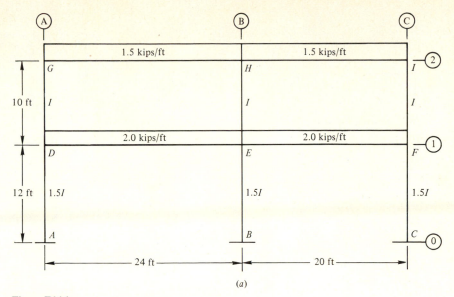

(a)

Figure E14.1

At column line B

$$N_{HE} = (1.5)\left(\tfrac{24}{2}\right) + (1.5)\left(\tfrac{20}{2}\right) = 33.0 \text{ kips} \qquad (\text{comp})$$

$$N_{EB} = 33.0 + (2.0)\left(\tfrac{24}{2}\right) + (2.0)\left(\tfrac{20}{2}\right) = 77.0 \text{ kips} \qquad (\text{comp})$$

At column line C

$$N_{IF} = (1.5)\left(\tfrac{20}{2}\right) = 15.0 \text{ kips} \qquad (\text{comp})$$

$$N_{FC} = 15 + (2.0)\left(\tfrac{20}{2}\right) = 35.0 \text{ kips} \qquad (\text{comp})$$

Beam moments The maximum positive moment at center span (bottom fibers in tension) is

$$M = 0.08wl^2 \tag{14.1}$$

$$M_{GH} = (0.08)(1.5)(24)^2 = 69.1 \text{ ft·kips}$$

$$M_{HI} = (0.08)(1.5)(20)^2 = 48.0 \text{ ft·kips}$$

$$M_{DE} = (0.08)(2.0)(24)^2 = 92.2 \text{ ft·kips}$$

$$M_{EF} = (0.08)(2.0)(20)^2 = 64.0 \text{ ft·kips}$$

Support moments (top fibers in tension)

$$M = 0.045wl^2 \tag{14.2}$$

$$M_{GH} = M_{HG} = (0.045)(1.5)(24)^2 = 38.9 \text{ ft·kips}$$

$$M_{HI} = M_{IH} = (0.045)(1.5)(20)^2 = 27.0 \text{ ft·kips}$$

$$M_{DE} = M_{ED} = (0.045)(20)(24)^2 = 51.8 \text{ ft·kips}$$

$$M_{EF} = M_{FE} = (0.045)(20)(20)^2 = 36.0 \text{ ft·kips}$$

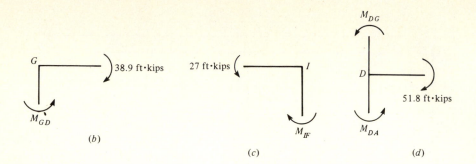

(b)

(c)

(d)

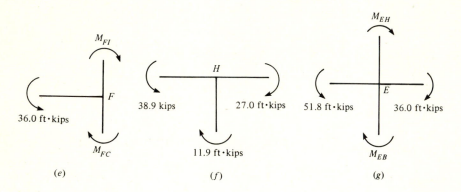

(e)

(f)

(g)

Figure E14.1 (*Cont'd.*)

Column moments The moments have to balance about joints G and I (Fig. E14.1b and c)

$$\widehat{\Sigma M_G} = 0: \qquad\qquad 38.9 - M_{GD} = 0$$

$$M_{GD} = 38.9 \text{ ft·kips}$$

$$\widehat{\Sigma M_I} = 0: \qquad\qquad -27 + M_{IF} = 0$$

$$M_{IF} = 27.0 \text{ ft·kips}$$

Consider the distribution of moments at joint D (Fig. E14.1d). From Eq. (14.3)

$$M_{DG} = 51.8 \frac{I_{DG}/10}{1.5I_{DG}/12 + I_{DG}/10} = 51.8 \frac{0.100}{0.255} = 23.0 \text{ ft·kips}$$

$$M_{DA} = 51.8 \frac{1.5I_{DG}/12}{1.5I_{DG}/12 + I_{DG}/10} = 51.8 \frac{0.125}{0.225} = 28.8 \text{ ft·kips}$$

Check:

$$\widehat{\Sigma M_D} = 0: \qquad\qquad 51.8 - 23.0 - 28.8 = 0$$

Consider joint F (Fig. E14.1e). From Eq. (14.3)

$$M_{FI} = 36 \frac{I_{FI}/10}{1.5I_{FI}/12 + I_{FI}/10} = 36 \frac{0.10}{0.225} = 16.0 \text{ ft·kips}$$

$$M_{FC} = 36 \frac{1.5I_{FI}/12}{1.5I_{FI}/12 + I_{FI}/10} = 36 \frac{0.125}{0.225} = 20 \text{ ft·kips}$$

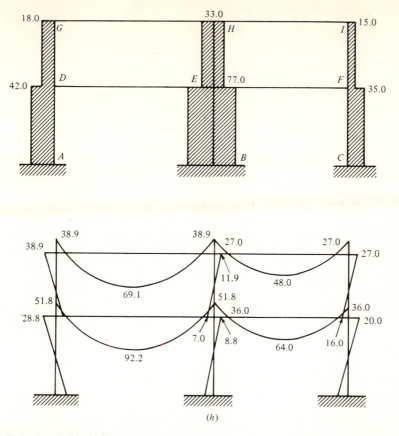

(h)

Figure E14.1 (*Cont'd.*)

Check:

$$\widehat{\Sigma M_F} = 0: \qquad\qquad -36 + 20 + 16 = 0$$

In general, the end girder moments at an interior column tend to cancel. In this case, however, because of unequal bay sizes and constant loading, there is an unbalanced moment of $38.9 - 27 = 11.9$ ft·kips at joint H. This moment must be carried by column EH (Fig. E14.1f).

For joint E, the unbalanced girder moment of $51.8 - 36 = 15.8$ ft must be carried by columns EH and EB on the basis of their relative stiffness (Fig. E14.1g)

$$M_{EH} = 15.8 \frac{0.100}{0.225} = 7.0 \text{ ft·kips}$$

$$M_{EB} = 15.8 \frac{0.125}{0.225} = 8.8 \text{ ft·kips}$$

The resulting force diagrams are developed in Fig. 14.1h. Another convention used in plotting bending moments is to plot the positive bending moments on the tension side of members.

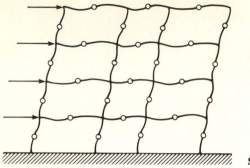

Sketch 14.21

14.8.3 Analysis of Lateral Loading

A rigid frame responds quite differently to lateral loads than to vertical loads. As a result, the assumptions made for the reaction of a frame to vertical loads do not apply to the lateral-loading condition. A rigid frame deformed by lateral loads is shown in Sketch 14.21. The lateral forces are resisted entirely by the frame. Any stiffness due to floors and walls is neglected. The deflected shape indicates that inflection (contraflexure) points form near the midspan of beams and columns. The *portal method* of analysis for lateral loads makes use of this pattern of behavior. This method is considered to be applicable to buildings of regular shape up to 25 stories high with height-to-width ratios of less than 5. Variations in bay size and story height should be small.

The portal method makes the following assumptions:

1. The frame deforms in such a way that it forms inflection points at the midspan of beams and midheight of columns.
2. Lateral forces are resisted completely by the frame. The bent of a frame acts as a series of individual portals. Each portal is assumed to carry a lateral load proportional to its span (Sketch 14.22).

$$\frac{P_1}{L_1} = \frac{P_2}{L_2} = \frac{P_3}{L_3} = \frac{P}{B}$$

or
$$P_1 = \frac{PL_1}{B} \qquad P_2 = \frac{PL_2}{B} \qquad P_3 = \frac{PL_3}{B} \qquad (14.4)$$

In order to calculate the axial forces in the columns, consider the free-body diagram of the upper portions of the portal (cut at the inflection point) in Sketch 14.23. Consider the first portal and take moments about the hinge of the right column

$$\frac{P_1 h}{2} - N_1 L_1 = 0$$

$$N_1 = \frac{P_1 h}{L_1 2}$$

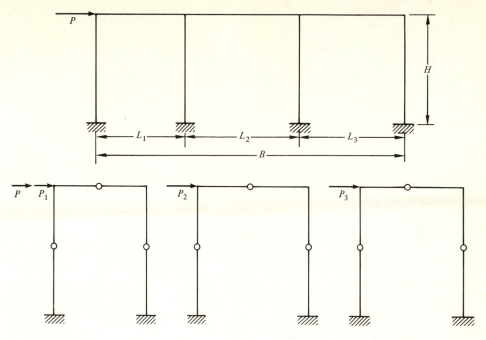

Sketch 14.22

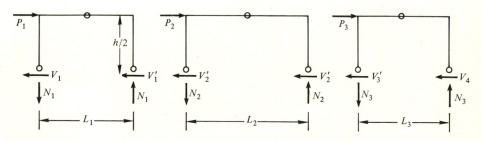

Sketch 14.23

Replacing P_1 by PL_1/B [(Eq. (14.4)] gives

$$N_1 = \frac{Ph}{2B} \tag{14.5}$$

In a similar manner

$$N_2 = \frac{Ph}{2B} \quad \text{and} \quad N_3 = \frac{Ph}{2B}$$

Since $N_1 = N_2$, interior columns do not carry axial forces due to wind loads. In general the wind moment is resisted by a couple composed of the axial forces of the exterior columns.

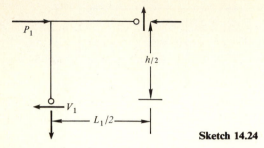

Sketch 14.24

To determine the column shears consider the free-body diagram of a portal cut at the inflection points in the column and beam (Sketch 14.24). Summing up moments about the midspan of the beam gives

$$\frac{V_1 h}{2} - \frac{N_1 L_1}{2}$$

$$V_1 = \frac{N_1 L_1}{h} \tag{14.6}$$

Substituting Eq. (14.5) in (14.6) gives

$$V_1 = \frac{Ph}{2B}\frac{L_1}{h} = \frac{PL_1}{2B}$$

Similarly $\qquad V_2' = \dfrac{PL_2}{2B} \qquad$ and $\qquad V_3' = \dfrac{PL_3}{2B}$

As a result, the column shears at midstory height are

$$V_1 = \frac{WL_1}{2B} \qquad V_2 = V_1' + V_2' = \frac{W(L_1 + L_2)}{2B}$$

$$V_3 = V_2' + V_3' = \frac{W(L_2 + L_3)}{2B} \qquad V_4 = \frac{WL_3}{2B} \tag{14.7}$$

In other words, the total wind shear is distributed to the columns in proportion to the width, i.e., floor area, each column is supporting.

It has been assumed that column sizes are approximately equal at any floor level. The assumed location of the inflection points is reasonable for the middle portion of a building, where the stiffness of the columns is approximately equal to the stiffness of the beams. However, in the upper two or three stories the stiffness of the columns is relatively small compared with that of the beams. As a result the inflection points move downward. They are often assumed to be at about $0.4h$. In the lowest three stories the opposite will be true, and the inflection point moves up to about $0.6h$. An increase in column height in the ground floor will also move the inflection point upward. The restraint provided by the foundation also influences the location of the inflection point. As the foundation provides less restraint, the inflection point moves down, eventually reaching the base of the column for a hinge condition. The portal method also assumes rigid connections even though true continuity may not exist. It does not take into account the secondary stresses resulting from frame-member deformations. The last two factors can normally be neglected in buildings up to 25 stories with slenderness ratios less than 5. Example 14.2 illustrates the application of the portal method to a simple bent.

Example 14.2 Determine the distribution of moments and axial forces in the two-story rigid frame due to the lateral forces shown in Fig. E14.2a.

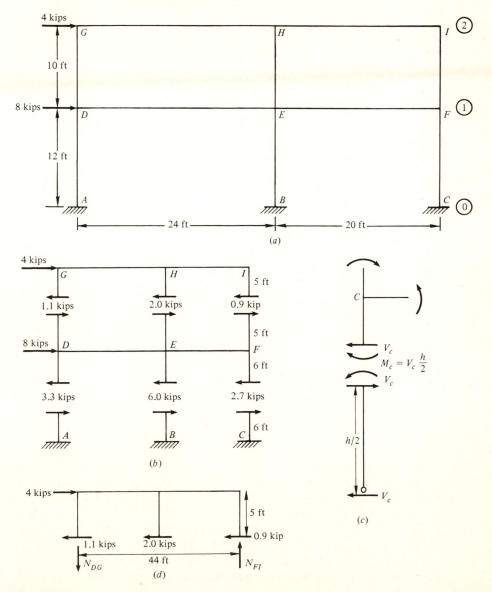

Figure E14.2

SOLUTION *Column shear forces* According to Eq. (14.7), the total shear is distributed to the columns in direct proportion to the floor area each column is supporting (Fig. E14.2b).

$$V_{DG} = \frac{(4)(24)}{(2)(44)} = 1.1 \text{ kip} \qquad V_{EH} = \frac{(4)(24 + 20)}{(2)(44)} = 2.0 \text{ kips}$$

$$V_{FI} = \frac{(4)(20)}{(2)(44)} = 0.9 \text{ kip} \qquad V_{AD} = \frac{(4 + 8)(24)}{(2)(44)} = 3.3 \text{ kips}$$

$$V_{BE} = \frac{(4 + 8)(24 + 20)}{(2)(44)} = 6.0 \text{ kips} \qquad V_{CF} = \frac{(4 + 8)(20)}{(2)(44)} = 2.7 \text{ kips}$$

Column moments at joints From the rotational equilibrium of the column free-body diagram, the column moments can be found by multiplying the column shear by one-half the column length (Fig. E14.2c):

$$M_{DG} = -M_{GD} = (1.1)(5) = 5.5 \text{ ft·kips}$$

$$M_{EH} = -M_{HE} = (2.0)(5) = 10.0 \text{ ft·kips}$$

$$M_{FI} = -M_{IF} = (0.9)(5) = 4.5 \text{ ft·kips}$$

$$M_{AD} = -M_{DA} = (3.3)(6) = 19.8 \text{ ft·kips}$$

$$M_{BE} = -M_{EB} = (6.0)(6) = 36.0 \text{ ft·kips}$$

$$M_{CF} = -M_{FC} = (2.7)(6) = 16.2 \text{ ft·kips}$$

Axial column forces The moments due to lateral forces are resisted by a couple composed of the axial forces of the exterior columns. As a result of the lateral forces, the interior columns do not carry axial loads. The exterior axial column forces for the upper bent, cutting the columns at the inflection points, can be obtained by taking moments about an exterior-column inflection point (Fig. E14.2d):

$$\overset{\curvearrowright}{\Sigma M_x} = 0: \qquad (4)(5) - 44N_{FI}$$

$$N_{FI} = \frac{(4)(5)}{44} = 0.45 \text{ kip}$$

$$\uparrow \Sigma V = 0: \qquad -N_{DG} + 0.45 = 0$$

$$N_{DG} = 0.45 \text{ kip}$$

Similarly (Fig. E14.2e)

$$\overset{\curvearrowright}{\Sigma M_x} = 0: \qquad (4)(16)(8)(6) - 44N_{CF} = 0$$

$$N_{CF} = 2.55 \text{ kips}$$

$$\uparrow \Sigma V = 0: \qquad -N_{AD} + 2.55 = 0$$

$$N_{AD} = 2.55 \text{ kips}$$

Beam shear and axial forces The shear forces in the beams are found by considering the vertical equilibrium of forces at each joint. In a similar manner the beam axial forces can be determined by considering the horizontal equilibrium of forces at each joint (Fig. E14.2f and g). Consider equilibrium of joint G:

$$\uparrow \Sigma V = 0: \qquad -0.45 + V_{GH} = 0$$

$$V_{GH} = 0.45 \text{ kip} = -V_{HG}$$

$$\overrightarrow{\Sigma H} = 0: \qquad 4.0 - 1.1 - N_{GH} = 0$$

$$N_{GH} = 2.9 \text{ kips} = -N_{HG}$$

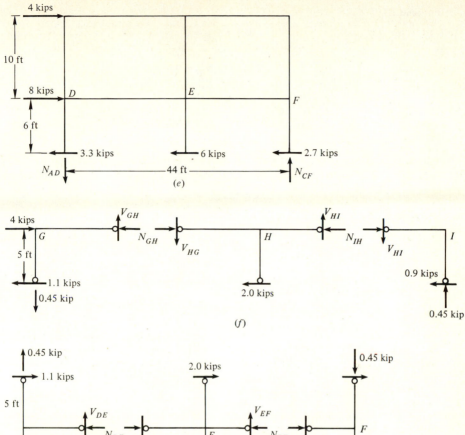

(e)

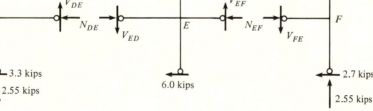

(f)

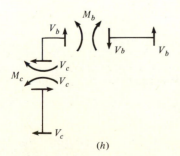

(g)

(h)

Figure E14.2 (*Cont'd.*)

For joint H

$\uparrow \Sigma V = 0$:
$$-0.45 + V_{HI} = 0$$
$$V_{HI} = 0.45 \text{ kip} = -V_{IH}$$

$\overrightarrow{\Sigma H} = 0$:
$$2.9 - 2.0 - N_{HI} = 0$$
$$N_{HI} = 0.9 \text{ kip} = -N_{IH}$$

For joint I

$\uparrow \Sigma V = 0$: $\qquad -0.45 + 0.45 = 0 \qquad$ check

$\overrightarrow{\Sigma H} = 0$: $\qquad 0.9 - 0.9 = 0 \qquad$ check

For joint D

$\uparrow \Sigma V = 0$:
$$0.45 - 2.55 + V_{DE} = 0$$
$$V_{DE} = 2.10 \text{ kips} = -V_{ED}$$

$\overrightarrow{\Sigma H} = 0$:
$$1.1 - 3.3 - N_{DE} = 0$$
$$N_{DE} = 2.2 \text{ kips} - N_{ED}$$

For joint E

$\uparrow \Sigma V = 0$:
$$-2.10 + V_{EF} = 0$$
$$V_{EF} = 2.10 \text{ kips} = -V_{FE}$$

$\overrightarrow{\Sigma H} = 0$:
$$2.2 + 2.0 - 6.0 - N_{EF} = 0$$
$$N_{EF} = 1.8 \text{ kips} = -N_{FE}$$

For joint F

$\uparrow \Sigma V = 0$: $\qquad -2.10 - 0.45 + 2.55 = 0 \qquad$ check

$\overrightarrow{\Sigma H} = 0$: $\qquad 1.8 + 0.9 - 2.7 = 0 \qquad$ check

Beam moments at joints The beam moments can now be found by multiplying the beam shear by half the beam span (Fig. E14.2h):

$$M_{GH} = -M_{HG} = (0.45)(12) = 5.4 \text{ ft·kips}$$
$$M_{HI} = -M_{IH} = (0.45)(10) = 4.5 \text{ ft·kips}$$
$$M_{DE} = -M_{ED} = (2.10)(12) = 25.2 \text{ ft·kips}$$
$$M_{EF} = -M_{FE} = (2.10)(10) = 21.0 \text{ ft·kips}$$

The moments obtained by the columns and beams can be checked by considering the rotational equilibrium of the various joints (Fig. E14.2i):

$\overset{\curvearrowright}{\Sigma M_G} = 0$: $\qquad -5.4 + 5.5 = -0.1 \approx 0 \qquad$ OK

$\overset{\curvearrowright}{\Sigma M_H} = 0$: $\qquad +10.0 - 5.4 - 4.5 = 0.1 \approx 0 \qquad$ OK

The resulting force diagrams for the rigid frame are given in Fig. E14.2j.

14.8.4 Approximate Design of Rigid-Frame Buildings

For the design of rigid-frame members two loading conditions are normally considered:

1. Only gravity loads (dead plus live)
2. Gravity loads acting together with wind or earthquake forces

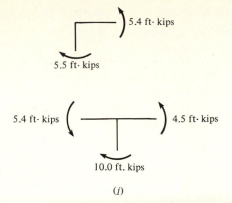

5.4 ft· kips

5.5 ft· kips

5.4 ft· kips   4.5 ft· kips

10.0 ft. kips

(*i*)

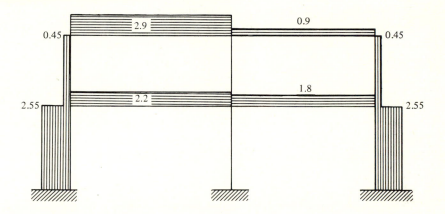

0.45   2.9   0.9   0.45

2.55   2.2   1.8   2.55

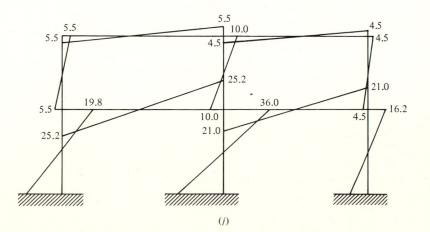

5.5
5.5   10.0   4.5
     4.5   4.5

25.2

5.5   19.8   36.0   21.0

   10.0   4.5   16.2
25.2   21.0

(*j*)

Figure E14.3 (*Cont'd.*)

Typical frame members must resist both compression and bending. The method of design has been discussed in Sec. 12.9.

Design for gravity loads Beams and girders must be proportioned to carry dead and live loads at the basic allowable stress. In some cases, the live loads on girders supporting a large floor area may be reduced, thanks to the low probability that the whole area will be loaded at once. Since the complete compression flange will be supported by the concrete floor slab, the basic allowable bending stresses for beams with fully supported compression flanges may be used. For both simple and rigid framing a more economical design is often made possible by taking advantage of composite action by the use of shear connectors in regions of positive moment (Sec. 14.9).

Design for wind plus gravity loads With combinations of wind and other loads, a $33\frac{1}{3}$ percent increase in allowable stress is permitted by AISC specifications [1]. This is usually accomplished by multiplying loads by $\frac{3}{4}$ and designing with the basic stress.

In rigid framing the negative moments at the supports may be checked by adding the moments due to gravity and wind. For this value to control design, three-quarters of the moment must exceed $0.080wl^2$ for the gravity loads alone.

The maximum positive moment due to gravity plus wind will vary in magnitude and location depending on their relative values. The moments and reactions of a girder are shown in Sketch 14.25. The positive moment M at any

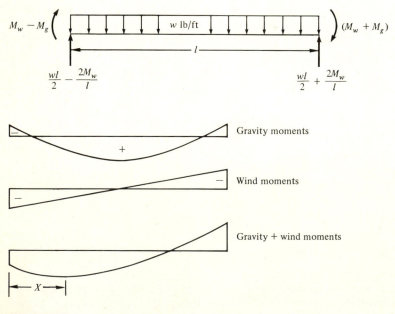

Sketch 14.25

point x in the girder is equal to

$$M = (M_w - M_g) + x\left(\frac{wl}{2} - \frac{2Mw}{l}\right) - \frac{wx^2}{2}$$

where M_w = wind moment
  M_g = negative gravity moment at support
  w = uniform load
  l = length of girder

The location of the maximum positive moment can be obtained by differentiating the moment equation with respect to x and equating to zero

$$x = \frac{l}{2} - \frac{2M_w}{wl}$$

The magnitude of the maximum positive moment becomes

$$M = \frac{wl^2}{8} + \frac{2M_w^2}{wl^2} - M_g \tag{14.8}$$

The design moment using normal working stresses is three-quarters of the value and it must exceed $0.080wl^2$ for gravity loads alone.

The design of a column in a rigid frame requires knowing the allowable compressive stress F_a, which is directly related to its slenderness Kl/r. The difficulty lies in the evaluation of the end restraint K provided by the girders attached to the upper and lower end of the column. For the ideal case in which infinite restraint is provided by floor girders, the stiffness factor is $K = 1$. As the restraint against end rotation is reduced, the effective column length increases. The amount of end restraint is also dependent on the type of girder-to-column connection (hinged, semirigid, or rigid).

A rigid-frame building that is laterally stabilized by shear walls does not develop lateral sidesway under the action of lateral loads. Depending on the rotational rigidity of the girders and their connections, the stiffness factor K for a column will be in the range of 0.5 to 1.0. The value is often safely assumed to be equal to 1.0.

Rigid-frame buildings with only light curtain walls and no lateral bracing develop sidesway under lateral loads. For these conditions the column stiffness factors are always larger than 1.

The typical stiffness factor for the lower stories of a high-rise rigid-frame building is in the range of 1.60, increasing slightly toward the upper stories. A conservative value of $K = 2$ can be used as a rough first approximation. It is important to consider the restraint relative to both axes.

14.9 COMPOSITE DESIGN

When a concrete slab is supported by a steel beam and there is no provision for the development of shear at their interface, noncomposite action occurs under the action of loads. This is not the most efficient use of the material in the beam

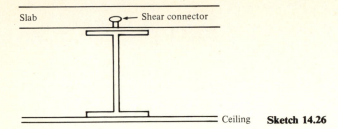

Slab   ← Shear connector

Ceiling   **Sketch 14.26**

and slab. The first approval for composite building floors was given by the 1952 AISC specification, and such floors have grown in popularity since then. They may either be encased in concrete or nonencased with shear connectors (Sketch 14.26), the latter being more common today.

In composite action the concrete slab acts as an integral part of the beam. A particular advantage of this type of construction is that it can take advantage of the high compressive strength of concrete along with the tensile strength of the steel. The result is a saving in steel. Composite sections have greater stiffness and therefore smaller deflections. A further advantage of composite construction is the possibility of smaller floor depths, which permit smaller story height and saving in materials. A disadvantage of this type of construction is the additional cost of providing and installing the shear connector.

Composite construction can be accomplished with or without shoring. Once the steel beams have been erected, the concrete slab can be placed on the beam. The formwork, fresh concrete, and other construction loads must be supported by the steel beam or by temporary shoring. If shoring is used, lighter beams become possible, but this saving in cost must be compared with the cost of the shoring. When the shoring is removed, the concrete slab participates in composite action by supporting the dead loads. Under these long-term loads, however, the creep and shrinkage of the concrete will transfer a considerable amount of the stress to the steel beam. This means that in the long run, the steel beam probably carries the dead load in either case and that only the live load is carried by composite action. As a result, the usual decision is to use the heavier steel beams and to do without shoring.

14.9.1 Effective Flange Widths

The portion of the slab that can be considered to participate in the beam action is controlled by specifications. For buildings AISC uses the same criteria as ACI [3] for reinforced concrete tee beams.

The maximum effective flange width is the least value of b determined according to the following criteria (Fig. 14.9):

1. b not greater than one-fourth the span of the beam
2. b' not greater than one-half the clear distance to the adjacent beam
3. b' not greater than 8 times the slab thickness

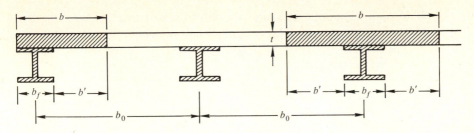

Figure 14.9 Effective flange width of composite section.

Should the slab exist only on one side of the beam, the following requirements control:

1. b not greater than one-twelfth the beam span
2. b' not greater than one-half the clear distance to the adjacent beam
3. b' not greater than 6 times the slab thickness

14.9.2 Analysis of Composite Section

For stress calculations the properties of a composite section are based on a transformed area. In this method the cross-sectional area of one of the two materials is transformed into an equivalent area of the other. For composite design of this kind it is customary to replace the concrete with an equivalent area of steel. The theory of composite action was developed in Sec. 11.10. If the neutral axis (centroidal axis) falls within the concrete flange, the concrete below the neutral axis should not be considered in calculating the moment of inertia of the section because it is assumed to be cracked. AISC specifications require 75 percent of the compressive strength f_c' of the concrete to be developed before composite action may be assumed. The modulus of elasticity of concrete in pounds per square inch may be taken as

$$E_c = w^{1.5}\left(33\sqrt{f_c'}\,\right) \qquad \text{according to ACI}$$

where w is the weight of concrete in pounds per cubic foot and f_c' is taken with units of pounds per square inch.

The AISC method of design can be summarized as follows. The composite section modulus S_{tr} must be capable of supporting all the loads without exceeding the allowable stresses, $F_b = 0.66F_y$ or $0.60F_y$, depending upon whether the beam is compact or not. The section modulus S_{tr} can be used for stress calculations in the steel section under total dead-load and live moment, providing its value does not exceed

$$S_{tr} = \left(1.35 + 0.35\frac{M_L}{M_D}\right)S_s$$

where M_L = moment caused by loads to concrete after hardening
M_D = moment caused by loads applied before concrete hardening
S_s = section modulus of steel beam referred to flange where stress is being computed

At sections subject to positive bending moment the stress should be computed for the steel tension, and at sections subject to negative bending moment the stress should be computed for both the tension and compression flanges. The section modulus of the transformed section is used to calculate the concrete flexural compressive stress, which should not exceed $0.45f_c'$.

The entire horizontal shear at the interface between the steel beam and the concrete slab is assumed to be transferred by shear connectors welded to the top flange of the beam and embedded in the concrete slab. According to AISC, this horizontal shear is estimated at ultimate load conditions (Sketch 14.27), where $0.85f_c'$ is assumed to be the average stress at failure, A_c is the actual area of effective concrete flange, and A_s is the area of the steel beam. These values are divided by 2 to estimate conditions at working loads. The critical value is

$$V_{hc} = 0.85f_c' A_c$$

$$V_{hs} = A_s F_y$$

Sketch 14.27

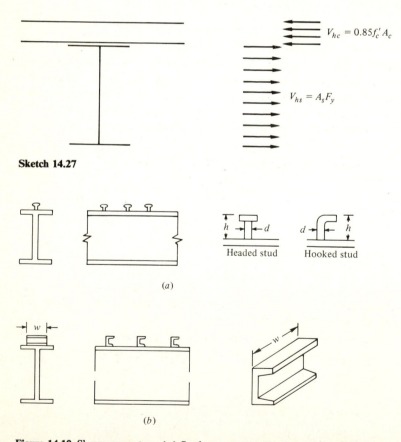

Headed stud   Hooked stud

(a)

(b)

Figure 14.10 Shear connectors. (a) Stud connectors; (b) channel connectors.

Table 14.1 Recommended shear values for shear connectors [1]

Allowable horizontal shear load for one connector q, kips[†]

| Connector[‡] | Specified compressive strength of concrete f_c', kips/in^2 | | |
	3.0	3.5	≥ 4.0
Hooked or headed stud:			
$\frac{1}{2}$-in-diam by 2 in	5.1	5.5	5.9
$\frac{5}{8}$-in-diam by $2\frac{1}{2}$ in	8.0	8.6	9.2
$\frac{3}{4}$-in-diam by 3 in	11.5	12.5	13.3
$\frac{7}{8}$-in-diam by $3\frac{1}{2}$ in	15.6	16.8	18.0
Channel:			
C3 × 4.1	4.3w[§]	4.7w[§]	5.0w[§]
C4 × 5.4	4.6w[§]	5.0w[§]	5.3w[§]
C5 × 6.7	4.9w[§]	5.3w[§]	5.6w[§]

   [†] Applicable only to concrete made with ASTM C33 aggregates.
   [‡] The allowable horizontal loads tabulated may also be used for studs longer than shown.
   [§] w = length of channel, in.

assumed to be the smaller of the two. The number of shear connectors required on each side of the point of maximum positive moment can be obtained by dividing the smaller value by the shear capacity q of one connector. AISC permits a uniform spacing of connectors. Table 14.1 gives recommended allowable shear values for different types of connectors, illustrated in Fig. 14.10.

Coefficients for use with concrete made with C33 aggregates

| Specified compressive strength of concrete f_c', kips/in^2 | Air-dry initial weight of concrete, lb/ft^2 | | | | | | |
	90	95	100	105	110	115	120
≤ 4.0	0.73	0.76	0.78	0.81	0.83	0.86	0.88
≥ 5.0	0.82	0.85	0.87	0.91	0.93	0.96	0.99

Example 14.3 Determine the service-load stresses for the structural system shown in Fig. E14.3a. Construction is without temporary shoring, and the subsequent live load is 100 lb/ft^2. Use AISC specifications. Assume simple spans and the following data:

$$\text{Partitions} = 20 \text{ lb/ft}^2 \quad \text{4-in slab} = \left(\tfrac{4}{12}\right)(150) = 50 \text{ lb/ft}^2$$

$$f_c' = 3000 \text{ lb/in}^2 \quad f_c = 0.45 f_c' = 1350 \text{ lb/in}^2 \quad \text{A36 steel}$$

SOLUTION

$$E_c = w^{1.5}\left(33\sqrt{f_c'}\,\right) = (150)^{1.5}(33\sqrt{3000}\,) = 3320 \text{ kips/in}^2$$

$$n = \frac{29,500}{3320} = 8.9 \quad \text{use } 9$$

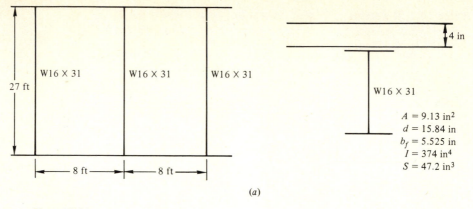

(a)

Figure E14.3

Calculations of moments The moment due to load applied before concrete hardens is

$$\text{Slab} = (8)(50) = 400 \text{ lb/ft}$$

$$\text{Beam} = \underline{31 \text{ lb/ft}}$$

$$\text{Total} = 431 \text{ lb/ft}$$

$$M_D = \frac{(0.431)(27)^2}{8} = 39.3 \text{ ft·kips}$$

and the moment due to loads applied after concrete hardens is

$$\text{Partitions } (8)(20) = 160 \text{ lb/ft}$$

$$\text{Live load } (8)(100) = \underline{800 \text{ lb/ft}}$$

$$\text{Total} = 960 \text{ lb/ft}$$

$$M_L = \frac{(0.960)(27)^2}{8} = 87.5 \text{ ft·kips}$$

$$M_T = M_D + M_L = 127.8 \text{ ft·kips}$$

Effective width of flange

$$b = \left(\tfrac{1}{4}\right)[(27)(12)] = 81 \text{ in}$$

$$b' = \left(\tfrac{1}{2}\right)[(8)(12) - 5.525] = 45.2 \text{ in}$$

or

$$b = (2)(45.2) + 5.525 = 96 \text{ in}$$

$$b' = (8)(4) = 32 \text{ in}$$

or

$$b = (2)(32) + 5.525 = 69.525 \text{ in} \qquad \text{controls}$$

Properties of composite section (Fig. E14.3b)

$$A = 9.13 + (4)(7.72) = 40.0 \text{ in}^2$$

$$40\bar{y} = 9.13\frac{15.84}{2} + (4)(7.72)(17.84) \qquad \bar{y} = 15.57 \text{ in}$$

$$I = 374 + 9.13(15.57 - 7.92)^2 + \frac{(7.72)(4)^3}{12} + (30.88)(17.84 - 15.57)^2$$

$$= 1108 \text{ in}^4$$

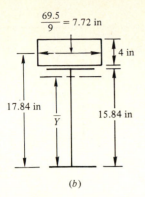

$$\frac{69.5}{9} = 7.72 \text{ in}$$

4 in

17.84 in

$\overline{Y}$

15.84 in

(b)

Figure E14.3 (*Cont'd.*)

Review of stresses Before concrete hardens

$$f_s = \frac{(39.3)(12)(7.92)}{374} = 9.99 < 24 \text{ kips/in}^2 \ (0.66 F_y)$$

The maximum permissible S_{tr}, according to AISC, is

$$\left(1.35 + 0.35 \frac{87.4}{39.3}\right)(47.2) = 100.5 \text{ in}^3$$

$$S_{tr} = \frac{1108}{15.57} = 71.2 < 100.5 \text{ in}^3 \quad \text{OK}$$

After concrete hardens

$$f_c = \frac{(126.8)(12)(4.27)}{(1108)(9)} = 0.651 < 1.350 \text{ kips/in}^2 \quad \text{OK}$$

$$f_s = \frac{(126.8)(12)(15.57)}{1108} = 21.4 < 24 \text{ kips/in}^2 \quad \text{OK}$$

Example 14.4 Design a composite section using A36 steel and AISC specifications for the situation in Fig. E14.4a. No shoring is to be used. Assume simple spans. The following

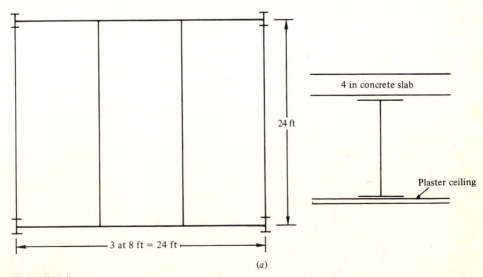

4 in concrete slab

24 ft

Plaster ceiling

3 at 8 ft = 24 ft

(a)

Figure E14.4

information is available:

$$\text{Live load} = 100 \text{ lb/ft}^2 \quad \text{Ceiling} = 8 \text{ lb/ft}^2 \quad \text{Partitions} = 16 \text{ lb/ft}^2$$

$$\text{4-in concrete slab} = 50 \text{ lb/ft}^2 \quad \text{Formwork} = 15 \text{ lb/ft}^2$$

$$f_c' = 3600 \text{ lb/in}^2 \quad f_c = 1620 \text{ lb/in}^2 \quad n = 9 \quad \text{Use A36 steel.}$$

SOLUTION

Calculation of moments Moments due to loads before concrete hardens are

$$\text{Slab} = (8)(50) = 400 \text{ lb/ft}$$

$$\text{Beam (assumed)} = \underline{25 \text{ lb/ft}}$$

$$\text{Total} = 425 \text{ lb/ft}$$

$$M_D = \frac{(0.452)(24)^2}{8} = 30.6 \text{ ft·kips}$$

and due to loads after concrete hardens are

$$\text{Ceiling} = (8)(8) \quad = 64 \text{ lb/ft}$$

$$\text{Partitions} = (16)(8) \quad = 128 \text{ lb/ft}$$

$$\text{Live load} = (8)(100) = \underline{800 \text{ lb/ft}}$$

$$\text{Total} = 992 \text{ lb/ft}$$

$$M_L = \frac{(0.992)(24)^2}{8} = 71.4 \text{ ft·kips}$$

$$M_T = M_D + M_L = 102 \text{ ft·kips}$$

Selection of trial section The AISC Steel Manual [1] table dealing with composite sections facilitates the selection of a steel section. Without such help the design process involves trial and error in selecting a steel section capable of supporting the dead loads during construction and of supporting the dead plus live load once composite action is possible. The composite section is usually critical. Cover plates on the tension flange are often an effective design.

Required section moduli

For M_D:

$$S_s = \frac{(30.6)(12)}{24} = 15.3 \text{ in}^3$$

For M_T:

$$S_{tr} = \frac{(102)(12)}{24} = 51 \text{ in}^3$$

Try a W12 × 27

$$S_s = 34.2 \text{ in}^3$$

(halfway between calculated S_s and S_{tr} values)

$$A = 7.95 \text{ in}^2 \quad b_f = 6.5 \text{ in} \quad d = 11.96 \text{ in}$$

$$t_f = 0.400 \text{ in} \quad t_w = 0.240 \text{ in} \quad I = 204 \text{ in}^4$$

$$b = \left(\tfrac{24}{4}\right)(12) = 72 \text{ in}$$

or

$$b = (8)(12) = 96 \text{ in}$$

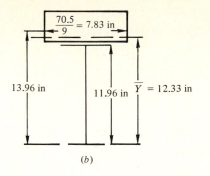

(b)

Figure E14.4 (*Cont'd.*)

or $\qquad b = 6.5 + (2)(8)(4) = 70.5$ in $\qquad$ controls (Fig. E14.4b)

$$\bar{y}[(7.83)(4) + 7.95] = 7.95\frac{11.96}{2} + (7.83)(4)(13.96)$$

$$\bar{y} = 12.33 \text{ in}$$

$$I = 204 + (7.95)(5.98 + 0.37)^2 + \frac{(7.83)(4)^3}{12} + (7.83)(4)(1.63)^2$$

$$= 649.6 \text{ in}^4$$

$$S_{tr} = \begin{cases} \dfrac{649.6}{12.33} = 52.7 \text{ in}^3 & \text{tension flange} \\[2mm] \dfrac{649.6}{3.67} = 177.0 \text{ in}^3 & \text{compression flange} \end{cases}$$

Review of stresses Before concrete hardness

$$f_s = \frac{(30.6)(12)}{34.2} = 10.7 < 24 \text{ kips/in}^2 \qquad \text{OK}$$

The maximum permissible S_{tr}, according to AISC, is

$$\left(1.35 + 0.35\frac{71.4}{30.6}\right)(34.2) = 74 > 52.7 \text{ in}^3 \qquad \text{OK}$$

After concrete hardens

$$f_s = \frac{(102)(12)}{52.7} = 23.2 < 24 \text{ kips/in}^2 \qquad \text{OK}$$

$$f_c = \frac{(102)(12)}{(177.0)(9)} = 0.77 < 1.35 \text{ kips/in}^2 \qquad \text{OK}$$

Design of shear connectors

$$V_{hc} = \frac{0.85f_c'}{2}A_c = \frac{(0.85)(3000)(3.67)(70.5)}{2} = 329.9 \text{ kips}$$

$$V_{hs} = \frac{A_s F_y}{2} = \frac{(7.95)(36)}{2} = 143.1 \text{ kips} \qquad \text{controls}$$

Assuming $\frac{3}{4}$-in diameter by 3-in hooked or headed stud,

$$q = 11.5 \text{ kips} \qquad \text{(Table 14.1)}$$

$$\text{Number required} = \frac{143.1}{11.5} = 12.4$$

Use 13 connectors, each half of beam length or a total of 26 connectors equally spaced over beam length

$$\text{Spacing} = \frac{(24)(12)}{26} = 11 \text{ in}$$

14.10 SUMMARY

This chapter covered some of the concepts involved in the design of steel buildings. The first portion discussed briefly the types of construction, including bearing-wall, steel-skeleton, and long-span structures. This was followed by a brief review of the associated types of floor construction. The main types are (1) concrete slabs supported on beams and open-web joists and (2) precast-concrete floor units. Roof construction involves precast and cast-in-place slabs that are normally made of lightweight concrete. The insulating property of a roof slab is an important consideration.

Frames used in building construction involve single-story bents that may be statically determinate or indeterminate. Because multistory frames must resist significant lateral forces, they either have rigid connections or simple connections with appropriate lateral bracing.

Approximate solutions to statically indeterminate structures are possible by making reasonable assumptions relative to the location of points of inflection in the beams and columns. This approach is referred to as the portal method for multistory buildings. Gravity loads and lateral loads (wind and earthquake) are considered separately and then combined, keeping in mind that allowable stresses can be increased $33\frac{1}{3}$ percent when wind or earthquake loads are involved.

Concrete slabs supported by steel beams make possible composite design, a condition in which the concrete slab functions with the steel beam. Such action necessitates the development of shear at the concrete-steel interface by means of shear connectors.

PROBLEMS

14.1 Analyze the bent shown in Fig. P14.1. Draw shear and bending-moment diagrams for each of the members.

14.2 (*a*) Analyze the steel frame in Fig. P14.2.

(*b*) Determine the additional wind moments due to a lateral load of 6 kips at *B*. Check whether they become critical when combined with gravity loads.

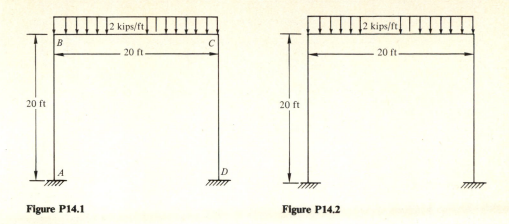

Figure P14.1   Figure P14.2

14.3 A building bent has three equal bays of 20 ft each and two stories of 12 ft each. The columns of the first story are fixed at their bases. Each girder carries a combined dead and live load of 2 kips/ft. Determine the axial column loads and moment diagrams for girders and columns (Fig. P14.3).

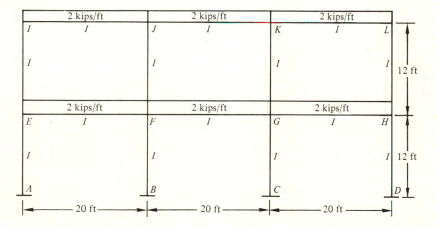

Figure P14.3

14.4 The building bent of Prob. 14.3 is acted upon by a horizontal force of 5 kips applied at each girder level on the left exterior columns. Determine the bending-moment diagrams for all members by the portal method.

14.5 Design column AE and girder EF of the building bent analyzed in Probs. 14.3 and 14.4. Consider the possible combination of lateral (wind) and gravity loads.

14.6 Determine the force distributions (axial and moments) due to wind for the six-story rigid-frame building shown in Fig. P14.6. The frame bents are spaced 20 ft apart.

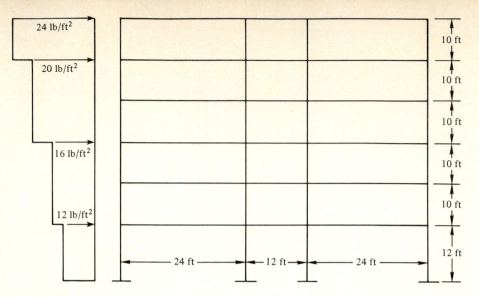

24 lb/ft²

20 lb/ft²

16 lb/ft²

12 lb/ft²

10 ft

10 ft

10 ft

10 ft

10 ft

12 ft

24 ft — 12 ft — 24 ft

Figure P14.6

4 in slab

W14 × 34

Beams 8 ft oc

Figure P14.7

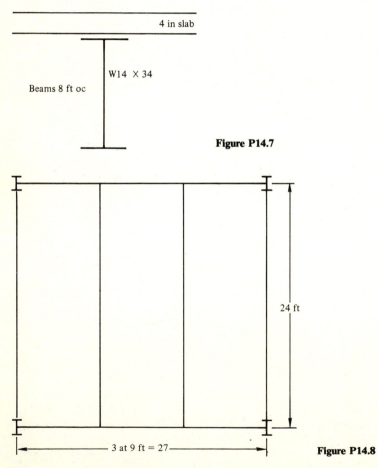

24 ft

— 3 at 9 ft = 27 —

Figure P14.8

14.7 Calculate the bending stresses for the section in Fig. P14.7. The section is used for a simple span of 24 ft and is to have a live uniform load of 150 lb/ft^2 applied after composite action develops. Assume no shoring and use AISC specifications; $n = 9$.

14.8 For the floor plan in Fig. P14.8 design a composite section for the simple span beams assuming a 5-in concrete slab. Use A36 steel and concrete with $f_c' = 3000$ lb/in^2; $n = 9$. The live load applied after composite action is 150 lb/ft^2. No shoring is to be used. Design the shear connectors.

REFERENCES

1. American Institute of Steel Construction, "Manual of Steel Construction," 8th ed., Chicago, 1978.
2. Building Officials and Code Administration International, Inc., "BOAC Basic Building Code/1978," Homewood, Ill., 1978.
3. American Concrete Institute, "Building Code Requirements for Reinforced Concrete (ACI 318)," Detroit, 1976.

APPENDIX

Table A.1 Allowable stresses for visual grading of structural lumber per normal load duration [1]

Species and commercial grade	Size classification	Design values, in lb/in²						
		Extreme fiber in bending, F_b		Tension parallel to grain, F_t	Horizontal shear, F_v	Compression perpendicular to grain, $F_{c\perp}$	Compression parallel to grain, F_c	Modulus of elasticity, E
		Single member uses	Repetitive member uses					
Douglas fir-larch (surfaced dry or surfaced green, used at 19 percent maximum moisture content)								
Dense select structural		2450	2800	1400	95	455	1850	1,900,000
Select structural		2100	2400	1200	95	385	1600	1,800,000
Dense no. 1		2050	2400	1200	95	455	1450	1,900,000
No. 1	2 in to 4 in thick;	1750	2050	1050	95	385	1250	1,800,000
Dense no. 2		1700	1950	1000	95	455	1150	1,700,000
No. 2	2 in to 4 in wide	1450	1650	850	95	385	1000	1,700,000
No. 3		800	925	475	95	385	600	1,500,000
Appearance		1750	2050	1050	95	385	1500	1,800,000
Stud		800	925	475	95	385	600	1,500,000
Construction	2 in to 4 in thick;	1050	1200	625	95	385	1150	1,500,000
Standard	4 in wide	600	675	350	95	385	925	1,500,000
Utility		275	325	175	95	385	600	1,500,000
Dense select structural		2100	2400	1400	95	455	1650	1,900,000
Select structural		1800	2050	1200	95	385	1400	1,800,000
Dense no. 1	2 in to 4 in thick;	1800	2050	1200	95	455	1450	1,900,000
No. 1	5 in and wider	1500	1750	1000	95	385	1250	1,800,000
Dense no. 2		1450	1700	775	95	455	1250	1,700,000
No. 2		1250	1450	650	95	385	1050	1,700,000
No. 3		725	850	375	95	385	675	1,500,000
Appearance		1500	1750	1000	95	385	1500	1,800,000
Stud		725	850	375	95	385	675	1,500,000
Dense select structural		1900	...	1100	85	455	1300	1,700,000
Select structural	Beams and stringers	1600	...	950	85	385	1100	1,600,000
Dense no. 1		1550	...	775	85	455	1100	1,700,000
No. 1		1300	...	675	85	385	925	1,600,000

Table A.1 (*Cont'd.*)

Species and commercial grade	Size classification	Design values, in lb/in²						
		Extreme fiber in bending, F_b		Tension parallel to grain, F_t	Horizontal shear, F_v	Compression perpendicular to grain, $F_{c\perp}$	Compression parallel to grain, F_c	Modulus of elasticity, E
		Single member uses	Repetitive member uses					
Dense select structural	Posts and timbers	1750	...	1150	85	455	1350	1,700,000
Select structural		1500	...	1000	85	385	1150	1,600,000
Dense no. 1		1400	...	950	85	455	1200	1,700,000
No. 1		1200	...	825	85	385	1000	1,600,000
Select dex	Decking	1750	2000	...	...	385	...	1,800,000
Commercial dex		1450	1650	...	...	385	...	1,700,000
Dense select structural	Beams and stringers	1900	...	1250	85	455	1300	1,700,000
Select structural		1600	...	1050	85	385	1100	1,600,000
Dense no. 1		1550	...	1050	85	455	1100	1,700,000
No. 1		1350	...	900	85	385	925	1,600,000
Dense select structural	Posts and timbers	1750	...	1150	85	455	1350	1,700,000
Select structural		1500	...	1000	85	385	1150	1,600,000
Dense no. 1		1400	...	950	85	455	1200	1,700,000
No. 1		1200	...	825	85	385	1000	1,600,000
Selected decking	Decking	...	2000	...	...	...	...	1,800,000
Commercial decking		...	1650	...	...	...	...	1,700,000
Selected decking	Decking	...	2150	(Surfaced at 15 percent maximum moisture content (mcc) and used at 15 percent mcc)				1,900,000
Commercial decking		...	1800					1,700,000

Southern pine (Surfaced green; used any condition)

Grade	Size							
Select structural	2½ in to 4 in thick; 2½ in to 4 in wide	1600	1850	925	95	270	1050	1,500,000
Dense select structural		1850	2150	1100	95	315	1200	1,600,000
No. 1		1350	1550	800	95	270	825	1,500,000
No. 1 dense		1600	1800	925	95	315	950	1,600,000
No. 2		1150	1300	675	85	270	650	1,400,000
No. 2 dense		1350	1500	775	85	315	750	1,400,000
No. 3		625	725	375	85	270	400	1,200,000
No. 3 dense		725	850	425	85	315	450	1,300,000
Stud		625	725	375	85	270	400	1,200,000
Construction	2½ in to 4 in thick; 4 in wide	825	925	475	95	270	725	1,200,000
Standard		475	525	275	85	270	600	1,200,000
Utility		200	250	125	85	270	400	1,200,000
Select structural	2½ in to 4 in thick; 5 in and wider	1400	1600	900	85	270	900	1,500,000
Dense select structural		1600	1850	1050	85	315	1050	1,600,000
No. 1		1200	1350	775	85	270	825	1,500,000
No. 1 dense		1400	1600	925	85	315	950	1,600,000
No. 2		975	1100	500	85	270	675	1,400,000
No. 2 dense		1150	1300	600	85	315	800	1,400,000
No. 3		550	650	300	85	270	425	1,200,000
No. 3 dense		650	750	350	85	315	475	1,300,000
Stud		575	675	300	85	270	425	1,200,000
Dense standard decking	2 in and wider; Decking	1600	1800	…	…	315	…	1,600,000
Select decking		1150	1300	…	…	270	…	1,400,000
Dense select decking		1350	1500	…	…	315	…	1,400,000
Commercial decking		1150	1300	…	…	270	…	1,400,000
Dense commercial decking		1350	1500	…	…	315	…	1,400,000
No. 1 SR	5 in and thicker	1350	…	875	110	270	775	1,500,000
No. 1 dense SR		1550	…	1050	110	315	925	1,600,000
No. 2 SR		1100	…	725	95	270	625	1,400,000
No. 2 dense SR		1250	…	850	95	315	725	1,400,000
Dense structural 86	2½ in and thicker	2100	2400	1400	145	315	1300	1,600,000
Dense structural 72		1750	2050	1200	120	315	1100	1,600,000
Dense structural 65		1600	1800	1050	110	315	1000	1,600,000

Table A.2 Properties of sections [2]

Nominal size, in b d	Actual size, in b d	Area, in^2	Axis xx S, in^3	Axis xx I, in^4	Axis yy S, in^3	Axis yy I, in^4	Board measure per lineal foot	Weight per lineal foot, lb
2×2	$1\text{-}1/2 \times 1\text{-}1/2$	2.25	.56	.42	.56	.42	.33	.63
3	$2\text{-}1/2$	3.75	1.56	1.95	.94	.70	.50	1.05
4	$3\text{-}1/2$	5.25	3.06	5.36	1.31	.99	.67	1.46
6	$5\text{-}1/2$	8.25	7.56	20.80	2.06	1.55	1.00	2.29
8	$7\text{-}1/4$	10.88	13.14	47.63	2.72	2.06	1.33	2.98
10	$9\text{-}1/4$	13.88	21.39	98.93	3.57	2.62	1.67	3.87
12	$11\text{-}1/4$	16.88	31.64	177.98	4.23	3.18	2.00	4.68
14	$13\text{-}1/4$	19.88	43.89	290.77	4.97	3.75	2.33	5.50
3×3	$2\text{-}1/2 \times 2\text{-}1/2$	6.25	2.61	3.25	2.6	3.24	.75	1.73
4	$3\text{-}1/2$	8.75	5.10	8.93	3.64	4.56	1.00	2.43
6	$5\text{-}1/2$	13.75	12.60	34.66	5.73	7.16	1.50	3.82
8	$7\text{-}1/4$	18.13	21.90	79.39	7.56	9.53	2.00	5.03
10	$9\text{-}1/4$	23.13	35.65	164.88	9.63	12.16	2.50	6.44
12	$11\text{-}1/4$	28.13	52.73	296.63	11.75	14.79	3.00	7.83
14	$13\text{-}1/4$	33.13	73.15	484.63	14.91	17.34	3.50	9.18
4×4	$3\text{-}1/2 \times 3\text{-}1/2$	12.25	7.15	12.50	7.14	12.52	1.33	3.39
6	$5\text{-}1/2$	19.25	17.65	48.53	11.23	19.64	2.00	5.34
8	$7\text{-}1/4$	25.38	30.66	111.15	14.82	26.15	2.67	7.05
10	$9\text{-}1/4$	32.38	49.91	230.84	18.97	33.23	3.33	8.98
12	$11\text{-}1/4$	39.38	73.82	415.28	23.03	40.30	4.00	10.91
14	$13\text{-}1/2$	46.38	106.31	717.61	27.07	47.59	4.67	12.90
$6 \times 6^*$	$5\text{-}1/2 \times 5\text{-}1/2$	30.25	27.73	76.25	27.73	76.25	3.00	8.40
8	$7\text{-}1/2$	41.25	51.56	193.35	37.81	103.98	4.00	11.46
10	$9\text{-}1/2$	52.25	82.73	392.96	47.89	131.71	5.00	14.51
12	$11\text{-}1/2$	63.25	121.23	697.07	57.98	159.44	6.00	17.57
14	$13\text{-}1/2$	74.25	167.06	1127.67	68.06	187.17	7.00	20.62
$8 \times 8^*$	$7\text{-}1/2 \times 7\text{-}1/2$	56.25	70.31	263.67	70.31	263.67	5.33	15.62
10	$9\text{-}1/2$	71.25	112.81	535.86	89.06	333.98	6.67	19.79
12	$11\text{-}1/2$	86.25	165.31	950.55	107.81	404.30	8.00	23.96
14	$13\text{-}1/2$	101.25	227.81	1537.73	126.56	474.61	9.33	28.12
$10 \times 10^*$	$9\text{-}1/2 \times 9\text{-}1/2$	90.25	142.89	678.75	142.89	678.75	8.33	25.07
12	$11\text{-}1/2$	109.25	209.39	1204.03	172.98	821.65	10.00	30.35
14	$13\text{-}1/2$	128.25	288.56	1947.80	203.06	964.25	11.67	35.62
$12 \times 12^*$	$11\text{-}1/2 \times 11\text{-}1/2$	132.25	253.48	1457.51	253.48	1457.51	12.00	36.74
14	$13\text{-}1/2$	155.25	349.31	2357.86	297.56	1710.98	14.00	43.12
$14 \times 14^*$	$13\text{-}1/2 \times 13\text{-}1/2$	182.25	410.06	2767.92	410.06	2767.92	16.33	50.62

*Note: Properties are based on minimum dressed green size which is 1/2 in off nominal in both b and d dimensions.

Table A.3 Unit properties of sections [3]

To determine weight per lineal foot (in pounds) divide the area by 4

SECTION PROPERTIES

No. of Lams 1½″	No. of Lams ¾″	Depth, d (inches)	Size Factor, C_F	Area, A (inches²)	Section Modulus, S (inches³)	Moment of Inertia, I (inches⁴)
3⅛″ WIDTH						
2	4	3.00	1.00	9.4	4.7	7.0
	5	3.75	1.00	11.7	7.3	13.7
3	6	4.50	1.00	14.1	10.5	23.7
	7	5.25	1.00	16.4	14.4	37.7
4	8	6.00	1.00	18.8	18.8	56.3
	9	6.75	1.00	21.1	23.7	80.1
5	10	7.50	1.00	23.4	29.3	109.9
	11	8.25	1.00	25.8	35.4	146.2
6	12	9.00	1.00	28.1	42.2	189.8
	13	9.75	1.00	30.5	49.5	241.4
7	14	10.50	1.00	32.8	57.4	301.5
	15	11.25	1.00	35.2	65.9	370.8
8	16	12.00	1.00	37.5	75.0	450.0
	17	12.75	0.99	39.8	84.7	539.8
9	18	13.50	0.99	42.2	94.9	640.7
	19	14.25	0.98	44.5	105.8	753.6
10	20	15.00	0.98	46.9	117.2	878.9
	21	15.75	0.97	49.2	129.2	1,017.4
11	22	16.50	0.97	51.6	141.8	1,169.8
	23	17.25	0.96	53.9	155.0	1,336.7
12	24	18.00	0.96	56.3	168.8	1,518.8
	25	18.75	0.95	58.6	183.1	1,716.6
13	26	19.50	0.95	60.9	198.0	1,931.0
	27	20.25	0.94	63.3	213.6	2,162.4
14	28	21.00	0.94	65.6	229.7	2,411.7
	29	21.75	0.94	68.0	246.4	2,679.5
15	30	22.50	0.93	70.3	263.7	2,966.3
	31	23.25	0.93	72.7	281.5	3,272.9
16	32	24.00	0.93	75.0	300.0	3,600.0
5⅛″ WIDTH						
3	6	4.50	1.00	23.1	17.3	38.9
	7	5.25	1.00	26.9	23.5	61.8
4	8	6.00	1.00	30.8	30.8	92.3
	9	6.75	1.00	34.6	38.9	131.3
5	10	7.50	1.00	38.4	48.0	180.2
	11	8.25	1.00	42.3	58.1	239.8
6	12	9.00	1.00	46.1	69.2	311.3
	13	9.75	1.00	50.0	81.2	395.8
7	14	10.50	1.00	53.8	94.2	494.4
	15	11.25	1.00	57.7	108.1	608.1
8	16	12.00	1.00	61.5	123.0	738.0
	17	12.75	0.99	65.3	138.9	885.2
9	18	13.50	0.99	69.2	155.7	1,050.8
	19	14.25	0.98	73.0	173.4	1,235.8
10	20	15.00	0.98	76.9	192.2	1,441.4
	21	15.75	0.97	80.7	211.9	1,668.6
11	22	16.50	0.97	84.6	232.5	1,918.5
	23	17.25	0.96	88.4	254.2	2,192.2
12	24	18.00	0.96	92.3	276.8	2,490.8
	25	18.75	0.95	96.1	300.3	2,815.2
13	26	19.50	0.95	99.9	324.8	3,166.8
	27	20.25	0.94	103.8	350.3	3,546.4
14	28	21.00	0.94	107.6	376.7	3,955.2
	29	21.75	0.94	111.5	404.1	4,394.3
15	30	22.50	0.93	115.3	432.4	4,864.7
	31	23.25	0.93	119.2	461.7	5,367.6
16	32	24.00	0.93	123.0	492.0	5,904.0
	33	24.75	0.92	126.8	523.2	6,475.0
17	34	25.50	0.92	130.7	555.4	7,081.6
	35	26.25	0.92	134.5	588.6	7,725.0
18	36	27.00	0.91	138.4	622.7	8,406.3
	37	27.75	0.91	142.2	657.8	9,126.4
19	38	28.50	0.91	146.1	693.8	9.886.6
	39	29.25	0.91	149.9	730.8	10,687.8
20	40	30.00	0.90	153.8	768.8	11,531.3
	41	30.75	0.90	157.6	807.7	12,417.9
21	42	31.50	0.90	161.4	847.5	13,348.9
	43	32.25	0.90	165.3	888.4	14,325.2
22	44	33.00	0.89	169.1	930.2	15,348.1
	45	33.75	0.89	173.0	972.9	16,418.5
23	46	34.50	0.89	176.8	1,016.7	17,537.6
	47	35.25	0.89	180.7	1,061.4	18,706.4
24	48	36.00	0.88	184.5	1,107.0	19,926.0
6¾″ WIDTH						
4	8	6.00	1.00	40.5	40.5	121.5
	9	6.75	1.00	45.6	51.3	173.0
5	10	7.50	1.00	50.6	63.3	237.3
	11	8.25	1.00	55.7	76.6	315.9
6	12	9.00	1.00	60.8	91.1	410.1
	13	9.75	1.00	65.8	106.9	521.4
7	14	10.50	1.00	70.9	124.0	651.2
	15	11.25	1.00	75.9	142.4	800.9
8	16	12.00	1.00	81.0	162.0	972.0
	17	12.75	0.99	86.1	182.9	1,165.9
9	18	13.50	0.99	91.1	205.0	1,384.0
	19	14.25	0.98	96.2	228.4	1,627.7
10	20	15.00	0.98	101.3	253.1	1,898.4
	21	15.75	0.97	106.3	279.1	2,197.7
11	22	16.50	0.97	111.4	306.3	2,526.8
	23	17.25	0.96	116.4	334.8	2,887.3
12	24	18.00	0.96	121.5	364.5	3,280.5
	25	18.75	0.95	126.6	395.5	3,707.9
13	26	19.50	0.95	131.6	427.8	4,170.9
	27	20.25	0.94	136.7	461.3	4,670.9
14	28	21.00	0.94	141.8	496.1	5,209.3
	29	21.75	0.94	146.8	532.2	5,787.6
15	30	22.50	0.93	151.9	569.5	6,407.2
	31	23.25	0.93	156.9	608.1	7,069.5
16	32	24.00	0.93	162.0	648.0	7,776.0
	33	24.75	0.92	167.1	689.1	8,528.0
17	34	25.50	0.92	172.1	731.5	9,327.0
	35	26.25	0.92	177.2	775.2	10,174.4
18	36	27.00	0.91	182.3	820.1	11,071.7
	37	27.75	0.91	187.3	866.3	12,020.2
19	38	28.50	0.91	192.4	913.8	13,021.4
	39	29.25	0.91	197.4	962.5	14,076.7
20	40	30.00	0.90	202.5	1,012.5	15,187.5
	41	30.75	0.90	207.6	1,063.8	16,355.3
21	42	31.50	0.90	212.6	1,116.3	17,581.4
	43	32.25	0.90	217.7	1,170.1	18,867.4
22	44	33.00	0.89	222.8	1,225.1	20,214.6
	45	33.75	0.89	227.8	1,281.4	21,624.4
23	46	34.50	0.89	232.9	1,339.0	23,098.3
	47	35.25	0.89	237.9	1,397.9	24,637.7
24	48	36.00	0.88	243.0	1,458.0	26,244.0
	49	36.75	0.88	248.1	1,519.4	27,918.7
25	50	37.50	0.88	253.1	1,582.0	29,663.1
	51	38.25	0.88	258.2	1,645.9	31,478.7
26	52	39.00	0.88	263.3	1,711.1	33,366.9
	53	39.75	0.88	268.3	1,777.6	35,329.2
27	54	40.50	0.87	273.4	1,845.3	37,367.0
	55	41.25	0.87	278.4	1,914.3	39,481.6
28	56	42.00	0.87	283.5	1,984.5	41,674.5
	57	42.75	0.87	288.6	2,056.0	43,947.2
29	58	43.50	0.87	293.6	2,128.8	46,301.0
	59	44.25	0.87	298.7	2,202.8	48,737.4
30	60	45.00	0.86	303.8	2,278.1	51,257.8
	61	45.75	0.86	308.8	2,354.7	53,863.7
31	62	46.50	0.86	313.9	2,432.5	56,556.4
	63	47.25	0.86	318.9	2,511.6	59,337.3
32	64	48.00	0.86	324.0	2,592.0	62,208.0
8¾″ WIDTH						
6	12	9.00	1.00	78.8	118.1	531.6
	13	9.75	1.00	85.3	138.6	675.8
7	14	10.50	1.00	91.9	160.8	844.1
	15	11.25	1.00	98.4	184.6	1,038.2
8	16	12.00	1.00	105.0	210.0	1,260.0
	17	12.75	0.99	111.6	237.1	1,511.3
9	18	13.50	0.99	118.1	265.8	1,794.0
	19	14.25	0.98	124.7	296.1	2,109.9
10	20	15.00	0.98	131.3	328.1	2,460.9
	21	15.75	0.97	137.8	361.8	2,848.8
11	22	16.50	0.97	144.4	397.0	3,275.5
	23	17.25	0.96	150.9	433.9	3,742.8
12	24	18.00	0.96	157.5	472.5	4,252.5
	25	18.75	0.95	164.1	512.7	4.806.5
13	26	19.50	0.95	170.6	554.5	5,406.7
	27	20.25	0.94	177.2	598.0	6,054.8
14	28	21.00	0.94	183.8	643.1	6,752.8
	29	21.75	0.94	190.3	689.9	7,502.5
15	30	22.50	0.93	196.9	738.3	8,305.7
	31	23.25	0.93	203.4	788.3	9,164.2
16	32	24.00	0.93	210.0	840.0	10,080.0
	33	24.75	0.92	216.6	893.3	11,054.8
17	34	25.50	0.92	223.1	948.3	12,090.6
	35	26.25	0.92	229.7	1,004.9	13,189.1
18	36	27.00	0.91	236.3	1,063.1	14,352.2
	37	27.75	0.91	242.8	1,123.0	15,581.7
19	38	28.50	0.91	249.4	1,184.5	16,879.6
	39	29.25	0.91	255.9	1,247.7	18,247.5
20	40	30.00	0.90	262.5	1,312.5	19,687.5
	41	30.75	0.90	269.1	1,378.9	21,201.3
21	42	31.50	0.90	275.6	1,447.0	22,790.7
	43	32.25	0.90	282.2	1,516.8	24,457.7
22	44	33.00	0.89	288.8	1,588.1	26,204.1
	45	33.75	0.89	295.3	1,661.1	28,031.6
23	46	34.50	0.89	301.9	1,735.8	29,942.2
	47	35.25	0.89	308.4	1,812.1	31,937.7
24	48	36.00	0.88	315.0	1,890.0	34,020.0
	49	36.75	0.88	321.6	1,969.6	36,190.9
25	50	37.50	0.88	328.1	2,050.8	38,452.2
	51	38.25	0.88	334.7	2,133.6	40,805.7
26	52	39.00	0.88	341.3	2,218.1	43,253.4
	53	39.75	0.88	347.8	2,304.3	45,797.1
27	54	40.50	0.87	354.4	2,392.0	48,438.6
	55	41.25	0.87	360.9	2,481.4	51,179.8
28	56	42.00	0.87	367.5	2,572.5	54,022.5
	57	42.75	0.87	374.1	2,665.2	56,968.6
29	58	43.50	0.87	380.6	2,759.5	60,019.8
	59	44.25	0.87	387.2	2,855.5	63,178.1
30	60	45.00	0.86	393.8	2,953.1	66,445.3
	61	45.75	0.86	400.3	3,052.4	69.823.3
31	62	46.50	0.86	406.9	3,153.3	73,313.8
	63	47.25	0.86	413.4	3,255.8	76,918.8
32	64	48.00	0.86	420.0	3,360.0	80,640.0
	65	48.75	0.86	426.6	3,465.8	84,479.4
33	66	49.50	0.85	433.1	3,573.3	88,438.7
	67	50.25	0.85	439.7	3,682.4	92,519.9
34	68	51.00	0.85	446.3	3,793.1	96,724.7
	69	51.75	0.85	452.8	3,905.5	101,055.0
35	70	52.50	0.85	459.4	4,019.5	105,512.7
	71	53.25	0.85	465.9	4,135.2	110,099.6
36	72	54.00	0.85	472.5	4,252.5	114,817.5
	73	54.75	0.85	479.1	4,371.4	119,668.3
37	74	55.50	0.84	485.6	4,492.0	124,653.9
	75	56.25	0.84	492.2	4,614.3	129,776.0
38	76	57.00	0.84	498.8	4,738.1	135,036.6
	77	57.75	0.84	505.3	4,863.6	140,437.4
39	78	58.50	0.84	511.9	4,990.8	145,980.4
	79	59.25	0.84	518.4	5,119.6	151,667.3
40	80	60.00	0.84	525.0	5,250.0	157,500.0
	81	60.75	0.84	531.6	5,382.1	163,480.4
41	82	61.50	0.83	538.1	5,515.8	169,610.3
	83	62.25	0.83	544.7	5,651.1	175,891.5
42	84	63.00	0.83	551.3	5,788.1	182,326.0
10¾″ WIDTH						
7	14	10.50	1.00	112.9	197.5	1,037.0
	15	11.25	1.00	120.9	226.8	1,275.5
8	16	12.00	1.00	129.0	258.0	1,548.0
	17	12.75	0.99	137.1	291.3	1,856.8
9	18	13.50	0.99	145.1	326.5	2,204.1
	19	14.25	0.98	153.2	363.8	2,592.2
10	20	15.00	0.98	161.3	403.1	3,023.4
	21	15.75	0.97	169.3	444.4	3,500.0
11	22	16.50	0.97	177.4	487.8	4,024.2
	23	17.25	0.96	185.4	533.1	4,598.3
12	24	18.00	0.96	193.5	580.5	5,224.5
	25	18.75	0.95	201.6	629.9	5,905.2
13	26	19.50	0.95	209.6	681.3	6,642.5
	27	20.25	0.94	217.7	734.7	7,438.8
14	28	21.00	0.94	225.8	790.1	8,296.3
	29	21.75	0.94	233.8	847.6	9,217.3
15	30	22.50	0.93	241.9	907.0	10,204.1
	31	23.25	0.93	249.9	968.5	11,258.9
16	32	24.00	0.93	258.0	1,032.0	12,384.0
	33	24.75	0.92	266.1	1,097.5	13,581.7
17	34	25.50	0.92	274.1	1,165.0	14,854.1
	35	26.25	0.92	282.2	1,234.6	16,203.7
18	36	27.00	0.91	290.3	1,306.1	17,632.7
	37	27.75	0.91	298.3	1,379.7	19,143.3
19	38	28.50	0.91	306.4	1,455.3	20,737.8
	39	29.25	0.91	314.4	1,532.9	22,418.4
20	40	30.00	0.90	322.5	1,612.5	24,187.5
	41	30.75	0.90	330.6	1,694.1	26,047.3
21	42	31.50	0.90	338.6	1,777.8	28,000.1
	43	32.25	0.90	346.7	1,863.4	30,048.1
22	44	33.00	0.89	354.8	1,951.1	32,193.6
	45	33.75	0.89	362.8	2,040.8	34,438.8
23	46	34.50	0.89	370.9	2,132.5	36,786.2
	47	35.25	0.89	378.9	2,226.3	39,237.8
24	48	36.00	0.88	387.0	2,322.0	41,796.0
	49	36.75	0.88	395.1	2,419.8	44,463.1
25	50	37.50	0.88	403.1	2,519.5	47,241.2
	51	38.25	0.88	411.2	2,621.3	50,132.8
26	52	39.00	0.88	419.3	2,725.1	53,139.9
	53	39.75	0.88	427.3	2,830.9	56,265.0
27	54	40.50	0.87	435.4	2,938.8	59,510.3
	55	41.25	0.87	443.4	3,048.6	62,878.1
28	56	42.00	0.87	451.5	3,160.5	66,370.5
	57	42.75	0.87	459.6	3,274.4	69,989.9
29	58	43.50	0.87	467.6	3,390.3	73,738.6
	59	44.25	0.87	475.7	3,508.2	77,618.8
30	60	45.00	0.86	483.8	3,628.1	81,632.8
	61	45.75	0.86	491.8	3,750.1	85,782.9
31	62	46.50	0.86	499.9	3,874.0	90,071.2
	63	47.25	0.86	507.9	4,000.0	94,500.2
32	64	48.00	0.86	516.0	4,128.0	99,072.0
	65	48.75	0.86	524.1	4,258.0	103,789.0
33	66	49.50	0.85	532.1	4,390.0	108.653.3
	67	50.25	0.85	540.2	4,524.1	113,667.3
34	68	51.00	0.85	548.3	4,660.1	118,833.2
	69	51.75	0.85	556.3	4,798.2	124,153.3
35	70	52.50	0.85	564.4	4,938.3	129,629.9
	71	53.25	0.85	572.4	5,080.4	135,265.2
36	72	54.00	0.85	580.5	5,224.5	141,061.5
	73	54.75	0.85	588.6	5,370.6	147,021.1
37	74	55.50	0.84	596.6	5,518.8	153,146.2
	75	56.25	0.84	604.7	5,668.9	159,439.1
38	76	57.00	0.84	612.8	5,821.1	165.902.1
	77	57.75	0.84	620.8	5,975.3	172,537.4
39	78	58.50	0.84	628.9	6,131.5	179,347.3
	79	59.25	0.84	636.9	6,289.8	186,334.1
40	80	60.00	0.84	645.0	6,450.0	193,500.0
	81	60.75	0.84	653.1	6,612.3	200,847.4
41	82	61.50	0.83	661.1	6,776.5	208,378.4
	83	62.25	0.83	669.2	6,942.8	216,095.3
42	84	63.00	0.83	677.3	7,111.1	224,000.5
	85	63.75	0.83	685.3	7,281.4	232,096.1
43	86	64.50	0.83	693.4	7,453.8	240,384.5
	87	65.25	0.83	701.4	7,628.1	248,867.9
44	88	66.00	0.83	709.5	7,804.5	257,548.5
	89	66.75	0.83	717.6	7,982.9	266,428.8
45	90	67.50	0.83	725.6	8,163.3	275,510.8
	91	68.25	0.82	733.7	8,345.7	284,796.9
46	92	69.00	0.82	741.8	8,530.1	294,289.3
	93	69.75	0.82	749.8	8,716.6	303,990.5
47	94	70.50	0.82	757.9	8,905.0	313,902.4
	95	71.25	0.82	765.9	9,095.5	324,027.5
48	96	72.00	0.82	774.0	9,288.0	334,368.0
	97	72.75	0.82	782.1	9,482.5	344,926.3
49	98	73.50	0.82	790.1	9,679.0	355,704.5
	99	74.25	0.82	798.2	9,877.6	366,704.8
50	100	75.00	0.82	806.3	10,078.1	377,929.7

†Data for $1\frac{1}{2}$-in lamination thickness may be obtained by entering the table with double the number of laminations or with the actual depth of the section.

Table A.4 Allowable stresses for laminated timbers [3]

Allowable Unit Stresses (psi) For Normal Conditions Of Loading, Members Stressed Principally In Bending, Loaded Perpendicular To The Wide Face of the Laminations[1, 2, 3, 4]

Combination Symbol	Allowable Unit Stresses, psi						Modulus of Elasticity E psi
	Extreme Fiber in Bending F_b	Tension Parallel to Grain F_t	Compression Parallel to Grain F_c	Compression Perpendicular to Grain $F_{c\perp}$		Horizontal Shear F_v	
				Tension Face	Comp. Face		
DOUGLAS FIR AND LARCH							
16F	1600	1600	1500	385	385	165	1,600,000
18F	1800	1600	1500	385	385	165	1,700,000
20F	2000	1600	1500	385	385	165	1,700,000
	2000	1600	1500	410	410	165	1,700,000
	2000	1600	1500	450	450	165	1,700,000
22F	2200	1600	1500	410	410	165	1,800,000
	2200	1600	1500	450	385	165	1,800,000
24F	2400	1600	1500	450	385	165	1,800,000
NOTE: The 26F combination is only available by special inquiry to the manufacturer. Other combinations are generally available from all laminators.							
26F	2600	1600	1500	450	410	165	1,800,000
HEM-FIR							
18F	1800	1300	1250	245		155	1,600,000
20F	2000	1300	1250	245		155	1,600,000
24F	2400	1300	1250	245		155	1,600,000
SOUTHERN PINE							
16F	1600	500	700	385		140	1,500,000
18F	1800	1600	1500	385		200	1,600,000
20F	2000	500	700	385		140	1,600,000
	2000	1600	1500	385		200	1,700,000
	2000	1600	1500	450		200	1,700,000
22F	2200	1600	1500	385		200	1,700,000
	2200	1600	1500	450		200	1,700,000
24F	2400	1100	1000	385		140	1,700,000
	2400	1600	1500	385		200	1,800,000
	2400	1600	1500	450		200	1,800,000
NOTE: The 26F combination is only available by special inquiry to the manufacturer. Other combinations are generally available from all laminators.							
26F	2600	1600	1500	385		200	1,800,000
	2600	1600	1500	450		200	1,800,000
CALIFORNIA REDWOOD							
16F	1600	1600	1600	325		125	1,400,000
NOTE: The 16F combination is generally available. The 22F combinations are generally available only in members without end joints and the designer should check with the laminator prior to specifying this stress level.							
22F	2200	2000	2000	325		125	1,400,000

1. The tabulated stresses in this table are primarily applicable to members stressed in bending due to a load applied perpendicular to the wide face of the laminations. For combinations and stresses applicable to members loaded primarily axially or parallel to wide face of the laminations, see Table 2.

2. The tabulated bending stresses are applicable to members 12 inches or less in depth. For members greater than 12 inches in depth, the size factor modifications apply.

3. The tabulated combinations are applicable to arches, compression members, tension members and also bending members less than 16¼ inches in depth. For bending members 16¼ inches or more in depth, AITC tension lamination restrictions apply.

4. To obtain wet-use stresses, multiply dry-use stresses by the following factors:

Type of stress	:	Wet-use factor
Bending and tension parallel to grain	:	0.80
Compression parallel to grain	:	0.73
Compression perpendicular to grain	:	0.67
Shear	:	0.88
Modulus of elasticity	:	0.83

Allowable Unit Stresses (psi) For Normal Conditions of Loading. Members Stressed Principally In Axial Tension, Axial Compression Or Loaded In Bending Parallel Or Perpendicular To the Wide Face [4, 5]

Combination Symbol	Tension Parallel to Grain[6] F_t	Compression Parallel to Grain[6] F_c	Extreme Fiber in Bending F_b When Loaded		Compression Perpendicular to Grain[7] $F_{c\perp}$	Horizontal Shear F_v, When Loaded		Modulus of Elasticity E psi
			Parallel to Wide Face[7]	Perpendicular to Wide Face[6, 8]		Parallel to Wide Face[7]	Perpendicular to Wide Face[6]	
DOUGLAS FIR AND LARCH								
1	1200	1500	900	1200	385	145	165	1,600,000
2	1800	1800	1500	1800	385	145	165	1,800,000
3	2200	2100	1900	2200	450	145	165	1,900,000
4	2400	2000	2100	2400	410	145	165	2,000,000
5	2600	2200	2300	2600	450	145	165	2,100,000
HEM-FIR								
1	800	1250	700	1000	245	125	155	1,300,000
2	1150	1500	1200	1400	245	125	155	1,400,000
3	1450	1600	1550	1800	245	125	155	1,600,000
4	1700	1800	1800	2400	245	125	155	1,700,000
SOUTHERN PINE								
1	1600	1400	900	1100	385	165	200	1,500,000
2	2200	1900	1550	1800	385	165	200	1,700,000
3	2600	2200	1800	2100	450	165	200	1,800,000
4	2400	2100	1900	2400	385	165	200	1,900,000
5	2600	2200	2200	2600	450	165	200	2,000,000
CALIFORNIA REDWOOD								
1	1800	1800	1000	1400	325	115	125	1,300,000
2	1800	1800	1000	1400	325	115	125	1,300,000
3	2000	2000	1400	2000	325	125	125	1,400,000
4	2200	2200	2200	2200	325	125	125	1,400,000
5	2200	2200	2200	2200	325	125	125	1,400,000

5. The tabulated stresses in this table are primarily applicable to members loaded axially or parallel to the wide face of the laminations. For combinations and stresses applicable to members stressed principally in bending due to a load applied perpendicular to the wide face of the laminations, see Table 1.

6. The tabulated stresses are applicable to members containing four (4) or more laminations.

7. The tabulated stresses are applicable to members containing three (3) or more laminations.

8. It is not intended that these combinations be used for deep bending members, but if bending members 16¼ inches or deeper are used, AITC tension lamination restrictions apply.

WIDE FLANGE SHAPES

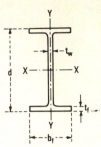

Theoretical Dimensions and Properties for **Designing**

Section Number	Weight per Foot	Area of Section	Depth of Section	Flange Width	Flange Thickness	Web Thickness	Axis X-X I_x	Axis X-X S_x	Axis X-X r_x	Axis Y-Y I_y	Axis Y-Y S_y	Axis Y-Y r_y	r_T
		A	d	b_f	t_f	t_w							
	lb	in.²	in.	in.	in.	in.	in.⁴	in.³	in.	in.⁴	in.³	in.	in.
W36 x 300	300	88.3	36.74	16.655	1.680	0.945	20300	1110	15.2	1300	156	3.83	4.39
	280	82.4	36.52	16.595	1.570	0.885	18900	1030	15.1	1200	144	3.81	4.37
	260	76.5	36.26	16.550	1.440	0.840	17300	953	15.0	1090	132	3.78	4.34
	245	72.1	36.08	16.510	1.350	0.800	16100	895	15.0	1010	123	3.75	4.32
	230	67.6	35.90	16.470	1.260	0.760	15000	837	14.9	940	114	3.73	4.30
W36 x 210	210	61.8	36.69	12.180	1.360	0.830	13200	719	14.6	411	67.5	2.58	3.09
	194	57.0	36.49	12.115	1.260	0.765	12100	664	14.6	375	61.9	2.56	3.07
	182	53.6	36.33	12.075	1.180	0.725	11300	623	14.5	347	57.6	2.55	3.05
	170	50.0	36.17	12.030	1.100	0.680	10500	580	14.5	320	53.2	2.53	3.04
	160	47.0	36.01	12.000	1.020	0.650	9750	542	14.4	295	49.1	2.50	3.02
	150	44.2	35.85	11.975	0.940	0.625	9040	504	14.3	270	45.1	2.47	2.99
	135	39.7	35.55	11.950	0.790	0.600	7800	439	14.0	225	37.7	2.38	2.93
W33 x 241	241	70.9	34.18	15.860	1.400	0.830	14200	829	14.1	932	118	3.63	4.17
	221	65.0	33.93	15.805	1.275	0.775	12800	757	14.1	840	106	3.59	4.15
	201	59.1	33.68	15.745	1.150	0.715	11500	684	14.0	749	95.2	3.56	4.12
W33 x 152	152	44.7	33.49	11.565	1.055	0.635	8160	487	13.5	273	47.2	2.47	2.94
	141	41.6	33.30	11.535	0.960	0.605	7450	448	13.4	246	42.7	2.43	2.92
	130	38.3	33.09	11.510	0.855	0.580	6710	406	13.2	218	37.9	2.39	2.88
	118	34.7	32.86	11.480	0.740	0.550	5900	359	13.0	187	32.6	2.32	2.84
W30 x 211	211	62.0	30.94	15.105	1.315	0.775	10300	663	12.9	757	100	3.49	3.99
	191	56.1	30.68	15.040	1.185	0.710	9170	598	12.8	673	89.5	3.46	3.97
	173	50.8	30.44	14.985	1.065	0.655	8200	539	12.7	598	79.8	3.43	3.94
W30 x 132	132	38.9	30.31	10.545	1.000	0.615	5770	380	12.2	196	37.2	2.25	2.68
	124	36.5	30.17	10.515	0.930	0.585	5360	355	12.1	181	34.4	2.23	2.66
	116	34.2	30.01	10.495	0.850	0.565	4930	329	12.0	164	31.3	2.19	2.64
	108	31.7	29.83	10.475	0.760	0.545	4470	299	11.9	146	27.9	2.15	2.61
	99	29.1	29.65	10.450	0.670	0.520	3990	269	11.7	128	24.5	2.10	2.57

All shapes on these pages have parallel-faced flanges.

WIDE FLANGE SHAPES

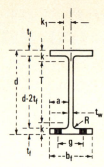

Approximate Dimensions for **Detailing**

Section Number	Weight per Foot	Depth of Section	Flange Width	Flange Thickness	Web Thickness	Half Web Thickness	d-2t_f	a	T	k	k₁	R	Usual Flange Gage
		d	b_f	t_f	t_w	$\frac{t_w}{2}$							g
	lb	in.	in.	in.	in.	in.	in.	in.	in.	in.	in.	in.	in.
W36 x	300	36¾	16⅝	1¹¹⁄₁₆	¹⁵⁄₁₆	½	33⅜	7⅞	31⅛	2¹³⁄₁₆	1½	0.95	5½
	280	36½	16⅝	1⁹⁄₁₆	⅞	⁷⁄₁₆	33⅜	7⅞	31⅛	2¹¹⁄₁₆	1½	0.95	5½
	260	36¼	16½	1⁷⁄₁₆	¹³⁄₁₆	⁷⁄₁₆	33⅜	7⅞	31⅛	2⁹⁄₁₆	1½	0.95	5½
	245	36⅛	16½	1⅜	¹³⁄₁₆	⁷⁄₁₆	33⅜	7⅞	31⅛	2½	1⁷⁄₁₆	0.95	5½
	230	35⅞	16½	1¼	¾	⅜	33⅜	7⅞	31⅛	2⅜	1⁷⁄₁₆	0.95	5½
W36 x	210	36¾	12⅛	1⅜	¹³⁄₁₆	⁷⁄₁₆	34	5⅝	32⅛	2⁵⁄₁₆	1¼	0.75	5½
	194	36½	12⅛	1¼	¾	⅜	34	5⅝	32⅛	2³⁄₁₆	1³⁄₁₆	0.75	5½
	182	36⅜	12⅛	1³⁄₁₆	¾	⅜	34	5⅝	32⅛	2⅛	1³⁄₁₆	0.75	5½
	170	36⅛	12	1⅛	¹¹⁄₁₆	⅜	34	5⅝	32⅛	2	1³⁄₁₆	0.75	5½
	160	36	12	1	⅝	⁵⁄₁₆	34	5⅝	32⅛	1¹⁵⁄₁₆	1⅛	0.75	5½
	150	35⅞	12	¹⁵⁄₁₆	⅝	⁵⁄₁₆	34	5⅝	32⅛	1⅞	1⅛	0.75	5½
	135	35½	12	¹³⁄₁₆	⅝	⁵⁄₁₆	34	5⅝	32⅛	1¹¹⁄₁₆	1⅛	0.75	5½
W33 x	241	34⅛	15⅞	1⅜	¹³⁄₁₆	⁷⁄₁₆	31⅜	7½	29¾	2³⁄₁₆	1³⁄₁₆	0.70	5½
	221	33⅞	15¾	1¼	¾	⅜	31⅜	7½	29¾	2¹⁄₁₆	1³⁄₁₆	0.70	5½
	201	33⅝	15¾	1⅛	¹¹⁄₁₆	⅜	31⅜	7½	29¾	1¹⁵⁄₁₆	1⅛	0.70	5½
W33 x	152	33½	11⅝	1¹⁄₁₆	⅝	⁵⁄₁₆	31⅜	5½	29¾	1⅞	1⅛	0.70	5½
	141	33¼	11½	¹⁵⁄₁₆	⅝	⁵⁄₁₆	31⅜	5½	29¾	1¾	1¹⁄₁₆	0.70	5½
	130	33⅛	11½	⅞	⁹⁄₁₆	⁵⁄₁₆	31⅜	5½	29¾	1¹¹⁄₁₆	1¹⁄₁₆	0.70	5½
	118	32⅞	11½	¾	⁹⁄₁₆	⁵⁄₁₆	31⅜	5½	29¾	1⁹⁄₁₆	1¹⁄₁₆	0.70	5½
W30 x	211	31	15⅛	1⁵⁄₁₆	¾	⅜	28⁵⁄₁₆	7⅛	26¾	2⅛	1⅛	0.65	5½
	191	30⅝	15	1³⁄₁₆	¹¹⁄₁₆	⅜	28⁵⁄₁₆	7⅛	26¾	1¹⁵⁄₁₆	1¹⁄₁₆	0.65	5½
	173	30½	15	1¹⁄₁₆	⅝	⁵⁄₁₆	28⁵⁄₁₆	7⅛	26¾	1⅞	1¹⁄₁₆	0.65	5½
W30 x	132	30¼	10½	1	⅝	⁵⁄₁₆	28⁵⁄₁₆	5	26¾	1¾	1¹⁄₁₆	0.65	5½
	124	30⅛	10½	¹⁵⁄₁₆	⁹⁄₁₆	⁵⁄₁₆	28⁵⁄₁₆	5	26¾	1¹¹⁄₁₆	1	0.65	5½
	116	30	10½	⅞	⁹⁄₁₆	⁵⁄₁₆	28⁵⁄₁₆	5	26¾	1⅝	1	0.65	5½
	108	29⅞	10½	¾	⁹⁄₁₆	⁵⁄₁₆	28⁵⁄₁₆	5	26¾	1⁹⁄₁₆	1	0.65	5½
	99	29⅝	10½	¹¹⁄₁₆	½	¼	28⁵⁄₁₆	5	26¾	1⁷⁄₁₆	1	0.65	5½

WIDE FLANGE SHAPES

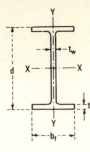

Theoretical Dimensions and Properties for **Designing**

Section Number	Weight per Foot	Area of Section A	Depth of Section d	Flange		Web Thickness t_w	Axis X-X			Axis Y-Y			r_T
				Width b_f	Thickness t_f		I_x	S_x	r_x	I_y	S_y	r_y	
	lb	in.²	in.	in.	in.	in.	in.⁴	in.³	in.	in.⁴	in.³	in.	in.
W27 x 178	178	52.3	27.81	14.085	1.190	0.725	6990	502	11.6	555	78.8	3.26	3.72
	161	47.4	27.59	14.020	1.080	0.660	6280	455	11.5	497	70.9	3.24	3.70
	146	42.9	27.38	13.965	0.975	0.605	5630	411	11.4	443	63.5	3.21	3.68
W27 x 114	114	33.5	27.29	10.070	0.930	0.570	4090	299	11.0	159	31.5	2.18	2.58
	102	30.0	27.09	10.015	0.830	0.515	3620	267	11.0	139	27.8	2.15	2.56
	94	27.7	26.92	9.990	0.745	0.490	3270	243	10.9	124	24.8	2.12	2.53
	84	24.8	26.71	9.960	0.640	0.460	2850	213	10.7	106	21.2	2.07	2.49
W24 x 162	162	47.7	25.00	12.955	1.220	0.705	5170	414	10.4	443	68.4	3.05	3.45
	146	43.0	24.74	12.900	1.090	0.650	4580	371	10.3	391	60.5	3.01	3.43
	131	38.5	24.48	12.855	0.960	0.605	4020	329	10.2	340	53.0	2.97	3.40
	117	34.4	24.26	12.800	0.850	0.550	3540	291	10.1	297	46.5	2.94	3.37
	104	30.6	24.06	12.750	0.750	0.500	3100	258	10.1	259	40.7	2.91	3.35
W24 x 94	94	27.7	24.31	9.065	0.875	0.515	2700	222	9.87	109	24.0	1.98	2.33
	84	24.7	24.10	9.020	0.770	0.470	2370	196	9.79	94.4	20.9	1.95	2.31
	76	22.4	23.92	8.990	0.680	0.440	2100	176	9.69	82.5	18.4	1.92	2.29
	68	20.1	23.73	8.965	0.585	0.415	1830	154	9.55	70.4	15.7	1.87	2.26
W24 x 62	62	18.2	23.74	7.040	0.590	0.430	1550	131	9.23	34.5	9.80	1.38	1.71
	55	16.2	23.57	7.005	0.505	0.395	1350	114	9.11	29.1	8.30	1.34	1.68
W21 x 147	147	43.2	22.06	12.510	1.150	0.720	3630	329	9.17	376	60.1	2.95	3.34
	132	38.8	21.83	12.440	1.035	0.650	3220	295	9.12	333	53.5	2.93	3.31
	122	35.9	21.68	12.390	0.960	0.600	2960	273	9.09	305	49.2	2.92	3.30
	111	32.7	21.51	12.340	0.875	0.550	2670	249	9.05	274	44.5	2.90	3.28
	101	29.8	21.36	12.290	0.800	0.500	2420	227	9.02	248	40.3	2.89	3.27
W21 x 93	93	27.3	21.62	8.420	0.930	0.580	2070	192	8.70	92.9	22.1	1.84	2.17
	83	24.3	21.43	8.355	0.835	0.515	1830	171	8.67	81.4	19.5	1.83	2.15
	73	21.5	21.24	8.295	0.740	0.455	1600	151	8.64	70.6	17.0	1.81	2.13
	68	20.0	21.13	8.270	0.685	0.430	1480	140	8.60	64.7	15.7	1.80	2.12
	62	18.3	20.99	8.240	0.615	0.400	1330	127	8.54	57.5	13.9	1.77	2.10
W21 x 57	57	16.7	21.06	6.555	0.650	0.405	1170	111	8.36	30.6	9.35	1.35	1.64
	50	14.7	20.83	6.530	0.535	0.380	984	94.5	8.18	24.9	7.64	1.30	1.60
	44	13.0	20.66	6.500	0.450	0.350	843	81.6	8.06	20.7	6.36	1.26	1.57

All shapes on these pages have parallel-faced flanges.

WIDE FLANGE SHAPES

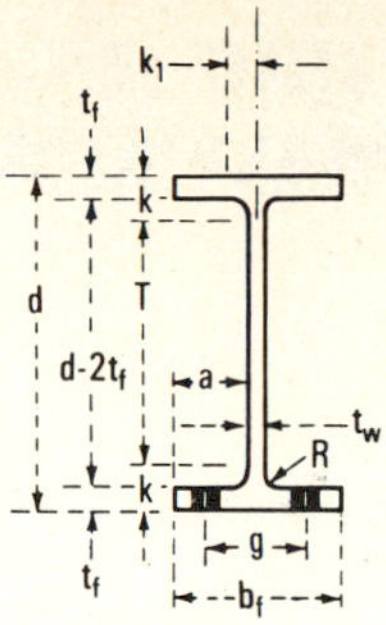

Approximate Dimensions for **Detailing**

Section Number	Weight per Foot	Depth of Section d	Flange Width b_f	Flange Thickness t_f	Web Thickness t_w	Half Web Thickness $\frac{t_w}{2}$	$d-2t_f$	a	T	k	k_1	R	Usual Flange Gage g
	lb	in.	in.	in.	in.	in.	in.	in.	in.	in.	in.	in.	in.
W27 x	178	27 3/4	14 1/8	1 3/16	3/4	3/8	25 7/16	6 5/8	24	1 7/8	1 1/16	0.60	5 1/2
	161	27 5/8	14	1 1/16	11/16	3/8	25 7/16	6 5/8	24	1 13/16	1	0.60	5 1/2
	146	27 3/8	14	1	5/8	5/16	25 7/16	6 5/8	24	1 11/16	1	0.60	5 1/2
W27 x	114	27 1/4	10 1/8	15/16	9/16	5/16	25 7/16	4 3/4	24	1 5/8	15/16	0.60	5 1/2
	102	27 1/8	10	13/16	1/2	1/4	25 7/16	4 3/4	24	1 9/16	15/16	0.60	5 1/2
	94	26 7/8	10	3/4	1/2	1/4	25 7/16	4 3/4	24	1 7/16	15/16	0.60	5 1/2
	84	26 3/4	10	5/8	7/16	1/4	25 7/16	4 3/4	24	1 3/8	15/16	0.60	5 1/2
W24 x	162	25	13	1 1/4	11/16	3/8	22 9/16	6 1/8	21	2	1 1/16	0.50	5 1/2
	146	24 3/4	12 7/8	1 1/16	5/8	5/16	22 9/16	6 1/8	21	1 7/8	1 1/16	0.50	5 1/2
	131	24 1/2	12 7/8	15/16	5/8	5/16	22 9/16	6 1/8	21	1 3/4	1 1/16	0.50	5 1/2
	117	24 1/4	12 3/4	7/8	9/16	5/16	22 9/16	6 1/8	21	1 5/8	1	0.50	5 1/2
	104	24	12 3/4	3/4	1/2	1/4	22 9/16	6 1/8	21	1 1/2	1	0.50	5 1/2
W24 x	94	24 1/4	9 1/8	7/8	1/2	1/4	22 9/16	4 1/4	21	1 5/8	1	0.50	5 1/2
	84	24 1/8	9	3/4	1/2	1/4	22 9/16	4 1/4	21	1 9/16	15/16	0.50	5 1/2
	76	23 7/8	9	11/16	7/16	1/4	22 9/16	4 1/4	21	1 7/16	15/16	0.50	5 1/2
	68	23 3/4	9	9/16	7/16	1/4	22 9/16	4 1/4	21	1 3/8	15/16	0.50	5 1/2
W24 x	62	23 3/4	7	9/16	7/16	1/4	22 9/16	3 1/4	21	1 3/8	15/16	0.50	3 1/2
	55	23 5/8	7	1/2	3/8	3/16	22 9/16	3 1/4	21	1 5/16	15/16	0.50	3 1/2
W21 x	147	22	12 1/2	1 1/8	3/4	3/8	19 3/4	5 7/8	18 1/4	1 7/8	1 1/16	0.50	5 1/2
	132	21 7/8	12 1/2	1 1/16	5/8	5/16	19 3/4	5 7/8	18 1/4	1 13/16	1	0.50	5 1/2
	122	21 5/8	12 3/8	15/16	5/8	5/16	19 3/4	5 7/8	18 1/4	1 11/16	1	0.50	5 1/2
	111	21 1/2	12 3/8	7/8	9/16	5/16	19 3/4	5 7/8	18 1/4	1 5/8	15/16	0.50	5 1/2
	101	21 3/8	12 1/4	13/16	1/2	1/4	19 3/4	5 7/8	18 1/4	1 9/16	15/16	0.50	5 1/2
W21 x	93	21 5/8	8 3/8	15/16	9/16	5/16	19 3/4	3 7/8	18 1/4	1 11/16	1	0.50	5 1/2
	83	21 3/8	8 3/8	13/16	1/2	1/4	19 3/4	3 7/8	18 1/4	1 9/16	15/16	0.50	5 1/2
	73	21 1/4	8 1/4	3/4	7/16	1/4	19 3/4	3 7/8	18 1/4	1 1/2	15/16	0.50	5 1/2
	68	21 1/8	8 1/4	11/16	7/16	1/4	19 3/4	3 7/8	18 1/4	1 7/16	7/8	0.50	5 1/2
	62	21	8 1/4	5/8	3/8	3/16	19 3/4	3 7/8	18 1/4	1 3/8	7/8	0.50	5 1/2
W21 x	57	21	6 1/2	5/8	3/8	3/16	19 3/4	3 1/8	18 1/4	1 3/8	7/8	0.50	3 1/2
	50	20 7/8	6 1/2	9/16	3/8	3/16	19 3/4	3 1/8	18 1/4	1 5/16	7/8	0.50	3 1/2
	44	20 5/8	6 1/2	7/16	3/8	3/16	19 3/4	3 1/8	18 1/4	1 3/16	7/8	0.50	3 1/2

WIDE FLANGE SHAPES

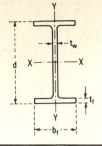

Theoretical Dimensions and Properties for **Designing**

Section Number	Weight per Foot	Area of Section	Depth of Section	Flange		Web Thickness	Axis X-X			Axis Y-Y			
				Width	Thickness		I_x	S_x	r_x	I_y	S_y	r_y	r_T
		A	d	b_f	t_f	t_w							
	lb	in.²	in.	in.	in.	in.	in.⁴	in.³	in.	in.⁴	in.³	in.	in.
W18 x	119	35.1	18.97	11.265	1.060	0.655	2190	231	7.90	253	44.9	2.69	3.02
	106	31.1	18.73	11.200	0.940	0.590	1910	204	7.84	220	39.4	2.66	3.00
	97	28.5	18.59	11.145	0.870	0.535	1750	188	7.82	201	36.1	2.65	2.99
	86	25.3	18.39	11.090	0.770	0.480	1530	166	7.77	175	31.6	2.63	2.97
	76	22.3	18.21	11.035	0.680	0.425	1330	146	7.73	152	27.6	2.61	2.95
W18 x	71	20.8	18.47	7.635	0.810	0.495	1170	127	7.50	60.3	15.8	1.70	1.98
	65	19.1	18.35	7.590	0.750	0.450	1070	117	7.49	54.8	14.4	1.69	1.97
	60	17.6	18.24	7.555	0.695	0.415	984	108	7.47	50.1	13.3	.1.69	1.96
	55	16.2	18.11	7.530	0.630	0.390	890	98.3	7.41	44.9	11.9	1.67	1.95
	50	14.7	17.99	7.495	0.570	0.355	800	88.9	7.38	40.1	10.7	1.65	1.94
W18 x	46	13.5	18.06	6.060	0.605	0.360	712	78.8	7.25	22.5	7.43	1.29	1.54
	40	11.8	17.90	6.015	0.525	0.315	612	68.4	7.21	19.1	6.35	1.27	1.52
	35	10.3	17.70	6.000	0.425	0.300	510	57.6	7.04	15.3	5.12	1.22	1.49
W16 x	100	29.4	16.97	10.425	0.985	0.585	1490	175	7.10	186	35.7	2.52	2.81
	89	26.2	16.75	10.365	0.875	0.525	1300	155	7.05	163	31.4	2.49	2.79
	77	22.6	16.52	10.295	0.760	0.455	1110	134	7.00	138	26.9	2.47	2.77
	67	19.7	16.33	10.235	0.665	0.395	954	117	6.96	119	23.2	2.46	2.75
W16 x	57	16.8	16.43	7.120	0.715	0.430	758	92.2	6.72	43.1	12.1	1.60	1.86
	50	14.7	16.26	7.070	0.630	0.380	659	81.0	6.68	37.2	10.5	1.59	1.84
	45	13.3	16.13	7.035	0.565	0.345	586	72.7	6.65	32.8	9.34	1.57	1.83
	40	11.8	16.01	6.995	0.505	0.305	518	64.7	6.63	28.9	8.25	1.57	1.82
	36	10.6	15.86	6.985	0.430	0.295	448	56.5	6.51	24.5	7.00	1.52	1.79
W16 x	31	9.12	15.88	5.525	0.440	0.275	375	47.2	6.41	12.4	4.49	1.17	1.39
	26	7.68	15.69	5.500	0.345	0.250	301	38.4	6.26	9.59	3.49	1.12	1.36

All shapes on these pages have parallel-faced flanges.

WIDE FLANGE SHAPES

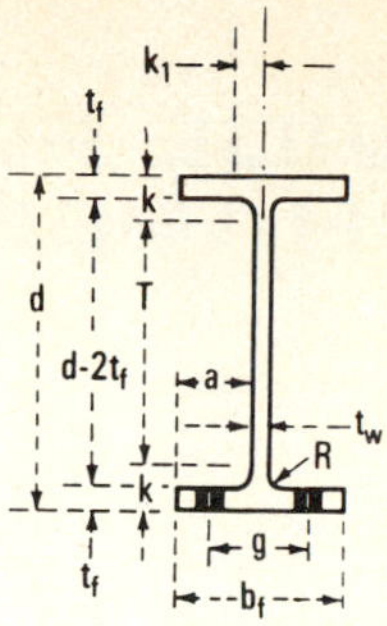

Approximate Dimensions for **Detailing**

Section Number	Weight per Foot	Depth of Section	Flange Width	Flange Thickness	Web Thickness	Half Web Thickness	$d-2t_f$	a	T	k	k_1	R	Usual Flange Gage
		d	b_f	t_f	t_w	$\frac{t_w}{2}$							g
	lb	in.	in.	in.	in.	in.	in.	in.	in.	in.	in.	in.	in.
W18 x	119	19	11¼	1 1/16	⅝	5/16	16⅞	5¼	15½	1¾	15/16	0.40	5½
	106	18¾	11¼	15/16	9/16	5/16	16⅞	5¼	15½	1⅝	15/16	0.40	5½
	97	18⅝	11⅛	⅞	9/16	5/16	16⅞	5¼	15½	1 9/16	⅞	0.40	5½
	86	18⅜	11⅛	¾	½	¼	16⅞	5¼	15½	1 7/16	⅞	0.40	5½
	76	18¼	11	11/16	7/16	¼	16⅞	5¼	15½	1⅜	13/16	0.40	5½
W18 x	71	18½	7⅝	13/16	½	¼	16⅞	3⅝	15½	1½	⅞	0.40	3½
	65	18⅜	7⅝	¾	7/16	¼	16⅞	3⅝	15½	1 7/16	⅞	0.40	3½
	60	18¼	7½	11/16	7/16	¼	16⅞	3⅝	15½	1⅜	13/16	0.40	3½
	55	18⅛	7½	⅝	⅜	3/16	16⅞	3⅝	15½	1 5/16	13/16	0.40	3½
	50	18	7½	9/16	⅜	3/16	16⅞	3⅝	15½	1¼	13/16	0.40	3½
W18 x	46	18	6	⅝	⅜	3/16	16⅞	2⅞	15½	1¼	13/16	0.40	3½
	40	17⅞	6	½	5/16	3/16	16⅞	2⅞	15½	1 3/16	13/16	0.40	3½
	35	17¾	6	7/16	5/16	3/16	16⅞	2⅞	15½	1⅛	¾	0.40	3½
W16 x	100	17	10⅜	1	9/16	5/16	15	4⅞	13⅝	1 11/16	15/16	0.40	5½
	89	16¾	10⅜	⅞	½	¼	15	4⅞	13⅝	1 9/16	⅞	0.40	5½
	77	16½	10¼	¾	7/16	¼	15	4⅞	13⅝	1 7/16	⅞	0.40	5½
	67	16⅜	10¼	11/16	⅜	3/16	15	4⅞	13⅝	1⅜	13/16	0.40	5½
W16 x	57	16⅜	7⅛	11/16	7/16	¼	15	3⅜	13⅝	1⅜	⅞	0.40	3½
	50	16¼	7⅛	⅝	⅜	3/16	15	3⅜	13⅝	1 5/16	13/16	0.40	3½
	45	16⅛	7	9/16	⅜	3/16	15	3⅜	13⅝	1¼	13/16	0.40	3½
	40	16	7	½	5/16	3/16	15	3⅜	13⅝	1 3/16	13/16	0.40	3½
	36	15⅞	7	7/16	5/16	3/16	15	3⅜	13⅝	1⅛	¾	0.40	3½
W16 x	31	15⅞	5½	7/16	¼	⅛	15	2⅝	13⅝	1⅛	¾	0.40	2¾
	26	15¾	5½	⅜	¼	⅛	15	2⅝	13⅝	1 1/16	¾	0.40	2¾

WIDE FLANGE SHAPES

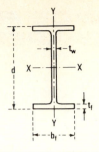

Theoretical Dimensions and Properties for **Designing**

Section Number	Weight per Foot	Area of Section	Depth of Section	Flange		Web Thick- ness	Axis X-X			Axis Y-Y			
				Width	Thick- ness		I_x	S_x	r_x	I_y	S_y	r_y	r_T
		A	d	b_f	t_f	t_w							
	lb	in.²	in.	in.	in.	in.	in.⁴	in.³	in.	in.⁴	in.³	in.	in.
W14 x	730*	215	22.42	17.890	4.910	3.070	14300	1280	8.17	4720	527	4.69	4.99
	665*	196	21.64	17.650	4.520	2.830	12400	1150	7.98	4170	472	4.62	4.92
	605*	178	20.92	17.415	4.160	2.595	10800	1040	7.80	3680	423	4.55	4.85
	550*	162	20.24	17.200	3.820	2.380	9430	931	7.63	3250	378	4.49	4.79
	500*	147	19.60	17.010	3.500	2.190	8210	838	7.48	2880	339	4.43	4.73
	455*	134	19.02	16.835	3.210	2.015	7190	756	7.33	2560	304	4.38	4.68
W14 x	426	125	18.67	16.695	3.035	1.875	6600	707	7.26	2360	283	4.34	4.64
	398	117	18.29	16.590	2.845	1.770	6000	656	7.16	2170	262	4.31	4.61
	370	109	17.92	16.475	2.660	1.655	5440	607	7.07	1990	241	4.27	4.57
	342	101	17.54	16.360	2.470	1.540	4900	559	6.98	1810	221	4.24	4.54
	311	91.4	17.12	16.230	2.260	1.410	4330	506	6.88	1610	199	4.20	4.50
	283	83.3	16.74	16.110	2.070	1.290	3840	459	6.79	1440	179	4.17	4.46
	257	75.6	16.38	15.995	1.890	1.175	3400	415	6.71	1290	161	4.13	4.43
	233	68.5	16.04	15.890	1.720	1.070	3010	375	6.63	1150	145	4.10	4.40
	211	62.0	15.72	15.800	1.560	0.980	2660	338	6.55	1030	130	4.07	4.37
	193	56.8	15.48	15.710	1.440	0.890	2400	310	6.50	931	119	4.05	4.35
	176	51.8	15.22	15.650	1.310	0.830	2140	281	6.43	838	107	4.02	4.32
	159	46.7	14.98	15.565	1.190	0.745	1900	254	6.38	748	96.2	4.00	4.30
	145	42.7	14.78	15.500	1.090	0.680	1710	232	6.33	677	87.3	3.98	4.28
W14 x	132	38.8	14.66	14.725	1.030	0.645	1530	209	6.28	548	74.5	3.76	4.05
	120	35.3	14.48	14.670	0.940	0.590	1380	190	6.24	495	67.5	3.74	4.04
	109	32.0	14.32	14.605	0.860	0.525	1240	173	6.22	447	61.2	3.73	4.02
	99	29.1	14.16	14.565	0.780	0.485	1110	157	6.17	402	55.2	3.71	4.00
	90	26.5	14.02	14.520	0.710	0.440	999	143	6.14	362	49.9	3.70	3.99
W14 x	82	24.1	14.31	10.130	0.855	0.510	882	123	6.05	148	29.3	2.48	2.74
	74	21.8	14.17	10.070	0.785	0.450	796	112	6.04	134	26.6	2.48	2.72
	68	20.0	14.04	10.035	0.720	0.415	723	103	6.01	121	24.2	2.46	2.71
	61	17.9	13.89	9.995	0.645	0.375	640	92.2	5.98	107	21.5	2.45	2.70
W14 x	53	15.6	13.92	8.060	0.660	0.370	541	77.8	5.89	57.7	14.3	1.92	2.15
	48	14.1	13.79	8.030	0.595	0.340	485	70.3	5.85	51.4	12.8	1.91	2.13
	43	12.6	13.66	7.995	0.530	0.305	428	62.7	5.82	45.2	11.3	1.89	2.12

*These shapes have a 1°-00′ (1.75%) flange slope. Flange thicknesses shown are average thicknesses.
Properties shown are for a parallel flange section.

All other shapes on these pages have parallel-faced flanges.

WIDE FLANGE SHAPES

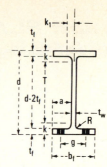

Approximate Dimensions for **Detailing**

Section Number	Weight per Foot	Depth of Section	Flange Width	Flange Thickness	Web Thickness	Half Web Thickness	$d-2t_f$	a	T	k	k_1	R	Usual Flange Gage
		d	b_f	t_f	t_w	$\frac{t_w}{2}$							g
	lb	in.	in.	in.	in.	in.	in.	in.	in.	in.	in.	in.	in.
W14 x	730	22⅜	17⅞	4¹⁵/₁₆	3¹/₁₆	1⁹/₁₆	12⅝	7⅜	11¼	5⁹/₁₆	2³/₁₆	0.60	3-(7½)-3
	665	21⅝	17⅝	4½	2¹³/₁₆	1⁷/₁₆	12⅝	7⅜	11¼	5³/₁₆	2¹/₁₆	0.60	3-(7½)-3
	605	20⅞	17⅜	4³/₁₆	2⅝	1⁵/₁₆	12⅝	7⅜	11¼	4¹³/₁₆	1¹⁵/₁₆	0.60	3-(7½)-3
	550	20¼	17¼	3¹³/₁₆	2⅜	1³/₁₆	12⅝	7⅜	11¼	4½	1¹³/₁₆	0.60	3-(7½)-3
	500	19⅝	17	3½	2³/₁₆	1⅛	12⅝	7⅜	11¼	4³/₁₆	1¾	0.60	3-(7½)-3
	455	19	16⅞	3³/₁₆	2	1	12⅝	7⅜	11¼	3⅞	1⅝	0.60	3-(7½)-3
W14 x	426	18⅝	16¾	3¹/₁₆	1⅞	¹⁵/₁₆	12⅝	7⅜	11¼	3¹¹/₁₆	1⁹/₁₆	0.60	3-(5½)-3
	398	18¼	16⅝	2⅞	1¾	⅞	12⅝	7⅜	11¼	3½	1½	0.60	3-(5½)-3
	370	17⅞	16½	2¹¹/₁₆	1⅝	¹³/₁₆	12⅝	7⅜	11¼	3⁵/₁₆	1⁷/₁₆	0.60	3-(5½)-3
	342	17½	16⅜	2½	1⁹/₁₆	¹³/₁₆	12⅝	7⅜	11¼	3⅛	1⅜	0.60	3-(5½)-3
	311	17⅛	16¼	2¼	1⁷/₁₆	¾	12⅝	7⅜	11¼	2¹⁵/₁₆	1⁵/₁₆	0.60	3-(5½)-3
	283	16¾	16⅛	2¹/₁₆	1⁵/₁₆	¹¹/₁₆	12⅝	7⅜	11¼	2¾	1¼	0.60	3-(5½)-3
	257	16⅜	16	1⅞	1³/₁₆	⅝	12⅝	7⅜	11¼	2⁹/₁₆	1³/₁₆	0.60	3-(5½)-3
	233	16	15⅞	1¾	1¹/₁₆	⁹/₁₆	12⅝	7⅜	11¼	2⅜	1³/₁₆	0.60	3-(5½)-3
	211	15¾	15¾	1⁹/₁₆	1	½	12⅝	7⅜	11¼	2¼	1⅛	0.60	3-(5½)-3
	193	15½	15¾	1⁷/₁₆	⅞	⁷/₁₆	12⅝	7⅜	11¼	2⅛	1¹/₁₆	0.60	3-(5½)-3
	176	15¼	15⅝	1⁵/₁₆	¹³/₁₆	⁷/₁₆	12⅝	7⅜	11¼	2	1¹/₁₆	0.60	3-(5½)-3
	159	15	15⅝	1³/₁₆	¾	⅜	12⅝	7⅜	11¼	1⅞	1	0.60	3-(5½)-3
	145	14¾	15½	1¹/₁₆	¹¹/₁₆	⅜	12⅝	7⅜	11¼	1¾	1	0.60	3-(5½)-3
W14 x	132	14⅝	14¾	1	⅝	⁵/₁₆	12⅝	7	11¼	1¹¹/₁₆	¹⁵/₁₆	0.60	5½
	120	14½	14⅝	¹⁵/₁₆	⁹/₁₆	⁵/₁₆	12⅝	7	11¼	1⅝	¹⁵/₁₆	0.60	5½
	109	14⅜	14⅝	⅞	½	¼	12⅝	7	11¼	1⁹/₁₆	⅞	0.60	5½
	99	14⅛	14⅝	¾	½	¼	12⅝	7	11¼	1⁷/₁₆	⅞	0.60	5½
	90	14	14½	¹¹/₁₆	⁷/₁₆	¼	12⅝	7	11¼	1⅜	⅞	0.60	5½
W14 x	82	14¼	10⅛	⅞	½	¼	12⅝	4¾	11	1⅝	1	0.60	5½
	74	14⅛	10⅛	¹³/₁₆	⁷/₁₆	¼	12⅝	4¾	11	1⁹/₁₆	¹⁵/₁₆	0.60	5½
	68	14	10	¾	⁷/₁₆	¼	12⅝	4¾	11	1½	¹⁵/₁₆	0.60	5½
	61	13⅞	10	⅝	⅜	³/₁₆	12⅝	4¾	11	1⁷/₁₆	¹⁵/₁₆	0.60	5½
W14 x	53	13⅞	8	¹¹/₁₆	⅜	³/₁₆	12⅝	3⅞	11	1⁷/₁₆	¹⁵/₁₆	0.60	5½
	48	13¾	8	⅝	⁵/₁₆	³/₁₆	12⅝	3⅞	11	1⅜	⅞	0.60	5½
	43	13⅝	8	½	⁵/₁₆	³/₁₆	12⅝	3⅞	11	1⁵/₁₆	⅞	0.60	5½

WIDE FLANGE SHAPES

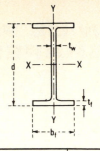

Theoretical Dimensions and Properties for **Designing**

Section Number	Weight per Foot	Area of Section	Depth of Section	Flange		Web Thickness	Axis X-X			Axis Y-Y			r_T
		A	d	Width b_f	Thickness t_f	t_w	I_x	S_x	r_x	I_y	S_y	r_y	
	lb	in.²	in.	in.	in.	in.	in.⁴	in.³	in.	in.⁴	in.³	in.	in.
W14 x	**38**	11.2	14.10	6.770	0.515	0.310	385	54.6	5.88	26.7	7.88	1.55	1.77
	34	10.0	13.98	6.745	0.455	0.285	340	48.6	5.83	23.3	6.91	1.53	1.76
	30	8.85	13.84	6.730	0.385	0.270	291	42.0	5.73	19.6	5.82	1.49	1.74
W14 x	**26**	7.69	13.91	5.025	0.420	0.255	245	35.3	5.65	8.91	3.54	1.08	1.28
	22	6.49	13.74	5.000	0.335	0.230	199	29.0	5.54	7.00	2.80	1.04	1.25
W12 x	**190**	55.8	14.38	12.670	1.735	1.060	1890	263	5.82	589	93.0	3.25	3.50
	170	50.0	14.03	12.570	1.560	0.960	1650	235	5.74	517	82.3	3.22	3.47
	152	44.7	13.71	12.480	1.400	0.870	1430	209	5.66	454	72.8	3.19	3.44
	136	39.9	13.41	12.400	1.250	0.790	1240	186	5.58	398	64.2	3.16	3.41
	120	35.3	13.12	12.320	1.105	0.710	1070	163	5.51	345	56.0	3.13	3.38
	106	31.2	12.89	12.220	0.990	0.610	933	145	5.47	301	49.3	3.11	3.36
	96	28.2	12.71	12.160	0.900	0.550	833	131	5.44	270	44.4	3.09	3.34
	87	25.6	12.53	12.125	0.810	0.515	740	118	5.38	241	39.7	3.07	3.32
	79	23.2	12.38	12.080	0.735	0.470	662	107	5.34	216	35.8	3.05	3.31
	72	21.1	12.25	12.040	0.670	0.430	597	97.4	5.31	195	32.4	3.04	3.29
	65	19.1	12.12	12.000	0.605	0.390	533	87.9	5.28	174	29.1	3.02	3.28
W12 x	**58**	17.0	12.19	10.010	0.640	0.360	475	78.0	5.28	107	21.4	2.51	2.72
	53	15.6	12.06	9.995	0.575	0.345	425	70.6	5.23	95.8	19.2	2.48	2.71
W12 x	**50**	14.7	12.19	8.080	0.640	0.370	394	64.7	5.18	56.3	13.9	1.96	2.17
	45	13.2	12.06	8.045	0.575	0.335	350	58.1	5.15	50.0	12.4	1.94	2.15
	40	11.8	11.94	8.005	0.515	0.295	310	51.9	5.13	44.1	11.0	1.93	2.14
W12 x	**35**	10.3	12.50	6.560	0.520	0.300	285	45.6	5.25	24.5	7.47	1.54	1.74
	30	8.79	12.34	6.520	0.440	0.260	238	38.6	5.21	20.3	6.24	1.52	1.73
	26	7.65	12.22	6.490	0.380	0.230	204	33.4	5.17	17.3	5.34	1.51	1.72
W12 x	**22**	6.48	12.31	4.030	0.425	0.260	156	25.4	4.91	4.66	2.31	0.848	1.02
	19	5.57	12.16	4.005	0.350	0.235	130	21.3	4.82	3.76	1.88	0.822	0.997
	16	4.71	11.99	3.990	0.265	0.220	103	17.1	4.67	2.82	1.41	0.773	0.963
	14	4.16	11.91	3.970	0.225	0.200	88.6	14.9	4.62	2.36	1.19	0.753	0.946

All shapes on these pages have parallel-faced flanges.

WIDE FLANGE SHAPES

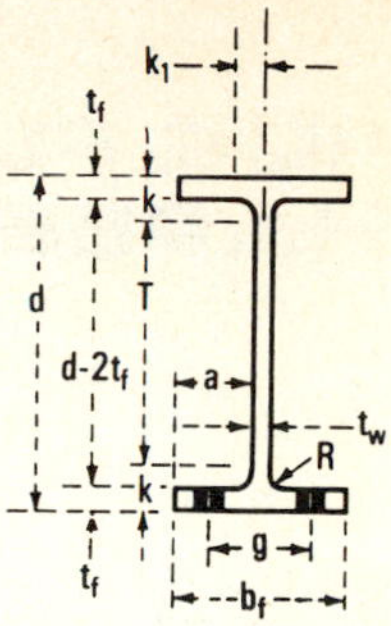

Approximate Dimensions for **Detailing**

Section Number	Weight per Foot	Dept of Section d	Flange Width b_f	Flange Thickness t_f	Web Thickness t_w	Half Web Thickness $\frac{t_w}{2}$	$d-2t_f$	a	T	k	k_1	R	Usual Flange Gage g
	lb	in.	in.	in.	in.	in.	in.	in.	in.	in.	in.	in.	in.
W14 x	38	14 1/8	6 3/4	1/2	5/16	3/16	13 1/16	3 1/4	12	1 1/16	5/8	0.40	3 1/2
	34	14	6 3/4	7/16	5/16	3/16	13 1/16	3 1/4	12	1	5/8	0.40	3 1/2
	30	13 7/8	6 3/4	3/8	1/4	1/8	13 1/16	3 1/4	12	15/16	5/8	0.40	3 1/2
W14 x	26	13 7/8	5	7/16	1/4	1/8	13 1/16	2 3/8	12	15/16	9/16	0.40	2 3/4
	22	13 3/4	5	5/16	1/4	1/8	13 1/16	2 3/8	12	7/8	9/16	0.40	2 3/4
W12 x	190	14 3/8	12 5/8	1 3/4	1 1/16	9/16	10 15/16	5 3/4	9 1/2	2 7/16	1 3/16	0.60	5 1/2
	170	14	12 5/8	1 9/16	15/16	1/2	10 15/16	5 3/4	9 1/2	2 1/4	1 1/8	0.60	5 1/2
	152	13 3/4	12 1/2	1 3/8	7/8	7/16	10 15/16	5 3/4	9 1/2	2 1/8	1 1/16	0.60	5 1/2
	136	13 3/8	12 3/8	1 1/4	13/16	7/16	10 15/16	5 3/4	9 1/2	1 15/16	1	0.60	5 1/2
	120	13 1/8	12 3/8	1 1/8	11/16	3/8	10 15/16	5 3/4	9 1/2	1 13/16	1	0.60	5 1/2
	106	12 7/8	12 1/4	1	5/8	5/16	10 15/16	5 3/4	9 1/2	1 11/16	15/16	0.60	5 1/2
	96	12 3/4	12 1/8	7/8	9/16	5/16	10 15/16	5 3/4	9 1/2	1 5/8	7/8	0.60	5 1/2
	87	12 1/2	12 1/8	13/16	1/2	1/4	10 15/16	5 3/4	9 1/2	1 1/2	7/8	0.60	5 1/2
	79	12 3/8	12 1/8	3/4	1/2	1/4	10 15/16	5 3/4	9 1/2	1 7/16	7/8	0.60	5 1/2
	72	12 1/4	12	11/16	7/16	1/4	10 15/16	5 3/4	9 1/2	1 3/8	7/8	0.60	5 1/2
	65	12 1/8	12	5/8	3/8	3/16	10 15/16	5 3/4	9 1/2	1 5/16	13/16	0.60	5 1/2
W12 x	58	12 1/4	10	5/8	3/8	3/16	10 15/16	4 7/8	9 1/2	1 3/8	13/16	0.60	5 1/2
	53	12	10	9/16	3/8	3/16	10 15/16	4 7/8	9 1/2	1 1/4	13/16	0.60	5 1/2
W12 x	50	12 1/4	8 1/8	5/8	3/8	3/16	10 15/16	3 7/8	9 1/2	1 3/8	13/16	0.60	5 1/2
	45	12	8	9/16	5/16	3/16	10 15/16	3 7/8	9 1/2	1 1/4	13/16	0.60	5 1/2
	40	12	8	1/2	5/16	3/16	10 15/16	3 7/8	9 1/2	1 1/4	3/4	0.60	5 1/2
W12 x	35	12 1/2	6 1/2	1/2	5/16	3/16	11 7/16	3 1/8	10 1/2	1	9/16	0.30	3 1/2
	30	12 3/8	6 1/2	7/16	1/4	1/8	11 7/16	3 1/8	10 1/2	15/16	1/2	0.30	3 1/2
	26	12 1/4	6 1/2	3/8	1/4	1/8	11 7/16	3 1/8	10 1/2	7/8	1/2	0.30	3 1/2
W12 x	22	12 1/4	4	7/16	1/4	1/8	11 7/16	1 7/8	10 1/2	7/8	1/2	0.30	2 1/4
	19	12 1/8	4	3/8	1/4	1/8	11 7/16	1 7/8	10 1/2	13/16	1/2	0.30	2 1/4
	16	12	4	1/4	1/4	1/8	11 7/16	1 7/8	10 1/2	3/4	1/2	0.30	2 1/4
	14	11 7/8	4	1/4	3/16	1/8	11 7/16	1 7/8	10 1/2	11/16	1/2	0.30	2 1/4

WIDE FLANGE SHAPES

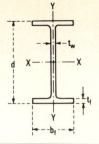

Theoretical Dimensions and Properties for **Designing**

Section Number	Weight per Foot	Area of Section A	Depth of Section d	Flange Width b_f	Flange Thickness t_f	Web Thickness t_w	Axis X-X I_x	Axis X-X S_x	Axis X-X r_x	Axis Y-Y I_y	Axis Y-Y S_y	Axis Y-Y r_y	r_T
	lb	in.²	in.	in.	in.	in.	in.⁴	in.³	in.	in.⁴	in.³	in.	in.
W10 x 112	112	32.9	11.36	10.415	1.250	0.755	716	126	4.66	236	45.3	2.68	2.88
100	100	29.4	11.10	10.340	1.120	0.680	623	112	4.60	207	40.0	2.65	2.85
88	88	25.9	10.84	10.265	0.990	0.605	534	98.5	4.54	179	34.8	2.63	2.83
77	77	22.6	10.60	10.190	0.870	0.530	455	85.9	4.49	154	30.1	2.60	2.80
68	68	20.0	10.40	10.130	0.770	0.470	394	75.7	4.44	134	26.4	2.59	2.79
60	60	17.6	10.22	10.080	0.680	0.420	341	66.7	4.39	116	23.0	2.57	2.77
54	54	15.8	10.09	10.030	0.615	0.370	303	60.0	4.37	103	20.6	2.56	2.75
49	49	14.4	9.98	10.000	0.560	0.340	272	54.6	4.35	93.4	18.7	2.54	2.74
W10 x 45	45	13.3	10.10	8.020	0.620	0.350	248	49.1	4.33	53.4	13.3	2.01	2.18
39	39	11.5	9.92	7.985	0.530	0.315	209	42.1	4.27	45.0	11.3	1.98	2.16
33	33	9.71	9.73	7.960	0.435	0.290	170	35.0	4.19	36.6	9.20	1.94	2.14
W10 x 30	30	8.84	10.47	5.810	0.510	0.300	170	32.4	4.38	16.7	5.75	1.37	1.55
26	26	7.61	10.33	5.770	0.440	0.260	144	27.9	4.35	14.1	4.89	1.36	1.54
22	22	6.49	10.17	5.750	0.360	0.240	118	23.2	4.27	11.4	3.97	1.33	1.51
W10 x 19	19	5.62	10.24	4.020	0.395	0.250	96.3	18.8	4.14	4.29	2.14	0.874	1.03
17	17	4.99	10.11	4.010	0.330	0.240	81.9	16.2	4.05	3.56	1.78	0.845	1.01
15	15	4.41	9.99	4.000	0.270	0.230	68.9	13.8	3.95	2.89	1.45	0.810	0.987
12	12	3.54	9.87	3.960	0.210	0.190	53.8	10.9	3.90	2.18	1.10	0.785	0.965
W8 x 67	67	19.7	9.00	8.280	0.935	0.570	272	60.4	3.72	88.6	21.4	2.12	2.28
58	58	17.1	8.75	8.220	0.810	0.510	228	52.0	3.65	75.1	18.3	2.10	2.26
48	48	14.1	8.50	8.110	0.685	0.400	184	43.3	3.61	60.9	15.0	2.08	2.23
40	40	11.7	8.25	8.070	0.560	0.360	146	35.5	3.53	49.1	12.2	2.04	2.21
35	35	10.3	8.12	8.020	0.495	0.310	127	31.2	3.51	42.6	10.6	2.03	2.20
31	31	9.13	8.00	7.995	0.435	0.285	110	27.5	3.47	37.1	9.27	2.02	2.18
W8 x 28	28	8.25	8.06	6.535	0.465	0.285	98.0	24.3	3.45	21.7	6.63	1.62	1.77
24	24	7.08	7.93	6.495	0.400	0.245	82.8	20.9	3.42	18.3	5.63	1.61	1.76
W8 x 21	21	6.16	8.28	5.270	0.400	0.250	75.3	18.2	3.49	9.77	3.71	1.26	1.41
18	18	5.26	8.14	5.250	0.330	0.230	61.9	15.2	3.43	7.97	3.04	1.23	1.39
W8 x 15	15	4.44	8.11	4.015	0.315	0.245	48.0	11.8	3.29	3.41	1.70	0.876	1.03
13	13	3.84	7.99	4.000	0.255	0.230	39.6	9.91	3.21	2.73	1.37	0.843	1.01
10	10	2.96	7.89	3.940	0.205	0.170	30.8	7.81	3.22	2.09	1.06	0.841	0.994

WIDE FLANGE SHAPES

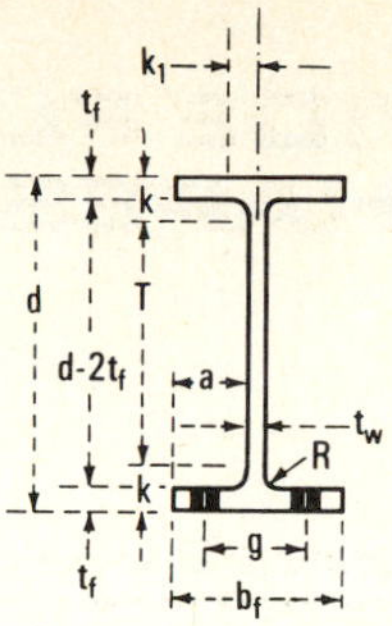

Approximate Dimensions for **Detailing**

Section Number	Weight per Foot	Depth of Section d	Flange Width b_f	Flange Thickness t_f	Web Thickness t_w	Half Web Thickness $\frac{t_w}{2}$	$d-2t_f$	a	T	k	k_1	R	Usual Flange Gage g
	lb	in.	in.	in.	in.	in.	in.	in.	in.	in.	in.	in.	in.
W10 x	112	11⅜	10⅜	1¼	¾	⅜	8⅞	4⅞	7⅝	1⅞	15/16	0.50	5½
	100	11⅛	10⅜	1⅛	11/16	⅜	8⅞	4⅞	7⅝	1¾	⅞	0.50	5½
	88	10⅞	10¼	1	⅝	5/16	8⅞	4⅞	7⅝	1⅝	13/16	0.50	5½
	77	10⅝	10¼	⅞	½	¼	8⅞	4⅞	7⅝	1½	13/16	0.50	5½
	68	10⅜	10⅛	¾	½	¼	8⅞	4⅞	7⅝	1⅜	¾	0.50	5½
	60	10¼	10⅛	11/16	7/16	¼	8⅞	4⅞	7⅝	1 5/16	¾	0.50	5½
	54	10⅛	10	⅝	⅜	3/16	8⅞	4⅞	7⅝	1¼	11/16	0.50	5½
	49	10	10	9/16	5/16	3/16	8⅞	4⅞	7⅝	1 3/16	11/16	0.50	5½
W10 x	45	10⅛	8	⅝	⅜	3/16	8⅞	3⅞	7⅝	1¼	11/16	0.50	5½
	39	9⅞	8	½	5/16	3/16	8⅞	3⅞	7⅝	1⅛	11/16	0.50	5½
	33	9¾	8	7/16	5/16	3/16	8⅞	3⅞	7⅝	1 1/16	11/16	0.50	5½
W10 x	30	10½	5¾	½	5/16	3/16	9 7/16	2¾	8⅝	15/16	½	0.30	2¾
	26	10⅜	5¾	7/16	¼	⅛	9 7/16	2¾	8⅝	⅞	½	0.30	2¾
	22	10⅛	5¾	⅜	¼	⅛	9 7/16	2¾	8⅝	¾	½	0.30	2¾
W10 x	19	10¼	4	⅜	¼	⅛	9 7/16	1⅞	8⅝	13/16	½	0.30	2¼
	17	10⅛	4	5/16	¼	⅛	9 7/16	1⅞	8⅝	¾	½	0.30	2¼
	15	10	4	¼	¼	⅛	9 7/16	1⅞	8⅝	11/16	7/16	0.30	2¼
	12	9⅞	4	3/16	3/16	⅛	9 7/16	1⅞	8⅝	⅝	7/16	0.30	2¼
W8 x	67	9	8¼	15/16	9/16	5/16	7⅛	3⅞	6⅛	1 7/16	11/16	0.40	5½
	58	8¾	8¼	13/16	½	¼	7⅛	3⅞	6⅛	1 5/16	11/16	0.40	5½
	48	8½	8⅛	11/16	⅜	3/16	7⅛	3⅞	6⅛	1 3/16	⅝	0.40	5½
	40	8¼	8⅛	9/16	⅜	3/16	7⅛	3⅞	6⅛	1 1/16	⅝	0.40	5½
	35	8⅛	8	½	5/16	3/16	7⅛	3⅞	6⅛	1	9/16	0.40	5½
	31	8	8	7/16	5/16	3/16	7⅛	3⅞	6⅛	15/16	9/16	0.40	5½
W8 x	28	8	6½	7/16	5/16	3/16	7⅛	3⅛	6⅛	15/16	9/16	0.40	3½
	24	7⅞	6½	⅜	¼	⅛	7⅛	3⅛	6⅛	⅞	9/16	0.40	3½
W8 x	21	8¼	5¼	⅜	¼	⅛	7½	2½	6⅝	13/16	½	0.30	2¾
	18	8⅛	5¼	5/16	¼	⅛	7½	2½	6⅝	¾	7/16	0.30	2¾
W8 x	15	8⅛	4	5/16	¼	⅛	7½	1⅞	6⅝	¾	½	0.30	2¼
	13	8	4	¼	¼	⅛	7½	1⅞	6⅝	11/16	7/16	0.30	2¼
	10	7⅞	4	3/16	3/16	⅛	7½	1⅞	6⅝	⅝	7/16	0.30	2¼

WIDE FLANGE SHAPES

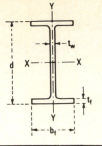

Theoretical Dimensions and Properties for **Designing**

Section Number	Weight per Foot	Area of Section A	Depth of Section d	Flange			Axis X-X			Axis Y-Y			r_T
				Width b_f	Thickness t_f	Web Thickness t_w	I_x	S_x	r_x	I_y	S_y	r_y	
	lb	in.²	in.	in.	in.	in.	in.⁴	in.³	in.	in.⁴	in.³	in.	in.
W6 x	25	7.34	6.38	6.080	0.455	0.320	53.4	16.7	2.70	17.1	5.61	1.52	1.66
	20	5.87	6.20	6.020	0.365	0.260	41.4	13.4	2.66	13.3	4.41	1.50	1.64
	15	4.43	5.99	5.990	0.260	0.230	29.1	9.72	2.56	9.32	3.11	1.45	1.61
W6 x	16	4.74	6.28	4.030	0.405	0.260	32.1	10.2	2.60	4.43	2.20	0.967	1.08
	12	3.55	6.03	4.000	0.280	0.230	22.1	7.31	2.49	2.99	1.50	0.918	1.05
	9	2.68	5.90	3.940	0.215	0.170	16.4	5.56	2.47	2.20	1.11	0.905	1.03
W5 x	19	5.54	5.15	5.030	0.430	0.270	26.2	10.2	2.17	9.13	3.63	1.28	1.38
	16	4.68	5.01	5.000	0.360	0.240	21.3	8.51	2.13	7.51	3.00	1.27	1.37
†W4 x	13	3.83	4.16	4.060	0.345	0.280	11.3	5.46	1.72	3.86	1.90	1.00	1.10

MISCELLANEOUS SHAPE

Theoretical Dimensions and Properties for **Designing**

Section Number	Weight per Foot	Area of Section A	Depth of Section d	Flange			Axis X-X			Axis Y-Y			r_T
				Width b_f	Thickness t_f	Web Thickness t_w	I_x	S_x	r_x	I_y	S_y	r_y	
	lb	in.²	in.	in.	in.	in.	in.⁴	in.³	in.	in.⁴	in.³	in.	in.
†M5 x	18.9	5.55	5.00	5.003	0.416	0.316	24.1	9.63	2.08	7.86	3.14	1.19	1.32

†W4 x 13 and M5 x 18.9 have flange slopes of 2.0 and 7.4 pct respectively. Flange thickness shown for these sections are average thicknesses. Properties are the same as if flanges were parallel.

All other shapes on these pages have parallel-faced flanges.

WIDE FLANGE SHAPES

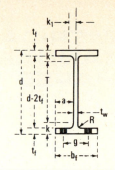

Approximate Dimensions for **Detailing**

Section Number	Weight per Foot	Depth of Section	Flange		Web Thickness	Half Web Thickness	d-2t_f	a	T	k	k₁	R	Usual Flange Gage
			Width	Thickness									
		d	b_f	t_f	t_w	$\frac{t_w}{2}$	d-2t$_f$	a	T	k	k_1	R	g
	lb	in.	in.	in.	in.	in.	in.	in.	in.	in.	in.	in.	in.
W6 x	25	6⅜	6⅛	⁷⁄₁₆	⁵⁄₁₆	³⁄₁₆	5½	2⅞	4¾	¹³⁄₁₆	⁷⁄₁₆	0.25	3½
	20	6¼	6	⅜	¼	⅛	5½	2⅞	4¾	¾	⁷⁄₁₆	0.25	3½
	15	6	6	¼	¼	⅛	5½	2⅞	4¾	⅝	⅜	0.25	3½
W6 x	16	6¼	4	⅜	¼	⅛	5½	1⅞	4¾	¾	⁷⁄₁₆	0.25	2¼
	12	6	4	¼	¼	⅛	5½	1⅞	4¾	⅝	⅜	0.25	2¼
	9	5⅞	4	³⁄₁₆	³⁄₁₆	⅛	5½	1⅞	4¾	⁹⁄₁₆	⅜	0.25	2¼
W5 x	19	5⅛	5	⁷⁄₁₆	¼	⅛	4⁵⁄₁₆	2⅜	3½	¹³⁄₁₆	⁷⁄₁₆	0.30	2¾
	16	5	5	⅜	¼	⅛	4⁵⁄₁₆	2⅜	3½	¾	⁷⁄₁₆	0.30	2¾
W4 x	13	4⅛	4	⅜	¼	⅛	3½	1⅞	2¾	¹¹⁄₁₆	⁷⁄₁₆	0.25	2¼

MISCELLANEOUS SHAPE

Theoretical Dimensions and Properties for **Detailing**

Section Number	Weight per Foot	Depth of Section	Flange		Web Thickness	Half Web Thickness	d-2t_f	a	T	k	k₁	R	Usual Flange Gage
			Width	Thickness									
		d	b_f	t_f	t_w	$\frac{t_w}{2}$	d-2t$_f$	a	T	k	k_1	R	g
	lb	in.	in.	in.	in.	in.	in.	in.	in.	in.	in.	in.	in.
M5 x	18.9	5	5	⁷⁄₁₆	⁵⁄₁₆	³⁄₁₆	4³⁄₁₆	2⅜	3¼	⅞	½	0.313	2¾

AMERICAN STANDARD SHAPES

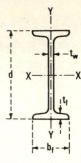

Theoretical Dimensions and Properties for **Designing**

Section Number	Weight per Foot	Area of Section	Depth of Section	Flange		Web Thick-ness	Axis X-X			Axis Y-Y			
				Width	Average Thick-ness								r_T
		A	d	b_f	t_f	t_w	I_x	S_x	r_x	I_y	S_y	r_y	
	lb	in.²	in.	in.	in.	in.	in.⁴	in.³	in.	in.⁴	in.³	in.	in.
S24 x	**121.0**	35.6	24.50	8.050	1.090	0.800	3160	258	9.43	83.3	20.7	1.53	1.86
	106.0	31.2	24.50	7.870	1.090	0.620	2940	240	9.71	77.1	19.6	1.57	1.86
S24 x	**100.0**	29.3	24.00	7.245	0.870	0.745	2390	199	9.02	47.7	13.2	1.27	1.59
	90.0	26.5	24.00	7.125	0.870	0.625	2250	187	9.21	44.9	12.6	1.30	1.60
	80.0	23.5	24.00	7.000	0.870	0.500	2100	175	9.47	42.2	12.1	1.34	1.61
S20 x	**96.0**	28.2	20.30	7.200	0.920	0.800	1670	165	7.71	50.2	13.9	1.33	1.63
	86.0	25.3	20.30	7.060	0.920	0.660	1580	155	7.89	46.8	13.3	1.36	1.63
S20 x	**75.0**	22.0	20.00	6.385	0.795	0.635	1280	128	7.62	29.8	9.32	1.16	1.43
	66.0	19.4	20.00	6.255	0.795	0.505	1190	119	7.83	27.7	8.85	1.19	1.44
S18 x	**70.0**	20.6	18.00	6.251	0.691	0.711	926	103	6.71	24.1	7.72	1.08	1.40
	54.7	16.1	18.00	6.001	0.691	0.461	804	89.4	7.07	20.8	6.94	1.14	1.40
S15 x	**50.0**	14.7	15.00	5.640	0.622	0.550	486	64.8	5.75	15.7	5.57	1.03	1.30
	42.9	12.6	15.00	5.501	0.622	0.411	447	59.6	5.95	14.4	5.23	1.07	1.30
S12 x	**50.0**	14.7	12.00	5.477	0.659	0.687	305	50.8	4.55	15.7	5.74	1.03	1.31
	40.8	12.0	12.00	5.252	0.659	0.462	272	45.4	4.77	13.6	5.16	1.06	1.28
S12 x	**35.0**	10.3	12.00	5.078	0.544	0.428	229	38.2	4.72	9.87	3.89	0.980	1.20
	31.8	9.35	12.00	5.000	0.544	0.350	218	36.4	4.83	9.36	3.74	1.00	1.20

All shapes on these pages have a flange slope of 16⅔ pct.

AMERICAN STANDARD SHAPES

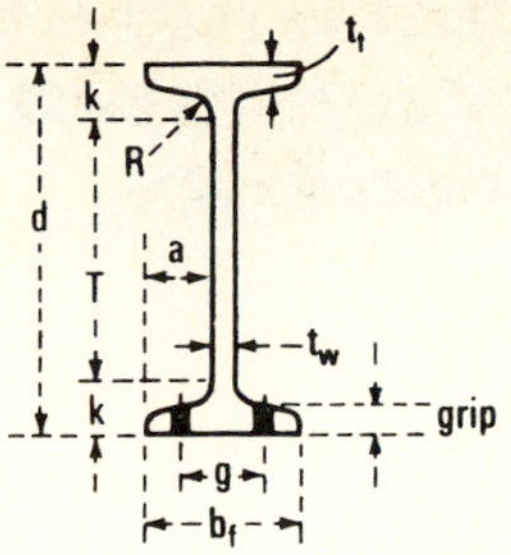

Approximate Dimensions for **Detailing**

Section Number	Weight per Foot	Depth of Section d	Flange Width b_f	Flange Average Thickness t_f	Web Thickness t_w	Half Web Thickness $\frac{t_w}{2}$	a	T	k	R	Grip	Max Flange Fastener	Usual Flange Gage g
	lb	in.	in.	in.	in.	in.	in.	in.	in.	in.	in.	in.	in.
S24 x	**121.0**	24½	8	1¹⁄₁₆	¹³⁄₁₆	⅜	3⅝	20½	2	.60	1⅛	1	4
	106.0	24½	7⅞	1¹⁄₁₆	⅝	⁵⁄₁₆	3⅝	20½	2	.60	1⅛	1	4
S24 x	**100.0**	24	7¼	⅞	¾	⅜	3¼	20½	1¾	.60	⅞	1	4
	90.0	24	7⅛	⅞	⅝	⁵⁄₁₆	3¼	20½	1¾	.60	⅞	1	4
	80.0	24	7	⅞	½	¼	3¼	20½	1¾	.60	⅞	1	4
S20 x	**96.0**	20¼	7¼	¹⁵⁄₁₆	¹³⁄₁₆	⅜	3¼	16¾	1¾	.60	¹⁵⁄₁₆	1	4
	86.0	20¼	7	¹⁵⁄₁₆	¹¹⁄₁₆	⁵⁄₁₆	3¼	16¾	1¾	.60	¹⁵⁄₁₆	1	4
S20 x	**75.0**	20	6⅜	¹³⁄₁₆	⅝	⁵⁄₁₆	2⅞	16¾	1⅝	.60	¹³⁄₁₆	⅞	3½
	66.0	20	6¼	¹³⁄₁₆	½	¼	2⅞	16¾	1⅝	.60	¹³⁄₁₆	⅞	3½
S18 x	**70.0**	18	6¼	¹¹⁄₁₆	¹¹⁄₁₆	⅜	2¾	15	1½	.56	¹¹⁄₁₆	⅞	3½
	54.7	18	6	¹¹⁄₁₆	⁷⁄₁₆	¼	2¾	15	1½	.56	¹¹⁄₁₆	⅞	3½
S15 x	**50.0**	15	5⅝	⅝	⁹⁄₁₆	¼	2½	12¼	1⅜	.51	⁹⁄₁₆	¾	3½
	42.9	15	5½	⅝	⁷⁄₁₆	³⁄₁₆	2½	12¼	1⅜	.51	⁹⁄₁₆	¾	3½
S12 x	**50.0**	12	5½	¹¹⁄₁₆	¹¹⁄₁₆	⁵⁄₁₆	2⅜	9⅛	1⁷⁄₁₆	.56	¹¹⁄₁₆	¾	3
	40.8	12	5¼	¹¹⁄₁₆	⁷⁄₁₆	¼	2⅜	9⅛	1⁷⁄₁₆	.56	⅝	¾	3
S12 x	**35.0**	12	5⅛	⁹⁄₁₆	⁷⁄₁₆	³⁄₁₆	2⅜	9⅝	1³⁄₁₆	.45	½	¾	3
	31.8	12	5	⁹⁄₁₆	⅜	³⁄₁₆	2⅜	9⅝	1³⁄₁₆	.45	½	¾	3

AMERICAN STANDARD SHAPES

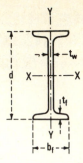

Theoretical Dimensions and Properties for **Designing**

Section Number	Weight per Foot	Area of Section A	Depth of Section d	Flange Width b_f	Flange Average Thickness t_f	Web Thickness t_w	Axis X-X I_x	Axis X-X S_x	Axis X-X r_x	Axis Y-Y I_y	Axis Y-Y S_y	Axis Y-Y r_y	r_T
	lb	in.²	in.	in.	in.	in.	in.⁴	in.³	in.	in.⁴	in.³	in.	in.
S10 x	**35.0**	10.3	10.00	4.944	0.491	0.594	147	29.4	3.78	8.36	3.38	0.901	1.14
	25.4	7.46	10.00	4.661	0.491	0.311	124	24.7	4.07	6.79	2.91	0.954	1.12
S8 x	**23.0**	6.77	8.00	4.171	0.425	0.441	64.9	16.2	3.10	4.31	2.07	0.798	0.984
	18.4	5.41	8.00	4.001	0.425	0.271	57.6	14.4	3.26	3.73	1.86	0.831	0.969
S7 x	**15.3**	4.50	7.00	3.662	0.392	0.252	36.7	10.5	2.86	2.64	1.44	0.766	0.891
S6 x	**17.25**	5.07	6.00	3.565	0.359	0.465	26.3	8.77	2.28	2.31	1.30	0.675	0.842
	12.5	3.67	6.00	3.332	0.359	0.232	22.1	7.37	2.45	1.82	1.09	0.705	0.814
S5 x	**10.0**	2.94	5.00	3.004	0.326	0.214	12.3	4.92	2.05	1.22	0.809	0.643	0.739
S4 x	**ʸ9.5**	2.79	4.00	2.796	0.293	0.326	6.79	3.39	1.56	0.903	0.646	0.569	0.682
	7.7	2.26	4.00	2.663	0.293	0.193	6.08	3.04	1.64	0.764	0.574	0.581	0.661
S3 x	**7.5**	2.21	3.00	2.509	0.260	0.349	2.93	1.95	1.15	0.586	0.468	0.516	0.620
	5.7	1.67	3.00	2.330	0.260	0.170	2.52	1.68	1.23	0.455	0.390	0.522	0.584

All shapes on these pages have a flange slope of 16⅔ pct.
ʸAvailable subject to inquiry.

AMERICAN STANDARD SHAPES

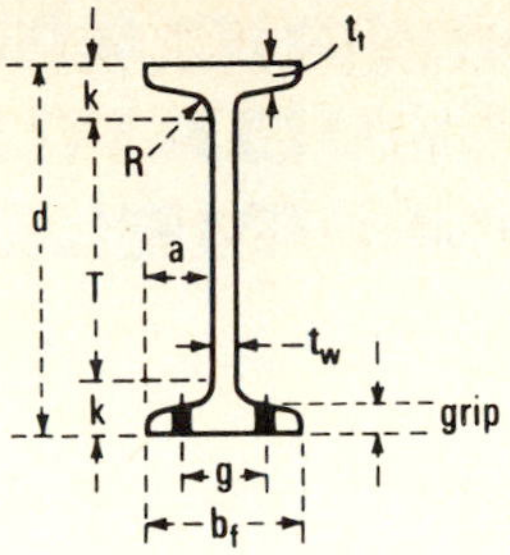

Approximate Dimensions for **Detailing**

Section Number	Weight per Foot	Depth of Section d	Flange Width b_f	Flange Average Thickness t_f	Web Thickness t_w	Half Web Thickness $\frac{t_w}{2}$	a	T	k	R	Grip	Max Flange Fastener	Usual Flange Gage g
	lb	in.	in.	in.	in.	in.	in.	in.	in.	in.	in.	in.	in.
S10 x	**35.0**	10	5	1/2	5/8	5/16	2 1/8	7 3/4	1 1/8	.41	1/2	3/4	2 3/4
	25.4	10	4 5/8	1/2	5/16	1/8	2 1/8	7 3/4	1 1/8	.41	1/2	3/4	2 3/4
S8 x	**23.0**	8	4 1/8	7/16	7/16	1/4	1 7/8	6	1	.37	7/16	3/4	2 1/4
	18.4	8	4	7/16	1/4	1/8	1 7/8	6	1	.37	7/16	3/4	2 1/4
S7 x	**15.3**	7	3 5/8	3/8	1/4	1/8	1 3/4	5 1/4	7/8	.35	3/8	5/8	2 1/4
S6 x	**17.25**	6	3 5/8	3/8	7/16	1/4	1 1/2	4 3/8	13/16	.33	3/8	5/8	2
	12.5	6	3 3/8	3/8	1/4	1/8	1 1/2	4 3/8	13/16	.33	5/16	—	—
S5 x	**10.0**	5	3	5/1	3/16	1/8	1 3/8	3 1/2	3/4	.31	5/16	—	—
S4 x	**y9.5**	4	2 3/4	5/16	5/16	3/16	1 1/4	2 5/8	11/16	.29	5/16	—	—
	7.7	4	2 5/8	5/16	3/16	1/8	1 1/4	2 5/8	11/16	.29	5/16	—	—
S3 x	**7.5**	3	2 1/2	1/4	3/8	3/16	1 1/8	1 3/4	5/8	.27	1/4	—	—
	5.7	3	2 3/8	1/4	3/16	1/16	1 1/8	1 3/4	5/8	.27	1/4	—	—

AMERICAN STANDARD CHANNELS

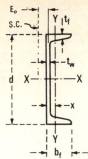

Theoretical Dimensions and Properties for **Designing**

Section Number	Weight per Foot	Area of Section	Depth of Section	Flange		Web Thickness	Axis X-X			Axis Y-Y				Shear Center Location
		A	d	Width b_f	Average Thickness t_f	t_w	I_x	S_x	r_x	I_y	S_y	r_y	x	E_o
	lb	in.²	in.	in.	in.	in.	in.⁴	in.³	in.	in.⁴	in.³	in.	in.	in.
C15 x	50.0	14.7	15.00	3.716	0.650	0.716	404	53.8	5.24	11.0	3.78	0.867	0.799	0.941
	40.0	11.8	15.00	3.520	0.650	0.520	349	46.5	5.44	9.23	3.36	0.886	0.778	1.03
	33.9	9.96	15.00	3.400	0.650	0.400	315	42.0	5.62	8.13	3.11	0.904	0.787	1.10
C12 x	30.0	8.82	12.00	3.170	0.501	0.510	162	27.0	4.29	5.14	2.06	0.763	0.674	0.873
	25.0	7.35	12.00	3.047	0.501	0.387	144	24.1	4.43	4.47	1.88	0.780	0.674	0.940
	20.7	6.09	12.00	2.942	0.501	0.282	129	21.5	4.61	3.88	1.73	0.799	0.698	1.01
C10 x	30.0	8.82	10.00	3.033	0.436	0.673	103	20.7	3.42	3.94	1.65	0.669	0.649	0.705
	25.0	7.35	10.00	2.886	0.436	0.526	91.2	18.2	3.52	3.36	1.48	0.676	0.617	0.757
	20.0	5.88	10.00	2.739	0.436	0.379	78.9	15.8	3.66	2.81	1.32	0.691	0.606	0.826
	15.3	4.49	10.00	2.600	0.436	0.240	67.4	13.5	3.87	2.28	1.16	0.713	0.634	0.916
C9 x	15.0	4.41	9.00	2.485	0.413	0.285	51.0	11.3	3.40	1.93	1.01	0.661	0.586	0.824
	13.4	3.94	9.00	2.433	0.413	0.233	47.9	10.6	3.48	1.76	0.962	0.668	0.601	0.859
C8 x	18.75	5.51	8.00	2.527	0.390	0.487	44.0	11.0	2.82	1.98	1.01	0.599	0.565	0.674
	13.75	4.04	8.00	2.343	0.390	0.303	36.1	9.03	2.99	1.53	0.853	0.615	0.553	0.756
	11.5	3.38	8.00	2.260	0.390	0.220	32.6	8.14	3.11	1.32	0.781	0.625	0.571	0.807
C7 x	12.25	3.60	7.00	2.194	0.366	0.314	24.2	6.93	2.60	1.17	0.702	0.571	0.525	0.695
	9.8	2.87	7.00	2.090	0.366	0.210	21.3	6.08	2.72	0.968	0.625	0.581	0.541	0.752
C6 x	13.0	3.83	6.00	2.157	0.343	0.437	17.4	5.80	2.13	1.05	0.642	0.525	0.514	0.599
	10.5	3.09	6.00	2.034	0.343	0.314	15.2	5.06	2.22	0.865	0.564	0.529	0.500	0.643
	8.2	2.40	6.00	1.920	0.343	0.200	13.1	4.38	2.34	0.692	0.492	0.537	0.512	0.699
C5 x	9.0	2.64	5.00	1.885	0.320	0.325	8.90	3.56	1.83	0.632	0.449	0.489	0.478	0.590
	6.7	1.97	5.00	1.750	0.320	0.190	7.49	3.00	1.95	0.478	0.378	0.493	0.484	0.647
C4 x	7.25	2.13	4.00	1.721	0.296	0.321	4.59	2.29	1.47	0.432	0.343	0.450	0.459	0.546
	5.4	1.59	4.00	1.584	0.296	0.184	3.85	1.93	1.56	0.319	0.283	0.449	0.458	0.594
C3 x	5.0	1.47	3.00	1.498	0.273	0.258	1.85	1.24	1.12	0.247	0.233	0.410	0.438	0.521
	4.1	1.21	3.00	1.410	0.273	0.170	1.66	1.10	1.17	0.197	0.202	0.404	0.437	0.546

All shapes on these pages have a flange slope of 16⅔ pct.

AMERICAN STANDARD CHANNELS

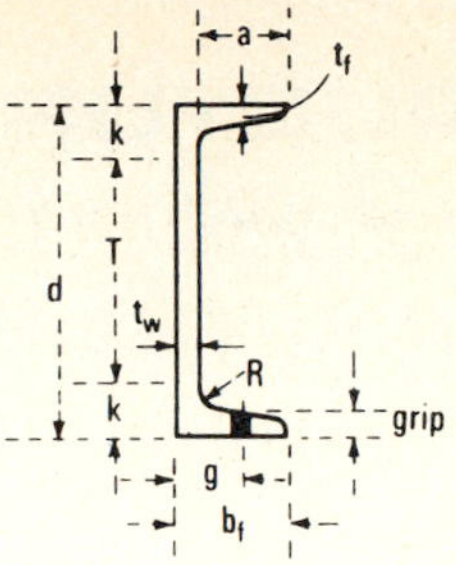

Approximate Dimensions for **Detailing**

Section Number	Weight per Foot	Depth of Section d	Flange Width b_f	Flange Average Thickness t_f	Web Thickness t_w	Half Web Thickness $\frac{t_w}{2}$	a	T	k	R	Grip	Max Flange Fastener	Usual Flange Gage g
	lb	in.	in.	in.	in.	in.	in.	in.	in.	in.	in.	in.	in.
C15 x	**50.0**	15	3¾	⅝	11/16	⅜	3	12⅛	1 7/16	0.50	⅝	1	2¼
	40.0	15	3½	⅝	½	¼	3	12⅛	1 7/16	0.50	⅝	1	2
	33.9	15	3⅜	⅝	⅜	3/16	3	12⅛	1 7/16	0.50	⅝	1	2
C12 x	**30.0**	12	3⅛	½	½	¼	2⅝	9¾	1⅛	0.38	½	⅞	1¾
	25.0	12	3	½	⅜	3/16	2⅝	9¾	1⅛	0.38	½	⅞	1¾
	20.7	12	3	½	5/16	⅛	2⅝	9¾	1⅛	0.38	½	⅞	1¾
C10 x	**30.0**	10	3	7/16	11/16	5/16	2⅜	8	1	0.34	7/16	¾	1¾
	25.0	10	2⅞	7/16	½	¼	2⅜	8	1	0.34	7/16	¾	1¾
	20.0	10	2¾	7/16	⅜	3/16	2⅜	8	1	0.34	7/16	¾	1½
	15.3	10	2⅝	7/16	¼	⅛	2⅜	8	1	0.34	7/16	¾	1½
C9 x	**15.0**	9	2½	7/16	5/16	⅛	2¼	7⅛	15/16	0.33	7/16	¾	1⅜
	13.4	9	2⅜	7/16	¼	⅛	2¼	7⅛	15/16	0.33	7/16	¾	1⅜
C8 x	**18.75**	8	2½	⅜	½	¼	2	6⅛	15/16	0.32	⅜	¾	1½
	13.75	8	2⅜	⅜	5/16	⅛	2	6⅛	15/16	0.32	⅜	¾	1⅜
	11.5	8	2¼	⅜	¼	⅛	2	6⅛	15/16	0.32	⅜	¾	1⅜
C7 x	**12.25**	7	2¼	⅜	5/16	3/16	1⅞	5¼	⅞	0.31	⅜	⅝	1¼
	9.8	7	2⅛	⅜	3/16	⅛	1⅞	5¼	⅞	0.31	⅜	⅝	1¼
C6 x	**13.0**	6	2⅛	5/16	7/16	3/16	1¾	4⅜	13/16	0.30	5/16	⅝	1⅜
	10.5	6	2	5/16	5/16	3/16	1¾	4⅜	13/16	0.30	⅜	⅝	1⅛
	8.2	6	1⅞	5/16	3/16	⅛	1¾	4⅜	13/16	0.30	5/16	⅝	1⅛
C5 x	**9.0**	5	1⅞	5/16	5/16	3/16	1½	3½	¾	0.29	5/16	⅝	1⅛
	6.7	5	1¾	5/16	3/16	⅛	1½	3½	¾	0.29	5/16	–	–
C4 x	**7.25**	4	1¾	5/16	5/16	3/16	1⅜	2⅝	11/16	0.28	5/16	⅝	1
	5.4	4	1⅝	5/16	3/16	1/16	1⅜	2⅝	11/16	0.28	¼	–	–
C3 x	**5.0**	3	1½	¼	¼	⅛	1¼	1⅝	11/16	0.27	¼	–	–
	4.1	3	1⅜	¼	3/16	1/16	1¼	1⅝	11/16	0.27	¼	–	–

STRUCTURAL TEES (Cut from W Shapes)

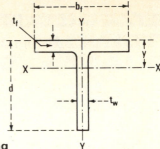

Theoretical Dimensions and Properties for **Designing**

Section Number	Weight per Foot	Area of Section A	Depth of Section d	Flange Width b_f	Flange Thickness t_f	Stem Thickness t_w	Axis X-X I_x	S_x	r_x	y	Axis Y-Y I_y	S_y	r_y
	lb	in.²	in.	in.	in.	in.	in.⁴	in.³	in.	in.⁴	in.³	in.³	in.
WT18 x **150**		44.1	18.370	16.655	1.680	0.945	1230	86.1	5.27	4.13	648	77.8	3.83
140		41.2	18.260	16.595	1.570	0.885	1140	80.0	5.25	4.07	599	72.2	3.81
130		38.2	18.130	16.550	1.440	0.840	1060	75.1	5.26	4.05	545	65.9	3.78
122.5		36.0	18.040	16.510	1.350	0.800	995	71.0	5.26	4.03	507	61.4	3.75
115		33.8	17.950	16.470	1.260	0.760	934	67.0	5.25	4.01	470	57.1	3.73
WT18 x **105**		30.9	18.345	12.180	1.360	0.830	985	73.1	5.65	4.87	206	33.8	2.58
97		28.5	18.245	12.115	1.260	0.765	901	67.0	5.62	4.80	187	30.9	2.56
91		26.8	18.165	12.075	1.180	0.725	845	63.1	5.62	4.77	174	28.8	2.55
85		25.0	18.085	12.030	1.100	0.680	786	58.9	5.61	4.73	160	26.6	2.53
80		23.5	18.005	12.000	1.020	0.650	740	55.8	5.61	4.74	147	24.6	2.50
75		22.1	17.925	11.975	0.940	0.625	698	53.1	5.62	4.78	135	22.5	2.47
67.5		19.9	17.775	11.950	0.790	0.600	636	49.7	5.66	4.96	113	18.9	2.38
WT16.5 x **120.5**		35.4	17.090	15.860	1.400	0.830	871	65.8	4.96	3.85	466	58.8	3.63
110.5		32.5	16.965	15.805	1.275	0.775	799	60.8	4.96	3.81	420	53.2	3.59
100.5		29.5	16.840	15.745	1.150	0.715	725	55.5	4.95	3.78	375	47.6	3.56
WT16.5 x **76**		22.4	16.745	11.565	1.055	0.635	592	47.4	5.14	4.26	136	23.6	2.47
70.5		20.8	16.650	11.535	0.960	0.605	552	44.7	5.15	4.29	123	21.3	2.43
65		19.2	16.545	11.510	0.855	0.580	513	42.1	5.18	4.36	109	18.9	2.39
59		17.3	16.430	11.480	0.740	0.550	469	39.2	5.20	4.47	93.6	16.3	2.32
WT15 x **105.5**		31.0	15.470	15.105	1.315	0.775	610	50.5	4.43	3.40	378	50.1	3.49
95.5		28.1	15.340	15.040	1.185	0.710	549	45.7	4.42	3.35	336	44.7	3.46
86.5		25.4	15.220	14.985	1.065	0.655	497	41.7	4.42	3.31	299	39.9	3.43
WT15 x **66**		19.4	15.155	10.545	1.000	0.615	421	37.4	4.66	3.90	98.0	18.6	2.25
62		18.2	15.085	10.515	0.930	0.585	396	35.3	4.66	3.90	90.4	17.2	2.23
58		17.1	15.005	10.495	0.850	0.565	373	33.7	4.67	3.94	82.1	15.7	2.19
54		15.9	14.915	10.475	0.760	0.545	349	32.0	4.69	4.01	73.0	13.9	2.15
49.5		14.5	14.825	10.450	0.670	0.520	322	30.0	4.71	4.09	63.9	12.2	2.10
WT13.5 x **89**		26.1	13.905	14.085	1.190	0.725	414	38.2	3.98	3.05	278	39.4	3.26
80.5		23.7	13.795	14.020	1.080	0.660	372	34.4	3.96	2.99	248	35.4	3.24
73		21.5	13.690	13.965	0.975	0.605	336	31.2	3.95	2.95	222	31.7	3.21
WT13.5 x **57**		16.8	13.645	10.070	0.930	0.570	289	28.3	4.15	3.42	79.4	15.8	2.18
51		15.0	13.545	10.015	0.830	0.515	258	25.3	4.14	3.37	69.6	13.9	2.15
47		13.8	13.460	9.990	0.745	0.490	239	23.8	4.16	3.41	62.0	12.4	2.12
42		12.4	13.355	9.960	0.640	0.460	216	21.9	4.18	3.48	52.8	10.6	2.07

Properties shown in this table are for the full center split section.

STRUCTURAL TEES (Cut from W Shapes)

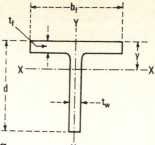

Theoretical Dimensions and Properties for **Designing**

Section Number	Weight per Foot	Area of Section	Depth of Section	Flange		Stem Thickness	Axis X-X				Axis Y-Y		
				Width	Thickness		I_x	S_x	r_x	y	I_y	S_y	r_y
		A	d	b_f	t_f	t_w							
	lb	in.²	in.	in.	in.	in.	in.⁴	in.³	in.	in.⁴	in.³	in.³	in.
WT12 x 81		23.9	12.500	12.955	1.220	0.705	293	29.9	3.50	2.70	221	34.2	3.05
73		21.5	12.370	12.900	1.090	0.650	264	27.2	3.50	2.66	195	30.3	3.01
65.5		19.3	12.240	12.855	0.960	0.605	238	24.8	3.52	2.65	170	26.5	2.97
58.5		17.2	12.130	12.800	0.850	0.550	212	22.3	3.51	2.62	149	23.2	2.94
52		15.3	12.030	12.750	0.750	0.500	189	20.0	3.51	2.59	130	20.3	2.91
WT12 x 47		13.8	12.155	9.065	0.875	0.515	186	20.3	3.67	2.99	54.5	12.0	1.98
42		12.4	12.050	9.020	0.770	0.470	166	18.3	3.67	2.97	47.2	10.5	1.95
38		11.2	11.960	8.990	0.680	0.440	151	16.9	3.68	3.00	41.3	9.18	1.92
34		10.0	11.865	8.965	0.585	0.415	137	15.6	3.70	3.06	35.2	7.85	1.87
WT12 x 31		9.11	11.870	7.040	0.590	0.430	131	15.6	3.79	3.46	17.2	4.90	1.38
27.5		8.10	11.785	7.005	0.505	0.395	117	14.1	3.80	3.50	14.5	4.15	1.34
WT10.5 x 73.5		21.6	11.030	12.510	1.150	0.720	204	23.7	3.08	2.39	188	30.0	2.95
66		19.4	10.915	12.440	1.035	0.650	181	21.1	3.06	2.33	166	26.7	2.93
61		17.9	10.840	12.390	0.960	0.600	166	19.3	3.04	2.28	152	24.6	2.92
55.5		16.3	10.755	12.340	0.875	0.550	150	17.5	3.03	2.23	137	22.2	2.90
50.5		14.9	10.680	12.290	0.800	0.500	135	15.8	3.01	2.18	124	20.2	2.89
WT10.5 x 46.5		13.7	10.810	8.420	0.930	0.580	144	17.9	3.25	2.74	46.4	11.0	1.84
41.5		12.2	10.715	8.355	0.835	0.515	127	15.7	3.22	2.66	40.7	9.75	1.83
36.5		10.7	10.620	8.295	0.740	0.455	110	13.8	3.21	2.60	35.3	8.51	1.81
34		10.0	10.565	8.270	0.685	0.430	103	12.9	3.20	2.59	32.4	7.83	1.80
31		9.13	10.495	8.240	0.615	0.400	93.8	11.9	3.21	2.58	28.7	6.97	1.77
WT10.5 x 28.5		8.37	10.530	6.555	0.650	0.405	90.4	11.8	3.29	2.85	15.3	4.67	1.35
25		7.36	10.415	6.530	0.535	0.380	80.3	10.7	3.30	2.93	12.5	3.82	1.30
22		6.49	10.330	6.500	0.450	0.350	71.1	9.68	3.31	2.98	10.3	3.18	1.26
WT9 x 59.5		17.5	9.485	11.265	1.060	0.655	119	15.9	2.60	2.03	126	22.5	2.69
53		15.6	9.365	11.200	0.940	0.590	104	14.1	2.59	1.97	110	19.7	2.66
48.5		14.3	9.295	11.145	0.870	0.535	93.8	12.7	2.56	1.91	100	18.0	2.65
43		12.7	9.195	11.090	0.770	0.480	82.4	11.2	2.55	1.86	87.6	15.8	2.63
38		11.2	9.105	11.035	0.680	0.425	71.8	9.83	2.54	1.80	76.2	13.8	2.61
WT9 x 35.5		10.4	9.235	7.635	0.810	0.495	78.2	11.2	2.74	2.26	30.1	7.89	1.70
32.5		9.55	9.175	7.590	0.750	0.450	70.7	10.1	2.72	2.20	27.4	7.22	1.69
30		8.82	9.120	7.555	0.695	0.415	64.7	9.29	2.71	2.16	25.0	6.63	1.69
27.5		8.10	9.055	7.530	0.630	0.390	59.5	8.63	2.71	2.16	22.5	5.97	1.67
25		7.33	8.995	7.495	0.570	0.355	53.5	7.79	2.70	2.12	20.0	5.35	1.65
WT9 x 23		6.77	9.030	6.060	0.605	0.360	52.1	7.77	2.77	2.33	11.3	3.72	1.29
20		5.88	8.950	6.015	0.525	0.315	44.8	6.73	2.76	2.29	9.55	3.17	1.27
17.5		5.15	8.850	6.000	0.425	0.300	40.1	6.21	2.79	2.39	7.67	2.56	1.22

Properties shown in this table are for the full center split section.

STRUCTURAL
TEES (Cut from W Shapes)

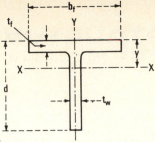

Theoretical Dimensions and Properties for **Designing**

Section Number	Weight per Foot	Area of Section	Depth of Section	Flange		Stem Thickness	Axis X-X				Axis Y-Y		
				Width	Thickness		I_x	S_x	r_x	y	I_y	S_y	r_y
		A	d	b_f	t_f	t_w							
	lb	in.²	in.	in.	in.	in.	in.⁴	in.³	in.	in.	in.⁴	in.³	in.
WT8 x	**50**	14.7	8.485	10.425	0.985	0.585	76.8	11.4	2.28	1.76	93.1	17.9	2.52
	44.5	13.1	8.375	10.365	0.875	0.525	67.2	10.1	2.27	1.70	81.3	15.7	2.49
	38.5	11.3	8.260	10.295	0.760	0.455	56.9	8.59	2.24	1.63	69.2	13.4	2.47
	33.5	9.84	8.165	10.235	0.665	0.395	48.6	7.36	2.22	1.56	59.5	11.6	2.46
WT8 x	**28.5**	8.38	8.215	7.120	0.715	0.430	48.7	7.77	2.41	1.94	21.6	6.06	1.60
	25	7.37	8.130	7.070	0.630	0.380	42.3	6.78	2.40	1.89	18.6	5.26	1.59
	22.5	6.63	8.065	7.035	0.565	0.345	37.8	6.10	2.39	1.86	16.4	4.67	1.57
	20	5.89	8.005	6.995	0.505	0.305	33.1	5.35	2.37	1.81	14.4	4.12	1.57
	18	5.28	7.930	6.985	0.430	0.295	30.6	5.05	2.41	1.88	12.2	3.50	1.52
WT8 x	**15.5**	4.56	7.940	5.525	0.440	0.275	27.4	4.64	2.45	2.02	6.20	2.24	1.17
	13	3.84	7.845	5.500	0.345	0.250	23.5	4.09	2.47	2.09	4.80	1.74	1.12
WT7 x	**365**	107	11.210	17.890	4.910	3.070	739	95.4	2.62	3.47	2360	264	4.69
	332.5	97.8	10.820	17.650	4.520	2.830	622	82.1	2.52	3.25	2080	236	4.62
	302.5	88.9	10.460	17.415	4.160	2.595	524	70.6	2.43	3.05	1840	211	4.55
	275	80.9	10.120	17.200	3.820	2.380	442	60.9	2.34	2.85	1630	189	4.49
	250	73.5	9.800	17.010	3.500	2.190	375	52.7	2.26	2.67	1440	169	4.43
	227.5	66.9	9.510	16.835	3.210	2.015	321	45.9	2.19	2.51	1280	152	4.38
WT7 x	**213**	62.6	9.335	16.695	3.035	1.875	287	41.4	2.14	2.40	1180	141	4.34
	199	58.5	9.145	16.590	2.845	1.770	257	37.6	2.10	2.30	1090	131	4.31
	185	54.4	8.960	16.475	2.660	1.655	229	33.9	2.05	2.19	994	121	4.27
	171	50.3	8.770	16.360	2.470	1.540	203	30.4	2.01	2.09	903	110	4.24
	155.5	45.7	8.560	16.230	2.260	1.410	176	26.7	1.96	1.97	807	99.4	4.20
	141.5	41.6	8.370	16.110	2.070	1.290	153	23.5	1.92	1.86	722	89.7	4.17
	128.5	37.8	8.190	15.995	1.890	1.175	133	20.7	1.88	1.75	645	80.7	4.13
	116.5	34.2	8.020	15.890	1.720	1.070	116	18.2	1.84	1.65	576	72.5	4.10
	105.5	31.0	7.860	15.800	1.560	0.098	102	16.2	1.81	1.57	515	65.0	4.07
	96.5	28.4	7.740	15.710	1.440	0.890	89.8	14.4	1.78	1.49	466	59.3	4.05
	88	25.9	7.610	15.650	1.310	0.830	80.5	13.0	1.76	1.43	419	53.5	4.02
	79.5	23.4	7.490	15.565	1.190	0.745	70.2	11.4	1.73	1.35	374	48.1	4.00
	72.5	21.3	7.390	15.500	1.090	0.680	62.5	10.2	1.71	1.29	338	43.7	3.98

Properties shown in this table are for the full center split section.

STRUCTURAL TEES (Cut from W Shapes)

Theoretical Dimensions and Properties for **Designing**

Section Number	Weight per Foot	Area of Section	Depth of Section	Flange		Stem Thickness	Axis X-X				Axis Y-Y		
				Width	Thickness		I_x	S_x	r_x	y	I_y	S_y	r_y
		A	d	b_f	t_f	t_w							
	lb	in.²	in.	in.	in.	in.	in.⁴	in.³	in.	in.	in.⁴	in.³	in.
WT7 x	66	19.4	7.330	14.725	1.030	0.645	57.8	9.57	1.73	1.29	274	37.2	3.76
	60	17.7	7.240	14.670	0.940	0.590	51.7	8.61	1.71	1.24	247	33.7	3.74
	54.5	16.0	7.160	14.605	0.860	0.525	45.3	7.56	1.68	1.17	223	30.6	3.73
	49.5	14.6	7.080	14.565	0.780	0.485	40.9	6.88	1.67	1.14	201	27.6	3.71
	45	13.2	7.010	14.520	0.710	0.440	36.4	6.16	1.66	1.09	181	25.0	3.70
WT7 x	41	12.0	7.155	10.130	0.855	0.510	41.2	7.14	1.85	1.39	74.2	14.6	2.48
	37	10.9	7.085	10.070	0.785	0.450	36.0	6.25	1.82	1.32	66.9	13.3	2.48
	34	9.99	7.020	10.035	0.720	0.415	32.6	5.69	1.81	1.29	60.7	12.1	2.46
	30.5	8.96	6.945	9.995	0.645	0.375	28.9	5.07	1.80	1.25	53.7	10.7	2.45
WT7 x	26.5	7.81	6.960	8.060	0.660	0.370	27.6	4.94	1.88	1.38	28.8	7.16	1.92
	24	7.07	6.895	8.030	0.595	0.340	24.9	4.48	1.87	1.35	25.7	6.40	1.91
	21.5	6.31	6.830	7.995	0.530	0.305	21.9	3.98	1.86	1.31	22.6	5.65	1.89
WT7 x	19	5.58	7.050	6.770	0.515	0.310	23.3	4.22	2.04	1.54	13.3	3.94	1.55
	17	5.00	6.990	6.745	0.455	0.285	20.9	3.83	2.04	1.53	11.7	3.45	1.53
	15	4.42	6.920	6.730	0.385	0.270	19.0	3.55	2.07	1.58	9.79	2.91	1.49
WT7 x	13	3.85	6.955	5.025	0.420	0.255	17.3	3.31	2.12	1.72	4.45	1.77	1.08
	11	3.25	6.870	5.000	0.335	0.230	14.8	2.91	2.14	1.76	3.50	1.40	1.04
WT6 x	95	27.9	7.190	12.670	1.735	1.060	79.0	14.2	1.68	1.62	295	46.5	3.25
	85	25.0	7.015	12.570	1.560	0.960	67.8	12.3	1.65	1.52	259	41.2	3.22
	76	22.4	6.855	12.480	1.400	0.870	58.5	10.8	1.62	1.43	227	36.4	3.19
	68	20.0	6.705	12.400	1.250	0.790	50.6	9.46	1.59	1.35	199	32.1	3.16
	60	17.6	6.560	12.320	1.105	0.710	43.4	8.22	1.57	1.28	172	28.0	3.13
	53	15.6	6.445	12.220	0.990	0.610	36.3	6.91	1.53	1.19	151	24.7	3.11
	48	14.1	6.355	12.160	0.900	0.550	32.0	6.12	1.51	1.13	135	22.2	3.09
	43.5	12.8	6.265	12.125	0.810	0.515	28.9	5.60	1.50	1.10	120	19.9	3.07
	39.5	11.6	6.190	12.080	0.735	0.470	25.8	5.03	1.49	1.06	108	17.9	3.05
	36	10.6	6.125	12.040	0.670	0.430	23.2	4.54	1.48	1.02	97.5	16.2	3.04
	32.5	9.54	0.060	12.000	0.605	0.390	20.6	4.06	1.47	0.985	87.2	14.5	3.02

Properties shown in this table are for the full center split section.

STRUCTURAL TEES (Cut from W Shapes)

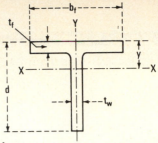

Theoretical Dimensions and Properties for **Designing**

Section Number	Weight per Foot	Area of Section	Depth of Section	Flange Width	Flange Thickness	Stem Thickness	Axis X-X				Axis Y-Y		
		A	d	b_f	t_f	t_w	I_x	S_x	r_x	y	I_y	S_y	r_y
	lb	in.²	in.	in.	in.	in.	in.⁴	in.³	in.	in.	in.⁴	in.³	in.
WT6 x	29	8.52	6.095	10.010	0.640	0.360	19.1	3.76	1.50	1.03	53.5	10.7	2.51
	26.5	7.78	6.030	9.995	0.575	0.345	17.7	3.54	1.51	1.02	47.9	9.58	2.48
WT6 x	25	7.34	6.095	8.080	0.640	0.370	18.7	3.79	1.60	1.17	28.2	6.97	1.96
	22.5	6.61	6.030	8.045	0.575	0.335	16.6	3.39	1.58	1.13	25.0	6.21	1.94
	20	5.89	5.970	8.005	0.515	0.295	14.4	2.95	1.57	1.08	22.0	5.51	1.93
WT6 x	17.5	5.17	6.250	6.560	0.520	0.300	16.0	3.23	1.76	1.30	12.2	3.73	1.54
	15	4.40	6.170	6.520	0.440	0.260	13.5	2.75	1.75	1.27	10.2	3.12	1.52
	13	3.82	6.110	6.490	0.380	0.230	11.7	2.40	1.75	1.25	8.66	2.67	1.51
WT6 x	11	3.24	6.155	4.030	0.425	0.260	11.7	2.59	1.90	1.63	2.33	1.16	0.848
	9.5	2.79	6.080	4.005	0.350	0.235	10.1	2.28	1.90	1.65	1.88	0.939	0.822
	8	2.36	5.995	3.990	0.265	0.220	8.70	2.04	1.92	1.74	1.41	0.706	0.773
	7	2.08	5.955	3.970	0.225	0.200	7.67	1.83	1.92	1.76	1.18	0.594	0.753
WT5 x	56	16.5	5.680	10.415	1.250	0.755	28.6	6.40	1.32	1.21	118	22.6	2.68
	50	14.7	5.550	10.340	1.120	0.680	24.5	5.56	1.29	1.13	103	20.0	2.65
	44	12.9	5.420	10.265	0.990	0.605	20.8	4.77	1.27	1.06	89.3	17.4	2.63
	38.5	11.3	5.300	10.190	0.870	0.530	17.4	4.05	1.24	0.990	76.8	15.1	2.60
	34	9.99	5.200	10.130	0.770	0.470	14.9	3.49	1.22	0.932	66.8	13.2	2.59
	30	8.82	5.110	10.080	0.680	0.420	12.9	3.04	1.21	0.884	58.1	11.5	2.57
	27	7.91	5.045	10.030	0.615	0.370	11.1	2.64	1.19	0.836	51.7	10.3	2.56
	24.5	7.21	4.990	10.000	0.560	0.340	10.0	2.39	1.18	0.807	46.7	9.34	2.54
WT5 x	22.5	6.63	5.050	8.020	0.620	0.350	10.2	2.47	1.24	0.907	26.7	6.65	2.01
	19.5	5.73	4.960	7.985	0.530	0.315	8.84	2.16	1.24	0.876	22.5	5.64	1.98
	16.5	4.85	4.865	7.960	0.435	0.290	7.71	1.93	1.26	0.869	18.3	4.60	1.94
WT5 x	15	4.42	5.235	5.810	0.510	0.300	9.28	2.24	1.45	1.10	8.35	2.87	1.37
	13	3.81	5.165	5.770	0.440	0.260	7.86	1.91	1.44	1.06	7.05	2.44	1.36
	11	3.24	5.085	5.750	0.360	0.240	6.88	1.72	1.46	1.07	5.71	1.99	1.33
WT5 x	9.5	2.81	5.120	4.020	0.395	0.250	6.68	1.74	1.54	1.28	2.15	1.07	0.874
	8.5	2.50	5.055	4.010	0.330	0.240	6.06	1.62	1.56	1.32	1.78	0.888	0.845
	7.5	2.21	4.995	4.000	0.270	0.230	5.45	1.50	1.57	1.37	1.45	0.723	0.810
	6	1.77	4.935	3.960	0.210	0.190	4.35	1.22	1.57	1.36	1.09	0.551	0.785

Properties shown in this table are for the full center split section.

STRUCTURAL
TEES (Cut from W Shapes)

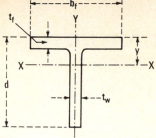

Theoretical Dimensions and Properties for **Designing**

Section Number	Weight per Foot	Area of Section	Depth of Section	Flange		Stem Thickness	Axis X-X				Axis Y-Y		
				Width	Thickness		I_x	S_x	r_x	y	I_y	S_y	r_y
		A	d	b_f	t_f	t_w							
	lb	in.²	in.	in.	in.	in.	in.⁴	in.³	in.	in.	in.⁴	in.³	in.
WT4 x	33.5	9.84	4.500	8.280	0.935	0.570	10.9	3.05	1.05	0.936	44.3	10.7	2.12
	29	8.55	4.375	8.220	0.810	0.510	9.12	2.61	1.03	0.874	37.5	9.13	2.10
	24	7.05	4.250	8.110	0.685	0.400	6.85	1.97	0.986	0.777	30.5	7.52	2.08
	20	5.87	4.125	8.070	0.560	0.360	5.73	1.69	0.988	0.735	24.5	6.08	2.04
	17.5	5.14	4.060	8.020	0.495	0.310	4.81	1.43	0.968	0.688	21.3	5.31	2.03
	15.5	4.56	4.000	7.995	0.435	0.285	4.28	1.28	0.968	0.668	18.5	4.64	2.02
WT4 x	14	4.12	4.030	6.535	0.465	0.285	4.22	1.28	1.01	0.734	10.8	3.31	1.62
	12	3.54	3.965	6.495	0.400	0.245	3.53	1.08	0.999	0.695	9.14	2.81	1.61
WT4 x	10.5	3.08	4.140	5.270	0.400	0.250	3.90	1.18	1.12	0.831	4.89	1.85	1.26
	9	2.63	4.070	5.250	0.330	0.230	3.41	1.05	1.14	0.834	3.98	1.52	1.23
WT4 x	7.5	2.22	4.055	4.015	0.315	0.245	3.28	1.07	1.22	0.998	1.70	0.849	0.876
	6.5	1.92	3.995	4.000	0.255	0.230	2.89	0.974	1.23	1.03	1.37	0.683	0.844
	5	1.48	3.945	3.940	0.205	0.170	2.15	0.717	1.20	0.953	1.05	0.532	0.841
WT3 x	12.5	3.67	3.190	6.080	0.455	0.320	2.29	0.886	0.789	0.610	8.53	2.81	1.52
	10	2.94	3.100	6.020	0.365	0.260	1.76	0.693	0.774	0.560	6.64	2.21	1.50
	7.5	2.21	2.995	5.990	0.260	0.230	1.41	0.577	0.797	0.558	4.66	1.56	1.45
WT3 x	8	2.37	3.140	4.030	0.405	0.260	1.69	0.685	0.844	0.677	2.21	1.10	0.967
	6	1.78	3.015	4.000	0.280	0.230	1.32	0.564	0.862	0.677	1.50	0.748	0.918
	4.5	1.34	2.950	3.940	0.215	0.170	0.950	0.408	0.842	0.623	1.10	0.557	0.905
WT2.5 x	9.5	2.77	2.575	5.030	0.430	0.270	1.01	0.485	0.605	0.487	4.56	1.82	1.28
	8	2.34	2.505	5.000	0.360	0.240	0.845	0.413	0.601	0.458	3.75	1.50	1.27
WT2 x	6.5	1.91	2.080	4.060	0.345	0.280	0.526	0.321	0.524	0.440	1.93	0.950	1.00

Properties shown in this table are for the full center split section.

ANGLES
equal legs

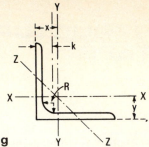

Theoretical Dimensions and Properties for **Designing**

Section Number and Size	Thick-ness	Weight per Foot	Area of Section	k	Axis X-X and Axis Y-Y				Axis Z-Z
					$I_{x,y}$	$S_{x,y}$	$r_{x,y}$	y or x	r_z
in.	in.	lb	in.²	in.	in.⁴	in.³	in.	in.	in.
L8 x 8 x	**1⅛**	56.9	16.7	1¾	98.0	17.5	2.42	2.41	1.56
R=⅝	**1**	51.0	15.0	1⅝	89.0	15.8	2.44	2.37	1.56
	⅞	45.0	13.2	1½	79.6	14.0	2.45	2.32	1.57
	¾	38.9	11.4	1⅜	69.7	12.2	2.47	2.28	1.58
	⅝	32.7	9.61	1¼	59.4	10.3	2.49	2.23	1.58
	ʸ⁹⁄₁₆	29.6	8.68	1³⁄₁₆	54.1	9.34	2.50	2.21	1.59
	½	26.4	7.75	1⅛	48.6	8.36	2.50	2.19	1.59
L6 x 6 x	**1**	37.4	11.0	1½	35.5	8.57	1.80	1.86	1.17
R=½	⅞	33.1	9.73	1⅜	31.9	7.63	1.81	1.82	1.17
	¾	28.7	8.44	1¼	28.2	6.66	1.83	1.78	1.17
	⅝	24.2	7.11	1⅛	24.2	5.66	1.84	1.73	1.18
	ʸ⁹⁄₁₆	21.9	6.43	1¹⁄₁₆	22.1	5.14	1.85	1.71	1.18
	½	19.6	5.75	1	19.9	4.61	1.86	1.68	1.18
	ʸ⁷⁄₁₆	17.2	5.06	¹⁵⁄₁₆	17.7	4.08	1.87	1.66	1.19
	⅜	14.9	4.36	⅞	15.4	3.53	1.88	1.64	1.19
L5 x 5 x	⅞	27.2	7.98	1⅜	17.8	5.17	1.49	1.57	0.973
R=½	¾	23.6	6.94	1¼	15.7	4.53	1.51	1.52	0.975
	ˣ ʸ⅝	20.0	5.86	1⅛	13.6	3.86	1.52	1.48	0.978
	½	16.2	4.75	1	11.3	3.16	1.54	1.43	0.983
	ʸ⁷⁄₁₆	14.3	4.18	¹⁵⁄₁₆	10.0	2.79	1.55	1.41	0.986
	⅜	12.3	3.61	⅞	8.74	2.42	1.56	1.39	0.990
	⁵⁄₁₆	10.3	3.03	¹³⁄₁₆	7.42	2.04	1.57	1.37	0.994
L4 x 4 x	¾	18.5	5.44	1⅛	7.67	2.81	1.19	1.27	0.778
R=⅜	⅝	15.7	4.61	1	6.66	2.40	1.20	1.23	0.779
	½	12.8	3.75	⅞	5.56	1.97	1.22	1.18	0.782
	ʸ⁷⁄₁₆	11.3	3.31	¹³⁄₁₆	4.97	1.75	1.23	1.16	0.785
	⅜	9.8	2.86	¾	4.36	1.52	1.23	1.14	0.788
	⁵⁄₁₆	8.2	2.40	¹¹⁄₁₆	3.71	1.29	1.24	1.12	0.791
	¼	6.6	1.94	⅝	3.04	1.05	1.25	1.09	0.795
L3½ x 3½ x	⅜	8.5	2.48	¾	2.87	1.15	1.07	1.01	0.687
R=⅜	⁵⁄₁₆	7.2	2.09	¹¹⁄₁₆	2.45	0.976	1.08	0.990	0.690
	¼	5.8	1.69	⅝	2.01	0.794	1.09	0.968	0.694

ˣPrice—Subject to Inquiry.
ʸAvailability—Subject to Inquiry.

ANGLES
equal legs

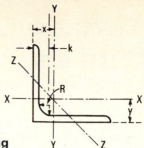

Theoretical Dimensions and Properties for **Designing**

Section Number and Size	Thick-ness	Weight per Foot	Area of Section	k	Axis X-X and Axis Y-Y				Axis Z-Z
					$I_{x,y}$	$S_{x,y}$	$r_{x\,y}$	y or x	r_z
in.	in.	lb	in.²	in.	in.⁴	in.³	in.	in.	in.
L3 x 3 x	½	9.4	2.75	¹³⁄₁₆	2.22	1.07	0.898	0.932	0.584
R=⁵⁄₁₆	x y⁷⁄₁₆	8.3	2.43	¾	1.99	0.954	0.905	0.910	0.585
	⅜	7.2	2.11	¹¹⁄₁₆	1.76	0.833	0.913	0.888	0.587
	⁵⁄₁₆	6.1	1.78	⅝	1.51	0.707	0.922	0.865	0.589
	¼	4.9	1.44	⁹⁄₁₆	1.24	0.577	0.930	0.842	0.592
	³⁄₁₆	3.71	1.09	½	0.962	0.441	0.939	0.820	0.596
L2½ x 2½ x	z½	7.7	2.25	¹³⁄₁₆	1.23	0.724	0.739	0.806	0.487
R=¼	z⅜	5.9	1.73	¹¹⁄₁₆	0.984	0.566	0.753	0.762	0.487
	z⁵⁄₁₆	5.0	1.46	⅝	0.849	0.482	0.761	0.740	0.489
	z¼	4.1	1.19	⁹⁄₁₆	0.703	0.394	0.769	0.717	0.491
	z³⁄₁₆	3.07	0.902	½	0.547	0.303	0.778	0.694	0.495
L2 x 2 x	z⅜	4.7	1.36	¹¹⁄₁₆	0.479	0.351	0.594	0.636	0.389
R=⁹⁄₃₂	z⁵⁄₁₆	3.92	1.15	⅝	0.416	0.300	0.601	0.614	0.390
	z¼	3.19	0.938	⁹⁄₁₆	0.348	0.247	0.609	0.592	0.391
	z³⁄₁₆	2.44	0.715	½	0.272	0.190	0.617	0.569	0.394
	z⅛	1.65	0.484	⁷⁄₁₆	0.190	0.131	0.626	0.546	0.398
L1¾ x 1¾ x	z¼	2.77	0.813	½	0.227	0.186	0.529	0.529	0.341
R=¼	z³⁄₁₆	2.12	0.621	⁷⁄₁₆	0.179	0.144	0.537	0.506	0.343
	z⅛	1.44	0.422	⅜	0.126	0.099	0.546	0.484	0.347
L1½ x 1½ x	z¼	2.34	0.688	⁷⁄₁₆	0.139	0.134	0.449	0.466	0.292
R=³⁄₁₆	z³⁄₁₆	1.80	0.527	⅜	0.110	0.104	0.457	0.444	0.293
	z⁵⁄₃₂	1.52	0.444	⅜	0.094	0.088	0.461	0.433	0.295
	z⅛	1.23	0.359	⁵⁄₁₆	0.078	0.072	0.465	0.421	0.296
L1¼ x 1¼ x	z¼	1.92	0.563	⁷⁄₁₆	0.077	0.091	0.369	0.403	0.243
R=³⁄₁₆	z³⁄₁₆	1.48	0.434	⅜	0.061	0.071	0.377	0.381	0.244
	z⅛	1.01	0.297	⁵⁄₁₆	0.044	0.049	0.385	0.359	0.246
L1 x 1 x	z¼	1.49	0.438	⅜	0.037	0.056	0.290	0.339	0.196
R=⅛	z³⁄₁₆	1.16	0.340	⁵⁄₁₆	0.030	0.044	0.297	0.318	0.195
	z⅛	0.80	0.234	¼	0.022	0.031	0.304	0.296	0.196

ˣPrice—Subject to Inquiry.
ʸAvailability—Subject to Inquiry.
ᶻWest Coast Mills Only.

ANGLES
unequal legs

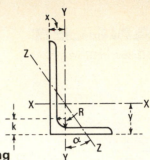

Theoretical Dimensions and Properties for **Designing**

Section Number and Size	Thickness	Weight per Foot	Area of Section	k	Axis X-X				Axis Y-Y				Axis Z-Z	
					I_x	S_x	r_x	y	I_y	S_y	r_y	x	r_z	Tan α
in.	in.	lb	in.²	in.	in.⁴	in.³	in.	in.	in.⁴	in.³	in.	in.	in.	
L9 x 4 x	z⅝	26.3	7.73	1⅛	64.9	11.5	2.90	3.36	8.32	2.65	0.858	1.04	0.847	0.216
R=½	z⁹⁄₁₆	23.8	7.00	1¹⁄₁₆	59.1	10.4	2.91	3.33	7.63	2.41	0.834	1.04	0.850	0.218
	z½	21.3	6.25	1	53.2	9.34	2.92	3.31	6.92	2.17	0.810	1.05	0.854	0.220
L8 x 6 x	1	44.2	13.0	1½	80.8	15.1	2.49	2.65	38.8	8.92	1.73	1.65	1.28	0.543
R=½	x y⅞	39.1	11.5	1⅜	72.3	13.4	2.51	2.61	34.9	7.94	1.74	1.61	1.28	0.547
	¾	33.8	9.94	1¼	63.4	11.7	2.53	2.56	30.7	6.92	1.76	1.56	1.29	0.551
	⁹⁄₁₆	25.7	7.56	1¹⁄₁₆	49.3	8.95	2.55	2.50	24.0	5.34	1.78	1.50	1.30	0.556
	½	23.0	6.75	1	44.3	8.02	2.56	2.47	21.7	4.79	1.79	1.47	1.30	0.558
	y⁷⁄₁₆	20.2	5.93	¹⁵⁄₁₆	39.2	7.07	2.57	2.45	19.3	4.23	1.80	1.45	1.31	0.640
L8 x 4 x	1	37.4	11.0	1½	69.6	14.1	2.52	3.05	11.6	3.94	1.03	1.05	0.846	0.247
R=½	x y⅞	33.1	9.73	1⅜	62.5	12.5	2.53	3.00	10.5	3.51	1.04	0.999	0.848	0.253
	¾	28.7	8.44	1¼	54.9	10.9	2.55	2.95	9.36	3.07	1.05	0.953	0.852	0.258
	x y⅝	24.2	7.11	1⅛	46.9	9.21	2.57	2.91	8.10	2.62	1.07	0.906	0.857	0.262
	⁹⁄₁₆	21.9	6.43	1¹⁄₁₆	42.8	8.35	2.58	2.88	7.43	2.38	1.07	0.882	0.861	0.265
	½	19.6	5.75	1	38.5	7.49	2.59	2.86	6.74	2.15	1.08	0.859	0.865	0.267
	y⁷⁄₁₆	17.2	5.06	¹⁵⁄₁₆	34.1	6.60	2.60	2.83	6.02	1.90	1.09	0.835	0.869	0.269
L7 x 4 x	¾	26.2	7.69	1¼	37.8	8.42	2.22	2.51	9.05	3.03	1.09	1.01	0.860	0.324
R=½	x y⅝	22.1	6.48	1⅛	32.4	7.14	2.24	2.46	7.84	2.58	1.10	0.963	0.865	0.329
	½	17.9	5.25	1	26.7	5.81	2.25	2.42	6.53	2.12	1.11	0.917	0.872	0.335
	y⁷⁄₁₆	15.8	4.62	¹⁵⁄₁₆	23.7	5.13	2.26	2.39	5.83	1.88	1.12	0.893	0.876	0.337
	⅜	13.6	3.98	⅞	20.6	4.44	2.27	2.37	5.10	1.63	1.13	0.870	0.880	0.340

ˣPrice—Subject to Inquiry
ʸAvailability—Subject to Inquiry.
ᶻWest Coast Mills Only.

ANGLES

unequal legs

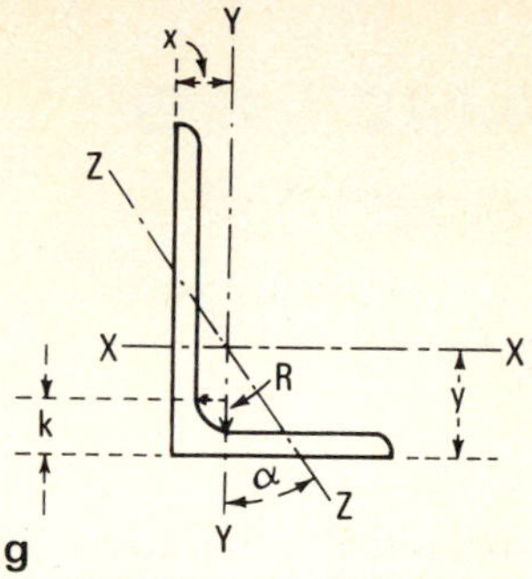

Theoretical Dimensions and Properties for **Designing**

Section Number and Size	Thick-ness	Weight per Foot	Area of Section	k	Axis X-X				Axis Y-Y				Axis Z-Z	
					I_x	S_x	r_x	y	I_y	S_y	r_y	x	r_z	Tan α
in.	in.	lb	in.²	in.	in.⁴	in.³	in.	in.	in.⁴	in.³	in.	in.	in.	
L6 x 4 x	¾	23.6	6.94	1¼	24.5	6.25	1.88	2.08	8.68	2.97	1.12	1.08	0.860	0.428
R=½	⅝	20.0	5.86	1⅛	21.1	5.31	1.90	2.03	7.52	2.54	1.13	1.03	0.864	0.435
	[x,y] 9/16	18.1	5.31	1 1/16	19.3	4.83	1.90	2.01	6.91	2.31	1.14	1.01	0.866	0.438
	½	16.2	4.75	1	17.4	4.33	1.91	1.99	6.27	2.08	1.15	0.987	0.870	0.440
	[y] 7/16	14.3	4.18	15/16	15.5	3.83	1.92	1.96	5.60	1.85	1.16	0.964	0.873	0.443
	⅜	12.3	3.61	⅞	13.5	3.32	1.93	1.94	4.90	1.60	1.17	0.941	0.877	0.446
	[y] 5/16	10.3	3.03	13/16	11.4	2.79	1.94	1.92	4.18	1.35	1.17	0.918	0.882	0.448
L6 x 3½ x	⅜	11.7	3.42	⅞	12.9	3.24	1.94	2.04	3.34	1.23	0.988	0.787	0.767	0.350
R=½	5/16	9.8	2.87	13/16	10.9	2.73	1.95	2.01	2.85	1.04	0.996	0.763	0.772	0.352
	[y] ¼	7.9	2.31	¾	8.86	2.21	1.96	1.99	2.34	0.847	1.01	0.740	0.777	0.355
L5 x 3½ x	¾	19.8	5.81	1¼	13.9	4.28	1.55	1.75	5.55	2.22	0.977	0.996	0.748	0.464
R=7/16	[x,y] ⅝	16.8	4.92	1⅛	12.0	3.65	1.56	1.70	4.83	1.90	0.991	0.951	0.751	0.472
	½	13.6	4.00	1	9.99	2.99	1.58	1.66	4.05	1.56	1.01	0.906	0.755	0.479
	[y] 7/16	12.0	3.53	15/16	8.90	2.64	1.59	1.63	3.63	1.39	1.01	0.883	0.758	0.482
	⅜	10.4	3.05	⅞	7.78	2.29	1.60	1.61	3.18	1.21	1.02	0.861	0.762	0.486
	5/16	8.7	2.56	13/16	6.60	1.94	1.61	1.59	2.72	1.02	1.03	0.838	0.766	0.489
	[y] ¼	7.0	2.06	¾	5.39	1.57	1.62	1.56	2.23	0.830	1.04	0.814	0.770	0.492
L5 x 3 x	½	12.8	3.75	1	9.45	2.91	1.59	1.75	2.58	1.15	0.829	0.750	0.648	0.357
R=⅜	[x,y] 7/16	11.3	3.31	15/16	8.43	2.58	1.60	1.73	2.32	1.02	0.837	0.727	0.651	0.361
	⅜	9.8	2.86	⅞	7.37	2.24	1.61	1.70	2.04	0.888	0.845	0.704	0.654	0.364
	5/16	8.2	2.40	13/16	6.26	1.89	1.61	1.68	1.75	0.753	0.853	0.681	0.658	0.368
	¼	6.6	1.94	¾	5.11	1.53	1.62	1.66	1.44	0.614	0.861	0.657	0.663	0.371
L4 x 3½ x	½	11.9	3.50	15/16	5.32	1.94	1.23	1.25	3.79	1.52	1.04	1.00	0.722	0.750
R=⅜	[x,y] 7/16	10.6	3.09	⅞	4.76	1.72	1.24	1.23	3.40	1.35	1.05	0.978	0.724	0.753
	⅜	9.1	2.67	13/16	4.18	1.49	1.25	1.21	2.95	1.17	1.06	0.955	0.727	0.755
	5/16	7.7	2.25	¾	3.56	1.26	1.26	1.18	2.55	0.994	1.07	0.932	0.730	0.757
	¼	6.2	1.81	11/16	2.91	1.03	1.27	1.16	2.09	0.808	1.07	0.909	0.734	0.759

[x] Price—Subject to Inquiry.
[y] Availability—Subject to Inquiry.

ANGLES
unequal legs

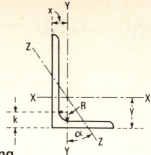

Theoretical Dimensions and Properties for **Designing**

Section Number and Size	Thick-ness	Weight per Foot	Area of Section	k	Axis X-X				Axis Y-Y				Axis Z-Z	
					I_x	S_x	r_x	y	I_y	S_y	r_y	x	r_z	Tan α
in.	in.	lb	in.²	in.	in.⁴	in.³	in.	in.	in.⁴	in.³	in.	in.	in.	
L4 x 3 x	½	11.1	3.25	¹⁵⁄₁₆	5.05	1.89	1.25	1.33	2.42	1.12	0.864	0.827	0.639	0.543
R=⅜	ˣ ʸ⁷⁄₁₆	9.8	2.87	⅞	4.52	1.68	1.25	1.30	2.18	0.992	0.871	0.804	0.641	0.547
	⅜	8.5	2.48	¹³⁄₁₆	3.96	1.46	1.26	1.28	1.92	0.866	0.879	0.782	0.644	0.551
	⁵⁄₁₆	7.2	2.09	¾	3.38	1.23	1.27	1.26	1.65	0.734	0.887	0.759	0.647	0.554
	¼	5.8	1.69	¹¹⁄₁₆	2.77	1.00	1.28	1.24	1.36	0.599	0.896	0.736	0.651	0.558
L3½ x 3 x	⅜	7.9	2.30	¹³⁄₁₆	2.72	1.13	1.09	1.08	1.85	0.851	0.897	0.830	0.625	0.721
R=⅜	⁵⁄₁₆	6.6	1.93	¾	2.33	0.954	1.10	1.06	1.58	0.722	0.905	0.808	0.627	0.724
	¼	5.4	1.56	¹¹⁄₁₆	1.91	0.776	1.11	1.04	1.30	0.589	0.914	0.785	0.631	0.727
L3½ x 2½ x	⅜	7.2	2.11	¹³⁄₁₆	2.56	1.09	1.10	1.16	1.09	0.592	0.719	0.660	0.537	0.496
R=⁵⁄₁₆	⁵⁄₁₆	6.1	1.78	¾	2.19	0.927	1.11	1.14	0.939	0.504	0.727	0.637	0.540	0.501
	¼	4.9	1.44	¹¹⁄₁₆	1.80	0.755	1.12	1.11	0.777	0.412	0.735	0.614	0.544	0.506
L3 x 2½ x	ᶻ⅜	6.6	1.92	¾	1.66	0.810	0.928	0.956	1.04	0.581	0.736	0.706	0.522	0.676
R=⁵⁄₁₆	ᶻ⁵⁄₁₆	5.6	1.62	¹¹⁄₁₆	1.42	0.688	0.937	0.933	0.898	0.494	0.744	0.683	0.525	0.680
	ᶻ¼	4.5	1.31	⅝	1.17	0.561	0.945	0.911	0.743	0.404	0.753	0.661	0.528	0.684
L3 x 2 x	ᶻ⅜	5.9	1.73	¹¹⁄₁₆	1.53	0.781	0.940	1.04	0.543	0.371	0.559	0.539	0.430	0.428
R=⁵⁄₁₆	ᶻ⁵⁄₁₆	5.0	1.46	⅝	1.32	0.664	0.948	1.02	0.470	0.317	0.567	0.516	0.432	0.435
	ᶻ¼	4.1	1.19	⁹⁄₁₆	1.09	0.542	0.957	0.993	0.392	0.260	0.574	0.493	0.435	0.440
	ʸᶻ³⁄₁₆	3.07	0.902	½	0.842	0.415	0.966	0.970	0.307	0.200	0.583	0.470	0.439	0.446
L2½ x 2 x	ᶻ⅜	5.3	1.55	¹¹⁄₁₆	0.912	0.547	0.768	0.831	0.514	0.363	0.577	0.581	0.420	0.614
R=¼	ᶻ⁵⁄₁₆	4.5	1.31	⅝	0.788	0.466	0.776	0.809	0.446	0.310	0.584	0.559	0.422	0.620
	ᶻ¼	3.62	1.06	⁹⁄₁₆	0.654	0.381	0.784	0.787	0.372	0.254	0.592	0.537	0.424	0.626
	ᶻ³⁄₁₆	2.75	0.809	½	0.509	0.293	0.793	0.764	0.291	0.196	0.600	0.514	0.427	0.631

ˣPrice—Subject to Inquiry.
ʸAvailability—Subject to Inquiry.
ᶻWest Coast Mills Only.

ANGLES
unequal legs

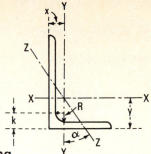

Theoretical Dimensions and Properties for **Designing**

Section Number and Size	Thick-ness	Weight per Foot	Area of Section	k	Axis X-X				Axis Y-Y				Axis Z-Z	
					I_x	S_x	r_x	y	I_y	S_y	r_y	x	r_z	Tan α
in.	in.	lb	in.²	in.	in.⁴	in.³	in.	in.	in.⁴	in.³	in.	in.	in.	
L2½ x 1½ x ᶻ ⁵⁄₁₆	3.92		1.15	⅝	0.711	0.444	0.785	0.898	0.191	0.174	0.408	0.398	0.322	0.349
R=³⁄₁₆ ᶻ ¼	3.19		0.938	⁹⁄₁₆	0.591	0.364	0.794	0.875	0.161	0.143	0.415	0.375	0.324	0.357
ᶻ ³⁄₁₆	2.44		0.715	½	0.461	0.279	0.803	0.852	0.127	0.111	0.422	0.352	0.327	0.364
L2 x 1½ x ᶻ ¼	2.77		0.813	½	0.316	0.236	0.623	0.663	0.151	0.139	0.432	0.413	0.320	0.543
R=³⁄₁₆ ᶻ ³⁄₁₆	2.12		0.621	⁷⁄₁₆	0.248	0.182	0.632	0.641	0.120	0.108	0.440	0.391	0.322	0.551
ᶻ ⅛	1.44		0.422	⅜	0.173	0.125	0.641	0.618	0.085	0.075	0.448	0.368	0.326	0.558
L1¾ x 1¼ x ᶻ ¼	2.34		0.688	⁷⁄₁₆	0.202	0.176	0.543	0.602	0.085	0.095	0.352	0.352	0.267	0.486
R=³⁄₁₆ ᶻ ³⁄₁₆	1.80		0.527	⅜	0.160	0.137	0.551	0.580	0.068	0.074	0.359	0.330	0.269	0.496
ᶻ ⅛	1.23		0.359	⁵⁄₁₆	0.113	0.094	0.560	0.557	0.049	0.051	0.368	0.307	0.271	0.506

ᶻWest Coast Mills Only.

529

REFERENCES

1. National Forest Products Association, Supplement of National Design Specifications for Wood Construction: Design Values for Wood Construction, Washington, 1977.
2. Southern Forest Products Association, *Tech. Bull.* 2, New Orleans, La., no date.
3. American Institute of Timber Construction, Glulam Systems, Washington, January 1975.
4. Bethlehem Steel Corporation, Structural Steel Data for Architectural and Engineering Students, Bethlehem, Pa., 1979.

INDEX